ANALOG MOS INTEGRATED CIRCUITS FOR SIGNAL PROCESSING

WILEY SERIES ON FILTERS:
Design, Manufacturing, and Applications

Editors: Robert A. Johnson and George Szentirmai

ANALOG MOS INTEGRATED CIRCUITS FOR SIGNAL PROCESSING

ROUBIK GREGORIAN

Sierra Semiconductor
San Jose, California

GABOR C. TEMES

Department of Electrical Engineering
University of California, Los Angeles

A Wiley-Interscience Publication

JOHN WILEY & SONS

New York   Chichester   Brisbane   Toronto   Singapore

Library of Congress Cataloging in Publication Data:

Gregorian, Roubik.
    Analog MOS integrated circuits for signal processing.

    (Wiley series on filters)
    "A Wiley-Interscience publication."
    Includes bibliographies and index.
    1. Switched capacitor circuits. 2. Linear integrated
circuits. 3. Metal oxide semiconductors. I. Temes,
Gabor C., 1929–   . II. Title. III. Series.
TK7868.S88G74   1986   621.381'73   85-31493
ISBN 0-471-09797-7

To our wives,

Ibi Temes

and Agnes Gregorian,

in gratitude for their patience and help

SERIES PREFACE

The primary objective of the Wiley Series On Filters is to bring together theory and industrial practice in a series of volumes written for filter users as well as those involved in filter design and manufacturing. Although this is a difficult task, the authors in this series are well qualified for the job. They bring both strong academic credentials and many years of industrial experience to their books. They have all designed filters, have been involved in manufacturing, and have had experience in interacting with the filter user.

Each of the books covers a wide range of subjects including filter specifications, design, theory, parts and materials, manufacturing, tuning, testing, specific applications, and help in using the filter in a circuit. The books also provide a broad view of each subject based on the authors' own work and involvement with filter experts from around the world.

The most outstanding feature of this series is the broad audience of filter builders and users, which it addresses. This includes filter research and development engineers, filter designers, and material specialists, as well as industrial, quality control, and sales engineers. On the filter user's side, the books are of help to the circuit designer, the system engineer, as well as applications, reliability, and component test experts, and specifications and standards engineers.

ROBERT A. JOHNSON
GEORGE SZENTIRMAI

PREFACE

The purpose of this book is to describe the operating principles of analog MOS integrated circuits and to teach the reader how to design and use such circuits. Examples of these devices include switched-capacitor filters, analog-to-digital and digital-to-analog converters, amplifiers, modulators, oscillators, and so on. The main emphasis is on the physical operation and on the design process. It is hoped that the book will be used as a senior- or graduate-level text in the electrical engineering curriculum of universities and also as training and reference material for industrial circuit designers. To increase the usefulness of the book as a text for classroom teaching, numerous problems are included at the end of each chapter; these problems may be used for homework assignments. To enhance its value as a design reference, tables and numerical design examples are included to clarify the step-by-step processes involved. The first three chapters provide a concise, basic-level, and (we hope) clear description of the general properties of analog MOS integrated circuits, and the required background in mathematics and semiconductor device physics. The remainder of the book is devoted to the design of the actual circuits, the practical problems encountered and their solutions, and some examples of system applications.

This book evolved from a set of lecture notes written originally for short courses presented several times annually since 1979 in the United States and in Western Europe, both as a public offering at UCLA, the Federal Institute of Technology of Switzerland, the University of Stuttgart, and so on, and as an in-house training course for high-technology semiconductor, communication, and computer companies, offered through the Continuing Education Institute of Los Angeles. Later, this material formed the basis of a graduate course offered on analog MOS integrated circuits at UCLA. The organization of the material was therefore influenced by the need to make the presentation suitable for audiences of widely varying backgrounds. Hence, we tried to make the book reasonably self-contained, and the presentation is at the simplest level afforded by the topics discussed. Only a limited amount of preparation was assumed on the part of the reader: mathematics on the junior level, and one or two introductory-level courses in electronics and semiconductor physics are the minimum requirements.

The origin of the book also influenced the detailed choice of its subject matter. Since the original short course was intended to train industrial engineers in the design of analog MOS circuits, the theoretical topics discussed were restricted to the minimum needed for the practical design process. Also, in those situations where a number of design techniques were available to accomplish a given task, we described only the one that was most extensively tested in practical applications. Hence many ingenious and effective design procedures were ignored.

Both authors have had considerable industrial experience and also extensive teaching background. We hope that this experience is detectable in our approach to the treatment of our subject.

The book contains eight chapters. Chapter 1 gives a basic introduction to switched-capacitor circuits, compares the analog MOS circuits with other signal processor implementations, and describes (but does not explain in any detail) some typical applications. This material can be covered in one lecture (two-hour lectures are assumed here and throughout this preface).

Chapter 2 describes the Laplace, Fourier, and z-transforms, and introduces the important s-to-z transformations needed to design a sampled-data system from an analog "model." Depending on the mathematical background of the students, this material may require two to three two-hour lectures.

Chapter 3 gives a brief description of the physics of MOS devices, discusses the linearized models of MOSFETs, and describes MOS capacitors and switches. The technology used to fabricate MOS devices is also briefly described. Once again, depending on the background of the audience, two or three lectures should suffice to cover the content of this chapter.

Chapter 4 discusses the circuit design techniques for realizing MOS operational amplifiers. The most common circuit configurations, as well as their design and limitations, are included, and a design example is worked out in detail. Complete coverage of all topics in this chapter requires about five lectures; this time can be reduced by restricting the discussions, leaving out some specialized subjects such as those discussed in Sections 4.10, 4.11, and 4.13, and assigning the design example for reading.

Chapter 5 deals with switched-capacitor filter design and hence represents the focal point of the book. As already mentioned, the design techniques discussed are restricted to the "mainstream" ones: those that have been most thoroughly tested in practical applications. The design of the two commonly used configurations—cascade and ladder circuits—is discussed in detail and illustrated with a numerical design example. Some special circuits, such as switched-capacitor N-path filters and simulated-resistor active-RC filters, are also described. A full coverage of all topics in this chapter needs about five lectures; by omitting Sections 5.8 and 5.9, this number can be reduced to three.

Chapter 6 deals with nonfiltering applications of switched-capacitor circuits. Such important circuits as voltage amplifiers, digital-to-analog and analog-to-digital converters, comparators, modulators, and oscillators are discussed on an introductory level. A complete coverage of all topics may require four lectures; it can be presented in two lectures if the detailed discussions of Sections 6.3 and 6.5 are condensed.

Chapter 7 contains a detailed discussion of the nonideal effects occurring in switched-capacitor circuits. This material is of utmost importance to the industrial designer of practical circuits and hence (in spite of its seemingly mundane subject matter) should be covered at least briefly even in an under-graduate lecture class. Two lectures should be sufficient for a brief presenta-tion.

Chapter 8 discusses some of the systems aspects of analog MOS signal processors and illustrates their use in commercial integrated systems. The first two sections, which deal with the prefiltering and postfiltering requirements of analog MOS circuits, should be discussed in the classroom. The remainder of the chapter discusses, on a descriptive level, some specific applications and can thus be assigned for reading. Hence the material in this chapter may be presented in one or two lectures, depending on how the application examples are treated.

Thus, depending on the depth of the presentation, the full coverage of all material in the book may require as many as 25 two-hour lectures or as few as 16. In intensive presentations (such as a short course or a training course), the complete book has been covered in four days, with six lecture hours per day.

We are grateful to our colleagues Drs. Du Xi-Yu, S. C. Fan, B. Fotouhi, B. Ghaderi, S. Law, K. Martin, T. Cataltepe, and H. J. Orchard, as well as our present and former students, J. N. Babanezhad, F. Dunlap, T. H. Hsu, L. Larson, J. B. Shyu, and F. J. Wang for discussions, review, and criticism. Most of the difficult typing task was done by Ms. Loetitia Loberman. We are grateful for her excellent and painstaking help. The artwork was done (excel-lently) by Mr. Kayvan Abolhassani of the Department of Electrical Engineer-ing at UCLA. Last, but not least, we would like to express our gratitude to our wives for graciously suffering neglect during the writing of this work.

ROUBIK GREGORIAN
GABOR C. TEMES

San Jose, California
Los Angeles, California
February 1986

CONTENTS

ANALOG MOS
INTEGRATED CIRCUITS
FOR SIGNAL PROCESSING

Chapter One ————————————

INTRODUCTION

In this chapter, the basic concept of a switched capacitor performing as a simulated resistor is introduced. Some of the physical properties of switched-capacitor circuits are also briefly discussed. Then, a comparison is made between signal processors using switched-capacitor circuits and some alternative implementations, such as discrete analog circuits, digital filters, and analog bipolar integrated circuits. Finally, a few representative examples are given of circuits and systems utilizing MOS analog signal processing techniques, to illustrate the great potential of these circuits in telecommunication systems and related applications.

1.1. THE USE OF ANALOG MOS INTEGRATED CIRCUITS FOR SIGNAL PROCESSING.[1-4]

Electrical signal processors are usually divided into two categories: analog and digital systems. An *analog system* carries signals in the form of voltages, currents, charges, and so on, which are *continuous* functions of the *continuous* time variable. Some typical examples of analog signal processors are audio amplifiers, passive- or active-RC filters, and so on. By contrast, in a *digital system* each signal is represented by a sequence of numbers. Since these numbers can only contain a finite number of digits (typically, coded in the form of binary digits or *bits*) they can only take on *discrete values*. Also, these numbers are the *sampled* values of the signal, taken at *discrete time* instances. Thus, both the dependent and independent variables of a digital signal are discrete. Since the processing of the digital bits is usually performed synchronously, a timing or *clock* circuit is an important part of the digital system. The clock provides one or more clock signals, each containing accurately timed pulses which operate or synchronize the operation of the components of the system. Typical examples of digital systems are a general-purpose digital

1

computer, or a special-purpose computer dedicated to (say) calculating the Fourier transform of a signal via the fast Fourier transform (FFT), or a digital filter used in speech analysis, and so on.

Most of the circuits considered in this book fall into a category which is in between the two main classifications described above. This is the category of *sampled-data analog systems*. For such systems, the signal is represented by the uncoded amplitude of an electrical quantity (normally, a voltage) as in an analog system. However, the system contains a clock, and the signal amplitude is sensed only at discrete time instances, as in a digital system. Prior to the development of the MOS circuits discussed in this work, the most important sampled-data analog systems were the charge-transfer devices, such as charge-coupled devices (CCDs) and bucket-brigade devices (BBDs). In these, the signal amplitude is represented by the amount of charge shifted from cell to cell. Since, with very few exceptions, these devices did not contain feedback loops, they were inherently nonrecursive in nature. Therefore, they were more suited for such applications as sampled-data delay lines, multiplexers, correlators, and so on which did not require accurately controlled poles as well as zeros, than for the commonly needed frequency-selective filtering tasks. Also, they require special fabrication technology, rather than the standard MOS process used to manufacture digital MOS circuits, and usually need some specialized peripheral (input and output) circuitry. For these reasons, their uses were restricted to a relatively few large companies, where the requisite special design background and technology could be developed and maintained.

By contrast, the circuits considered in this book can be fabricated utilizing standard digital MOS technology, and hence can also be placed on the same chip with digital circuitry. This latter aspect is of great importance, for example, in modern telecommunication systems, where both analog and digital functions are often needed within the same functional block. Furthermore, these circuits contain only a few standard building blocks: amplifiers, switches, capacitors, and, in rare instances, resistors. Once these have been developed and standardized in the locally available technology, a large number of applications can be accommodated using only slightly different configurations and/or dimensions.

To understand the basic concepts of the most commonly used configurations of analog MOS circuits, consider the simple analog transfer function

$$\frac{V_{out}(s)}{V_{in}(s)} = \frac{b}{s^2 + as + b}. \tag{1.1}$$

It is easy to verify that the RLC circuit shown in Fig. 1.1a can realize this function (Problem 1.1). While this circuit is easy to design, build, and test, the presence of the inductor in the circuit makes the fabrication in an integrated form impractical. In fact, for low-frequency applications, this circuit may well require a very large-valued, and hence bulky, inductor and capacitor. To

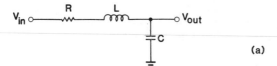

(a)

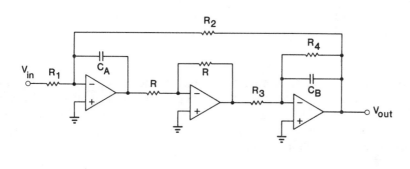

(b)

FIGURE 1.1.   Second-order filter realizations; (*a*) passive circuit; (*b*) active-RC circuit.

overcome this problem, the designer may decide instead to realize the desired transfer function using an active-RC circuit. It can readily be shown that the circuit of Fig. 1.1*b*, which utilizes three *operational amplifiers*, is capable of providing the transfer function specified in Eq. (1.1). This circuit needs no inductors and may be realized with small-sized discrete components for a wide variety of specifications (Problem 1.2). It turns out, however, that while the integration of this circuit on an MOS chip is, in principle, feasible (since the amplifiers, resistors, and capacitors needed can all be integrated), there are some major practical obstacles to integration. These include the very large chip area needed by the RC components, as well as the stringent accuracy and stability requirements for these elements. These requirements cannot be readily satisfied by integrated components, since neither the fabricated values nor the temperature-induced variations of the resistive and the capacitive elements track each other. The resulting pole/zero variations are too large for most applications. (This subject will be discussed in detail in Section 5.1 of Chapter 5.)

An effective strategy which can solve both the area and the matching problems is to replace each resistor in the circuit by a combination of a capacitor and a few switches. Consider the branch shown in Fig. 1.2. Here, the four switches S_1, S_2, S_3, and S_4 open and close periodically, at a rate which is much faster than that of the variations of the terminal voltages v_A and v_B. Switches S_1 and S_4 operate synchronously with each other, but in opposite phase with S_2 and S_3. Thus, when S_2 and S_3 are closed, S_1 and S_4 are open,

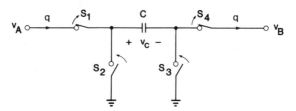

FIGURE 1.2.   Switched-capacitor realization of a resistive branch.

and vice versa. Now when S_2 and S_3 close, C is discharged. When next S_2 and S_3 open, and S_1 and S_4 close, C is recharged to the voltage $v_C = v_A - v_B$. This causes a charge $q = C(v_A - v_B)$ to flow through the branch of Fig. 1.2. Next, C is again discharged by S_2 and S_3, and so on. If this cycle is repeated every T seconds (where T is the *switching period* or *clock period*) then the average current through the branch is therefore

$$i_{av} = \frac{q}{T} = \frac{C}{T}(v_A - v_B). \tag{1.2}$$

Thus, i_{av} is *proportional* to the branch voltage $v_A - v_B$. Similarly, for a branch containing a resistor R, the branch current is $i = (1/R)(v_A - v_B)$. Thus, the average current flows in these two branches are the same if the relation $R = T/C$ holds.

It is plausible therefore that the branch of Fig. 1.2 can be used to replace all resistors in the circuit of Fig. 1.1*b*. The resulting stage[3] is shown in Fig. 1.3. In this circuit, switches which belong to different "resistors," but perform identical tasks, have been combined. Furthermore, the second operational amplifier

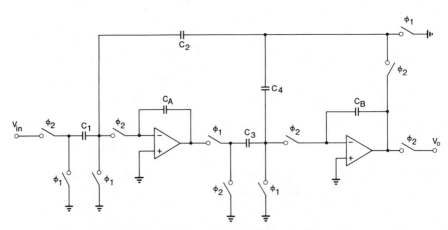

FIGURE 1.3.   Second-order switched-capacitor filter section.

(op-amp) in Fig. 1.1*b*, which acted merely as a phase inverter, has been eliminated. This was possible since by simply changing the phasing of two of the switches associated with capacitor C_3, the required phase inversion could be accomplished without an op-amp. The details of the transformation of the circuit of Fig. 1.1*b* to that of Fig. 1.3 are discussed in Chapter 5, Section 5.4.

As Fig. 1.3 illustrates, the transformed circuit contains only capacitors, switches, and op-amps. A major advantage of this new arrangement is that now all time constants, previously determined by the poorly controlled RC products, will be given by expressions of the form $(T/C_1)C_2 = T(C_2/C_1)$. Here, the clock period T is usually determined by a quartz-crystal-controlled clock circuit, and is hence very accurate and stable. The other factor of the time constant is C_2/C_1, that is, the *ratio* of two on-chip MOS capacitances. Using some simple rules in the layout of these elements (described in Section 3.5), it is possible to obtain an accuracy and stability of the order of 0.1% for this ratio. The resulting overall accuracy is at least a hundred times better than what can be achieved with an on-chip resistor and capacitor for the RC time constant.

A dramatic improvement is also achievable for the area required by the passive elements. To achieve a time constant in the audio-frequency range (say 10 krad/s), even with a large (10 pF) capacitor a resistance of 10 MΩ is required. Such a resistor will occupy an area of about 10^6 μm^2, which is prohibitively large; it is nearly 10% of the area of an average chip. By contrast, for a typical clock period of 10 μs, the capacitance of the switched capacitor realizing a 10-MΩ resistor is $C = T/R = 10^{-5}/10^7 = 10^{-12}$ F = 1 pF. The area required to realize this capacitance is about 2500 μm^2, or only 0.25% of that needed by the resistor which it replaces.

Using the three types of components (op-amps, capacitors, and switches) shown in Fig. 1.3, a large quantity of signal processing circuitry can be placed on a single chip. A high-quality op-amp can be fabricated on an area of 5×10^4 to 10^5 μm^2, while a switch needs typically only about 50 μm^2. Since the area of a large chip may be around 5×10^7 μm^2, such a chip can readily accommodate, say, 100 op-amps, 300 capacitors, and 500 switches. Extrapolating from the circuit of Fig. 1.3 which realizes the second-order transfer function given in Eq. (1.1), it can be seen that the signal processing capability of such a chip is sufficient to implement transfer functions with a combined order of 100. In fact, since the op-amps can be time shared (multiplexed) for low-frequency signals, even higher-order functions may be realized: a speech analyzer chip[5] implementing switched-capacitor filters with a total of 308 poles, in addition to a substantial quantity of on-chip digital circuitry was recently described!

In addition to frequency-selective filtering which has been the most common application of the switched-capacitor (SC) circuits introduced in Fig. 1.3, there are many other functions which such circuits can perform. These include analog-to-digital (A/D) and digital-to-analog (D/A) data conversion, programmable-gain amplification for AGC and other applications, as well as such nonlinear operations as multiplication, modulation, detection, rectification, zero-crossing detection, and so on. They have also been used extensively in large mixed analog–digital systems such as codecs, modems, and speech processors. It is expected that this range will expand further, as the quality (bandwidth, dynamic range, etc.) of the components, especially op-amps, improves, and as better circuit techniques are introduced.

1.2. COMPARISON OF ANALOG MOS SIGNAL PROCESSORS WITH OTHER IMPLEMENTATIONS

It is important to define the areas of applications in which analog MOS signal processors are competitive with, or even superior to, earlier implementations of signal processing systems. To do that, we list below some salient features of analog MOS (typically, switched-capacitor) circuits, and contrast them with those of alternative realizations.

1. *Switched-Capacitor Circuits Are Integrated Circuits.*   This property has a profound effect on the economy of its applications. The development (theoretical design, computer simulation, layout and fabrication, testing and troubleshooting) of such a circuit may require a combined initial design effort of one or more man-years, at a cost of \$50,000 or more. After this initial expenditure, the devices can be mass produced at a low per-unit cost, say \$5 or less. Hence, compared to a discrete implementation costing (say) \$15 and having a negligible design expenditure, the integrated realization is economical if the inequality

$$50,000 + 5N < 15N \qquad (1.3)$$

holds. Here, N is the total number of units required. For the values used, this gives $N > 5000$. Clearly, the actual figures depend on the experience, equipment, application, and so on; however, the orders of magnitude given are fairly typical.

Other features associated with the integrated-circuit (IC) character of SC circuits are small size, light weight, high reliability, and small dc bias power required. These may also have a great importance in, say, aerospace applications.

2. *Switched-Capacitor Circuits Are Sampled-Data Systems.*   As mentioned before, the signal values are evaluated only at periodic time instants in an SC circuit, and the sampling period is determined by a crystal-controlled clock. This feature makes it possible to have all pole and zero values dependent only on capacitance *ratios* (rather than on absolute values); it thus permits the realization of high-selectivity responses with good accuracy and stability. Furthermore, since only periodic samples of the signal are of interest, it is possible to time share (multiplex) the whole circuit, or parts such as the

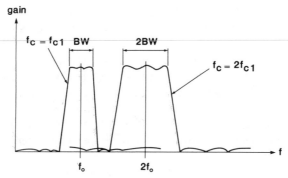

FIGURE 1.4.   Switched-capacitor bandpass filter responses for $f_c = f_{c1}$ and $f_c = 2f_{c1}$. Both the center frequency f_0 and the bandwidth BW are doubled when f_c doubles.

op-amps of the circuit, among several signal channels resulting in highly efficient multichannel systems.

Finally, all time constants of an SC circuit are proportional to the clock period T. As a result, the overall gain versus frequency response $H(f)$ can readily be scaled by changing the *clock frequency* $f_c = 1/T$. As an illustration, Fig. 1.4 shows the responses of an SC bandpass filter for two different clock frequencies. Clearly, changing the value of f_c from f_{c1} to $2f_{c1}$ simply expands the response curve horizontally by the same factor 2. This gives a valuable tool for the fine tuning of the response for applications such as voltage-controlled oscillators (VCOs), adaptive filters, tracking filters, and so on.

The above features (integrated realization, clock-controlled sampled-data operation) are shared with charge-transfer devices such as CCDs and BBDs, as well as with digital filters, but not with the other commonly used signal processor implementations.

3.   *Switched-Capacitor Circuits Are Analog Systems.*   In spite of their sampled-data characters, SC circuits handle signals in analog forms; thus, the amplitudes of the sampled voltages are the signal values, without the use of any encoding. This makes the basic operations (multiplication, addition, delay) needed in signal processing much simpler to perform than in digital circuits. Hence, the density of operations on the chip can be much higher than for digital signal processors. Without any multiplexing, linear filter sections realizing 100 or even more poles can be accommodated on a single chip.

Due to the basic simplicity of the circuits used in SC systems, the speed with which the signal processing tasks can be accomplished is much higher than for digital systems. The real-time filtering of signals with frequencies up to about 0.5 MHz is possible currently, and this figure is continuously rising as improved technology and design techniques become available.

Due to its much simpler structure, along with the chip area needed, the dc power required for a given signal processing task is also considerably less for a

SC implementation than for a digital one. A typical value is 1 mW dc power needed per filter pole realized, and this value can be drastically reduced if necessary.

In contrast to these advantages, when compared with digital signal processors, SC systems also have several important disadvantages. One is their limited accuracy. As mentioned earlier, the capacitance ratios (which determine the coefficients and thus the zeros and poles of the transfer function) can be made accurate and stable to about 0.1% of their nominal values. While this is a very impressive precision for an analog system, it only corresponds to 10-bit floating-point accuracy in digital terms. For some applications, this precision is not sufficient, and (say) a 16-bit accuracy is mandatory, precluding the use of analog methods.

Another potential problem area concerns dynamic range. Due to a relatively large level of noise originating from the operational amplifiers and switches, and also coupled from the clock and the supply lines, and so on, the dynamic range (basically, the maximum-signal-to-noise ratio) of SC circuits seldom exceeds 100,000 or 100 dB; values of 70 ~ 90 dB are common. The corresponding range of a digital filter is much larger.

Yet another advantage of digital signal processors over their analog counterparts is their superior flexibility and programmability. While SC circuits can also be made programmable, for a digital system the changing of the characteristics can be performed simply by selecting different coefficients from a ROM. This ease of operation holds also for the multiplexing of a digital signal processor among several signal channels.

Finally, an important practical advantage of digital signal processors is due to the fact that they are built using standard microprocessor technology. Hence, they can take advantage of the standard cells and other design information amassed for microprocessors. Because of the large volume and hence high priority of microprocessor manufacturing, there is usually ample available information concerning the technology, the basic circuits, accurate device models for computer simulation, and so on, which may or may not be the case for the analog circuits using the same technology. Also, the inevitable changes in the fabrication process, such as the scaling down of the minimum linewidth, reduction of threshold and supply voltages, and so on are aimed primarily at improving the performance of the higher-volume digital ICs. While changes aimed at the faster operation of digital circuits (such as scaling down their dimensions) tend to provide a similar improvement of the analog circuits sharing the same technology, they may also introduce or aggravate nonideal phenomena, such as short-channel and narrow-channel effects, which often affect the analog circuits more than the digital ones.

As the above arguments demonstrate, there is no clear-cut and general rule for choosing an analog SC implementation over a digital one, or vice versa. Whenever simplicity, speed of operation, small chip area, or small dc bias power are of prime importance, and the required input and output signals are inherently analog in nature, the analog implementation may be preferable over a digital one. Vice versa, if the input and/or output signals are already digital, or if the flexibility of the processor for programming or multiplexing is critical, or if the accuracy and/or dynamic range requirements are very stringent, or if the necessary components (ROMs, RAMs, arithmetic units, A/D and D/A converters) are already available, a digital signal processor may be the logical choice.

In recently developed signal processing chips (see, e.g., Ref. 5), the signal processing tasks are usually divided between analog and digital circuits. Analog circuits are in any case needed by the digital signal processors for peripheral tasks (A/D and D/A conversions, prefiltering and postfiltering, amplification, etc.), while digital circuits are needed by SC systems for clock signal generation, multiplexing and programming, calibration, testing, and so on. Thus, the decision between analog and digital circuits is not a binary one; usually, both are needed in a signal processor. The main issue is often the assignment of the different tasks to the most appropriate (digital or analog) functional blocks.

4. *Switched-Capacitor Circuits Are Realized in MOS Technology.*[1,6] An alternative technology available is, of course, the bipolar one. Prior to about 1977, there existed a clear separation of the bipolar and MOS technologies, according to the function required. MOS technology, with its superior device density, was used mostly for digital logic and memory applications, while all required analog functions (such as amplification, filtering, data conversion) were performed using bipolar integrated circuits, such as bipolar op-amps. Since that time, however, rapid progress made in MOS fabrication techniques

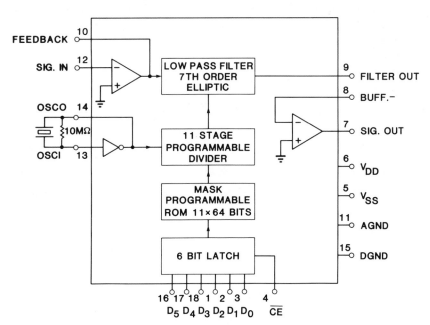

FIGURE 1.5. The block diagram of the AMI S3528 programmable low-pass filter.

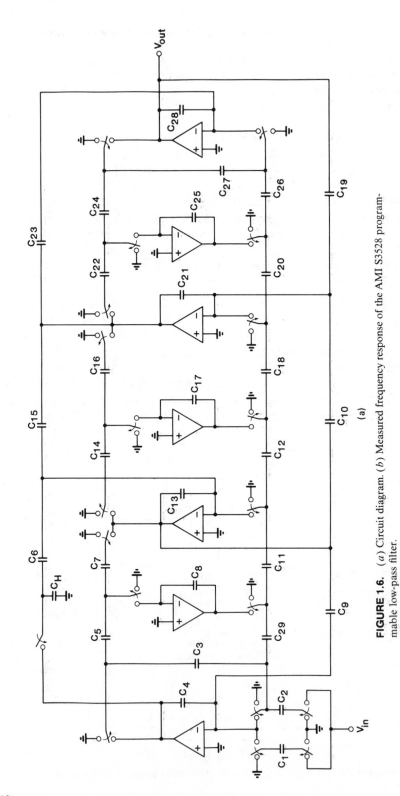

FIGURE 1.6. (*a*) Circuit diagram. (*b*) Measured frequency response of the AMI S3528 programmable low-pass filter.

(a)

made it possible to manufacture much more complex and flexible chips. In addition, new developments occurred in communication technology (such as digital telephony, data transmission via telephone lines, adaptive communication channels, etc.) which required analog and digital signal processing circuitry in the same functional blocks. The analog functions most often needed are filtering (for antialiasing, smoothing, band separation, etc.), amplification, sample-and-hold operations, voltage comparison, and the generation as well as precise scaling of voltages and currents for data conversion. The separation of these analog functions from the digital ones merely because of the different fabrication technologies used is undesirable, since it increases both the packaging costs and the space requirements and also, due to the additional interconnections required, degrades the performance. Hence, there was strong motivation to develop novel MOS circuits which can perform these analog functions and which can also share the area on the same chip with the digital circuitry.

The MOS technology has both advantages and disadvantages as compared with the bipolar one. An MOS device has an extremely high impedance at its input (gate) terminal, which enables it to sense the voltage across a capacitor without discharging it. Also, there is no inherent offset voltage across the MOS device when it is used as a conducting switch. Furthermore, high-quality capacitors can be fabricated reliably on an MOS chip. These features make the realization of such circuits as precision sample-and-hold stages feasible on an MOS chip.[1] This is usually not possible in bipolar technology.

On the negative side, the transconductance of MOS transistors is inherently lower than that of bipolar ones. A typical transconductance value for a

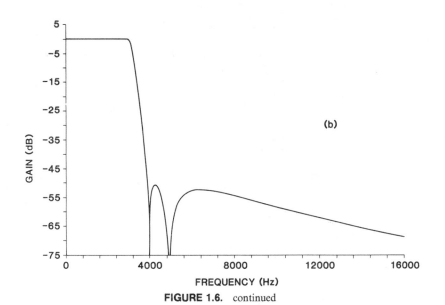

FIGURE 1.6.   continued

moderate-size MOS device is around 0.5 mA/V; for a bipolar transistor, it may be about a hundred times larger. This leads to a higher offset voltage for an MOS amplifier than for a bipolar one. (At the same time, however, the input capacitance of the MOS transistor is typically much smaller than that of a bipolar one.) Also, the noise generated in an MOS device is much higher, especially at low frequencies, than in a bipolar transistor. The conclusion is that the behavior of an amplifier realized on an MOS chip tends to be inferior to an equivalent bipolar realization in terms of offset voltage, noise, and dynamic range. However, it can have a much higher input impedance than its bipolar counterpart.

As a result of these properties, switched capacitor circuits are especially suitable for linear applications, where element-value accuracy is important, but the signal frequency is not too high, and the dynamic range required is not excessive. Voice- and audio-frequency filtering and data conversion are in this category, and they represent the bulk of past applications.

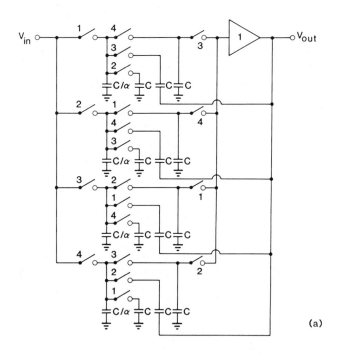

(a)

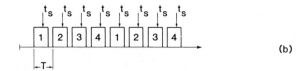

(b)

FIGURE 1.7.   Narrow-band switched-capacitor bandpass filter.

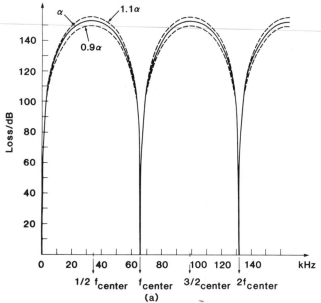

FIGURE 1.8.   Loss response of the narrow-band switched-capacitor filter of Fig. 1.7: (*a*) overall response; (*b*) passband response.

1.3.   EXAMPLES OF ANALOG MOS SIGNAL PROCESSING CIRCUITS AND SYSTEMS

In this section, a few selected examples of practical analog MOS signal processors are given. Of course, the reader should not expect to understand the details of these systems at this stage. However, the diagrams may give an idea of the potentials of these devices in modern telecommunication systems. In the later chapters of the book (Chapters 5–8), the design of such circuits will be discussed in detail.

Figure 1.5 shows the block diagram of a digitally programmable switched-capacitor low-pass filter (marketed as S3528 by American Microsystems, Inc.). By choosing the six-bit digital word (D_0 D_1 $\cdots$ D_5) at the digital input, the passband and stopband edge frequencies can be selected, in fairly small steps. The circuit diagram of the seventh-order switched-capacitor filter is shown in Fig. 1.6*a*; its measured response is shown in Fig. 1.6*b*. The measured passband gain variation was less than 0.06 dB. This represents a superior performance, which could not have been achieved without extensive trimming using any other analog filter technology.

Figure 1.7 illustrates the circuit diagram of a simple narrow-band switched-capacitor filter.[7] Figure 1.8*a* shows the overall loss response, and Fig. 1.8*b* the passband details. The relative bandwidth was less than 0.2%. Among analog circuits, only a carefully trimmed crystal or mechanical filter could parallel this performance.

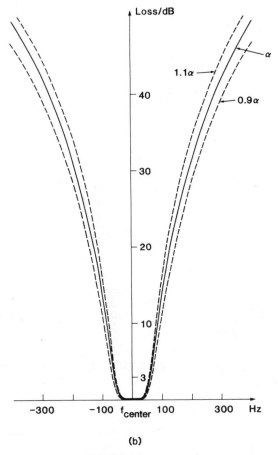

(b)

FIGURE 1.8. continued

Figure 1.9 shows the circuit diagram of a switched-capacitor sine-wave oscillator capable of generating low-distortion waveforms with very accurately controlled amplitude and frequency values.[8] Its performance is comparable to that of a high-quality tunable oscillator realized by a discrete circuit.

Figure 1.10 shows the block diagram of the AMI S3506 codec.[9] This circuit contains four filters as well as a nonlinear analog-to-digital converter, a nonlinear digital-to-analog converter, and an on-chip voltage reference circuit, in addition to a considerable amount of digital logic circuitry.

Figure 1.11 illustrates the block diagram of the dual-tone multifrequency (DTMF) receiver designed by Silicon Systems, Inc.[10] This chip contains a dozen filters, two multiplexed zero-crossing detectors, a power regulator, and voltage reference circuit, as well as extensive digital circuitry. The total number of operational amplifiers on the chip is 37.

Finally, Fig. 1.12 shows the diagram of a recent speech spectrum analyzer.[5] This circuit splits the speech spectrum into 20 frequency bands and measures

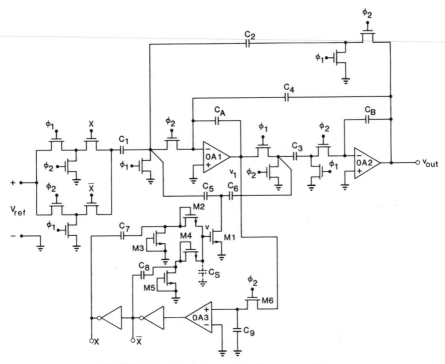

FIGURE 1.9. Switched-capacitor sine-wave oscillator.

the power in each band. The resulting 20 values are then converted into nine-bit digital words, and stored in the output buffer memory from which they can be asynchronously recalled. By using time-division multiplexing (made possible by the sampled-data character of the switched-capacitor filters), a total number of 308 filters poles is realized using "only" 30 operational amplifiers. In addition, the chip contains two voltage comparators, a 396-bit random-access memory (RAM) and 500 digital gates.

As the above examples illustrate, present-day analog MOS circuits have reached a certain amount of maturity. Already, almost any analog signal processing task in the voice- or audio-frequency range has a possible solution using such circuits. With the expected improvements in both fabrication technology and circuit design techniques, it is likely that the speed and dynamic range of these circuits will increase, allowing their use in such large-volume applications as video and radio systems, image processing, carrier-frequency telephony, and so on.

PROBLEMS

1.1. Show that the circuit of Fig. 1.1a can realize the transfer function of Eq. (1.1). What should be the element values R, L, and C?

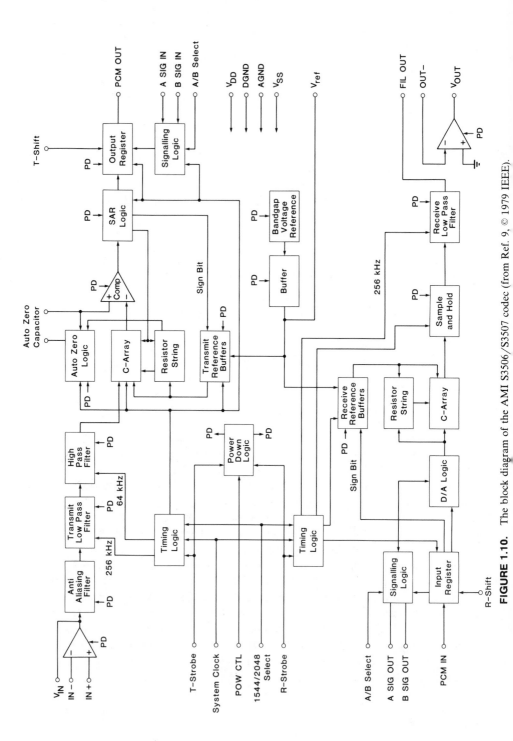

FIGURE 1.10. The block diagram of the AMI S3506/S3507 codec (from Ref. 9, © 1979 IEEE).

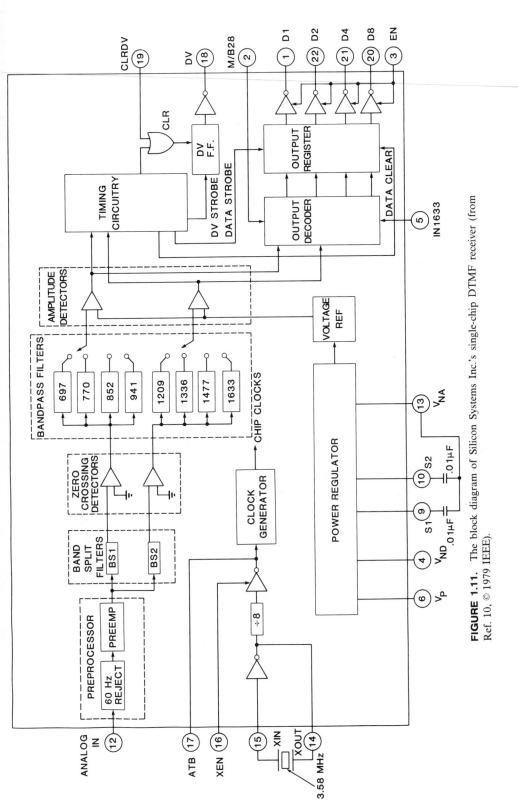

FIGURE 1.11. The block diagram of Silicon Systems Inc.'s single-chip DTMF receiver (from Ref. 10, © 1979 IEEE).

17

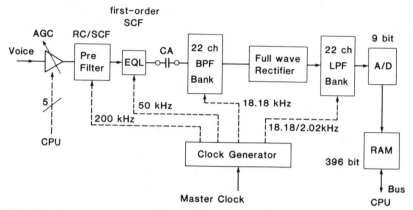

FIGURE 1.12. Speech spectrum analyzer block diagram (from Ref. 5, © 1984 IEEE).

1.2. Calculate the transfer function of the active-RC circuit of Fig. 1.1*b*. Assume that the circuit is to realize the transfer function of Eq. (1.1). Write the available equations for the element values. How many element values can be chosen arbitrarily?

REFERENCES

1. R. W. Brodersen, P. R. Gray, and D. A. Hodges, *Proc. IEEE*, **67**, 61–75 (1979).

2. Y. Tsividis, *Proc. IEEE*, **71**, 926–940 (1983).

3. R. Gregorian, K. W. Martin, and G. C. Temes, *Proc. IEEE*, **71**, 941–966 (1983).

4. D. J. Allstot and W. C. Black, Jr., *Proc. IEEE*, **71**, 967–986 (1983).

5. Y. Kuraishi, K. Nakayama, K. Miyadera, and T. Okamura, *IEEE J. Solid-State Circuits*, **SC-19**, 964–970 (1984).

6. P. R. Gray and R. G. Meyer, *Analysis and Design of Analog Integrated Circuits* (2nd ed.), Wiley, New York, 1984.

7. J. A. Nossek, *NTZ Archiv*, **3**, 351–358 (1981).

8. P. E. Fleischer, A. Ganesan and K. R. Laker, *IEEE J. Solid-State Circuits*, **SC-20**, 641–647 (1985).

9. R. Gregorian, G. A. Wegner, and W. E. Nicholson, *IEEE J. Solid-State Circuits*, **SC-14**, 65–73 (1979).

10. B. J. White, G. M. Jacobs, and G. F. Landsburg, *IEEE J. Solid-State Circuits*, 991–997 (1979).

Chapter Two _____

TRANSFORMATION METHODS

Switched-capacitor filters are *sampled-data* circuits, with *analog* signal representation. Hence, their analysis requires, in general, the mathematical tools of both analog signals (Laplace and Fourier transformations) and those of sampled signals (*z*-transformation). Furthermore, the relations between these two groups of transformations must be correctly formulated and used. For these reasons, this chapter gives a summary of the basic definitions of analog, digital, and sampled-analog systems, and then discusses the various transformation methods needed to analyze their time and frequency responses. Finally, design techniques will be described for obtaining the transfer function of a sampled-data system from that of a suitable analog "model" system.

2.1. ANALOG, DIGITAL, AND SAMPLED-ANALOG SIGNALS AND SYSTEMS[1-3]

A *signal* is a function—its independent variable is, in our applications, time; the dependent variable is a physical quantity such as voltage, charge, or current. A *continuous-time signal* is a signal which has a well-defined value at every point in the time interval of interest (Fig. 2.1a). A *discrete-time signal* has values only at discrete (usually equally spaced) time instances (Fig. 2.1b); it is unspecified at any other time. Often, the discrete-time signal is obtained by sampling a continuous-time one. Thus, the signal of Fig. 2.1b satisfies the relation

$$v(nT) = v(t)|_{t=nT}, \quad n = 0, 1, 2, \ldots$$

with the signal of Fig. 2.1a. Here, T is the *sampling interval*. A closely related

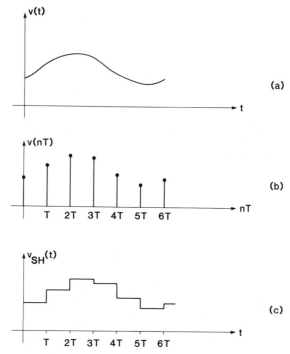

FIGURE 2.1. Signal types: (a) continuous-time signal; (b) discrete-time signal; (c) sampled-and-held signal.

signal type is the *sampled-and-held signal* illustrated in Fig. 2.1c. It satisfies

$$v_{SH}(t) = v(nT), \quad nT \leq t < (n + 1)T.$$

Thus, a sampled-and-held (S/H) signal is a continuous-time signal whose value changes only at discrete-time instances. It is completely defined by T and the values $v(nT)$ for $n = 0, 1, 2, \ldots$.

A *digital signal* is a sequence of numbers. Each number can be regarded as a representation of a value $v(nT)$ of a discrete-time signal. However, the numbers in the digital signal can only have a finite number of digits (bits, if the signal is binary coded). Hence, they can only assume discrete values, which are multiples of the smallest available digit. By contrast, the functions $v(t)$, $v(nT)$, and $v_{SH}(t)$ may have any values; they will be called *analog signals*. The last two functions, $v(nT)$ and $v_{SH}(t)$ are also often called *sampled-analog* signals, or *sampled-data analog* signals.

A *system* is a physical device capable of processing signals. It will be called analog or digital, continuous or discrete time, and so on, depending on the type of the signal which it can handle. An example of a digital system is a digital computer; an example of a continuous-time analog system is an amplifier. Most of the circuits discussed in this book process sampled-and-held analog signals such as that shown in Fig. 2.1c. They are thus continuous-time analog systems of a special type; they are used to transform a S/H input signal into another S/H signal, the output signal.

An analog system which processes electrical signals in the form of voltages, currents, fluxes, and charges is an *electric circuit* or *network*. Its analysis is performed by constructing and solving the *network equations*. These are typically obtained by using *Kirchhoff's voltage law* (KVL) which states that the sum of voltages in any closed loop of the circuit is zero; *Kirchhoff's current law* (KCL) which states that the sum of currents entering any node is zero; and the *branch relations* which are satisfied by the specific circuit elements (resistors, capacitors, amplifiers, etc.) used in the network. An excellent introduction to circuit analysis is given in Ref. 2.

The KVL and KCL involve only the addition and subtraction of voltages and currents. The branch relations, by contrast, may involve differentiation or integration of these quantities with respect to time. Thus, for an inductor L, the voltage $v_L(t)$ across the device and the current $i_L(t)$ flowing through it satisfy the equation

$$v_L(t) = L\frac{di_L(t)}{dt} \quad \text{or} \quad i_L(t) = \frac{1}{L}\int_0^t v_L(\tau)\,d\tau + i_L(0). \qquad (2.1)$$

Similarly, for a capacitor C, the relation

$$i_C(t) = C\frac{dv_C(t)}{dt} \quad \text{or} \quad v_C(t) = \frac{1}{C}\int_0^t i_C(\tau)\,d\tau + v_C(0) \qquad (2.2)$$

holds. Thus, the circuit equations are, in general, integro-differential relations. For the circuit shown in Fig. 2.2, for example, the current $i(t)$ may be found by solving the equation

$$L\frac{di(t)}{dt} + Ri(t) + \frac{1}{C}\int_0^t i(\tau)\,d\tau = v(t) - v_C(0). \qquad (2.3)$$

For the solution, the initial values $i(0)$ and $v_C(0)$ must be known.

The solution of such integro-differential equations can be performed using direct mathematical methods.[1] These tend to be complicated, however. It is

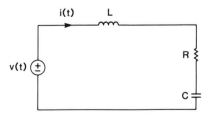

FIGURE 2.2. Continuous-time analog electric circuit.

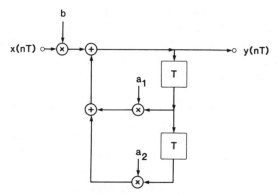

FIGURE 2.3. Digital system.

much more convenient to use solution techniques based on the *Laplace transformation*, to be discussed in the next section.

Similarly, the analysis of a discrete-time system involves the construction and solution of network equations. To construct these equations, topological relations which involve only additions and subtractions (and are hence analogous to the Kirchhoff laws) must be combined with branch relations.[3] The latter may involve multiplication by a constant, or a delay by the sampling interval T. Consider, as an example, the digital system shown in Fig. 2.3. It is easy to show that the input–output relation is

$$y(nT) - a_1 y(nT - T) - a_2 y(nT - 2T) = bx(nT). \qquad (2.4)$$

This is a *difference equation*; it can be solved by direct methods, if the initial conditions $y(0)$ and $y(-T)$ are given. It is much simpler, however, to use z-transformation, to be discussed in Section 2.4, in the process.

For a linear time-invariant continuous-time analog system, an important characteristic is its *impulse response*. The *impulse function* or *Dirac delta function* $\delta(t)$ can be regarded as the limiting case of the pulse function $p_\epsilon(t)$ shown in Fig. 2.4 as $\epsilon \to 0$. The impulse response $h(t)$ is the output of the system (with zero initial energy) if $\delta(t)$ is its input signal (Fig. 2.5a).

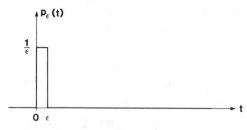

FIGURE 2.4. Pulse function. For $\epsilon \to 0$, $p_\epsilon(t)$ becomes the impulse function $\delta(t)$.

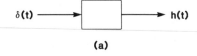

(a)

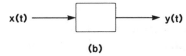

(b)

FIGURE 2.5. Continuous-time analog system with: (*a*) impulse-function input signal; (*b*) general input signal.

If now the same zero-energy system is driven by an input signal $x(t)$ (Fig. 2.5*b*), then it can be shown[1,2] that the output is given by

$$y(t) = \int_0^t x(t - \tau)h(\tau)\,d\tau, \tag{2.5}$$

where $x(t) = h(t) = 0$ for $t < 0$ is assumed. Thus, the impulse response defines the response of the system to other signals as well. The operation between $x(t)$ and $h(t)$, indicated in (2.5), is often denoted by $x(t) * h(t)$. It is called the (unilateral) *convolution* of $x(t)$ and $h(t)$. Clearly, except for very simple functions, it is a complicated process to carry out. Once again, however, the use of Laplace transformation can make the task much easier, as will be discussed in the next section.

Similar results can be derived for linear time-invariant discrete-time systems. The impulse function is now defined by

$$\delta(nT) = \begin{cases} 1, & n = 0 \\ 0, & n \neq 0. \end{cases} \tag{2.6}$$

The impulse response $h(nT)$ is the output of the system (with zero initial conditions) if $\delta(nT)$ is its input.

For a different input signal $x(nT)$, the output of the system with zero initial conditions is given by the *convolution* (or *convolution sum*[3])

$$y(nT) = x(nT) * h(nT) = \sum_{k=0}^{n} h(kT)x(nT - kT), \tag{2.7}$$

where $x(nT) = h(nT) = 0$ for $n < 0$ is assumed. This operation, performed directly, is tedious for all but the simplest $x(nT)$ and $h(nT)$. However, as will be shown later, the use of z-transformation reduces the necessary effort considerably.

We conclude that the analysis of linear time-invariant systems is made much easier by the use of transformation methods. Hence, in the next sections, these techniques will be explored.

2.2.   THE LAPLACE TRANSFORMATION[1,4]

The *Laplace transform* of a function $f(t)$ is defined by

$$F(s) = \int_0^\infty f(t)\, e^{-st}\, dt. \tag{2.8}$$

Equation (2.8) defines the *unilateral* Laplace transform; for the *bilateral* Laplace transform, the lower limit of the integration is $-\infty$. We will assume that s is a general complex variable, with values which allow the integral to converge. Also, it will be assumed that the time origin $t = 0$ is chosen such that $f(t) = 0$ for $t < 0$.

The Laplace transformation is a *linear* operation: if $f(t) = k_1 f_1(t) + k_2 f_2(t)$, where k_1, k_2 are constants, then its transform is $F(s) = k_1 F_1(s) + k_2 F_2(s)$, where F_1 and F_2 are the transforms of f_1 and f_2, respectively. In addition, it has three key properties:

1.   If the Laplace transform of $f(t)$ is $F(s)$, then that of $df(t)/dt$ is $sF(s) - f(0)$; or, in a self-explanatory shorthand notation,

$$\text{if } f(t) \leftrightarrow F(s), \text{ then } \frac{df(t)}{dt} \leftrightarrow sF(s) - f(0).$$

2.   In the same notation,

$$\text{if } f(t) \leftrightarrow F(s), \text{ then } \int_0^t f(\tau)\, d\tau \leftrightarrow \frac{F(s)}{s}.$$

3.   Finally,

$$\text{if } f_1(t) \leftrightarrow F_1(s) \text{ and } f_2(t) \leftrightarrow F_2(s), \text{ then}$$

$$f_1(t) * f_2(t) = \int_0^t f_1(t - \tau) f_2(\tau)\, d\tau \leftrightarrow F_1(s) F_2(s).$$

Clearly, properties (1) and (2) promise to simplify the solution of the integro-differential equation significantly. For example, applying them to Eq. (2.3), with $v_C(0) = 0$,

$$L \cdot [sI(s) - i(0)] + R \cdot I(s) + \frac{1}{C} \cdot \frac{I(s)}{s} = V(s) \tag{2.9}$$

results. Here, as elsewhere, capital letters denote transformed functions. This

TABLE 2.1.   Key Properties of the Laplace Transformation

Time Function	Laplace Transform
$f(t)$	$F(s)$
$df(t)/dt$	$sF(s) - f(0)$
$\int_0^t f(\tau)\, d\tau$	$\dfrac{1}{s} F(s)$
$f_1(t) * f_2(t)$	$F_1(s) F_2(s)$
$f(t - a)u(t - a), a \geq 0$	$e^{-as} F(s)$
$tf(t)$	$-\dfrac{dF(s)}{ds}$
$e^{at} f(t)$	$F(s - a)$
$k_1 f_1(t) + k_2 f_2(t)$	$k_1 F_1(s) + k_2 F_2(s)$

can easily be solved for $I(s)$:

$$I(s) = \frac{V(s) + Li(0)}{sL + R + 1/(sC)}. \qquad (2.10)$$

Hence, after the operation of Laplace transformation is inverted so that $i(t)$ is obtained from $I(s)$, the problem is solved.

Both Laplace transformation and inverse transformation can usually be carried out using tables listing the most commonly needed transforms and rules. Table 2.1 gives some key properties of the Laplace transformation; Table 2.2 the most commonly occurring functions. The reader is referred to Refs. 1, 2, and 4 for much more extensive discussions of these important topics.

TABLE 2.2.   Some Commonly Used Laplace-Transform Pairs

$f(t)$	$F(s)$
$\delta(t)$	1
K or $Ku(t)$	$\dfrac{K}{s}$
t^n	$\dfrac{n!}{s^{n+1}}$
e^{at}	$\dfrac{1}{s - a}$
$e^{at} \cos bt$	$\dfrac{s - a}{(s - a)^2 + b^2}$
$e^{at} \sin bt$	$\dfrac{b}{(s - a)^2 + b^2}$

The unilateral Laplace transform of a constant K is, as Table 2.2 indicates, K/s. This is the same as the Laplace transform of $Ku(t)$, where $u(t)$ is the *step function* given by

$$u(t) \triangleq \begin{cases} 1, & t \geq 0 \\ 0, & t < 0. \end{cases} \tag{2.11}$$

The reason is that the unilateral Laplace transformation of Eq. (2.8) takes into account only the behavior of $f(t)$ for $t \geq 0$. In that range $Ku(t) \equiv K$.

It should be noted that the time functions listed in Table 2.2 include all commonly occurring input signals for continuous-time circuits. Their Laplace transforms are all *rational functions* (i.e., ratios of polynomials) in s; their general form is

$$F(s) = \frac{a_m s^m + a_{m-1} s^{m-1} + \cdots + a_1 s + a_0}{b_n s^n + b_{n-1} s^{n-1} + \cdots + b_1 s + b_0}. \tag{2.12}$$

Property 3, also included as the third rule of Table 2.1, is of great importance for solving problems involving convolution. Applying it to Eq. (2.5)

$$Y(s) = X(s)H(s) \tag{2.13}$$

results. Here, $Y(s)$, $X(s)$, and $H(s)$ are the Laplace transforms of $y(t)$, $x(t)$, and $h(t)$, respectively. Thus, the integral relation of (2.5) becomes a simple multiplication in the s domain. $H(s)$ is often called the *transfer function* or *system function*.

In the Laplace transform domain, assuming zero initial conditions, that is, $i_L(0) = 0$ and $v_C(0) = 0$, the relations (2.1) and (2.2) become

$$V_L(s) = sL \cdot I_L(s)$$

and

$$V_C(s) = \frac{1}{sC} \cdot I_C(s). \tag{2.14}$$

This indicates that voltages and currents of the inductors and capacitors satisfy relations akin to Ohm's law in the Laplace transform domain. The role of resistance is played by the *impedance* Z; this is $Z_L = sL$ for an inductor, while $Z_C = 1/sC$ for a capacitor. The analysis of a continuous-time circuit with zero initial conditions and $X(s) = 1$ can be performed, for example, using nodal analysis. This involves the solution of a system of linear equations, where each coefficient is constructed from Z_L and Z_C by multiplication, division, addition, and subtraction.[5] Consequently, all coefficients are rational functions of s and hence (as can be deduced by considering Cramer's rule) so is $Y(s)$. Since for $X(s) = 1$ we have $Y(s) \equiv H(s)$, we conclude that $H(s)$ is also a rational function of s.

As mentioned earlier, all commonly used input functions $x(t)$ have rational Laplace transforms $X(s)$. Hence, all commonly occurring $Y(s)$ are also rational functions.

In view of the above, it is sufficient to discuss the inversion of the Laplace transformation for rational functions $F(s)$. The simplest technique in this case is based on *partial-fraction expansion*. The process will be first illustrated by an example, and subsequently generalized. Consider the inversion of

$$F(s) = \frac{s^2 + 3s + 7}{s^3 + 5s^2 + 12s + 8}. \tag{2.15}$$

We want to break $F(s)$ into a sum of simpler terms, preferably such as are contained in Table 2.2 that can hence be inverted by inspection. Since $s^3 + 5s^2 + 12s + 8 = (s^2 + 4s + 8)(s + 1) = (s + 2 - j2)(s + 2 + j2)(s + 1)$, we can write

$$F(s) = \frac{C_1}{s - s_1} + \frac{C_2}{s - s_2} + \frac{C_3}{s - s_3} \tag{2.16}$$

where the poles are $s_1 = -2 + j2$, $s_2 = -2 - j2$, and $s_3 = -1$. Clearly,

$$C_1 = (s - s_1)\left[F(s) - \frac{C_2}{s - s_2} - \frac{C_3}{s - s_3} \right]. \tag{2.17}$$

Note that since $F(s)$ contains $s - s_1$ in its denominator, this factor can be canceled in $(s - s_1)F(s)$. Hence, setting $s = s_1$ in (2.17),

$$C_1 = \left[(s - s_1)F(s) \right]_{s=s_1}$$

$$= \left[\frac{s^2 + 3s + 7}{(s + 2 + j2)(s + 1)} \right]_{s=-2+j2} = \frac{j}{4}. \tag{2.18}$$

Similarly,

$$C_2 = \left[(s - s_2)F(s) \right]_{s=s_2} = \left[\frac{s^2 + 3s + 7}{(s + 2 - j2)(s + 1)} \right]_{s=-2-j2} = \frac{-j}{4} \tag{2.19}$$

and

$$C_3 = \left[(s - s_3)F(s) \right]_{s=s_3} = \left[\frac{s^2 + 3s + 7}{s^2 + 4s + 8} \right]_{s=-1} = 1. \tag{2.20}$$

The first two terms can be combined to eliminate imaginary quantities:

$$\frac{j/4}{s + 2 - j2} - \frac{j/4}{s + 2 + j2} = \frac{-1}{(s + 2)^2 + 4} \tag{2.21}$$

and hence

$$F(s) = -\frac{1}{2}\frac{2}{(s + 2)^2 + 4} + \frac{1}{s + 1}. \tag{2.22}$$

From Table 2.2, therefore

$$f(t) = -\tfrac{1}{2}e^{-2t}\sin 2t + e^{-t}. \tag{2.23}$$

Clearly, in the general case of an $F(s)$ given by Eq. (2.12) with $m \leq n$, we can use the partial-fraction expansion

$$F(s) = C_0 + \sum_{i=1}^{n} \frac{C_i}{s - s_i}, \tag{2.24}$$

where the s_i are the zeros of the denominator of $F(s)$, and hence the poles of $F(s)$. The residues C_i are given by $C_0 = a_n/b_n$ for $m = n$, zero if $m < n$, and

$$C_i = \left[(s - s_i)F(s)\right]_{s=s_i}, \qquad i = 1, 2, \ldots, n. \tag{2.25}$$

Terms corresponding to complex conjugate poles s_i may be pairwise combined, and then Table 2.2 used to obtain the inverse transforms of all terms. Equations (2.24) and (2.25) hold for simple poles only.

It follows from the described procedure, and from the last three entries of Table 2.2, that corresponding to each pole $s_i = a_i + jb_i$ of $H(s)$ there will be a term of the form $e^{a_i t}$, or $e^{a_i t}\cos b_i t$, or $e^{a_i t}\sin b_i t$ in the impulse response $h(t)$ of the system. If the real part a_i of the pole is positive, then the magnitude of the corresponding term is exponentially increasing with time. Thus, for any input $x(t)$, $y(t) \to \infty$ as $t \to \infty$. We conclude therefore that the transfer function $H(s)$ of a *stable* system cannot have poles with positive real parts. Thus, all poles are in the left half of the s plane (Fig. 2.6). Any *simple* pole with zero real part results in a term $\cos b_i t$ or $\sin b_i t$ in the transient which continues with a constant amplitude. This is possible, for example, for a lossless circuit.[5] Multiple imaginary poles, however, are not possible.

There is also a *general* formula available[4] for finding the inverse Laplace transform of $F(s)$. Let the real part of the complex variable $s = \sigma + j\omega$ be chosen larger than the real parts of all singularities (e.g., poles) of $F(s)$; then $f(t)$ can be found from

$$f(t) = \frac{1}{2\pi j} \int_{\sigma - j\infty}^{\sigma + j\infty} e^{st}F(s)\, ds \tag{2.26}$$

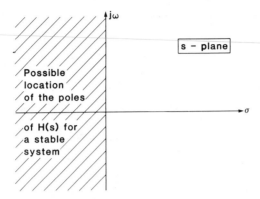

FIGURE 2.6. Pole location for the transfer function of a stable continuous-time system.

for $t > 0$. This formula is difficult to use for actual calculations; however, as will be shown in the next section, it has important theoretical implications.

The Laplace transformation, as shown above, is an effective tool for the mathematical analysis of continuous-time systems. It does not, however, provide much physical insight into the operation of the system. A slight modification of the defining formula, which allows a physical interpretation, results in another important transformation, to be discussed next.

2.3. THE FOURIER TRANSFORMATION[1,4]

The (bilateral) *Fourier transform* of a function $f(t)$ is defined by

$$F(j\omega) = \int_{-\infty}^{\infty} f(t) e^{-j\omega t} \, dt, \qquad (2.27)$$

while its inverse can be shown to be

$$f(t) = \frac{1}{2\pi} \int_{-\infty}^{\infty} F(j\omega) e^{j\omega t} \, d\omega. \qquad (2.28)$$

A comparison of these equations with (2.8) and (2.26) shows that the Fourier transform of $f(t)$ is its (bilateral) Laplace transform, evaluated for $s = j\omega$ (i.e., for $\sigma = 0$). In fact, these two transforms are essentially identical; they differ only in notation and application. However, the Fourier transform also has a more direct interpretation. Regarding the integral as the limiting form of summation, (2.28) shows that $f(t)$ can be regarded as a sum of sinusoidal functions $e^{j\omega t} = \cos \omega t + j \sin \omega t$; the weight (or amplitude) of each sinusoidal is $F(j\omega)/2\pi$. Hence, $F(j\omega)$ is often called the *spectrum* of $f(t)$.

TABLE 2.3. Key Properties of the Fourier Transformation

Time Function	Fourier Transform
$f(t)$	$F(j\omega)$
$df(t)/dt$	$j\omega F(j\omega)$
$\displaystyle\int_{-\infty}^{t} f(\tau)\,d\tau$	$\dfrac{1}{j\omega}F(j\omega) + \pi F(0)\delta(\omega)$
$f_1(t) * f_2(t)$	$F_1(j\omega)F_2(j\omega)$
$\displaystyle\int_{-\infty}^{\infty} |f(t)|^2\,dt$	$\dfrac{1}{2\pi}\displaystyle\int_{-\infty}^{\infty} |F(j\omega)|^2\,d\omega$
$f(t - t_0)$	$F(j\omega)e^{-j\omega t_0}$
$f(t)e^{j\omega_0 t}$	$F(j\omega - j\omega_0)$
$k_1 f_1(t) + k_2 f_2(t)$	$k_1 F_1(j\omega) + k_2 F_2(j\omega)$

Due to the basic similarities of the Laplace and Fourier transforms, many of the properties of the Laplace transformation, discussed in Section 2.2, have similar counterparts for the Fourier transformation. A number of key properties are listed in Table 2.3. Also, a few important transforms are included in Table 2.4. Note that the convolution $f_1(t) * f_2(t)$ indicated in Table 2.3 is the *bilateral* convolution defined by

$$f_1(t) * f_2(t) = \int_{-\infty}^{\infty} f_1(\tau)f_2(t - \tau)\,d\tau. \tag{2.29}$$

The formula given in Table 2.4 for the Fourier transform of $f_1 * f_2$ shows that

TABLE 2.4. Some Commonly Used Fourier-Transform Pairs

$f(t)$	$F(j\omega)$
$\delta(t)$	1
$u(t)$	$\pi\delta(\omega) + \dfrac{1}{j\omega}$
$e^{-a|t|}$	$\dfrac{2a}{\omega^2 + a^2}$
$\begin{cases} 1, & |t| < t_0 \\ 0, & |t| > t_0 \end{cases}$	$(2/\omega)\sin\omega t_0$
$\dfrac{\sin\omega_0 t}{\pi t}$	$\begin{cases} 1, & |\omega| < \omega_0 \\ 0, & |\omega| > \omega_0 \end{cases}$
$e^{j\omega_0 t}$	$2\pi\,\delta(\omega - \omega_0)$
1	$2\pi\,\delta(\omega)$
$\cos\omega_0 t$	$\pi\,\delta(\omega - \omega_0) + \pi\,\delta(\omega + \omega_0)$
$\sin\omega_0 t$	$-j\pi\,\delta(\omega - \omega_0) + j\pi\,\delta(\omega + \omega_0)$

if the spectrum of the input signal is $X(j\omega)$, then that of the output signal is

$$Y(j\omega) = H(j\omega)X(j\omega). \tag{2.30}$$

Here, $H(j\omega)$ is the spectrum of the impulse response $h(t)$ of the circuit. It is often called the *frequency response* of the network, and is usually its single most important characteristic. It determines, as Eq. (2.30) shows, how the circuit affects the various frequency components of the signal transmitted from its input to its output terminals.

In the next section, the use of transformation methods for sampled-data signals will be discussed. First, however, an interpretation of the sampled-data signal as a limiting case of a continuous-time one will be given.

2.4. SAMPLING AND THE z-TRANSFORM[1,3,4,6]

Consider the circuit shown in Fig. 2.7a. The switches S_1 and S_2 make momentary contact at the instances shown as t_1 and t_2, respectively, in Fig. 2.7b. Each time S_1 closes, C charges to the instantaneous value $f(nT)$ of $f(t)$. When τ seconds later S_2 closes, C is discharged. The voltage across C is amplified by the buffer amplifier with a voltage gain of k. As a result, the continuous input signal $f(t)$ is transformed into the pulse train $f^*(t)$. Clearly, the nth pulse, located between nT and $nT + \tau$, can be written in terms of the step function defined in (2.11) as

$$f_n(t) = kf(nT)[u(t - nT) - u(t - nT - \tau)]. \tag{2.31}$$

Hence, assuming $f(t) = 0$ for $t < 0$,

$$f^*(t) = \sum_{n=0}^{\infty} f_n(t) = k\sum_{n=0}^{\infty} f(nT)[u(t - nT) - u(t - nT - \tau)]. \tag{2.32}$$

Next, the Laplace transform $F^*(s)$ of $f^*(t)$ will be calculated. Using the relations $f(t - a)u(t - a) \leftrightarrow e^{-as}F(s)$ from Table 2.1, and $u(t) \leftrightarrow 1/s$ from Table 2.2,

$$F^*(s) = k\sum_{n=0}^{\infty} f(nT)[(1/s)e^{-snT} - (1/s)e^{-s(nT+\tau)}]$$

$$= k\frac{1 - e^{-s\tau}}{s}\sum_{n=0}^{\infty} f(nT)e^{-snT} \tag{2.33}$$

results.

The simple circuit of Fig. 2.7a is a *sampling stage*. It may or may not be useful for a given practical application; however, it can be used as a model for

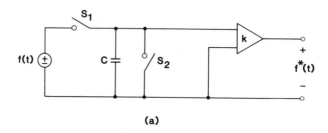

(a)

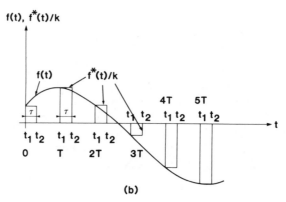

(b)

FIGURE 2.7. Sampling: (a) sampling stage; (b) continuous-time signal $f(t)$ and its sampled equivalent $f^*(t)$.

theoretical discussions. For example, for the generation of the sampled signal $v(nT)$ shown in Fig. 2.1b, we can assume that the interval τ between t_1 and t_2 is made very small. Then the factor preceding the summation in (2.33) satisfies the approximation

$$k\frac{1 - e^{-s\tau}}{s} \simeq k\frac{1 - (1 - s\tau)}{s} = k\tau. \tag{2.34}$$

To simplify the discussion, we choose $k = 1/\tau$. Then the *area*, rather than the amplitude, of each pulse of $f^*(t)$ equals $f(nT)$. With these assumptions, (2.33) becomes

$$F^*(s) = \sum_{n=0}^{\infty} f(nT)e^{-snT} = \sum_{n=0}^{\infty} f(nT)z^{-n}, \tag{2.35}$$

where the shorthand notation

$$z \triangleq e^{sT} \tag{2.36}$$

is used. It is customary to denote the right-hand side of (2.35) by $F(z)$; it is called the (unilateral) z-*transform* of the set of samples $f(nT)$. (For the bilateral z-transform, the lower limit of n is $-\infty$.) Note that the z-transform is related *only* to the samples $f(nT)$ of $f(t)$; it is independent of the function values between the samples. The set $\{f(nT)\}$ is often called a *sequence* or *time series*.

The process leading from $f(t)$ to $F(z)$ will be illustrated by the simple example $f(t) = e^{-at}u(t)$. Then the time series contains the samples

$$f(nT) = \begin{cases} e^{-anT}, & n \geq 0 \\ 0, & n < 0 \end{cases} \tag{2.37}$$

and hence, by (2.35), its z-transform is given by

$$F(z) = \sum_{n=0}^{\infty} f(nT)z^{-n} = \sum_{n=0}^{\infty} (e^{-aT}z^{-1})^n. \tag{2.38}$$

The infinite series in (2.38) converges only if $|e^{aT}z| > 1$, that is, if $|z| > |e^{-aT}|$. For such z values

$$F(z) = \frac{1}{1 - e^{-aT}z^{-1}} = \frac{z}{z - e^{-aT}}. \tag{2.39}$$

Clearly, the z-transform of a sampled function differs from its Laplace transform only by the use of the shorthand notation (2.36). Hence, the properties of z-transforms are readily deduced from those of Laplace transforms. For our purposes, the most important ones are the following:

1. If the z-transform of a sequence $f(nT)$ is $F(z)$, then that of the *delayed* sequence $f(nT - kT)$ is $z^{-k}F(z)$. Here, n and k are positive integers, and it is assumed that $f(nT) = 0$ for $n < 0$. Thus, if $f(nT) \leftrightarrow F(z)$, then $f(nT - kT) \leftrightarrow z^{-k}F(z)$.

2. If $f_1(nT) \leftrightarrow F_1(z)$ and $f_2(nT) \leftrightarrow F_2(z)$, then their convolution satisfies[3]

$$f_1(nT) * f_2(nT) \triangleq \sum_{k=0}^{n} f_1(kT)f_2(nT - kT)$$

$$\leftrightarrow F_1(z)F_2(z). \tag{2.40}$$

Property 1 can be derived directly from the definition (2.38) of the z transform. Alternatively, it can be seen from the shifting theorem $f(t - a) \leftrightarrow e^{-as}F(s)$ of the Laplace transformation, with $a = kT$ and $e^{-as} = e^{-sTk} = z^{-k}$. It is obviously useful for the analysis of sampled-data systems, such as that shown in Fig. 2.3. Applying it to Eq. (2.4), we get

$$Y(z) - a_1z^{-1}Y(z) - a_2z^{-2}Y(z) = bX(z) \tag{2.41}$$

TABLE 2.5. Key Properties of the z-Transformation

Sequence	z-Transform
$f(nT)$	$F(z)$
$f(nT - kT), \ k = 0,1,2,\dots$	$z^{-k}F(z)$
$f(nT + kT), \ k = 1,2,\dots$	$z^{k}F(z) - \displaystyle\sum_{m=0}^{k-1} f(mT)z^{-m}$
$\displaystyle\sum_{m=0}^{n-1} f(mT)$	$(z-1)^{-1}F(z)$
$a^{-n}f(nT)$	$F(az)$
$nf(nT)$	$-z\dfrac{dF(z)}{dz}$
$f(-nT)$	$F(1/z)$
$\displaystyle\sum_{k=0}^{n} f_1(kT)f_2(nT - kT)$	$F_1(z)F_2(z)$
$k_1 f_1(nT) + k_2 f_2(nT)$	$k_1 F_1(z) + k_2 F_2(z)$

which can be solved to give

$$Y(z) = \frac{bX(z)}{1 - a_1 z^{-1} - a_2 z^{-2}} \tag{2.42}$$

Property 2 is applicable to the input–output relations (2.7) of linear time-invariant sampled-data systems.

These, and some other key properties of the z-transformation are collected in Table 2.5. Some important sequences and their z-transforms are listed in Table 2.6. Since unilateral z-transforms are discussed, all sequences are assumed to be zero for negative values of n, or, equivalently, to be multiplied by the discrete-time unit-step sequence

$$u(nT) \triangleq \begin{cases} 1, & n \geq 0 \\ 0, & n < 0. \end{cases} \tag{2.43}$$

We note that all entries on the right-hand side of Table 2.6 are rational functions of z. Since the entries on the left-hand side include all commonly used input sequences, the z-transformed input $X(z)$ of a discrete-time system is usually a rational function. If the operations performed in the system are restricted to additions, multiplications by a constant, and delays (as is the case for linear filters), then all signals in the system will be rational if $X(z)$ is. Thus, $Y(z)$ and $H(z)$ are also rational.

TABLE 2.6. Some Commonly Used z-Transform Pairs

$f(nT)$	$F(z)$
$\delta(nT)$	1
$Ku(nT),\ K$	$\dfrac{Kz}{z-1}$
a^{nT}	$\dfrac{z}{z-a^T}$
$a^{nT}\sin n\omega_0 T$	$\dfrac{\left(a^T\sin\omega_0 T\right)z}{z^2-\left(2a^T\cos\omega_0 T\right)z+a^{2T}}$
$a^{nT}\cos n\omega_0 T$	$\dfrac{z\left(z-a^T\cos\omega_0 T\right)}{z^2-\left(2a^T\cos\omega_0 T\right)z+a^{2T}}$
nTa^{nT-T}	$\dfrac{zT}{\left(z-a^T\right)^2}$

Applying Eq. (2.40) of Property 2 to (2.7),

$$Y(z) = H(z)X(z) \tag{2.44}$$

results. If the input signal is the impulse function given in (2.6), then (from Table 2.6), $X(z) \equiv 1$ and hence $Y(z) = H(z)$. We thus confirm that $H(z)$ is the z-transform of the impulse response $h(nT)$ defined in Section 2.1.

Equation (2.44) gives an easy method for finding the z-transformed output signal of a discrete-time system if its input signal and impulse response are known. To obtain the output sequence in the time domain, however, the inverse operation of the z-transformation must be performed. This task is usually made simple by the rational character of the signal $F(z)$ at hand. As with the Laplace transforms, the simplest available technique is then partial-fraction expansion. The purpose is to break $F(z)$ into first- and second-order terms, which correspond directly to the entries given in Table 2.6. Since all of these (except the first one, which is constant) contain a factor z, it is more expedient to expand $F(z)/z$ rather than $F(z)$. The process will be illustrated by an example. Let $T = 1$ and

$$F(z) = \frac{8z^3 - 30z^2 + 19z}{(z-1)^2(z-4)}.$$

By partial-fraction expansion

$$\frac{F(z)}{z} = \frac{8z^2 - 30z + 19}{(z-1)^2(z-4)} = \frac{C_1}{z-1} + \frac{C_2}{(z-1)^2} + \frac{C_3}{z-4}.$$

To find the C_i, the terms on the right-hand side are recombined, and the numerator

$$C_1(z - 1)(z - 4) + C_2(z - 4) + C_3(z - 1)^2$$

is equated to $8z^2 - 30z + 19$. From the equality of the coefficients of like powers, the linear equations

$$C_1 + C_3 = 8,$$

$$-5C_1 + C_2 - 2C_3 = -30,$$

$$4C_1 - 4C_2 + C_3 = 19$$

result. Solution gives $C_1 = 5$, $C_2 = 1$, and $C_3 = 3$. Hence,

$$F(z) = \frac{5z}{z - 1} + \frac{z}{(z - 1)^2} + \frac{3z}{z - 4}.$$

Using Table 2.6, for $T = 1$ the solution

$$f(nT) = (5 + n + 3 \times 4^n)u(nT)$$

is obtained.

To illustrate the use of the z-transformation in the analysis of a discrete-time system, the system of Fig. 2.3 and its input–output relations will be revisited. The latter was given in the time domain by Eq. (2.4) and in the z domain by (2.42). Let $T = 2$, $b = 2$, $a_1 = 1$, and $a_2 = -0.5$. The *step response* $g(nT)$ will be calculated as follows.

Since now the z-transformed input is (from Table 2.6)

$$X(z) = \frac{z}{z - 1},$$

substitution in (2.42) gives

$$Y(z) = \frac{2z^3}{(z - 1)(z^2 - z + 0.5)}.$$

Partial-fraction expansion gives

$$\frac{Y(z)}{z} = \frac{2z^2}{(z - 1)(z^2 - z + 0.5)}$$

$$= \frac{4}{z - 1} - \frac{2z - 2}{z^2 - z + 0.5}.$$

For the second term, comparison of the coefficients with those of the fourth and fifth entries of Table 2.6 gives

$$a^{2T} = a^4 = 0.5$$

$$-2a^T \cos \omega_0 T = -2\sqrt{0.5} \cos \omega_0 T = -1.$$

Hence, $a = 2^{-1/4}$ and $\cos \omega_0 T = \sin \omega_0 T = \sqrt{2}/2$. Thus, $\omega_0 T = \pi/4$. Partitioning the second term in the form

$$-2 \frac{z - 0.5}{z^2 - z + 0.5} + 2 \frac{0.5}{z^2 - z + 0.5},$$

we get

$$Y(z) = \frac{4z}{z - 1} - \frac{2z(z - 0.5)}{z^2 - z + 0.5} + \frac{2z(0.5)}{z^2 - z + 0.5}.$$

Hence, from Table 2.6, the step response is

$$y(nT) = u(nT)\left[4 + (\sqrt{2}/2)^n(-2 \cos n\pi/4 + 2 \sin n\pi/4)\right].$$

From the last four entries of Table 2.6, it is clear that for any pole z_i of $H(z)$ there will be a factor $a^{nT} = |z_i|^n$ in the corresponding term of the transient response. If $|z_i| > 1$, then this term will increase exponentially with n. Thus, even for a finite (bounded) input, the output will be unbounded. It can therefore be concluded that a *stable* discrete-time system cannot have any pole z_i with a magnitude greater than 1. Thus, all poles of such a system must lie in the shaded area (often called the inside of the *unit circle* or the *unit disk*) in the z-plane (Fig. 2.8).

A simple pole for which $|z_i| = 1$ lies *on* the unit circle. By Table 2.6, such a pole gives rise to a transient of the form $\cos n\omega_0 T$ or $\sin n\omega_0 T$. Thus, it is possible for an undamped system. However, multiple poles with $|z_i| = 1$ give increasing transients and are thus not possible for a stable system.

The preceding discussion, starting with Eq. (2.34), referred to the *sampled* signal $v(nT)$ shown in Fig. 2.1b. However, the circuit of Fig. 2.7 can also be used to generate the *sampled-and-held* signal $v_{SH}(t)$ of Fig. 2.1c, merely by eliminating the switch S_2. Then $f^*(t)$ will remain constant between two adjacent sampling instances nT and $(n + 1)T$. The resulting time and s-domain functions can be obtained by substituting $\tau = T$ into Eqs. (2.32) and (2.33). Thus, using $k = 1$,

$$f_{SH}(t) = \sum_{n=0}^{\infty} f(nT)[u(t - nT) - u(t - nT - T)] \qquad (2.45)$$

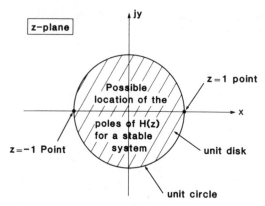

FIGURE 2.8. Pole location for the transfer function of a stable sampled-data system.

and

$$F_{SH}(s) = \frac{1 - e^{-sT}}{s} \sum_{n=0}^{\infty} f(nT)e^{-snT} \qquad (2.46)$$

results.

Comparing Eqs. (2.46) and (2.35), it is clear that for a given $f(t)$—and thus $f(nT)$—the Laplace transforms of the sampled and S/H signals differ only in the factor

$$H_{SH}(s) \triangleq \frac{1 - e^{-sT}}{s}. \qquad (2.47)$$

Hence, while $F_{SH}(s)$ cannot now be written simply as a rational function (or a power series) in z, it can be written as $H_{SH}(s)$ multiplied by such a function. Also, consider a linear system where the input signal $X(s)$ is in a sampled-and-held form, and the internal operations consist only of additions, multiplications by constants, and delaying the signal by T. For such a system, all signals, including the output Y, will be sampled-and-held ones. The transfer function $H = Y/X$ will not contain $H_{SH}(s)$ since that cancels in the ratio; also, by the same argument as for sampled-data systems, H will be a rational function of z. Then $Y = HX$ will be $H_{SH}(s)$ times a rational function of z, as will be all internal signals.

As was the case for continuous-time systems, the physical operation can be better understood using spectral analysis than general transformation methods. This will be discussed in the next section.

2.5.   THE SPECTRA OF SAMPLED-DATA SIGNALS

Consider the spectrum of the sampled signal described by (2.35). It is obtained by replacing s by $j\omega$ (or, equivalently, z by $e^{j\omega T}$). Then

$$F^*(j\omega) = \sum_{n=0}^{\infty} f(nT)e^{-jn\omega T}. \qquad (2.48)$$

To emphasize the fact that ω appears only through the exponentials, this spectrum is often written as $F^*(e^{j\omega T})$. We shall use both notations.

It is evident that $F^*(e^{j\omega T})$ is a *periodic* function of ω with a period $2\pi/T$, since if ω is replaced by $\omega + 2\pi/T$ in (2.48), each exponential factor satisfies

$$e^{-jn\omega T} \rightarrow e^{-jn(\omega + 2\pi/T)T} = e^{-jn\omega T}e^{-jn2\pi}. \qquad (2.49)$$

now since $e^{-jn2\pi} = 1$, $F^*(e^{j\omega T})$ remains unchanged. It hence has a period $2\pi/T$. In fact, it can be shown (Problem 2.20) that the spectrum of the sampled signal $f^*(t)$ is related to the spectrum of the continuous-time signal $f(t)$ from which it is obtained (Fig. 2.7) by

$$F^*(j\omega) = \frac{1}{T} \sum_{k=-\infty}^{\infty} F(j\omega - jk2\pi/T). \qquad (2.50)$$

Equation (2.50) has some very important implications. Consider the continuous-time signal spectrum $F_A(j\omega)$ in Fig. 2.9a. It is fully *band limited*, that is, it vanishes outside the bounds $-\omega_A$ and ω_A, where $\omega_A < \pi/T$. Hence, when $F_A(j\omega)$ is replicated with a repetition period $2\pi/T$ as dictated by (2.50), the replicas forming $F_A^*(j\omega)$ do *not* overlap (Fig. 2.9b). Thus, there is a one-to-one relation between the values of $F_A(j\omega)$ and $F_A^*(j\omega)$.

By contrast, the "tails" of the broader spectrum $F_B(j\omega)$ extend beyond the bounds $\pm\pi/T$ (Fig. 2.9a) to $\pm\omega_B$. Thus, when it is replicated (Fig. 2.9b), these tails overlap, and the value of $F_B^*(j\omega)$ at any frequency ω is influenced by the values of $F_B(j\omega)$ at several different frequencies. This phenomenon is called *aliasing* or *folding*. It is a nonlinear distortion.

For $F_A(j\omega)$, the spectrum and thus the time function $f_A(t)$ can be obtained from the sampled-signal spectrum uniquely, and in a straightforward way. Specifically, if an ideal low-pass filter with the transfer function

$$H(j\omega) = \begin{cases} 1, & |\omega| \leq \pi/T \\ 0, & |\omega| > \pi/T \end{cases} \qquad (2.51)$$

is used to process the sampled signal $f_A^*(t)$, then only the "main lobe" of

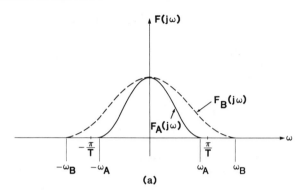

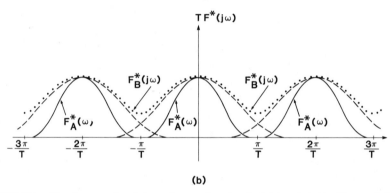

FIGURE 2.9. Signal spectra for: (*a*) continuous-time signals $f_A(t)$ and $f_B(t)$; (*b*) sampled-data signals obtained from $f_A(t)$ and $f_B(t)$.

$F_A^*(j\omega)$, lying in the $-\pi/T \leq \omega \leq \pi/T$ range, will pass through the filter. Since this main lobe is (apart from the unimportant factor T) identical to $F_A(j\omega)$, the continuous-time signal $f_A(t)$ is thus recovered. By contrast, no such operation can regain $F_B(j\omega)$ from $F_B^*(j\omega)$, since no part of the latter retains the original shape of $F_B(j\omega)$.

We can thus conclude that if the sampling frequency $2\pi/T$ is larger than twice the maximum frequency (say, ω_A) in the spectrum of the continuous-time signal, then no aliasing distortion will take place. The original spectrum $F_A(j\omega)$ can then be recovered undistorted from the sampled-signal spectrum by using a low-pass filter with a cutoff frequency π/T. If, on the other hand, $2\pi/T$ is less than twice the band limit of the original signal, then aliasing occurs, and the continuous-time signal is irretrievably lost.

This observation was first made by Nyquist who also showed that if $\pi/T > \omega_A$ then the formula for recovering $f(t)$ from the samples $f(nT)$ is

$$f(t) = \sum_{n=0}^{\infty} f(nT) \frac{\sin[(\pi/T)(t - nT)]}{(\pi/T)(t - nT)} . \qquad (2.52)$$

Equation (2.52) follows simply from Fig. 2.9 (cf. Problem 2.21).
    The condition on the sampling rate

$$\frac{2\pi}{T} > 2\omega_A \qquad (2.53)$$

is often called *Nyquist's criterion*, while the lower bound $2\omega_A$ on the sampling frequency is called the *Nyquist rate*. In fact, no physical signal is strictly band limited, and the "maximum frequency" ω_A may be arbitrarily designated such that the amplitude $|F_A(j\omega)|$ is, say, 60 dB below its low-frequency value for $|\omega| > \omega_A$.
    So far, the discussions involved a sampled signal with zero-width samples $(\tau \to 0)$. For sampled-and-held (S/H) signals, the spectrum is obtained by setting $s = j\omega$ in the Laplace transform (2.46):

$$F_{SH}(j\omega) = \frac{1 - e^{-j\omega T}}{j\omega} \sum_{n=0}^{\infty} f(nT) e^{-jn\omega T}. \qquad (2.54)$$

A comparison with (2.48) shows that $F_{SH}(j\omega) = H_{SH}(j\omega)F^*(j\omega)$, where the factor

$$H_{SH}(j\omega) \triangleq \frac{1 - e^{-j\omega T}}{j\omega} = Te^{-j\omega T/2} \frac{\sin(\omega T/2)}{\omega T/2} \qquad (2.55)$$

is due to the holding operation. $H_{SH}(j\omega)$ can be regarded as the transfer function of the holding device. Note that $H_{SH}(j\omega)$ has a linear phase. Its amplitude response is shown schematically in Fig. 2.10. It is often called the "$(\sin x)/x$ response," and is characteristic of sampled-and-held signal spectra.
    Next, the spectrum of a continuous-time signal $f(t)$ will be compared with that of the S/H signal obtained from it. Assuming that the Nyquist criterion (2.53) holds, the situation is as shown in Fig. 2.11. The original spectrum $F(j\omega)$ is changed into

$$F_{SH}(j\omega) = e^{-j\omega T/2} \frac{\sin(\omega T/2)}{\omega T/2} \sum_{k=-\infty}^{\infty} F\left(j\omega - jk\frac{2\pi}{T} \right), \qquad (2.56)$$

as a combination of Eqs. (2.54) and (2.50) shows. $F(j\omega)$ is thus replicated and

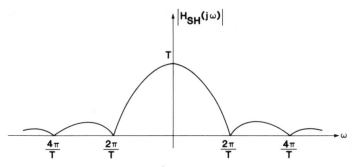

FIGURE 2.10. The amplitude response $|H_{SH}(j\omega)| = 2\sin(\omega T/2)/\omega$.

multiplied by the "$(\sin x)/x$" response. As a result, the main lobe in $-\pi/T < \omega < \pi/T$ is no longer quite the same as $F(j\omega)$, and the side lobes centered around $\pm 2\pi/T, \pm 4\pi/T, \ldots$ are greatly reduced.

The distortion of the main lobe is simply due to the factor $H_{SH}(j\omega)$. It is thus *linearly* distorted, as opposed to the *nonlinear* distortion which aliasing introduces. In fact, $F(j\omega)$ can be recovered from $F_{SH}(j\omega)$, if Nyquist's criterion is satisfied, in two steps. The ideal low-pass filter with the transfer function $H(j\omega)$ given in (2.51) can be used to select the main lobe and suppress all side lobes; then an "amplitude equalizer" stage with a transfer function $H_e(j\omega) = 1/H_{SH}(j\omega)$ will restore $F(j\omega)$.

In practice, Nyquist's criterion is often satisfied not simply by choosing the sampling rate $2\pi/T$ higher than $2\omega_A$, but rather by reducing ω_A. Thus, *before* sampling, the continuous-time signal $f(t)$ is passed through a low-pass filter

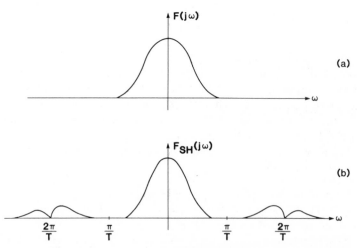

FIGURE 2.11. Signal spectra for: (*a*) a continuous-time signal $f(t)$; (*b*) a sampled-and-held signal obtained from $f(t)$.

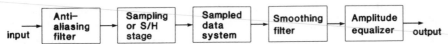

FIGURE 2.12. Block diagram of a sampled-data system with continuous-time input and output signals.

(called an *antialiasing* filter) which reduces its bandwidth to the $-\pi/T < \omega < \pi/T$ range. This can be accomplished by the ideal low-pass filter of Eq. (2.51). Having thus "cut off the tails" of the spectrum $F(j\omega)$ which extended beyond $\pm \pi/T$, $f(t)$ can be sampled (or sampled-and-held) without aliasing. The sampled signal can now be processed in a sampled-data system. Finally, if necessary, a continuous-time signal can again be regained by selecting the main lobe of the sampled-data output spectrum with another low-pass filter given by (2.51). This second filter is often called a *smoothing* filter, since it provides a smooth continuous-time signal $f(t)$ from a sampled (or S/H) one. If an S/H system is used, then (as discussed earlier) an equalizer may also be needed to reduce the $(\sin x)/x$ amplitude distortion.

A schematic representation of a sampled-data system with continuous-time input and output signals is therefore of the form shown in Fig. 2.12. The antialiasing and smoothing filters can be identical, and should ideally have the transfer function given in Eq. (2.51). In practice, of course, the infinitely sharp cutoff implied by (2.51) can only be approximated by a real filter. Hence, a margin of, say, 20% or more is allowed in the selection of the cutoff frequency of the filters so that $\omega_c \le 0.8 \, \pi/T$ can be chosen.

In the foregoing discussions, *analysis* techniques have been described for continuous-time and sampled-data systems. Next, the mathematical tools developed in the course of the discussions will be used for the *design* of sampled-data filters.

2.6. THE DESIGN OF SAMPLED-DATA FILTERS FROM CONTINUOUS-TIME MODELS

Consider the important case when the sampled-data system of Fig. 2.12 is to perform as a *filter*. Sampled-data filters can be divided into two categories: *finite-impulse-response* (*FIR*) and *infinite-impulse-response* (*IIR*) filters. As the names imply, if the impulse $\delta(nT)$ of Eq. (2.6) is applied to the input of an FIR filter, then its output signal—which is its impulse response $h(nT)$—will contain only a finite number of samples, so that $h(nT)$ is of finite duration. By contrast, the impulse response of IIR filters is infinitely long. Physically, an IIR filter has feedback loops, in which a signal, once injected by the input $\delta(nT)$, can circulate forever. An FIR filter typically has no such loops, although in principle it is possible to design an FIR filter which contains feedback, and relies on the cancellation of signals to limit the duration of $h(nT)$. Mathematically, as Table 2.6 shows, any terms in the partial-fraction expansion of $H(z)$

which are of the form $k/(z - a)$ or $k/(z^2 + az + b)$ correspond to sequences of infinite duration. Since such terms are introduced by those poles of $H(z)$ which lie at finite nonzero location, we conclude that all poles of the transfer function of an FIR filter must be either at $z = 0$ or $z \to \infty$. Hence, for such a filter $H(z)$ is a power series in z with a finite number of terms. Thus,

$$H(z) = \sum_{n=n_1}^{n_2} h_n z^{-n}, \qquad (2.57)$$

where $|n_1|, |n_2| < \infty$.

Since $H(z)$ is the z-transform of the impulse response $h(nT)$, that is,

$$H(z) = \sum_{n=0}^{\infty} h(nT) z^{-n}, \qquad (2.58)$$

clearly $h_n = h(nT)$, so the coefficients in (2.57) are the samples of the impulse response, and $n_2 > n_1 \geq 0$.

If a highly selective filter is required, then $H(z)$ needs poles (natural frequencies) near the cutoff of the filter. Hence, IIR filters are preferable in such applications. For an IIR filter, $H(z)$ is a *rational* function, that is, the ratio of two polynomials $N(z)$ and $D(z)$:

$$H(z) = \frac{N(z)}{D(z)} = \frac{a_N z^N + a_{N-1} z^{N-1} + \cdots + a_1 z + a_0}{b_M z^M + b_{M-1} z^{M-1} + \cdots + b_1 z + b_0}. \qquad (2.59)$$

Hence, some poles are finite and nonzero.

Usually, for a given specification, the transfer function $H(z)$ of an IIR filter is not found directly. Instead, advantage is taken of the fact that a large collection of transfer functions has been compiled over the years for *continuous-time* filters (see, e.g., Refs. 5 and 7–9). These functions $H(s)$ can also be used, with minor modifications, for the design of *sampled-data* filters using the transformation methods which will be described in this section.

All strategies which will be discussed below for obtaining $H(z)$ from a "model" transfer function are based on a transformation of the frequency variable. Thus, let the transfer function $H_a(s_a)$ of a continuous-time filter be given. (Here, and in what follows, the subscript a is used to distinguish the variables of the "model filter"; a stands for "analog" but for our purposes it implies "continuous time.") Then, we will obtain $H(z)$ simply by replacing s_a by some function

$$s_a = f(z) \qquad (2.60)$$

in H_a. Thus, the desired transfer function is

$$H(z) = H_a(s_a)\big|_{s_a = f(z)} = H_a[f(z)]. \qquad (2.61)$$

What are the requirements for $f(z)$? As mentioned earlier, $H_a(s_a)$ is a rational function of s_a, while $H(z)$ is a rational function of z. Hence, $f(z)$ must itself be a rational expression in z. Furthermore, for an analog filter, the frequency response $H_a(j\omega_a)$ of the model has some desirable properties (usually selectivity) which we would presumably like to obtain also for the frequency response $H(e^{j\omega T})$ of the sampled-data filter to be designed. Hence, the frequency axis $s_a = j\omega_a$ of the s_a-plane must be transformed into the unit circle $z = e^{j\omega T}$ of the z-plane. Finally, the $H_a(s_a)$ functions are tabulated only for stable (passive) filters, and (unless an oscillator is required) the final sampled-data filter designed from $H(z)$ must also be stable. From the stability conditions discussed earlier (cf. Figs. 2.6 and 2.8), we know therefore that the poles s_{ai} of $H_a(s_a)$ are inside the left half of the s_a plane, while the poles z_i of $H(z)$, given by the relation

$$s_{ai} = f(z_i), \tag{2.62}$$

must be inside the unit circle in the z-plane. We conclude that the mapping (2.60) must transform the left half of the s_a-plane and the z-plane unit disk into each other. The conditions of $f(z)$ are therefore the following:

1. $f(z)$ must be a rational function of z.
2. For $|z| = 1$, $f(z)$ must be pure imaginary; $f(e^{j\omega T}) = j\omega_a$. Vice versa, if $s_a = f(z)$ is imaginary, then $|z| = 1$ must hold.
3. For $|z| < 1$, the real part of $s_a = f(z)$ must be negative. Vice versa, if Re $s_a < 0$, then the corresponding z must have an absolute value less than 1.

Rational functions which satisfy these conditions can be found heuristically. However, a systematic derivation, based on numerical integration techniques, is also possible. The latter will be described here.

Consider a model continuous-time filter which has the transfer function $H_a(s_a)$. Its response can be determined from a system of first-order differential equations

$$\frac{dx_i(t)}{dt} = g_i(t), \qquad i = 1, 2, \ldots, N. \tag{2.63}$$

Here, the x_i are the *state variables* of the filter (Ref. 2, pp. 501–522), and the relations in (2.63) are its *state equations*. The $g_i(t)$ are linear functions of the state variables $x_i(t)$ and the input signal. Using Laplace transformation, (2.63) becomes

$$s_a X_i(s_a) = G_i(s_a), \qquad i = 1, 2, \ldots, N \tag{2.64}$$

where $x_i(t) = 0$ for $t \leq 0$ is assumed, and the subscript a used to indicate that the expression refers to the analog model.

Next, we will derive the state equations for some sampled-data systems which have approximately the same properties as the continuous-time filter. Since the required relations for the sampled-data systems are first-order *difference* equations, (2.63) is integrated between $(n - 1)T$ and nT:

$$\int_{nT-T}^{nT} \frac{dx_i(t)}{dt} dt = x_i(nT) - x_i(nT - T)$$

$$= \int_{nT-T}^{nT} g_i(t) dt, \qquad i = 1, 2, \ldots, N. \qquad (2.65)$$

This transforms the LHS of the equation into the proper difference form. To evaluate the integral on the RHS, numerical integration can be used. Several techniques are available.

Forward Euler Integration

This process is based on the approximation, due to Euler,

$$\int_{nT-T}^{nT} g_i(t) dt \cong Tg_i(nT - T) \qquad (2.66)$$

It is illustrated in Fig. 2.13a; the cross-hatched rectangular area is used as an estimate (in this case, a low estimate) of the area under the $g_i(t)$ curve between $t_1 = (n - 1)T$ and $t_2 = nT$.

Substituting the approximate value from (2.66) into (2.65), the difference equation approximating (2.63)

$$x_i(nT) - x_i(nT - T) = Tg_i(nT - T) \qquad (2.67)$$

results. Using z-transformation, (2.67) gives

$$X_i(z) - z^{-1}X_i(z) = Tz^{-1}G_i(z). \qquad (2.68)$$

{Note that $X_i(z) = X_i[f(z)]$ is a *different* function from $X_i(s_a)$, even though, for simplicity, the same notation is used.} From (2.68),

$$\frac{z - 1}{T} X_i(z) = G_i(z), \qquad i = 1, 2, \ldots, N. \qquad (2.69)$$

A comparison of (2.64) and (2.69) suggests the transformation

$$s_a = f(z) = \frac{z - 1}{T}. \qquad (2.70)$$

Next, the transformation of (2.70) will be tested with respect to Conditions 1–3. It is clearly a rational function of z as required by Condition 1. Testing

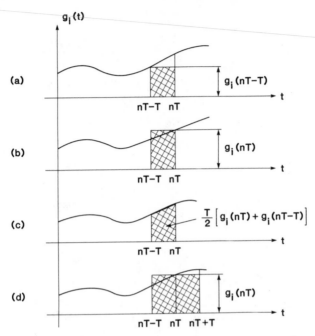

FIGURE 2.13. Numerical integration techniques: (*a*) forward Euler integration; (*b*) backward Euler integration; (*c*) trapezoidal integration; (*d*) midpoint integration.

for Condition 2, we set $s_a = j\omega_a$. Then

$$z = s_a T + 1 = 1 + j\omega_a T$$

and hence $|z| = 1$ holds only for $\omega_a = 0$. Figures 2.14*a* and 2.14*b* show the image of the $j\omega_a$-axis in the z-plane. Since the image is a vertical line tangential to the unit circle, for $|\omega_a T| \ll 1$ we have $|z| \simeq 1$. Thus, Condition 2 is met only approximately, and only for $|\omega_a| \ll 1/T$. The figure also shows three poles in the s_a-plane and their images in the z-plane. Clearly, all three poles belong to a stable continuous-time filter. However, in the z-plane, while z_1 is stable, z_2 is nearly on the unit circle and z_3 is outside it. Thus, the $H(z)$ obtained using the "forward-Euler" transformation of (2.70) from $H_a(s_a) = 1/[(s - s_{a1})(s - s_{a2})(s - s_{a3})]$ is in this example unstable. Hence, Condition 3 is violated for this mapping. Even if all poles of a $H_a(s_a)$ are mapped inside the unit circle, usually the "dominant" pole (i.e., that closest to the $j\omega_a$-axis) will move very close to the unit circle in the z-plane. Hence, a *peaking* of the passband gain occurs in the sampled-data filter (Figs. 2.15*a* and 2.15*b*). Also, if the zeros of $H_a(s_a)$ (i.e., the transmission zeros) are on the $j\omega_a$-axis, those of $H(z)$ will *not* be on the unit circle. Hence, $H(e^{j\omega T})$ will not be zero for any ω, and thus the stopband response will deteriorate (Figs. 2.15*a* and 2.15*b*). Of course,

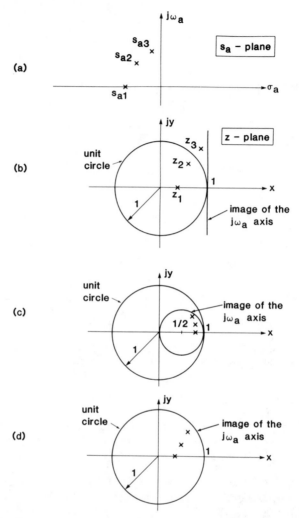

FIGURE 2.14. Illustration of the $s_a \leftrightarrow z$ mapping: (a) poles and frequency axis in the s_a-plane; (b) their images in the z-plane for the forward-Euler transformation; (c) the images for backward-Euler transformation; (d) the images for the bilinear transformation.

$H(e^{j\omega T})$ is periodic in ω, with a period $2\pi/T$. Only part of the $(0, 2\pi/T)$ period is shown in the Figure.

Backward Euler Integration

Using the rectangular approximation shown in Fig. 2.13b for the area under the $g_i(t)$ curve, we obtain

$$\int_{nT-T}^{nT} g_i(t)\, dt \cong Tg_i(nT). \qquad (2.71)$$

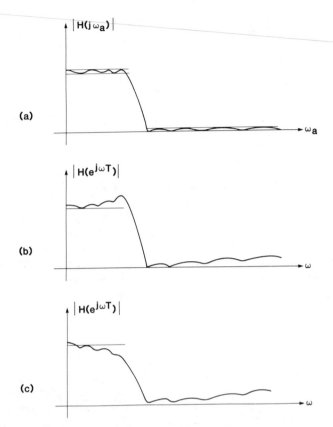

FIGURE 2.15. Frequency responses of low-pass filters: (a) continuous-time filter response with flat (equal-ripple) passband and stopband; (b) sampled-data response with peaking passband and deteriorated stopband, obtained by forward-Euler mapping; (c) response of filter obtained by the backward-Euler mapping.

Substitution into (2.65) gives now

$$x_i(nT) - x_i(nT - T) \cong Tg_i(nT). \qquad (2.72)$$

Hence, by z-transformation,

$$\frac{1 - z^{-1}}{T} X_i(z) = G_i(z) \qquad (2.73)$$

results. Comparison with (2.64) suggests the mapping

$$s_a = f(z) = \frac{1 - z^{-1}}{T} = \frac{z - 1}{Tz}. \qquad (2.74)$$

This is called the *backward-Euler transformation*. It is again rational, as required by Condition 1. However, for $s_a = j\omega_a$,

$$z = x + jy = \frac{1}{1 - j\omega_a T} = \frac{1 + j\omega_a T}{1 + \omega_a^2 T^2}. \tag{2.75}$$

Clearly, $|z| \neq 1$, and Condition 2 is violated. In fact, x and y satisfy

$$\left(x - \tfrac{1}{2}\right)^2 + y^2 = \left(\tfrac{1}{2}\right)^2 \tag{2.76}$$

as can readily be seen (Problem 2.23). Hence, the image of the $j\omega_a$-axis is now a *circle* with a radius of $\tfrac{1}{2}$. Its center lies at $z = \tfrac{1}{2}$ (Fig. 2.14c). The three mapped poles z_1, z_2, and z_3 are now within the mapped $j\omega_a$-axis, and hence inside the unit circle. Thus, a stable $H_a(s_a)$ is mapped into a stable $H(z)$, and Condition 3 holds. The dominant pole is now moved away from the unit circle. Hence, now a *rounding* of the passband response results (Fig. 2.15c). Since the zeros of $H(z)$ are again off the unit circle, the stopband response will also be adversely affected. $H(e^{j\omega T})$ is, of course, periodic in ω; the figure shows only part of one period.

Trapezoidal Integration

A better approximation than those based on the Euler formulas can be obtained if the area under the $g_i(t)$ curve for $(n-1)T \leq t < nT$ is approximated by that of a trapezoid (Fig. 2.13c). Then the approximating formula is

$$\int_{nT-T}^{nT} g_i(t)\, dt \cong \frac{T}{2}\left[g_i(nT - T) + g_i(nT)\right]. \tag{2.77}$$

Hence, from (2.65),

$$x_i(nT) - x_i(nT - T) \cong \frac{T}{2}\left[g_i(nT - T) + g_i(nT)\right] \tag{2.78}$$

results. Using z-transformation, (2.78) becomes

$$\frac{2}{T}\frac{z-1}{z+1}X_i(z) = G_i(z). \tag{2.79}$$

Comparison with (2.64), therefore, indicates the mapping

$$s_a = \frac{2}{T}\frac{z-1}{z+1}. \tag{2.80}$$

This is usually called the *bilinear s-to-z transformation*. It is, of course, rational.

Furthermore, for $s = j\omega_a$,

$$|z| = \left| \frac{1 + j\omega_a T/2}{1 - j\omega_a T/2} \right| \equiv 1. \tag{2.81}$$

Thus, the bilinear mapping satisfies Condition 2 (which neither Euler transformation could do). To test for Condition 3, let $s_a = \sigma_a + j\omega_a$ with $\sigma_a < 0$. Then

$$|z| = \left| \frac{(1 + \sigma_a T/2) + j\omega_a T/2}{(1 - \sigma_a T/2) - j\omega_a T/2} \right| < 1. \tag{2.82}$$

Hence, Condition 3 is also satisfied.

We saw from (2.81) that for the bilinear transformation the $j\omega_a$-axis maps to the unit circle, where $z = e^{j\omega T}$. Hence, from (2.80),

$$j\omega_a = \frac{2}{T} \frac{e^{j\omega T} - 1}{e^{j\omega T} + 1}. \tag{2.83}$$

Dividing numerator and denominator by $2e^{j\omega T/2}$, the RHS becomes $j(2/T)\tan(\omega T/2)$. Hence, (2.83) gives

$$\frac{\omega_a T}{2} = \tan \frac{\omega T}{2},$$

$$\frac{\omega T}{2} = \tan^{-1} \frac{\omega_a T}{2}. \tag{2.84}$$

Equation (2.84) gives the relation between any frequency of the model filter, and its image frequency for the sampled-data circuit. The relation is plotted in Fig. 2.16. Clearly, $\omega_a = 0$ is mapped into $\omega = 0$, $\pm 2\pi/T$, $\pm 4\pi/T,\dots$, while $\omega_a \to \infty$ is mapped into $\omega = \pm \pi/T$, $\pm 3\pi/T,\dots$. In the vicinity of $\omega = 0$ (and $\pm 2\pi/T$, etc.) the ω vs. ω_a relation is nearly linear. Elsewhere, however, it is not; the infinitely long ω_a-axis is compressed into the finite intervals $-\pi/T \leq \omega \leq \pi/T$, $\pi/T \leq \omega \leq 3\pi/T$, and so on of the ω-axis. Hence, the frequency scale of the ω-axis is *warped* as compared to that of the ω_a-axis. This must be taken into consideration in the design. However, since the ω_a- and ω-axes are mapped into each other, a continuous-time frequency response $H_a(j\omega_a)$ with *flat* passbands and stopbands (such as that shown in Fig. 2.15a) will be transformed into a sampled-data response $H(e^{j\omega T})$ which also has flat passbands and stopbands. Also, $H(e^{j\omega T})$ will show the same "ripple" amplitudes as those of $H_a(j\omega_a)$. Naturally, the sampled-data response will be periodic in ω, with a period $2\pi/T$.

On the basis of the above discussions, the following design procedure should be followed when a sampled-data filter transfer function $H(z)$ is to be found

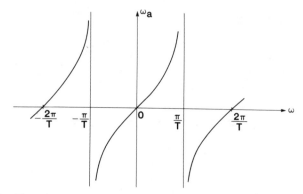

FIGURE 2.16. The relation between the continuous-time and sampled-data frequencies for the bilinear s-to-z mapping.

by the bilinear mapping from a continuous-time model transfer function $H_a(s_a)$:

1. The specifications of the sampled-data filter should be established. Usually, the following parameters are specified:

   (a) The passband limit frequency (ω_p or f_p) and the stopband limit frequency (ω_s or f_s). For bandpass and bandstop filters, there will be several such limit frequencies.

   (b) The maximum allowable variation of $|H(e^{j\omega T})|$ in the passband. This is usually given in decibels (dB) and denoted by α_p. It is called the *passband ripple* of the filter.

   (c) The maximum allowable value of $|H(e^{j\omega T})|$ in the stopband, relative to its maximum value in the passband. This is also usually given in dB's and is denoted by α_s. It is called the *minimum stopband loss*.

   (d) Less frequently, there may be requirements on the phase, delay, and/or transient response of the filter.

2. Since the transformation (2.84) between the frequency scales of the sampled-data filter and its continuous-time model is nonlinear, the specifications of the latter must be *prewarped*. Hence, instead of ω_p and ω_s, the model must be designed for the band-limit frequencies

$$\omega_{ap} = \frac{2}{T} \tan \frac{\omega_p T}{2},$$

$$\omega_{as} = \frac{2}{T} \tan \frac{\omega_s T}{2}.$$

(2.85)

The loss limits α_p and α_s remain, of course, the same since only the independent frequency variable is transformed when $H(e^{j\omega T})$ and $H_a(j\omega_a)$ are changed into each other.

3. A suitable continuous-time transfer function $H_a(s_a)$ is found such that the prewarped specifications represented by ω_{ap}, ω_{as}, α_p, and α_s are satisfied. Typically, this is done with the aid of design tables such as those given in Refs. 7–9. Often, the zeros and poles of $H_a(s_a)$ are thus obtained. From these, the factored form of $H_a(s_a)$ can readily be constructed.

4. To calculate $H(z)$, the variable s_a in $H_a(s_a)$ must be replaced, according to (2.80), by $(2/T)(z - 1)/(z + 1)$. This can be done in each factor of $H_a(s_a)$ to obtain the factored form of $H(z)$ which is often needed for realization. In that way, the zeros and poles in the z-plane can also easily be found.

Since this process is often needed in the design of switched-capacitor filters, it will be illustrated by a typical example. Let the specifications of the sampled-data low-pass filter be the following:

Passband:	0 to $f_p = 1$ kHz,
	$\alpha_p \leq 0.05$ dB.
Stopband:	$f_s \leq 1.5$ kHz to $f_c/2$,
	$\alpha_s \geq 38$ dB.
Sampling frequency:	$f_c = 1/T = 50$ kHz.

The order of the continuous-time model filter can be determined from α_p, α_s, and the *selectivity parameter* $k = \omega_{ap}/\omega_{as}$ (see Ref. 5, p. 518). Using (2.85), k is found from

$$k \triangleq \frac{\omega_{ap}}{\omega_{as}} = \frac{\tan(\omega_p T/2)}{\tan(\omega_s T/2)} = \frac{\tan(\pi f_p/f_c)}{\tan(\pi f_s/f_c)} \cong 0.6656.$$

Choosing an elliptic filter realization to minimize the order of the filter, the tables of Refs. 7–9 can be used to find the transfer function $H_a(s_a)$ of a suitable model. On p. 102 of Ref. 8, the zeros and poles of the continuous-time elliptic filter C05, $\theta = 42°$, $p = 10\%$ with the parameters

$$\hat{\omega}_{ap} = 1 \text{ rad/s},$$

$$\hat{\alpha}_p = 0.044 \text{ dB},$$

$$\hat{\omega}_{as} = 1.49448 \text{ rad/s},$$

$$\hat{\alpha}_s = 39.57 \text{ dB}$$

are given. For this filter, $\hat{\alpha}_p$ is slightly smaller, while $\hat{\alpha}_s$ and $\hat{k} \cong 0.669$ are slightly larger, than the specified values. Hence, it provides a suitable model. The corresponding transfer function is in the form

$$\hat{H}_a(s_a) = \hat{K} \frac{(s_a^2 + \hat{\omega}_1^2)(s_a^2 + \hat{\omega}_2^2)}{(s_a - \hat{a}_0)(s_a^2 - 2\hat{a}_1 s_a + \hat{a}_1^2 + \hat{b}_1^2)(s_a^2 - 2\hat{a}_2 s_a + \hat{a}_2^2 + \hat{b}_2^2)}$$

where, from the tables, the parameter values are

$$\hat{\omega}_1 = 1.5514948,$$

$$\hat{\omega}_2 = 2.32131474,$$

$$\hat{a}_0 = -0.78140011,$$

$$\hat{a}_1 = -0.48467278,$$

$$\hat{b}_1 = 0.82815049,$$

$$\hat{a}_2 = -0.128006731,$$

$$\hat{b}_2 = 1.100351473.$$

Here, the poles are $\hat{s}_1 = \hat{a}_0$, $\hat{s}_{2,3} = \hat{a}_1 \pm j\hat{b}_1$, and $\hat{s}_{4,5} = \hat{a}_2 \pm j\hat{b}_2$. The zeros are at $\pm j\hat{\omega}_1$, $\pm j\hat{\omega}_2$, and $s \to \infty$.

The constant $\hat{K}$ is chosen to assure $\hat{H}_a(0) = 1$. This gives $\hat{K} \cong 0.068$.

Note that the $\hat{H}_a(s_a)$ given above is frequency normalized,[5] so that $\omega_{ap} = 1$ rad/s. In fact, the desired passband limit is, from (2.85),

$$\omega_{ap} = \frac{2}{T} \tan \frac{\omega_p T}{2} = 2 f_c \tan \frac{\pi f_p}{f_c} \cong 6291.4667 \text{ rad/s.}$$

Multiplying therefore all zeros and poles by ω_{ap}, the model transfer function becomes

$$H_a(s_a) = K \frac{\left(s_a^2 + \omega_1^2\right)\left(s_a^2 + \omega_2^2\right)}{(s_a - a_0)\left(s_a^2 - 2a_1 s_a + a_1^2 + b_1^2\right)\left(s_a^2 - 2a_2 s_a + a_2^2 + b_2^2\right)}$$

where the denormalized parameter values are

$$K = 428.247646,$$

$$\omega_1 = 9.76117788 \times 10^3,$$

$$\omega_2 = 1.46044744 \times 10^4,$$

$$a_0 = -4.91615278 \times 10^3,$$

$$a_1 = -3.04930266 \times 10^3,$$

$$b_1 = 5.21028124 \times 10^3,$$

$$a_2 = -805.350086,$$

$$b_2 = 6.92282466 \times 10^3.$$

From $H_a(s_a)$, the desired transfer function $H(z)$ can now be obtained if we replace s_a by

$$f(z) = \frac{2}{T}\frac{z-1}{z+1} = 10^5\frac{z-1}{z+1}.$$

The resulting function is then

$$H(z) = C\frac{z+1}{z+d_0}\frac{z^2 + c_1z + 1}{z^2 + e_1z + f_1}\frac{z^2 + c_2z + 1}{z^2 + e_2z + f_2},$$

where

$$C = 3.8719271 \times 10^{-3},$$

$$c_1 = -1.962247471,$$

$$c_2 = -1.916465445,$$

$$d_0 = -0.906284158,$$

$$e_1 = -1.871739343,$$

$$f_1 = 0.88543246,$$

$$e_2 = -1.949416807,$$

$$f_2 = 0.968447477.$$

The frequency response $|H(e^{j\omega T})|$, calculated from $H(z)$, is shown in Fig. 2.17. It meets all specifications.

Midpoint Integration

In this process, the area under the $g_i(t)$ curve in the $(n-1)T \leq t \leq (n+1)T$ range is approximated by that of a rectangle of length $2T$ and height $g_i(nT)$, as shown in Fig. 2.13d. (Note that $t = nT$ is the *midpoint* of the range; hence the name of this numerical integration method.) Thus, from (2.63),

$$\int_{nT-T}^{nT+T}\frac{dx_i(t)}{dt}\,dt = x_i(nT+T) - x_i(nT-T)$$

$$= \int_{nT-T}^{nT+T}g_i(t)\,dt \cong 2Tg_i(nT). \qquad (2.86)$$

Using z-transformation,

$$zX_i(z) - z^{-1}X_i(z) = 2TG_i(z) \qquad (2.87)$$

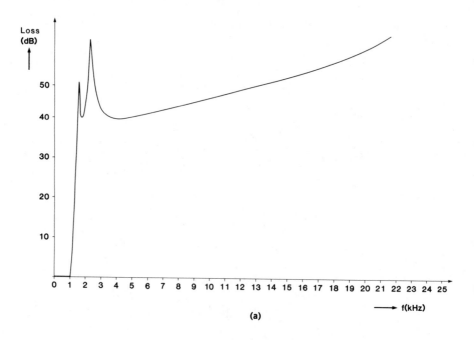

(a)

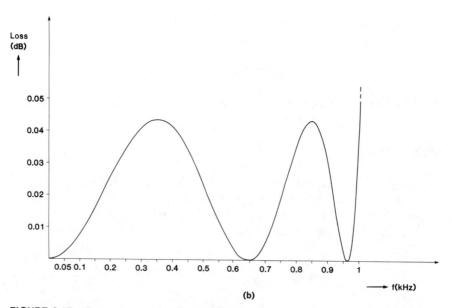

(b)

FIGURE 2.17. Frequency response of a sampled-data low-pass filter: (a) overall response for $0 \le f \le f_c/2$; (b) details of the passband response.

so that

$$\frac{z^2 - 1}{2Tz} X_i(z) = G_i(z). \tag{2.88}$$

Comparison with (2.64) shows that the transformation

$$s_a = f(z) = \frac{1}{2T}(z - z^{-1}) = \frac{z^2 - 1}{2Tz} \tag{2.89}$$

corresponds to this integration process. This mapping is rational, as required by Condition 1; furthermore, for $z = \exp(j\omega T)$,

$$s_a = \frac{1}{2T}(e^{j\omega T} - e^{-j\omega T}) = \frac{j}{T}\sin \omega T \tag{2.90}$$

so $s_a = j\omega_a$, with

$$\omega_a = \frac{1}{T}\sin \omega T. \tag{2.91}$$

Thus, as required by Condition 2, the unit circle of the z-plane is mapped on the $j\omega_a$-axis. Vice versa, if (2.89) is solved for z, we can find the image of the $j\omega_a$-axis. However, since $f(z)$ is a second-order rational expression, now *two z* values correspond to a single s_a. They are given by

$$z_{1,2} = s_a T \pm \sqrt{(s_a T)^2 + 1}. \tag{2.92}$$

If $s_a = j\omega_a$ and $|\omega_a| \leq 1/T$, then $s_a T$ is imaginary, while the square-root expression is real, and

$$|z_1| = |z_2| = (\omega_a T)^2 + \left[-(\omega_a T)^2 + 1\right] = 1. \tag{2.93}$$

Hence, both z values corresponding to such s_a are on the unit circle. However, if $|\omega_a| > 1/T$, then the square-root term is imaginary and hence so are both z_1 and z_2. Thus, $z_{1,2} = jy_{1,2}$ where

$$y_{1,2} = \omega_a T \pm \sqrt{(\omega_a T)^2 - 1}. \tag{2.94}$$

Thus, the image of the $j\omega_a$-axis has *two* branches. For $\omega_a \geq 1/T$, it is the imaginary (jy) axis (Fig. 2.18); for $\omega_a \leq 1/T$, it is the unit circle.

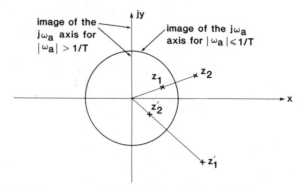

FIGURE 2.18.   The images of the $j\omega_a$-axis and two poles for the LDI mapping.

To check the validity of Condition 3, we note that by (2.92) the product of the two values z_1, z_2 corresponding to any single s_a satisfies

$$z_1 z_2 = \left[s_a T + \sqrt{(s_a T)^2 + 1} \right]\left[s_a T - \sqrt{(s_a T)^2 + 1} \right]$$

$$= (s_a T)^2 - \left[(s_a T)^2 + 1 \right] = -1. \tag{2.95}$$

Thus, $z_2 = -1/z_1$. Hence, if $|z_1| < 1$ then $|z_2| > 1$, and vice versa (Fig. 2.18). Thus, a "stable" pole s_a will transform into one stable and one unstable pole in the z-plane. Hence, a stable $H_a(s_a)$ will *not* transform into a stable $H(z)$. This, on the surface, seems to rule out any practical application of the mapping given in (2.89). Nevertheless, as will be shown in Chapter 5, the mapping in fact represents a useful tool in understanding the operation of switched-capacitor filters based on the lossless discrete integrator (LDI) concept. Hence, it is often called the *LDI transformation*.

As shown above, in the LDI transformation for $|\omega_a| < 1/T$ the $j\omega_a$ axis maps on the unit circle. The relations between the frequency variables is then given by (2.91). It is shown graphically in Fig. 2.19. Clearly, in contrast to the bilinear transformation (Fig. 2.16), the warping of the frequency scale is such that the ω_a axis is *expanded* by an average factor of $\pi/2$. For a low-pass filter, this means that the model filter must be *more* selective than the actual sampled-data one if the midpoint approximation is used. By contrast, the model needs to be less selective than the actual filter for the bilinear mapping. Hence, for low-pass filters, normally a lower-order model is needed if the bilinear (rather than the LDI) transformation is used.

Clearly, other numerical integration algorithms (Simpson's rule, etc.) can also be used to derive additional s-to-z transformations.[10] None of these, however, has been used in switched-capacitor filter design, while those discussed above do play important roles in practical design, as will be shown in later chapters.

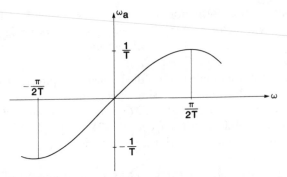

FIGURE 2.19. The relation between the continuous-time frequency ω_a and the sampled-data frequency ω for the LDI transformation.

A different method, not based on the numerical integration of Eq. (2.63), is the so-called *matched s-to-z transformation*.[11] In this technique, the zeros and poles of $H_a(s_a)$ are duplicated at the same location in the s-plane, the Laplace variable domain of the sampled-data system. Due to the periodic nature of $H(e^{sT})$, any zero (or pole) s_i will be accompanied by an infinite number of additional zeros (or poles) at $s_i + kj2\pi/T, \; k = \pm 1, \pm 2, \ldots$. Thus, for the s_a-plane pole/zero pattern shown in Fig. 2.20a, the corresponding pattern in the s-plane is that shown in Fig. 2.20b.

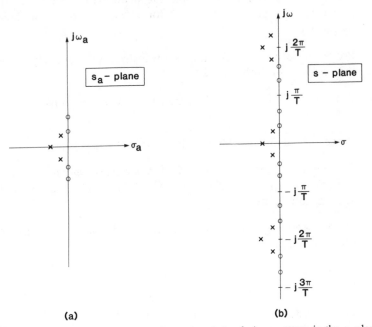

FIGURE 2.20. The matched s-to-z transformation: (a) pole/zero pattern in the s_a-plane; (b) transformed pattern in the z-plane.

To carry out the transformation of a given $H_a(s_a)$, each factor $s_a - a$ must be represented by a factor $z - e^{aT}$ in $H(z)$. A second-order factor $(s - \sigma_1)^2 + \omega_1^2 = (s - \sigma_1 - j\omega_1)(s - \sigma_1 + j\omega_1)$ will become $z^2 + a_1 z + a_2$, where $a_1 = -2e^{\sigma_1 T}\cos(\omega_1 T)$ and $a_2 = e^{2\sigma_1 T}$.

Unlike the transformation based on the bilinear mapping (2.80), the matched transformation is vulnerable to aliasing effects, since the frequency response $H(e^{j\omega T})$ is now related to $H_a(j\omega_a)$ as $F^*(j\omega)$ is to $F(j\omega)$ in (2.50). Hence, this transformation is useful only if $|H_a(j\omega_a)| \ll 1$ for $|\omega_a| \geq \pi/T$. The aliasing can be reduced somewhat by arbitrarily placing some zeros of $H(e^{sT})$ at $jk\pi/T$, $k = \pm 1, \pm 3, \pm 5, \ldots$.

An important advantage of the matched s-to-z transformation is that for narrow-band filters (low-pass or bandpass) it preserves both the loss and phase responses in and near the passband. Hence, unlike the bilinear transformation (2.80), it can be used to map a linear-phase continuous-time filter into a linear-phase discrete-time one.

PROBLEMS

2.1. Show that a continuous-time signal or system is always also an analog one.

2.2. From the definition of the Laplace transformation given in Eq. (2.8), prove the rules given in Table 2.1 for $df(t)/dt$. (*Hint:* Use integration by parts.)

2.3. Prove the formula for the Laplace transform of $\int_0^t f(\tau)\,d\tau$. (*Hint:* Use the result from Problem 2.2.)

2.4. Prove the rule for the Laplace transform of $f(t - a)u(t - a)$ given in Table 2.1.

2.5. Prove the rule for the Laplace transform of $e^{at}f(t)$.

2.6. Prove that the Laplace transformation is a *linear operation*, that is, that $k_1 f_1(t) + k_2 f_2(t) \leftrightarrow k_1 F_1(s) + k_2 F_2(s)$ for any constants k_1 and k_2 and any functions f_1 and f_2.

2.7. Prove the rule given in Table 2.2 for the Laplace transform of $\delta(t)$. (*Hint.* Use Fig. 2.4.)

2.8. Prove the rule for the formula given in Table 2.2 for the Laplace transform of a constant K.

2.9. Prove the formula for the Laplace transform of e^{at}.

2.10. Prove the formula for the Laplace transform of $e^{at}\cos bt$. [*Hint:* Use $\cos bt = (e^{jbt} + e^{-jbt})/2$.]

2.11. Calculate the Laplace and Fourier transforms of the pulse function

$$f(t) = \begin{cases} 1, & t_1 < t < t_2 \\ 0, & \text{otherwise} \end{cases}$$

where $t_1 > 0$. How are the two transforms related?

2.12. Find the Fourier transform of $f(t) = e^{-\alpha|t|}$.

2.13. Find the Fourier transform of

$$f(t) = \begin{cases} 1 - |t|, & |t| \le 1 \\ 0, & |t| > 1. \end{cases}$$

[*Hint:* Show first that $f(t) = g(t) * g(t)$, where

$$g(t) = \begin{bmatrix} 1, & |t| \le \frac{1}{2} \\ 0, & |t| > \frac{1}{2}. \end{bmatrix}$$

2.14. Verify the shifting theorem of the z-transformation. Show that if $f(nT) \leftrightarrow F(z)$, then $f(nT - kT) \leftrightarrow z^{-k}F(z)$.

2.15. Verify the entries given for the z-transforms of $\delta(nT)$, $Ku(nT)$, and a^{nT} in Table 2.6.

2.16. Verify the z-transforms given in Table 2.6 for $a^{nT}\sin n\omega_0 T$ and $a^{nT}\cos n\omega_0 T$. {*Hint:* Use $a^{nT}\sin n\omega_0 T = (1/2j)[(a^T e^{j\omega_0 T})^n - (a^T e^{-j\omega_0 T})^n]$, and the linearity of the z-transformation.}

2.17. Verify the input–output relation given in Eq. (2.4) for the system of Fig. 2.3.

2.18. Find the z-transform of $n^3 a^{nT}$. For what values of z does the z-transform converge?

2.19. Find the inverse z-transform of $z/(z - 1)^3$.

2.20. Prove Eq. (2.50). [*Hint:* Use the following basic relations of the Fourier transformation[1]

$$f(t) \sum_{n=-\infty}^{\infty} \delta(t - nT) \leftrightarrow \sum_{n=0}^{\infty} f(nT)e^{-jn\omega T},$$

$$f_1(t)f_2(t) \leftrightarrow \frac{1}{2\pi} F_1(j\omega) * F_2(j\omega),$$

$$\sum_{n=-\infty}^{\infty} \delta(t - nT) \leftrightarrow \frac{2\pi}{T} \sum_{n=-\infty}^{\infty} \delta\left(\omega - n\frac{2\pi}{T}\right),$$

$$F(\omega) * \sum_{n=-\infty}^{\infty} \delta\left(\omega - n\frac{2\pi}{T}\right) = \sum_{k=-\infty}^{\infty} F\left(\omega - k\frac{2\pi}{T}\right)$$

in the order given.]

2.21. Prove Eq. (2.52), Nyquist's interpolation formula. [*Hint:* If Nyquist's criterion (2.53) holds, then $F_A(\omega) = H(\omega)F_A^*(\omega)$, where $H(\omega)$ is given by (2.51).]

2.22. Where may the poles of $H_a(s_a)$ lie in order to give a stable $H(z)$ if the mapping of (2.70) is used?

2.23. Prove Eq. (2.76) for the backward-Euler transformation.

2.24. Derive $H(z)$ from the model transfer function

$$H_a(s_a) = \frac{1}{(s_a + a)(s_a^2 + bs_a + c)}$$

using (a) the forward Euler mapping; (b) the backward Euler mapping; (c) the bilinear s-to-z mapping.

2.25. What s_a is mapped by the bilinear transformation into $z = j$?

2.26. The three-point Simpson's integration formula is[10]

$$\int_{nT-T}^{nT+T} g_i(t)\, dt \cong \frac{T}{3}\left[g_i(nT - T) + 4g_i(nT) + g_i(nT + T)\right].$$

(a) Derive the corresponding $s_a = f(z)$ transformation.
(b) Does $f(z)$ meet Conditions 1–3 of Section 2.6?

2.27. Consider the mapping

$$s_a = f(z) = \frac{p_0 + p_1 z + \cdots + p_n z^n}{q_0 + q_1 z + \cdots + q_n z^n}.$$

Prove that if $f(z)$ satisfies Condition 2 then the coefficients satisfy[10]

$$\sum_{k=0}^{n-m}(p_k q_{k+m} + p_{k+m} q_k) = 0, \quad \text{for } m = 0, 1, 2, \ldots, n.$$

REFERENCES

1. A. Papoulis, *Circuits and Systems*, Holt, Rinehart & Winston, New York, 1980.
2. C. A. Desoer and E. S. Kuh, *Basic Circuit Theory*, McGraw-Hill, New York, 1969.
3. A. V. Oppenheim and R. W. Schafer, *Digital Signal Processing*, Prentice-Hall, Englewood Cliffs, N.J., 1975.
4. G. Doetsch, *Guide to the Applications of the Laplace and z-Transforms*, Van Nostrand, London, 1971.

5. G. C. Temes and J. LaPatra, *Introduction to Circuit Synthesis and Design*, McGraw-Hill, New York, 1977.

6. E. J. Angelo, Jr., *Bell Syst. Tech. J.*, **60**, 1499–1546 (1981).

7. R. Saal, *Handbook of Filter Design*, AEG-Telefunken, Berlin, 1979.

8. E. Christian and E. Eisenmann, *Filter Design Tables and Graphs*, Wiley, New York, 1966.

9. A. I. Zverev, *Handbook of Filter Synthesis*, Wiley, New York, 1967.

10. M. Jahanbegloo, New Approaches to Sampled-Data Analog Signal Processing, Ph.D. dissertation, UCLA, 1979.

11. G. C. Temes and S. K. Mitra (Eds.), *Modern Filter Theory and Design*, Wiley, New York, 1973.

Chapter Three ———————————————————————

MOS DEVICES AS
CIRCUIT ELEMENTS

In this chapter, the physics of MOS (metal oxide semiconductor) devices is discussed briefly. The most important and simplest current–voltage relations are given, and simple models introduced for MOS transistors in linear operation. The basic properties of MOS capacitors and switches are described, along with their imperfections and limitations. The fabrication process is also briefly described.

The discussions given here are in the simplest possible terms, and are aimed at providing some physical understanding of the highly complex device operation for the circuit designer. Precision and depth were regretfully sacrificed in the process. The ambitious reader is referred to the excellent specialized works listed as references at the end of the chapter.

3.1. SEMICONDUCTORS

In metals (e.g., aluminum, copper, silver, etc.) which are good electrical conductors the atoms are arranged in a regular crystal array. The electrons from the outer (valence) shell of the atoms are free to move within the material. Since the number of atoms, and thus the number of free electrons, is very large (of the order of 10^{23} cm^{-3}), even a small electric field results in a large electron current. Hence the high conductivity observed for these metals.

The picture is quite different for an insulator, such as silicon dioxide (SiO_2). Here, the valence electrons form the bonds between adjacent atoms, and are hence themselves tied to these atoms. Hence, there are no free electrons available for conduction, and the conductivity is very low.

Semiconductors (such as silicon or germanium) are in-between conductors and insulators in their electrical properties. At very low temperatures, the valence electrons are bound to their atoms which again form a regular lattice.

However, as the temperature is raised, due to the thermal vibrations of the atoms some bonds will be broken, and an electron escapes from each of these bonds. Such electrons are capable of conducting electricity. Furthermore, each fugitive electron leaves a charge deficit (called a *hole*) behind in the bond. A valence electron in a bond close to a hole can easily move over, filling the hole and leaving a new hole in its own bond. The effect is the same as if the hole had moved from one bond to the next. Since the hole "moves" in a direction opposite to that of the moving valence electron, in an electronic field it behaves like a positively charged particle.

Electrical conduction is thus possible for a semiconductor at room temperature. The density of thermally generated electrons and holes is, however, much smaller than that of the free electrons in metal. Typical numbers are 10^{10} charge carriers/cm^3 for silicon and 10^{13} cm^{-3} in germanium. In what follows, the currently dominant material, silicon, will be discussed exclusively.

The number of free charge carriers in a semiconductor can be raised by adding foreign elements (dopants) to the pure silicon. Silicon (and germanium) has *four* valence electrons. If an atom of an element with *five* valence electrons (such as arsenic, phosphorus, or antimony) is added to the semiconductor, then it may take the place of a silicon atom in the crystal lattice. Thus four of its valence electrons will participate in the four bonds tying the atom to adjacent semiconductor atoms in the lattice. The fifth valence electron of the foreign atom, however, will not have a place in any bond, and will thus be free to move away from its parent atom. Hence, such a dopant element (called a *donor*, since it contributes free electrons to the semiconductor) enhances the conductivity of the material.

Adding atoms of an element with *three* valence electrons will also contribute to the conductivity. Now there will be a bond *lacking* a valence electron for each dopant atom. Thus, one hole is created by each such atom. These dopants (e.g., boron, aluminum, gallium) are called *acceptors*, since the holes will propagate by accepting bound valence electrons from adjacent semiconductor atoms.

In doped semiconductors there will be carriers due to thermal effects as well as to the donor (or acceptor) atoms. Materials containing *donors* will thus have both free electrons and holes; but there will be more electrons than holes. Such semiconductors will be called *n-type*, where *n* stands for negative. Materials containing *acceptors* will have a majority of holes; they are called *p-type* semiconductors, where *p* stands for positive.

A semiconductor structure can also be fabricated which contains two adjacent regions of different types (Fig. 3.1). The surface joining the two regions is called a *p-n junction*. When the junction is fabricated, the random thermal motion of the majority carriers (electrons in the *n*-type region, holes in the *p*-type region) will cause electrons to spill over from the *n*-type region to the *p*-type region. Vice versa, holes will move from the *p*-type region to the *n*-type semiconductor. Thus, this random motion (called *diffusion*) results in the *p*-type semiconductor being charged negatively, while the *n*-type region is

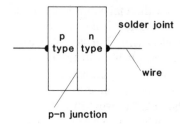

p–n junction

FIGURE 3.1.   A *p-n* junction diode.

charged positively. The effect will be strongest near the junction: there, in the *p*-type region, the negatively charged acceptor atoms will no longer be neutralized by holes; and (in the *n*-type region) the positively charged donor ions will no longer be surrounded by free electrons. Hence, in this area a dipole layer of fixed ions will be formed (Fig. 3.2). The electric field $\mathscr{E}$ created by the dipole opposes further majority-carrier diffusion. It helps, however, the thermally generated *minority carriers* (electrons in *p*-type regions, holes in *n*-type regions) to migrate from one region to another. Thus, after a short transient, an equilibrium will be obtained. Four different carrier currents will flow: majority carriers will move by diffusion from region to region *in spite* of $\mathscr{E}$, and minority carriers will flow *aided* by $\mathscr{E}$. These currents cancel each other in equilibrium, since the effects of $\mathscr{E}$ compensate for the larger number of available majority carriers.

The equilibrium will be upset if a voltage source is connected to the wires soldered to the semiconductor (Fig. 3.3). Assume first that the polarity of the source is such that it makes the *p*-region more positive with respect to the *n*-region, that is, that $v > 0$ in Fig. 3.3. Then v will reduce $\mathscr{E}$, and thus the current of majority carriers will become larger than that of the minority carriers. Since there are many majority carriers held back by $\mathscr{E}$ from spilling over the boundary, even a small reduction of $\mathscr{E}$ caused by, say, a battery of $v = 0.8$ V, can result in a large majority-carrier current (say, $i = 1$ A) in the circuit. Hence, v with the indicated polarity will be called *forward voltage* and i the *forward current*.

Let us now reverse the polarity of the voltage source, so that $v < 0$ in Fig. 3.3. Now v will aid $\mathscr{E}$ in obstructing the flow of majority carriers from region to region. If v is large enough, the majority current is essentially eliminated, and only the flow of minority carriers (electrons moving from the *p*-region to

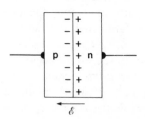

FIGURE 3.2.   Ion layers in a *p-n* junction.

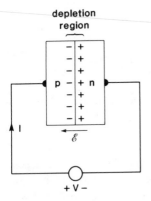

depletion
region

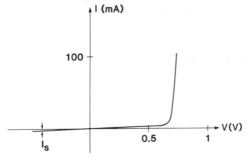

FIGURE 3.3. A circuit for testing a *p-n* diode.

the *n*-region, and holes moving in the opposite direction) remains. Since, however, the number of minority carriers is small and nearly independent of v, the resulting net current will be small and nearly constant. This is the case of *reverse voltage* and *current*. With the reference directions used in Fig. 3.3, now $i < 0$. Figure 3.4 illustrates the overall behavior of i as a function of v. A detailed theoretical analysis[1,2] reveals that the describing equation is, to a good approximation,

$$i = I_S(e^{qv/kT} - 1). \tag{3.1}$$

Here, I_S is the *saturation current*, determined by the geometry and the material properties of the device; $q \simeq 1.6 \times 10^{-19}$ C is the electron charge, and $k \simeq 1.38 \times 10^{-23}$ J/K Boltzmann's constant. T is the température of the semiconductor, in degrees Kelvin. At room temperature ($T = 300$ K), $kT/q \simeq 26$ mV. I_S is usually very small, of order 10^{-9} A or less. Thus, i increases exponentially with v for $v > 0$, while $i \simeq -I_S$ and is very small if $v < 0$ (Fig. 3.4).

The behavior of the region directly adjacent to the boundary between *p*- and *n*-regions is of prime importance. As mentioned earlier, the majority carriers are very sparse in this area; some have emigrated into the other region,

FIGURE 3.4. The current vs voltage characteristics of a *p-n* junction diode.

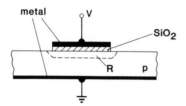

FIGURE 3.5.   A metal-oxide-semiconductor (MOS) structure.

and the others have been pushed back into the inside of their native region by the field $\mathscr{E}$. Hence, the border area contains only the fixed ions, charged negatively in the p-type region and positively in the n-type material (Fig. 3.3). This area is hence called the *depletion* region. Its width increases with increasing $\mathscr{E}$; hence, it will be greater (smaller) for reverse (forward) voltage v.

Due to the field $\mathscr{E}$, a voltage ϕ_i (often called the *built-in voltage*) appears across the depletion region for $v = 0$. The total potential across the junction, for $v \neq 0$, is thus $\phi_i - v$. Typically, $\phi_i = 0.5 \sim 1$ V.

For $v < 0$, the p-n junction can be regarded as a capacitor, since only a small saturation current I_S flows for a dc voltage v, and since adjacent positive $(+Q)$ and negative $(-Q)$ charges are stored in the depletion region (Fig. 3.3). Since the charge stored is a nonlinear function of v, the capacitance is nonlinear. We shall define the capacitance C by the incremental relation $C = dQ/dv$. It can then be shown (see Ref. 1, Section 3.3 and Ref. 2, Section 6.5) that for the device illustrated in Fig. 3.3,

$$C = \left(\frac{q\varepsilon_S / [2(1/N_a + 1/N_d)]}{\phi_i + |v|} \right)^{1/2} A \qquad (3.2)$$

holds. Here, $\varepsilon_S \cong 1.04$ pF/cm is the permittivity of the silicon: $\varepsilon_S = \varepsilon_0 K_S$, where ε_0 is the permittivity of free space ($\varepsilon_0 \simeq 8.86 \times 10^{-14}$ F/cm) and $K_S \simeq 11.7$ is the dielectric constant* of silicon. A is the area of the junction in cm^2, and $N_a(N_d)$ is the number of acceptor (donor) atoms per cm^3. Note that C decreases with $|v|$. It can be shown (see Ref. 1, Section 3.3) that the quantity under the square-root sign is $(\varepsilon_S/x_d)^2$, where x_d (in cm) is the width of the depletion region; hence, $C = \varepsilon_S A/x_d$ holds.

3.2.   THE MOS TRANSISTOR[1,3]

Consider next the structure shown in Fig. 3.5. It is a sandwich of several layers: from top to bottom, it contains layers of metal, silicon dioxide (SiO$_2$, an excellent insulator), p-type silicon, and a second metal layer connected to ground. It is called a *metal-oxide-semiconductor* (MOS) structure. Let v be

*Also called relative permittivity.

negative; then an electric field will be created across the dioxide layer, which will attract positive charges (holes) to the region R under the top metal electrode (Fig. 3.5). Thus, negative charges will be stored in the top metal electrode and positive charges in R. The device will thus behave as a capacitor C of magnitude

$$C = \varepsilon_{ox} \frac{A}{l} \qquad (3.3)$$

In (3.3), ε_{ox} is the permittivity of the SiO_2: $\varepsilon_{ox} = \varepsilon_0 K_{ox} \cong 0.35$ pF/cm, where K_{ox} is the dielectric constant of SiO_2 ($K_{ox} \simeq 3.9$). A is the area of the top electrode, and l the thickness of the SiO_2 layer.* The p-type Si layer between R and the bottom metal layer behaves as a resistor; hence, the overall structure simulates a lossy capacitor.

Next, let v be a small *positive* voltage in Fig. 3.5. The electric field will now repel holes. As a result, the fixed negatively charged ions in R will be abandoned by the mobile holes, and a net negative space charge will appear in R, which is now a depletion layer. Thus, again charge is stored in the top electrode and a capacitor is created. For very small values of v ($v \ll 1$ V), (3.3) will remain valid for the magnitude of the capacitance. As the value of v is increased, however, the charge in R becomes greater since the depletion region widens. Since the average ion is now farther away from the surface, the effective value of l in (3.3) increases and C decreases.

If v is increased even further, a new effect appears. Since the thermal generation of holes and electrons occurs continuously in the semiconductor, if the field created by a positive v is strong enough, it can attract thermal electrons to R; these will then move to the surface. When this occurs, the capacitor stores positive charges in the top electrode, while negative ones (electrons) are stored in the surface layer. Thus, in (3.3), l again becomes the SiO_2 thickness, and hence C has the same value as it had for negative voltage v. The overall behavior of C as a function of v is illustrated schematically in Fig. 3.6, which also gives the names of the three operating regions. The names of the first two are evident; the third one is called the *inversion region*, since (due to the high voltage v) mobile electrons are attracted into R which thus behaves as an n- (rather than p-) type material. It should be noted that since thermal electrons are generated only at a slow rate, the voltage v must be present for some time before the "inversion layer" is formed; hence, it will not appear if v is a high-frequency (say, $f > 1$ kHz) signal rather than a constant voltage.

Consider next the structure shown in Fig. 3.7. A new feature is the presence of two n^+- (i.e., heavily doped n-type) regions in the p-type material. The one on the left will be called the *source*; a voltage v_S is connected to it. The

---

*Often, the oxide thickness l is measured in angstroms (1 Å = 10^{-8} cm). Usual values of l are between 300 and 2000 Å.

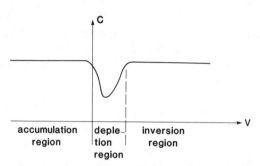

FIGURE 3.6. The capacitance vs voltage characteristics of an MOS structure.

n^+-region on the right will be called the *drain*; its voltage is denoted by v_D. The top metal electrode will be called the *gate*; its voltage is v_G. The body of the semiconductor is usually called the *substrate* or *bulk*. The overall device is the *MOS transistor*. Its operation will be briefly discussed next.

Let the source be grounded, so that $v_S = 0$. Also, let v_D have a small positive value, say 0.5 V. We will consider the behavior of the drain current i_D as v_G is raised from 0 to higher positive values. Since the gate is insulated from the rest of the device by the oxide layer, it will not conduct any current. The n^+-drain region and the surrounding p-type substrate form a p-n junction. Since the substrate is grounded, while $v_D > 0$, this junction is reverse biased. Hence, for $v_G = 0$, $i_D \approx 0$.

As v_G is increased, the region R under the gate will first be depleted, then inverted, as discussed earlier in connection with Figs. 3.5 and 3.6. When R is depleted, i_D remains zero, since the area around the drain is still reverse biased. However, the situation changes when v_G is so large that inversion occurs so that R is filled with electrons. Now, a layer containing mobile electrons, called an *inversion layer* or *channel*, connects the drain to the source. Since the drain is positive with respect to the source, electrons will flow from the source to the drain, and a positive current $i_D > 0$ will be observed. The smallest voltage v_G necessary to produce a channel is called the *threshold voltage* and is denoted by V_T. Usually, V_T is given as the v_G value needed for $i_D = 1\ \mu A$; it may range from a fraction of a volt to several volts.

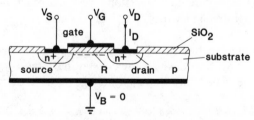

FIGURE 3.7. An MOS transistor.

It should be noted that for the structure of Fig. 3.7 most of the electrons in the channel do *not* originate from thermal effects in the bulk; instead, they are drawn by the electric field due to v_G out of the source. Some electrons are also drawn from the drain; however, since $v_D > 0$, the drain–substrate junction is more reverse biased, and hence it is harder for electrons to escape from the drain.

Since a potential difference v_D exists between the two ends of the channel, the electrons in the channel will be attracted to the drain. Therefore, in addition to the random thermal motion of the electrons, a steady motion (called *drift*) will occur which causes the current flow. For small v_D, the channel will therefore behave as a resistor, and hence $i_D \simeq v_D/R$, where the channel resistance R is given by

$$R = \frac{L}{W\mu_n|Q_n|}. \tag{3.4}$$

Here, L is the length and W the width of the channel, while μ_n is the mobility of the electrons in the channel,* defined by the relation (electron drift velocity) = (mobility) $\times$ (electric field). Finally, Q_n is the charge density (in C/cm^2) of the electrons in the channel. Since v_G can be considered as the sum of two terms, V_T (necessary to maintain the depletion region under the channel) and $v_G - V_T$ (necessary to maintain the channel), we have

$$Q_n = -C_{ox}(v_G - V_T), \tag{3.5}$$

where $C_{ox} = \varepsilon_{ox}/l$ is the capacitance (per unit area) of the oxide layer separating the gate from the channel. Hence, for small v_D (i.e., $v_D \ll v_G - V_T$), the relation

$$i_D = \mu_n C_{ox} \frac{W}{L}(v_G - V_T)v_D \tag{3.6}$$

holds. Thus, the transistor acts as a resistor, with resistance $R = [\mu_n C_{ox}(W/L)(v_G - V_T)]^{-1}$ controlled by v_G.

If v_D is increased, so that it is no longer negligible compared to v_G, (3.6) will become inaccurate. Since the potential of the channel at the grounded source is zero, while at the drain it is v_D, we can assume that its average potential is $v_D/2$. Hence, the average voltage between the gate and the channel is $(v_G - v_D/2)$. Replacing v_G by $(v_G - v_D/2)$ in (3.6) gives

$$i_D = \mu_n C_{ox} \frac{W}{L}(v_G - V_T - v_D/2)v_D. \tag{3.7}$$

---

*The mobility *in the bulk* of the semiconductor is higher, since μ_n decreases with the concentration of ionized impurities. Typical values are $\mu_n \simeq 1000$ $cm^2/V \cdot s$ for $N_D = 10^{16}$ cm^{-3}, while $\mu_n \simeq 100$ $cm^2/V \cdot s$ for $N_D = 10^{19}$ cm^{-3}.

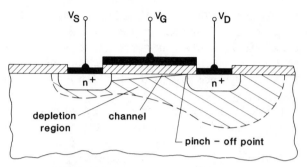

FIGURE 3.8. Pinch-off in an MOS transistor.

Equation (3.7) remains a good approximation for $v_D < v_G - V_T$. This range is called the *linear region** of operation of the MOS transistor.

When $v_D \geq v_G - V_T$, a new phenomenon occurs. Consider the situation illustrated in Fig. 3.8, where only the structure near the semiconductor surface is shown, magnified. As the figure indicates, due to the variation of the potential along the channel, the charge density Q_n decreases near the drain. If $v_D = v_G - V_T$, at the drain the gate-to-channel voltage is no longer sufficient to maintain the channel. Thus, the depletion region surrounding the source, the channel, and the drain extends all the way to the surface. This phenomenon is sometimes called *pinch-off*, and the region where it occurs is the *pinch-off point* (Fig. 3.8). If v_D is further increased, the pinch-off point will move toward the source, since the area where $v_G - v_D \leq V_T$ will increase. Hence, now the channel will extend only from the source to the pinch-off point, the latter being somewhere under the gate. The region between the pinch-off point and the drain is depleted. Electrons from the channel are injected into this depletion region at the pinch-off point, and are swept to the drain by the field created by the potential difference between the drain and the pinch-off point. The voltage $v_{DS} \triangleq v_D - v_S$ is thus divided between the two "series-connected" regions: the channel between source and pinch-off point, and the depletion region between pinch-off point and drain. Clearly, the latter has a higher resistance, and hence most of v_{DS} in fact appears across it. Any increase of v_D will, to a good approximation, result in an equal voltage increase across the depletion region and will hardly change i_D. Thus, for $v_D > v_G - V_T$, from (3.7),

$$i_D(v_D) \simeq i_{D\,\text{sat}} \triangleq i_D(v_{D\,\text{sat}})$$

$$= \frac{\mu_n C_{\text{ox}}}{2} \frac{W}{L} (v_G - V_T)^2. \tag{3.8}$$

*Or *triode region*.

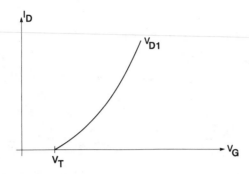

FIGURE 3.9.   The drain current vs gate voltage characteristics of an MOS transistor.

This phenomenon is called *saturation*; $v_{D\,sat} = v_G - V_T$ is the *drain saturation voltage* and $i_{D\,sat}$, as given by (3.8), is the *drain saturation current*.

The drain current does, in reality, increase somewhat with increasing v_D. This can be attributed to the move of the pinch-off point towards the source for increasing v_D, and hence to the shortened channel: as (3.8) indicates, i_D increases as L is reduced. As an approximation, this effect (often called *channel-length modulation*) can be included in the formula for $i_D(v_D)$ in the form of an added factor $(1 + \lambda v_D)$. Here, λ is a device constant which depends on L, on the doping concentration of the substrate, and on the substrate bias (to be discussed in the next section). For $L \simeq 10\ \mu m$, typically $\lambda \simeq 0.03\ V^{-1}$; generally, $\lambda \propto 1/L$.

It is usual to introduce the abbreviations $k' \triangleq \mu_n C_{ox}/2$ and $k \triangleq k'W/L$. Then the saturation current given by (3.8) becomes

$$i_D = k(v_G - V_T)^2(1 + \lambda v_D), \qquad v_G \geq V_T \tag{3.9}$$

which incorporates channel-length modulation. Figure 3.9 shows the variation of i_D with v_G for constant v_D; Fig. 3.10 illustrates its dependence on v_D for various v_G values, where $v_{G1} < v_{G2} < v_{G3} \cdots$.

All derivations of this section were performed for the structure of Fig. 3.7, whose source, drain, and channel were all *n*-type. This device is called an *n-channel* MOS, or *NMOS transistor*. A similar arrangement can be constructed by creating p^+ drain and source diffusions in an *n*-type substrate. Now a negative v_G is needed to create a *p*-type channel under the gate, and a negative v_D is used to attract the holes in the channel to the drain. Also, i_D will be negative if the reference direction of Fig. 3.7 is used. The resulting device is called a *p-channel* MOS or *PMOS transistor*. The formulas (3.3)–(3.9) remain valid if some small changes are made. The mobility μ_n of electrons must be replaced by μ_p, the hole mobility in the channel. As would be expected from the more elaborate mechanism of hole conduction, $\mu_p < \mu_n$: typical mobility values in the channel region, for an impurity concentration of

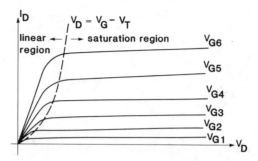

FIGURE 3.10.   The drain current vs drain-to-source voltage characteristics of an MOS transistor.

10^{16} cm^{-3} are $\mu_n = 1000$ cm^2/V · s and $\mu_p = 400$ cm^2/V · s. The electron charge density Q_n in the channel is to be replaced by Q_p, the hole charge density; also, a negative sign must be included in Eqs. (3.6)–(3.9) to account for the change in the charge of the carriers. Finally, v_D must be replaced by $|v_D|$ in (3.9), since now $v_D < 0$. In conclusion, (3.7) becomes

$$i_D = -2k(v_G - V_T - v_D/2)v_D. \tag{3.10}$$

Here, $k \triangleq \mu_p C_{\text{ox}} W/2L$ and $V_T < 0$. Equation (3.10) describes the drain current characteristics in the linear range. The behavior of i_D in the saturation region can be obtained by modifying (3.9):

$$i_D = -k(v_G - V_T)^2(1 + \lambda|v_D|). \tag{3.11}$$

The circuit symbols used for NMOS and PMOS transistors are shown in Figs. 3.11a and b, respectively. If the type is unimportant, the simplified symbol of Fig. 3.11c may be used for both NMOS and PMOS devices.

Since the operation of the devices described in this section is dependent upon the electric field induced by the gate voltage, they are called *field-effect devices* (*FETs*), or *MOSFETSs*.*

Since PMOS transistors are more easily fabricated than NMOS ones, they were initially predominant. However, later when the techniques for the reliable production of NMOS devices were developed, the latter became standard. The main reason for this is the higher mobility of electrons, which makes the NMOS transistors faster than PMOS ones.

---

*Since the charge carriers here are either electrons *or* holes (not both), FETs are also sometimes called *unipolar* devices, to contrast them with the bipolar transistors in which *both* electron and hole currents exist.

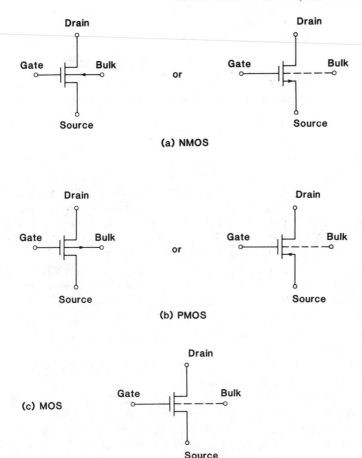

FIGURE 3.11. Transistor symbols.

3.3. MOS TRANSISTOR TYPES; BODY EFFECT

The MOS transistors described in Section 3.2, both NMOS and PMOS types, share several common features. In the structure, the gate is electrically insulated from the rest of the device by the SiO_2 layer under it. Hence, it is often called an *insulated-gate field-effect transistor*, or *IGFET*. Also, the voltage v_G induces and enhances the drain current. Thus, the described devices operate in the *enhancement mode*.

It is also possible to fabricate an MOS transistor which conducts drain current when $v_G = 0$. For example, an *n*-type layer can be introduced by doping which connects the source and drain of an NMOS device. With such a doped channel, the field of the gate is not needed to produce an inversion layer; the region R (Fig. 3.7) has now a "built-in" conducting *n*-type channel.

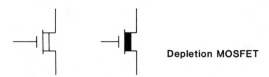

FIGURE 3.12. Symbols for depletion-mode transistors.

However, if a *negative* gate voltage is applied, the field thus created will repel electrons, and create a depletion layer in the channel adjacent to the SiO_2 surface, thereby reducing the conductivity and thus the drain current. If the magnitude of the negative gate voltage is sufficiently large, the channel becomes completely depleted and $i_D \simeq 0$ results. The value v_G at which this occurs is again called threshold voltage and is denoted by V_T. Now, however, $V_T < 0$. Such a device is called a *depletion-mode FET*.

It should be noted that even without a doped layer, an NMOS transistor can conduct for $v_G = 0$ due to oxide charges[4] if the bulk is very lightly doped. It is also of interest to note that if v_G is chosen positive in a depletion-mode device, it will attract additional electrons into the channel and i_D will increase. Thus, an NMOS transistor with $V_T < 0$ can be used either as a depletion- or as an enhancement-mode device.

It is also possible to create a depletion-mode PMOS device, with $V_T > 0$, by establishing a *p*-type doped channel.

The relations (3.6)–(3.11) remain valid for depletion-mode devices, if the value and sign of V_T is chosen appropriately, as described above. Two symbols often used to denote depletion-mode MOSFETs are shown in Fig. 3.12.

A totally different structure can also be used to produce a depletion-mode field-effect transistor (Fig. 3.13). Here, a lightly doped *n*-type layer (channel) connects the n^+-source and drain regions and the gate is a p^+-region implanted in this layer. Hence, for $v_S = v_G = 0$ and $v_D > 0$, a drain current will flow. If v_G is made *negative*, the p^+-implant acting as the gate will be surrounded by a depletion layer; the greater $|v_G|$, the deeper the layer. The mobile electrons in the channel cannot enter this depletion layer, nor the one along the *p-n* junction between the channel and the substrate. Hence, the

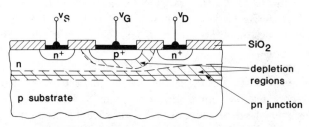

FIGURE 3.13. Junction field-effect transistor.

effective cross section of the channel is reduced as $|v_G|$ is increased. At some value $v_G = V_P$ ($V_P < 0$), i_D becomes zero (in practice, < 1 μA). Thus, V_P plays the same role as V_T for a depletion-mode MOSFET; it is called the *pinch-off* voltage. It can be visualized as the gate voltage which causes the two depletion regions in the channel to merge, leaving no conductive path between source and drain.

The device described and shown in Fig. 3.13 is called an *n*-channel *junction field-effect transistor* (*JFET*), since its gate is separated from the rest of the device by a reverse-biased *p-n* junction, rather than by an SiO_2 layer as for the MOSFET (IGFET).

Since the JFET is hardly ever used in analog MOS integrated circuits, it will not be discussed in detail here; Reference 1, Section 3.5 gives a clear description of its physics and current–voltage characteristics.

Next, a key limitation (called *body effect*) of MOSFETs used as analog circuit elements will be described. In the discussions of Section 3.2, it was always assumed that both the bulk and source are grounded, so that $v_B = v_S = 0$ held. Often, circuit considerations make this convenient arrangement impossible and $v_S \neq v_B$ must be used. Obviously, the voltage $v_S - v_B$ must be such that the source–bulk junction is *reverse* biased; otherwise, a large junction current will flow inside the transistor. This current may damage the device, and in any case will impede its proper operation. Thus, say, in an NMOS transistor, the bulk must be biased to make it negative with respect to both source and drain.

If the source potential is *not* zero, the voltages v_G and v_D must be replaced in all equations by $v_{GS} = v_G - v_S$ and $v_{DS} = v_D - v_S$, respectively. In addition, the depletion region around the channel (Fig. 3.8) will become wider if the reverse voltage between the bulk and the source (and hence the channel) is increased. Since the voltage $v_G = V_T$ is the gate voltage necessary to maintain the depletion region (without creating a channel), V_T will *increase* in magnitude. The dependence of V_T on the voltage $v_{SB} \triangleq v_S - v_B$ can be shown (see Ref. 1, Sec. 8.2) to be in the form

$$|V_T| = |V_{TO}| + \gamma\left(\sqrt{2|\phi_p| + |v_{SB}|} - \sqrt{2|\phi_p|}\right). \tag{3.12}$$

Here, V_{TO} is the threshold voltage for $v_{SB} = 0$ and γ is a device constant given by

$$\gamma = \frac{\sqrt{2\varepsilon_S q N_{imp}}}{C_{ox}}. \tag{3.13}$$

In (3.13), ε_S is the permittivity of silicon: $\varepsilon_S = \varepsilon_0 K_S$, $K_S \simeq 11.7$. Also, N_{imp} is the density of the impurity ions in the bulk. For NMOS, $N_{imp} = N_A$, the acceptor ion density; for PMOS, $N_{imp} = N_D$, the donor ion density. For example, for $N_{imp} = 10^{15}$ cm^{-3} and 800 Å oxide thickness (i.e., $C_{ox} \simeq 4.4 \times$

TABLE 3.1. Key Units and Constants for MOS Transistors

$1\ \mu m = 10^{-4}\ cm = 10^{4}\ \text{Å}$.

$1\ mil = 25.4\ \mu m = 0.0254\ mm$.

Electron charge (magnitude): $q = 1.6 \times 10^{-19}$ C.

Permittivity of free space: $\varepsilon_0 = 8.86 \times 10^{-14}$ F/cm.

Permittivity of silicon: $\varepsilon_{si} = \varepsilon_0 K_{si} = 1.04 \times 10^{-12}$ F/cm; $K_{si} = 11.7$.

Permittivity of silicon dioxide: $\varepsilon_{ox} = \varepsilon_0 K_{ox} = 3.5 \times 10^{-13}$ F/cm; $K_{ox} = 3.9$.

Oxide capacitance: $C_{ox} = \varepsilon_{ox}/t_{ox} = 3.5 \times 10^{-13}/t_{ox}^{(cm)}$ F/cm^2.

Intrinsic carrier concentration: $n_i = 1.5 \times 10^{10}$/cm^{-3} at $T = 300$K.

Boltzmann's constant: $k = 1.38 \times 10^{-23}$ J/K; kT/q (at $T = 300$K) $= 0.026$ V.

Electron mobility in Si ($N_{imp} = 10^{17}$ cm^{-3}, $T = 300$K): 670 cm^2/V · s.

Hole mobility in Si ($N_{imp} = 10^{17}$ cm^{-3}, $T = 300$K): 220 cm^2/V · s.

Body-effect coefficient: $\gamma = \sqrt{\dfrac{2qK_{si}N_{imp}}{\varepsilon_0}}\ \dfrac{t_{ox}}{K_{ox}} \simeq 1.67 \times 10^{-3} t_{ox}^{(cm)} \sqrt{N_{imp}^{(cm^{-3})}}$ V$^{1/2}$.

Bulk potential: $\phi_p = -\dfrac{kT}{q}\ln\dfrac{N_{imp}}{n_i} = 0.026\ln[0.67 \times 10^{-10}\ N_{imp}^{(cm^{-3})}]$

10^{-8} F/cm^2), $\gamma \simeq 0.423$ V$^{1/2}$. Finally, ϕ_p is a material constant* of the bulk; its value is around 0.3 V.

This phenomenon (called body effect) is a major limitation of MOS devices operated with $v_S \ne v_B$; its evil influence will be repeatedly lamented later in this work. As (3.12) and (3.13) show, to reduce the body effect N_{imp} should be made small. However, for very small N_A values (say, $N_A < 10^{13}$ cm^{-3}), an NMOS may behave as a depletion-mode device, as explained earlier. Thus, the body effect cannot be completely eliminated.

Some key constants and formulas on MOSFETs are summarized in Tables 3.1 and 3.2. They are based on some unpublished lecture notes of Professor K. W. Martin of UCLA (with his kind permission).

3.4. SMALL-SIGNAL OPERATION AND EQUIVALENT CIRCUIT OF MOSFETS

The previous discussion of MOS transistors assumed that all voltages and currents were either constant, or varied sufficiently slowly so that all capacitive currents could be neglected. On the other hand, the formulas were valid for small as well as large voltages and currents.

---

*$\phi_p \triangleq (E_i - E_f)/q$, where E_i is the intrinsic Fermi energy and E_f the Fermi energy of the semiconductor (see Ref. 1, p. 318).

TABLE 3.2. Drain–Current Relations for MOSFETs in Large-Signal, Low-Frequency Operation

	NMOS	PMOS
Triode region: $\|v_{GS}\| > \|V_T\|$; $\|v_{DS}\| < \|v_{GS}\| - \|V_T\|$	$i_D = \mu_n C_{ox} \dfrac{W}{L}\left(v_{GS} - V_T - \dfrac{v_{DS}}{2}\right) v_{DS}$	$-i_D = \mu_p C_{ox} \dfrac{W}{L}\left(v_{GS} - V_T - \dfrac{v_{DS}}{2}\right) v_{DS}$
Saturation region: $\|v_{GS}\| > \|V_T\|$; $\|v_{DS}\| > \|v_{GS}\| - \|V_T\|$	$i_D = \dfrac{\mu_n C_{ox} W}{2L}(v_{GS} - V_T)^2(1 + \lambda v_{DS})$, where $\lambda \propto L^{-1}(v_{DG} + V_T)^{-1/2}N_{\text{imp}}^{-1/2}$, $V_T = (V_T)_{v_{SB}=0} + \gamma\left(\sqrt{2\|\phi_p\| + v_{SB}} - \sqrt{2\|\phi_p\|}\right)$ (see Table 3.1 for the values of γ and ϕ_p)	$-i_D = \dfrac{\mu_p C_{ox} W}{2L}(v_{GS} - V_T)^2(1 - \lambda v_{DS})$, where $\lambda \propto L^{-1}(-v_{DG} - V_T)^{-1/2}N_{\text{imp}}^{-1/2}$ $V_T = (V_T)_{v_{SB}=0} - \gamma\left(\sqrt{2\|\phi_p\| - v_{SB}} - \sqrt{2\|\phi_p\|}\right)$

In some important linear applications (such as in operational amplifiers) the transistor voltages and currents vary so rapidly that capacitive effects cannot be ignored. At the same time, the device is used with sufficiently small signals, so that a linear approximation can be applied to all nonlinear relations. Since the MOSFET is usually biased in its saturation region for linear applications, the analysis will be illustrated for this case; however, it can equally well be applied to devices in their linear (i.e., nonsaturated) regions of operation.[12]

Assuming an NMOS transistor, and combining Eqs. (3.9) and (3.12), the relation

$$i_D = k\left(v_{GS} - V_{TO} - \gamma\sqrt{2\phi_p + v_{SB}} + \gamma\sqrt{2\phi_p}\right)^2(1 + \lambda v_{DS}) \qquad (3.14)$$

results. Here, we used $v_{GS} \triangleq v_G - v_S$ and $v_{DS} \triangleq v_D - v_S$ to replace v_G and v_D, since in general $v_S \neq 0$. For small variations of i_D, v_{GS}, v_{DS}, and v_{SB}, the nonlinear expression (3.14) can be replaced by a first-order Taylor approximation. Specifically, near a constant bias point $i_D^0 = f(v_{GS}^0, v_{DS}^0, v_{SB}^0)$ we can write

$$i_D^0 + \Delta i_D \simeq i_D^0 + \left(\frac{\partial i_D}{\partial v_{GS}}\right)^0 \Delta v_{GS} + \left(\frac{\partial i_D}{\partial v_{DS}}\right)^0 \Delta v_{DS} + \left(\frac{\partial i_D}{\partial v_{SB}}\right)^0 \Delta v_{SB} \qquad (3.15)$$

Here, $(\partial i_D/\partial v_{GS})^0$ and so on denotes the partial derivatives evaluated at the bias point. Δi_D is the deviation (increment) of i_D from its bias value; Δv_{GS}, Δv_{DS}, and Δv_{SB} are the increments of v_{GS}, v_{DS}, and v_{SB}. All deviations must be small for (3.15) to hold. If only the incremental (small-signal, ac) components are of interest then (3.15) can be written as

$$\Delta i_D \simeq g_m \Delta v_{GS} + g_d \Delta v_{DS} + g_{mb} \Delta v_{SB}, \qquad (3.16)$$

where

$$g_m \triangleq \left(\frac{\partial i_D}{\partial v_{GS}}\right)^0, \qquad g_d \triangleq \left(\frac{\partial i_D}{\partial v_{DS}}\right)^0, \qquad g_{mb} \triangleq \left(\frac{\partial i_D}{\partial v_{SB}}\right)^0. \qquad (3.17)$$

Here, g_d is the (incremental) drain conductance, while g_m and g_{mb} are transconductances which can be represented by voltage-controlled current sources (VCCSs). Hence, an equivalent circuit model, shown in Fig. 3.14, can

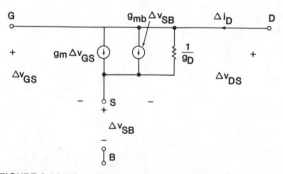

FIGURE 3.14. Low-frequency equivalent circuit of a MOSFET.

be constructed. The values of g_m, g_{mb}, and g_d can be found from (3.14):

$$g_m \triangleq \left(\frac{\partial i_D}{\partial v_{GS}}\right)^0 = 2k\left(v_{GS}^0 - V_{TO} - \gamma\sqrt{2\phi_p + v_{SB}^0} + \gamma\sqrt{2\phi_p}\right)\left(1 + \lambda v_{DS}^0\right)$$

$$= 2\sqrt{k\left(1 + \lambda v_{DS}^0\right)i_D^0}, \tag{3.18}$$

$$g_{mb} \triangleq \left(\frac{\partial i_D}{\partial v_{SB}}\right)^0 = -k\left(v_{GS}^0 - V_{TO} - \gamma\sqrt{2\phi_p + v_{SB}^0} + \gamma\sqrt{2\phi_p}\right)$$

$$\times\left(1 + \lambda v_{DS}^0\right)\frac{\gamma}{\sqrt{2\phi_p + v_{SB}^0}}$$

$$= \frac{-\gamma g_m/2}{\sqrt{2\phi_p + v_{SB}^0}}, \tag{3.19}$$

and

$$g_d \triangleq \left(\frac{\partial i_D}{\partial v_{DS}}\right)^0 = k\left(v_{GS}^0 - V_{TO} - \gamma\sqrt{2\phi_p + v_{SB}^0} + \gamma\sqrt{2\phi_p}\right)^2\lambda$$

$$= \frac{\lambda}{1 + \lambda v_{DS}^0}i_D^0. \tag{3.20}$$

Hence, to a good approximation, g_m and g_{mb} are proportional to $\sqrt{i_D^0}$, while g_d is proportional to i_D^0.

The other important components of the complete small-signal model of the MOSFET are the capacitors representing the incremental variations of stored charges with changing electrode voltages. These play an important role in the high-frequency operation of the device. The *intrinsic* components of the terminal capacitances of the MOSFET devices (associated with reverse-biased p-n junctions, channel and depletion regions) are strongly dependent on the region of operation, while the *extrinsic* components (due to layout parasitics, overlapping regions, etc.) are relatively constant. Assuming again that the transistor operates in the saturation region, it can be assumed that the channel begins at the source and extends over two-thirds of the distance to the drain. In this region of operation, the most important capacitances are the following:

C_{gd}: *Gate-to-Drain Capacitance.* This is due to the overlap of the gate and the drain diffusion. It is a thin-oxide capacitance, and hence, to a good approximation, it can be regarded as being voltage independent.

C_{gs}: *Gate-to-Source Capacitance.* This capacitance has two components: $C_{gs_{ov}}$, the gate-to-source thin-oxide overlap capacitance, and C_{gs}', the gate-to-channel capacitance. The latter (in the saturation region) is around $\frac{2}{3}C_{ox}$, where C_{ox} is the total thin-oxide capacitance between the gate and the surface of the substrate. C_{gs} is nearly voltage independent in the saturation region.

C_{sb}: *Source-to-Substrate Capacitance.*   This capacitance also has two components: C_{sbpn}, the *p-n* junction capacitance between the source diffusion and the substrate, and C'_{sb}, which can be estimated as two-thirds of the capacitance of the depletion region under the channel. The overall capacitance C_{sb} has a voltage dependence which is similar to that of an abrupt *p-n* junction.

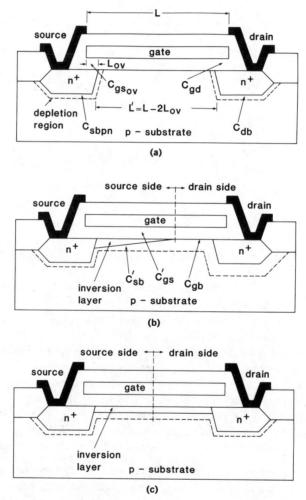

FIGURE 3.15. Parasitic capacitances in a MOSFET in the: (*a*) cutoff region; (*b*) saturation region; (*c*) triode region.

TABLE 3.3. The Terminal Capacitances of a MOSFET in the Three Main Regions of Operation (L' denotes $L - 2L_{ov}$)

Region of Operation	Capacitance				
	C_{gs}	C_{gd}	C_{gb}	C_{sb}	C_{db}
Cutoff region	$WL_{ov}C_{ox}$	$WL_{ov}C_{ox}$	$WL'C_{ox}$	$A_S C_{pn}(V_{sb})$	$A_d C_{pn}(V_{db})$
Saturation region	$WC_{ox}\left(L_{ov} + \frac{2}{3}L'\right)$	$WL_{ov}C_{ox}$	$\dfrac{1}{3}\dfrac{WL'C_{ox}C_{pn}(V_{db})}{C_{ox} + C_{pn}(V_{cb})}$	$A_s C_{pn}(V_{sb}) + \frac{2}{3}WL'C_{pn}(V_{sb})$	$A_d C_{pn}(V_{db})$
Nonsaturated (triode) region	$WL_{ov}C_{ox} + \frac{1}{2}WL'C_{ox}$	$WC_{ox}\left(L_{ov} + \frac{1}{2}L'\right)$	0	$A_s C_{pn}(V_{sb}) + \frac{1}{2}WL'C_{pn}(V_{sb})$	$A_d C_{pn}(V_{db}) + \frac{1}{2}WL'C_{pn}(V_{db})$

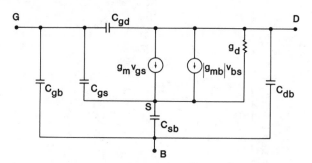

FIGURE 3.16.   High-frequency equivalent circuit of a MOSFET.

C_{db}: *Drain-to-Substrate Capacitance.*   This is a *p-n* junction capacitance and is thus voltage dependent.

C_{gb}: *Gate-to-Substrate Capacitance.*   This capacitance is usually small in the saturation region; its value is around $0.1C_{ox}$.

Figure 3.15 illustrates the physical structure of an NMOS transistor, and the locations of the capacitances in the cutoff (Fig. 15*a*), saturation (Fig. 15*b*), and nonsaturation or triode (Fig. 15*c*) regions. Table 3.3 lists the terminal capacitors of the NMOS device and their estimated values in the three regions of operation. The notations used are those shown in Figs. 3.15*a*–3.15*c*. Figure 3.16 depicts the complete high-frequency (ac) small-signal model of the MOSFET. In analyzing the small-signal behavior of MOSFETs, the model of Fig. 3.14 can be used if only low-frequency signals are present; if the capacitive currents are also of interest, then the circuit of Fig. 3.16 must be applied.

From the models of Figs. 3.14 and 3.16, and the accompanying discussions, a number of general statements can be made about the desirable construction of a MOSFET:

1.   For high ac gain, g_m should be large. This will be the case, by Eq. (3.18), if $k \triangleq (\mu_n C_{ox} W)/2L$ is large. Thus, the oxide should be thin to maximize C_{ox} (which is the oxide capacitance per unit area); also, W/L should be as large as possible. These measures, however, tend to increase the size and thus the cost of the integrated circuit. Also, by (3.18), the quiescent (bias) current i_D^0 should be as large as the allowable dc power dissipation permits.

2.   As the negative sign in Eq. (3.19) indicates, the body effect reduces the gain. To minimize g_{mb}, by Eqs. (3.19) and (3.13), we need large C_{ox}, small N_{imp} (i.e., lightly doped substrate), and a large bias voltage v_{SB}^0 for the source. (Of course, if v_{SB} is constant, then no incremental body effect occurs and these requirements are moot.)

3.   Ideally, the MOSFET in saturation should behave as a pure current source. Hence, as Fig. 3.14 illustrates, g_d should be small. By Eq. (3.20), this

TABLE 3.4. Small-Signal parameters of MOSFETs in Saturation (cf. Fig. 3.16) ($|\lambda v_{Ds}| \ll 1$ is assumed in all formulas)

	NMOS	PMOS
Transconductance: $g_m = \dfrac{\partial i_D}{\partial v_{GS}}$	$\dfrac{\mu_n C_{\text{ox}} W}{L}(v_{GS}^0 - V_T) = \sqrt{\dfrac{2\mu_n C_{\text{ox}} W i_D^0}{L}}$	$-\dfrac{\mu_n C_{\text{ox}} W}{L}(v_{GS}^0 - V_T) = -\sqrt{\dfrac{2\mu_p C_{\text{ox}} W(-i_D^0)}{L}}$
Body-effect transconductance: $g_{mb} = \dfrac{\partial i_D}{\partial v_{SB}}$	$-\dfrac{\gamma/2}{\sqrt{2\phi_p + v_{SB}^0}}\, g_m$	$-\dfrac{\gamma/2}{\sqrt{2\phi_p - v_{SB}^0}}\, g_m$
Drain conductance: $g_d = \dfrac{\partial i_D}{\partial v_{DS}}$	$\dfrac{\lambda i_D^0}{1 + \lambda v_{DS}^0}$	$\dfrac{\lambda i_D^0}{1 - \lambda v_{DS}^0}$
Gate-to-source capacitance C_{gs}	$\tfrac{2}{3} WLC_{\text{ox}}$	$\tfrac{2}{3} WLC_{\text{ox}}$
Gate-to-drain capacitance C_{gd}	$C_{gd\,\text{overlap}}$	$C_{gd\,\text{overlap}}$
Source- (or drain-) to-bulk capacitance C_{sb} (C_{db})	$\dfrac{C_{sb_0}}{\sqrt{1 + v_{SB}^0/2\phi_p}}\, ; \dfrac{C_{db_0}}{\sqrt{1 + v_{DB}^0/2\phi_p}}$	$\dfrac{C_{sb_0}}{\sqrt{1 - v_{SB}^0/2\phi_p}}\, ; \dfrac{C_{db_0}}{\sqrt{1 - v_{DB}^0/2\phi_p}}$

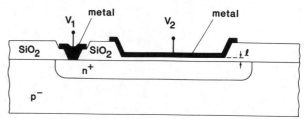

FIGURE 3.17.   Thin-oxide metal-to-n^+-region capacitor.

requires a small bias current i_D^0, a large bias voltage v_{DS}^0, and a small λ. Since λ is introduced by channel-length modulation, it can be reduced by increasing L and also (see Ref. 1, Sec. 8.4) by increasing N_{imp}.

A summary of the formulas derived in this section is given in Table 3.4.

3.5. MOS CAPACITORS[5,6,7]

Next to the MOSFETs, the key components of analog MOS circuits are the MOS capacitors. Their electrodes (i.e., conductive plates) can be realized in several different ways, to be discussed below; the dielectric, however, is almost always SiO$_2$, one of the most stable insulators known.* It has a relative permittivity $K_{ox} \simeq 3.9$, and a high breakdown field ($\mathscr{E}_b \simeq 8 \times 10^6$ V/cm). Depending on the technology available, the electrodes may be fabricated as follows:[5]

1. *Metal-or Polysilicon-over-Diffusion Structure (Fig. 3.17).* This structure is formed by growing a thin layer of SiO$_2$ over a heavily doped region (n^+-region in Fig. 3.17) in the substrate. The doped region thus serves as the bottom plate of the capacitor. The top plate is formed by covering the SiO$_2$ with a metal or polysilicon layer. In a metal-gate process this can be done in the same processing step which provides the metalization for the gates and leads (the "interconnects") of the integrated circuit. In a silicon-gate process, the conductive layers which form the MOSFET gates are not made of metal, but of deposited polycrystalline silicon (polysilicon). When it is heavily doped, this material is a good conductor, and thus suitable for use also as a capacitor electrode. Ideally, if the electrodes were perfect conductors, the capacitance formed would be given by Eq. (3.3). In fact, for a metal-over-diffusion structure, if $V_2 > V_1$ in Fig. 3.17, the charge at the bottom plate is negative, and is readily supplied by the n^+-layer. If, however, $V_2 < V_1$ then the bottom plate requires a positive charge. This can only be obtained by creating a

---

*A dielectric composed of SiO$_2$ and Si$_3$N$_4$ layers[8] can also be used. It has a higher permittivity and breakdown voltage than SiO$_2$ alone.

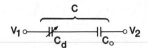

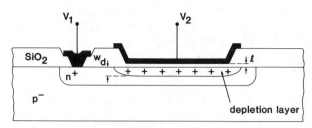

FIGURE 3.18.   MOS capacitor with negative applied voltage ($V_2 < V_1$).

depleted ion layer in the n^+-region under the thin oxide (Fig. 3.18). The capacitance C seen between the two terminals can then be regared as the series connection of two component capacitors: one is the oxide capacitance C_0 given by Eq. (3.3), the other the capacitance C_d of the depletion layer. Since the depth w_d of the depletion layer depends on the applied voltage $V_2 - V_1$, C_d is a voltage-variable capacitor. It can readily be shown (see Ref. 2, Sec. 9.2) that the voltage dependence of the total capacitance C per unit area is then

$$C = \frac{C_0 C_d}{C_0 + C_d} = C_0 \left[1 + \frac{2K_{ox}^2 \varepsilon_0 (V_1 - V_2)}{q N_D K_s l^2} \right]^{-1/2} \text{F/cm}^2. \quad (3.21)$$

Here, N_D is the uniform n^+-doping density in cm^{-3}, and K_s is the relative permittivity of silicon ($K_s \simeq 11.7$). As shown in Fig. 3.19 (curve 1), C decreases with increasing $V_1 - V_2$ in this region. When $V_1 - V_2$ reaches the threshold voltage $|V_T|$, strong inversion occurs on the silicon surface. Further increase of $V_1 - V_2$ will increase the capacitance, until C_0 is reached again (curve 2, Fig. 3.19). For a high-frequency signal, there will not be enough time for the minority carriers (holes) to form and dissolve the inversion layer, and hence Eq. (3.21) continues to hold (curve 1, Fig. 3.19).

For the high doping density normally used, $|V_T|$ is very large ($|V_T| > 100$ V) and the voltage dependence of C is slight for usual values of $|V_1 - V_2| \ll |V_T|$.

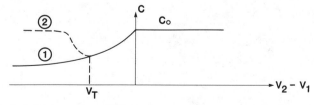

FIGURE 3.19.   The variation of a MOS capacitor with applied voltage.

In the region where $C \simeq C_0$, the voltage coefficient of C, defined as

$$\gamma_V^C \triangleq \frac{\Delta C / C}{\Delta V} \simeq \frac{1}{C} \frac{\partial C}{\partial V} \qquad (3.22)$$

is small; for $N_D = 5 \times 10^{20} \ \text{cm}^{-3}$ and $l = 1000 \ \text{Å}$, $\gamma_V^C \sim -10 \ \text{ppm/V}$ results. Here, "ppm" is short for "parts per million", that is, $10^{-4}\%$.

The temperature coefficient of C is defined similarly:

$$\gamma_T^C \triangleq \frac{\Delta C / C}{\Delta T} \simeq \frac{1}{C} \frac{\partial C}{\partial T}. \qquad (3.23)$$

The temperature dependence of C is due to the thermal expansion of the device, the change in K_{ox}, and the change in the depth of the depletion region. For $N_D = 5 \times 10^{20} \ \text{cm}^{-3}$, typical values for γ_T^C are around 20–50 ppm/°C.

The metal-over-diffusion capacitor construction is well suited for MOS processes used to fabricate metal-gate MOSFETs, since the steps needed to construct the source and drain diffusions and the gate can be used also to construct the capacitors. However, for the polysilicon-over-diffusion capacitor, the conventional self-aligned silicon-gate process cannot be used, because in such a process the polysilicon layer is deposited before the formation of the heavily doped source–drain regions. The construction of capacitors in a single polysilicon process involves one additional masking and processing step which inserts a heavily doped diffusion region under the polysilicon and thin-oxide layers.

Another important parameter in a monolithic capacitor is the tolerance of the absolute value of capacitance. The tolerance is primarily determined by the variation of the dielectric thickness and is usually within ±15%. The matching of two identical capacitors, however, within the same die area is very accurate and is in the range of 0.1–1%.

2. *Polysilicon-over-Polysilicon Structure (Fig. 3.20).* In a silicon-gate "double-poly" process, a second layer of low resistivity polysilicon is available for use as an interconnect or for the formation of a floating gate for memory applications. These two layers of polysilicon can also be used as the top and bottom electrodes of a monolithic capacitor. Figure 3.20 shows the construc-

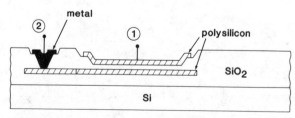

FIGURE 3.20.   Polysilicon-oxide-polysilicon capacitor structure.

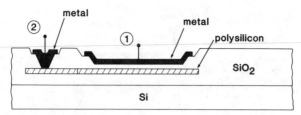

FIGURE 3.21. Metal-oxide-polysilicon capacitor structure.

tion of a capacitor with two polysilicon electrodes. The values of γ_V^C and γ_T^C are of the same order as for the metal- or polysilicon-over-diffusion structure. This structure can be fabricated using any MOSFET or charge-coupled device (CCD), or electrically programmable and erasable random access memory (EPROM or EEPROM) process in which two polysilicon layers are available. A disadvantage of this arrangement is that (due to the granularity of the polysilicon surface) the oxide thickness has a large random variation.

3. *Metal-over-Polysilicon Structure (Fig. 3.21).* This construction is similar to that shown in Fig. 3.20, and its properties are also nearly the same. Hence, it will not be discussed.

Naturally, other structures similar to those illustrated in Figs. 3.17–3.21 are also possible.

There are unavoidable parasitic capacitances associated with each of the described MOS capacitor types. In all cases, there is a large parasitic capacitance from the bottom plate of the capacitor to the substrate, and thus to the substrate bias supply. For a metal- or polysilicon-over-diffusion capacitor, where the bottom plate is embedded in the substrate, this stray capacitance is that of a reverse-biased *p-n* junction and can be 15 ~ 30% of *C* depending on the oxide thickness and the construction of the device. For the structures of Figs. 3.20 and 3.21, the bottom-plate stray capacitance is typically 5 ~ 20% of *C*.

Since both capacitor plates must be connected to the rest of the circuit, an additional stray capacitance to the substrate exists due to the connecting leads. Also, very often one or both plates of a capacitor are connected to the source or drain diffusion of a MOSFET used as a switch (Fig. 3.22). The *p-n* junction of the diffusion adds a parasitic space-charge (i.e., depletion layer) capacitance between the capacitor plate and the substrate. If the junction has a length L, width W, and depth x_j, then this added stray capacitance will be (Problem 3.7)

$$C_j \simeq C_a WL + C_s \pi (W + L) x_j. \tag{3.24}$$

Here, C_a is the unit-area capacitance of the bottom plate of the n^+-diffusion, and C_s is that of the side wall.

In addition to the junction capacitance C_j, the *p-n* junction also causes a leakage current between the capacitor plate and the substrate. This can be

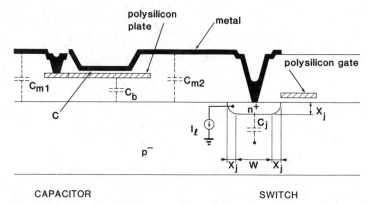

CAPACITOR   SWITCH

FIGURE 3.22. Metal-over-polysilicon capacitor and its corresponding strays.

represented by a constant current source (I_l in Fig. 3.22). The leakage allows the main capacitor C to lose charge, and hence it represents a *lower* limit on the frequency with which C may be charged or discharged without losing too much charge in one period. The leakage current is caused by the thermal generation and recombination of carriers in the depletion layer. Its value is usually around 1 ~ 10 pA per mil² junction area (1 mil = 10^{-3} in. ≈ 25.4 μm) at room temperature; it doubles for each 8–12°C rise in the junction temperature. It also depends somewhat on the junction shape and the voltage between the diffusion and the substrate.

As will be seen in later chapters, the response of a switched-capacitor circuit depends ideally only on the clock frequency and the ratios of the capacitances in the circuit. Hence, the accuracy with which these ratios can be controlled represents a major limitation on the accuracy of the frequency response. The achievable precision depends on many factors,[6,7] most of which are related to the fabrication process and are thus beyond the scope of this book. A few important general points will, however, be considered next.

The top view of an MOS capacitor is shown in Fig. 3.23. As indicated in the figure (grossly exaggerated), the edges of the electrodes are subject to a random

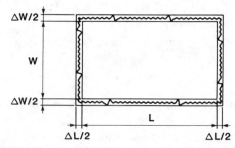

FIGURE 3.23. Random edge variations in an MOS capacitor.

variation. Hence, the area $A = WL$ in the basic formula Eq. (3.3) also varies, and thus so does C. From Eq. (3.3) and

$$\Delta C = \frac{\varepsilon}{l}[(W + \Delta W)(L + \Delta L) - WL] \qquad (3.25)$$

we obtain the relation

$$\frac{\Delta C}{C} \simeq \frac{\Delta W}{W} + \frac{\Delta L}{L} \qquad (3.26)$$

between the increments ΔW, ΔL, and ΔC.

Assuming that ΔW and ΔL are independent random variables with equal standard deviations $\sigma_L = \sigma_W$, the standard deviation of ΔC will be (Problem 3.8)

$$\sigma_c = C\sigma_L\sqrt{W^{-2} + L^{-2}}. \qquad (3.27)$$

Since C (and thus WL) is fixed, the relative error σ_c/C is minimized if $W = L$ is chosen. The relative capacitance error is then $\sqrt{2}\,\sigma_L/L$ (Problem 3.9). Thus, for the minimum possible relative capacitance error due to edge variations, the shape of the capacitors should be square. (Since the capacitance value is proportional to the area, while its error is proportional to the perimeter, a circular shape would be even better; however, it is impractical to fabricate.)

The deviation given above is valid for the absolute, rather than relative, value of C. However, it can readily be shown (Problem 3.10) that the same considerations and results apply to capacitance *ratios*. Specifically, assume that the nominal values of two capacitors with sides W_1, L_1 and W_2, L_2, respectively, are C_1 and C_2, so that the nominal value of the ratio is $\alpha = C_1/C_2 > 1$. If all dimensions have the same standard error σ, then it can be shown (Problem 3.10) that the relative error of α is

$$\frac{\sigma_\alpha}{\alpha} = \sigma_L\sqrt{L_1^{-2} + W_1^{-2} + L_2^{-2} + W_2^{-2}}. \qquad (3.28)$$

For prescribed $W_1 L_1$ and $W_2 L_2 = W_1 L_1/\alpha$, σ_α/α is minimum if $L_1 = W_1 = \sqrt{\alpha}\,L_2 = \sqrt{\alpha}\,W_2$. The minimum value is then

$$\left(\frac{\sigma_\alpha}{\alpha}\right)_{min} = \left(\frac{\sqrt{2}\,\sigma_L}{L_1}\right)(1 + \alpha)^{1/2}. \qquad (3.29)$$

Another source of error in the value of a capacitance ratio is the "undercut." This is due to an uncontrolled lateral etching of the plates of the capacitor along its perimeter during the fabrication. It results in a decrease of C, which is proportional to the perimeter of the device. Consider again two capacitors with

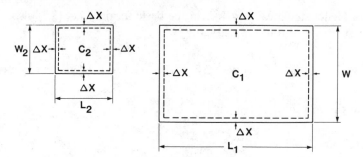

FIGURE 3.24.   Undercut error.

nominal dimensions W_1, L_1 and W_2, L_2, respectively, so that nominally

$$\alpha = \frac{W_1 L_1}{W_2 L_2}.$$

Then, due to the undercut, in fact

$$\alpha_{\text{actual}} \simeq \frac{W_1 L_1 - 2(W_1 + L_1)\,\Delta x}{W_2 L_2 - 2(W_2 + L_2)\,\Delta x} \tag{3.30}$$

results, where Δx is the undercut depth (Fig. 3.24), assumed to be constant along the perimeter.

If again $L_1 = W_1 = \sqrt{\alpha}\,L_2 = \sqrt{\alpha}\,W_2$ is chosen, then $\alpha_{\text{actual}} \neq \alpha$; however, it is readily shown (Problem 3.11) that for $W_2 = L_2$ and

$$W_1 = L_2 \left(\alpha - \sqrt{\alpha^2 - \alpha} \right) \tag{3.31}$$

and

$$L_1 = L_2 \left(\alpha + \sqrt{\alpha^2 - \alpha} \right)$$

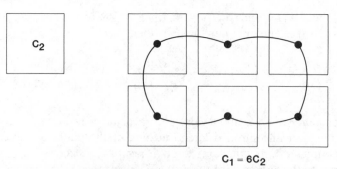

FIGURE 3.25.   Unit-capacitor structure.

the undercut causes no error in α. For this choice, the standard deviation of α is given by

$$\frac{\sigma_\alpha}{\alpha} = \left(\frac{\sigma_L}{L_2}\right)\sqrt{6 - 2\alpha^{-1}}. \tag{3.32}$$

For $\alpha \gg 1$, $\sigma_\alpha/\alpha \simeq \sqrt{6}\,\sigma_L/L_2$.

A more common technique[6] to avoid the undercut effect is to connect identically sized smaller capacitors ("unit capacitors") in parallel to construct a larger one (Fig. 3.25). Using this process, the area/perimeter ratio is nearly the same for any two capacitors, and hence $\alpha_{actual} \simeq \alpha$ can be obtained.

3.6. THE MOS SWITCH

Another important circuit element in switched-capacitor circuits is the MOS switch. The simplest realization of an on–off switch (Fig. 3.26a) in MOS technology is a single MOSFET (Fig. 3.26b). When the gate has a sufficiently high voltage of the appropriate polarity (positive for NMOS, negative for PMOS), the switch will be "on," and a current i_D will flow between nodes ① and ② in response to a potential difference v_{DS} between these nodes. Since the on-value of the gate voltage v_ϕ, derived from a clock (Fig. 3.27), is usually large (5 ~ 15 V) while the signal voltage between nodes ① and ② is typically of the order of 1 V or less, it can be assumed that the FET switch is in its linear region. Therefore, from Eq. (3.7), the current through an NMOS switch is given by

$$i_D = k\left[2(v_{GS} - V_T)v_{DS} - v_{DS}^2\right]. \tag{3.33}$$

For the usual case of $|v_{GS} - V_T| \gg |v_{DS}|$, the MOSFET behaves as a linear

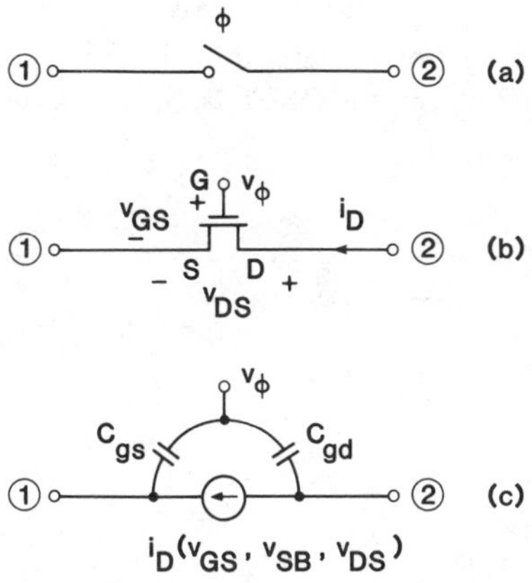

FIGURE 3.26. MOS switch: (a) symbol; (b) single-MOSFET realization; (c) equivalent circuit.

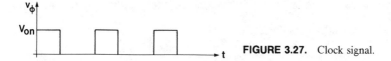

FIGURE 3.27. Clock signal.

resistor of value

$$R_{\text{on}} \approx \frac{1}{2k(v_{GS} - V_T)}.$$

(3.34)

(This approximation is, of course, not valid during the time the switch is off, or during the turn-on and turn-off transients.) Due to the nonzero value of R_{on}, it requires a nonzero time to charge or discharge a capacitor through the closed switch.

As discussed in Section 3.4, in addition to the channel current i_D, there are capacitive currents associated with the transient response of the MOSFET. Of the several stray capacitances present, the two gate capacitances C_{gd} and C_{gs} are of special importance in switching applications (Fig. 3.26c). Since the impedances which load nodes ① and ② in a switched-capacitor filter are usually capacitive, C_{gd} and C_{gs} form voltage dividers of transfer ratios $C_{gd}/(C_{gd} + C_2)$ and $C_{gs}/(C_{gs} + C_1)$, respectively. Here, C_1 is the capacitance loading node ① and C_2 that loading node ②. For typical values (e.g., C_{gs} and $C_{gd} \approx 0.02$ pF, C_1 and $C_2 \approx 2$ pF), the clock signal $v_\phi \approx 10$ V will be divided by about 100 when it is transmitted to nodes ① and ②. Thus, v_1 and v_2 will contain a "clock feedthrough" noise of the order of 0.1 V.

Very often, one of the nodes (say ②) is a virtual ground (Fig. 3.28). Then the charge entering into the circuit at node ② due to the transitions of v_ϕ is of interest. From Fig. 3.26c, the charge pumped into node ② as v_ϕ changes from 0 to V_{on} (Fig. 3.27) is $C_{gd}V_{\text{on}}$. For $C_{gd} \approx 0.02$ pF and $V_{\text{on}} = 10$ V, this charge pulse has a magnitude of 0.2 pC. If such pulse enters a feedback capacitor C of, say, 2 pF (Fig. 3.28), the resulting output voltage step is again 0.1 V, a significant value.

In most applications, noise voltages of such magnitude are not tolerable. To compensate for the clock feedthrough effect using only MOSFETs of the same

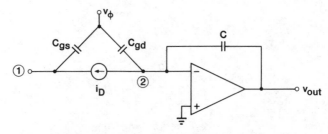

FIGURE 3.28. Clock feedthrough in an op-amp circuit.

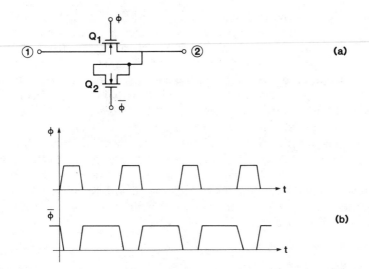

FIGURE 3.29. Clock feedthrough compensation scheme: (*a*) circuit; (*b*) clock signals.

type, the scheme of Fig. 3.29 may be used. Here, the feedthrough to node ② is reduced by having *two* feedthrough signals of nearly equal values but opposite signs. Since the capacitive coupling path from the short-circuited MOSFET Q_2 includes both C_{gd} and C_{gs}, Q_2 should have a narrower width W by a factor of about 2, and the same length L.

If both NMOS and PMOS transistors are available, then the circuit of Fig. 3.30, which compensates feedthroughs to both nodes ① and ②, can be used. This circuit also reduces the voltage drop across the switch, and thus increases the dynamic range since smaller signals may be used.

As Fig. 3.15 illustrates, C_{gs} contains a *nonlinear* component C'_{gs} which holds the channel charge. Hence, it is difficult to achieve perfect clock feedthrough compensation with *any* scheme. When the switch is turned off, the channel charge (usually a few pC) is swept in part into the drain and in part into the source. If the clock signal changes slowly, the channel charge drift can be analyzed; however, for fast clock transition, the distribution of the charge is unpredictable. Since this phenomenon is effectively the same as clock feedthrough via C'_{gs} (Fig. 3.15), it makes exact feedback cancellation even more complicated.

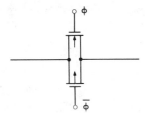

FIGURE 3.30. Clock feedthrough compensation using complementary-type transistors.

Due to the symmetric structure of the MOSFET (Fig. 3.8), the MOS switch is bidirectional. Also, there is no "offset" voltage (i.e., voltage appearing due to intrinsic phenomena) between the drain and source if $i_D = 0$.

After the switch is turned off, the drain (source) current consists of the leakage current of the reverse-biased drain (source) junction. While this current is very small ($I_l \ll 1$ pA) it can result in a steadily increasing drain or source charge and thus voltage, unless there is a dc path to ground connected at least intermittently to both switch terminals. Furthermore, as Fig. 3.15 indicates, the MOS switch introduces two nonlinear capacitances, C_{sb} and C_{db}, between these terminals and the substrate. These capacitances can cause harmonic distortion, and can also couple noise from the substrate into the signal path.

Another important nonideal effect associated with the MOS transistor used as a switch is the thermal noise associated with the conducting channel. This effect will be briefly discussed in the next section.

3.7.   NOISE IN MOSFETS

There are three distinct sources of noise in solid-state devices: shot noise, thermal noise, and flicker noise.

Shot Noise

Since electric currents are carried by randomly propagating individual charge carriers (electrons or holes), superimposed on the nominal (average) current I, there is always a random variation i_{ns}. This is due to the fluctuation in the number of carriers crossing a given surface in the conductor in any time interval. It can be shown that the mean square of i_{ns} is given by

$$\overline{i_{ns}^2} = 2qI\,\Delta f, \tag{3.35}$$

where $q = 1.6 \times 10^{-19}$ C is the magnitude of the electron charge and Δf the bandwidth. This formula only holds, however, if the density of the charge carriers is so low, and the external electric field is so high, that the interactions between the carriers is negligible. Otherwise, the randomness of their density and velocity is reduced due to the correlation introduced by the repulsion of their charges. The noise current is then much smaller than predicted by (3.35).

In a conducting MOSFET channel, the charge density is usually high and the electric field is low. Therefore, (3.35) does not hold. The noise current due to random carrier motions is hence better described as *thermal noise*, which is discussed next.

Thermal Noise

In a real resistor R, the electrons are in random thermal motion. As a result, a fluctuating voltage v_{nT} appears across the resistor even in the absence of a

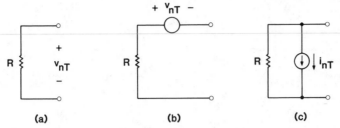

FIGURE 3.31. Thermal noise in a resistor: (a) noisy resistor; (b) and (c) equivalent circuits.

current from an external circuit (Fig. 3.31a). Thus, the Thévenin model of the real (noisy) resistor is that shown in Fig. 3.31b. Clearly, the higher the absolute temperature T of the resistor, the larger v_{nT} will be. In fact, it can be shown that the *mean square* of v_{nT} is given by

$$\overline{v_{nT}^2} = 4kTR\,\Delta f. \tag{3.36}$$

Here, k is the ubiquitous Boltzmann's constant, and Δf is the bandwidth in which the noise is measured, in Hz. (The value of $4kT$ at room temperature is about 1.66×10^{-20} V C.)

If Eq. (3.36) was true for any bandwidth, then the energy of the noise would be infinite. In fact, however, for very high frequencies ($\approx 10^{13}$ Hz) other physical phenomena enter, which cause $\overline{v_n^2}$ to decrease with increasing frequency so that the overall noise energy is finite.

The average value (dc component) of the thermal noise is zero. Since its spectral density $\overline{v_{nT}^2}/\Delta f$ is independent of frequency (at least for lower frequencies), it is a "white noise." Clearly, Fig. 3.31b may be redrawn in the form of a Norton equivalent, that is as a (noiseless) resistor R in parallel with a noise current source i_{nT} (Fig. 3.31c). The value of the latter is given by

$$\overline{i_{nT}^2} = 4kTG\,\Delta f \tag{3.37}$$

where $G = 1/R$.

Since the channel of a MOSFET in conduction contains free carriers, it is subject to thermal noise. Therefore, Eqs. (3.36) and (3.37) will hold, with R given by the *incremental channel resistance*. The noise can then be modeled by a current source, as shown in Fig. 3.32a. If the device is in saturation, its channel tapers off (Fig. 3.8), and the approximation $R \simeq 3/2g_m$ can be used in (3.37).

In most circuits, it is convenient to pretend that i_{nT} is caused by a voltage source connected to the gate of an (otherwise noiseless) MOSFET (Fig. 3.32b). This "gate referred" noise voltage source is then given by

$$\overline{v_{nT}^2} \simeq \overline{\left(i_{nT}/g_m\right)^2} = \frac{8}{3}\frac{kT}{g_m}\,\Delta f. \tag{3.38}$$

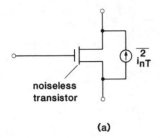

(a)

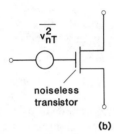

(b)

FIGURE 3.32. Equivalent models for the thermal noise in a MOSFET.

Both i_{nT} and v_{nT} depend thus on the dimensions, bias conditions, and temperature of the device. As an example of their orders of magnitude, for a transistor with $W = 200$ μm, $L = 10$ μm, and $C_{ox} = 4.34 \times 10^{-8}$ F/cm^2 (corresponding to an oxide thickness of 800 Å) which is operated in saturation at a drain current $i_D = 200$ μA, the gate-referred noise voltage at room temperature is about 9 nV/$\sqrt{\text{Hz}}$.

If the device is switched off, then R becomes very high, and the equivalent noise circuit will be a current source with a value given by Eq. (3.37). Clearly, $\overline{i_{nT}^2}$ is very small; hence, for usual (low or moderate) external impedance levels, the MOSFET can be regarded as a noiseless open circuit if it is turned off.

Flicker (1 / f) Noise

In an MOS transistor, extra electron energy states exist at the boundary between the Si and SiO$_2$. These can trap and release electrons from the channel, and hence introduce noise.[9,10] Since the process is relatively slow, most of the noise energy will be at low frequencies. As before, a possible model of this noise phenomenon is a current source in parallel with the channel resistance. The dc value of noise current is again zero. Its mean-square value *increases* with temperature and the density of the surface states; it *decreases* with the gate area $W \times L$ and the gate oxide capacitance per unit area C_{ox}. For devices fabricated with a "clean" process, the gate-referred noise voltage is nearly independent of the bias conditions and is given by the approximating formula

$$\overline{v_{nf}^2} = \frac{K}{C_{ox}WL} \frac{\Delta f}{f}.$$ (3.39)

Here, K depends on the temperature and the fabrication process; a typical value[11] is 3×10^{-24} $V^2 F$. For the transistor described in the preceding example, the formula gives a noise voltage of 83 nV/$\sqrt{Hz}$ at $f = 1$ kHz. As before, the equivalent channel current noise is related to $\overline{v_{nf}^2}$ by the formula $\overline{i_{nf}^2} = g_m^2 \overline{v_{nf}^2}$.

The noise process described is usually called *flicker noise* or (in reference to the $1/f$ factor in $\overline{v_{nf}^2}$ and $\overline{i_{nf}^2}$) $1/f$ *noise*. As the example given illustrates, at low frequencies (say, below 1 kHz) it is usually the dominant noise mechanism in a MOSFET.

In conclusion, the channel noise in a MOSFET can be modelled by an equivalent noise current generator, as in Fig. 3.32a. In the small-signal model, this generator will be in parallel with the current sources $g_m v_{gs}$ and $g_{mb} v_{bs}$ (Fig. 3.16). Its value can be chosen as the RMS (root mean square) noise current, which from (3.37)–(3.39) is

$$i_n = \sqrt{\overline{i_{nT}^2} + \overline{i_{nf}^2}}$$

$$= \sqrt{\left[4kTG + Kg_m^2/(C_{ox}WLf)\right] \Delta f}.$$ (3.40)

Note that the mean squares of the noise currents are added, since the different noise mechanisms are statistically independent.

Alternatively, the noise can be represented by its "gate-referred" voltage source (Fig. 3.32b), in series with the gate terminal. The value of the source is i_n/g_m, with i_n given by Eq. (3.40).

3.8. ANALOG MOS TECHNOLOGY [8, 10, 13, 14, 15]

As already discussed, the building blocks needed in an analog MOS circuit are capacitors, switches, op-amps, and (less frequently) resistors. Since switches and op-amps are realized by connecting appropriately MOS field-effect transistors (MOSFETs), the basic components are capacitors, MOSFETs, and (if needed) resistors. The properties, physical foundations, and limitations of the first two have already been discussed. In this section, we give a brief description of the actual devices, and the processes which are used to fabricate them.

The base material of the MOS chip is crystalline (single-crystal) silicon, which is lightly doped (10^{14}–10^{15} atoms/cm^3) with p- or n-type impurities. Circular slices, called *wafers*, are cut from cylindrical bars of this material. A wafer is normally 4–6 in. in diameter and approximately 0.01 in. thick. An important parameter of the wafer is its *orientation*. This describes the angle at which the wafer surface cuts across the crystal structure. Currently, the (111) cut, which exposes a triangular surface structure, and the (100) cut, which provides a rectangular pattern, are most widely used.

To create transistors, capacitors, and resistors on the wafer, the following operations are used:

1. Oxidation.
2. Photolithography (photomasking).
3. Etching.
4. Diffusion.
5. Deposition.
6. Ion implantation.

These operations will now be briefly discussed.

1. *Oxidation.* In this process, a protective layer of silicon dioxide (SiO_2, glass) is grown over the silicon surface. This is done by exposing the surface to oxygen (dry oxidation) or steam (wet oxidation) in a furnace at a temperature of 900–1200 °C, for a period which may range from a few minutes to several hours depending on the oxide thickness required. Dry oxidation provides a denser oxide layer, and allows more precise control over the oxide thickness. However, it is slower than wet oxidation, and is hence most often used to create thin oxides such as those needed under the gates. The resulting Si–SiO_2 interface will be *below* the original Si surface. For an oxide thickness t_{ox}, the shift is by $0.44t_{ox}$ (Fig. 3.33).[16]

2. *Photolithography.* This process is used for establishing patterns for the selective deposition or removal of material such as SiO_2. For the removal of SiO_2, its surface is first covered by a thin layer of a photosensitive material called *photoresist*. This is usually applied by *spinning*: a small amount of photoresist is placed on the center of the wafer, which is then rapidly spun around. The centrifugal force causes the photoresist to spread out and cover the wafer uniformly (Fig. 3.34a). Next, a hot air stream or infrared heating is

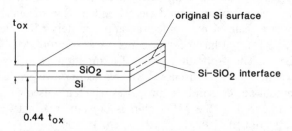

FIGURE 3.33. Oxidation: the Si surface shifts by $0.44t_{ox}$.

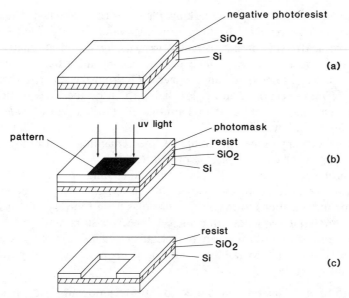

FIGURE 3.34. Photolithography: (*a*) wafer with photoresist; (*b*) masking and exposure; (*c*) wafer after development.

used to bake on the resist layer. The wafer is now covered by the *photomask*, a glass plate on which the desired pattern is defined (Fig. 3.34*b*). If a "negative" photoresist is used, then the mask will be opaque in those areas where the SiO₂ is to be removed and transparent elsewhere. (For a "positive" photoresist, the situation is reversed.) The mask has the size and shape of the wafer, and is in direct contact with it. Next, a collimated beam of ultraviolet (UV) light is used to illuminate the wafer through the mask. As a result, the photoresist is exposed to UV light at the regions where the mask was transparent. This causes it to harden at these areas. The wafer is now *developed*, that is immersed in a developer solution. This dissolves the photoresist in the unexposed areas, but retains the hardened layer in the exposed ones (Fig. 3.34*c*).

The described technique is called *contact printing*, since the mask is in close physical contact with the wafer. This allows good resolution, but it may lead to damage to the mask surface due, for example, to a speck of dust between the mask and the wafer. The probability of such damage can be reduced by allowing a small gap (10 ~ 30 μm wide) between the wafer and the mask (*proximity printing*), and can be eliminated by projecting the mask pattern optically on the wafer (*projection printing*). In the latter technique, in order to achieve high resolution, only a small part of the mask pattern is projected on the wafer, and this partial pattern is scanned or stepped across the wafer to obtain the complete desired repetitive exposure pattern.

Following the development step, the hardened resist usually must undergo a "postbake" process, which improves its adherence to the SiO₂ layer.

3. *Etching.*   After the photomasking, the wafer can be etched. The unprotected SiO_2 can be removed by immersing the wafer in an etchant (such as hydrofluoric acid). This process is called *wet etching*. In an alternative technique, called *plasma-assisted etching* or *dry etching*, a partially ionized gas (plasma) is used to remove the unprotected SiO_2 from the wafer. Dry etching can provide more accuracy, and is less liable to erode the sides of the layer under the hardened photoresist (a harmful effect often called *undercut* or *lateral etching*) than wet etching.

After the etching process is complete, the remaining photoresist can be removed using a solvent or a plasma process.

It should be noted that the photomasking and etching processes, which were discussed above in terms of removing a SiO_2 layer, can also be applied with only minor modifications to the etching of polysilicon (polycrystalline silicon) used as a material for gates, interconnect lines, resistors, and so on. Also, aluminum and silicon nitride can be selectively etched away using similar techniques. Finally, photomasking is also used in the selective deposition of material layers on the wafer.

4. *Diffusion.*   This process is used to introduce dopants into silicon. It is usually performed in two steps. In the first step, called *predeposition*, a specified density of dopant atoms is introduced into the silicon surface. In the second step, called *drive-in*, these impurities are caused to penetrate into the material deeper. Predeposition can be performed by exposing the silicon surface to a carrier gas at a high temperature. The dopant then leaves the gas and forms a deposit on the surface. Alternatively, the wafer can be coated by a doped layer of SiO_2 using chemical vapor deposition or a spin-on procedure, and the dopant then caused to diffuse into the silicon.

The drive-in is performed by heating the wafer for a predetermined length of time. This is usually done simultaneously with an oxidation process which protects the exposed diffusion region from contamination.

5. *Deposition.*   A number of materials can be deposited to form a thin film on wafers. These include polysilicon, SiO_2, silicon nitride (Si_3N_4), and aluminum. The most frequently used technique is chemical vapor deposition (CVD). In this process, a material is reacted at high temperature, and is deposited by way of a carrier gas (e.g., hydrogen or nitrogen). This procedure can be used to deposit polysilicon, SiO_2, and Si_3N_4. When (as is usually the case) a *doped* polysilicon layer is required, a layer of undoped polysilicon is usually deposited, and the dopant (boron, arsenic, or phosphorus) is added later using ion implantation or diffusion. Polysilicon can be deposited on thin oxide to form a gate, or on silicon to form a uniformly doped layer.

The deposition of aluminum or aluminum alloys (called *metalization*) is usually done using vacuum evaporation. The metal source and the wafers are placed in a vacuum chamber, and the source is heated or bombarded by an electron beam to cause some of the metal atoms to leave the source. An alternative technique is *sputtering*, in which positive ions are generated in a gas discharge, and are accelerated by a high voltage to a cathode which is coated by the source metal. This causes some of the metal atoms to leave and move to the wafers to form the deposited layer.

6. *Ion Implantation.* Like diffusion, ion implantation is a process used for introducing impurities into silicon. Here, ionized atoms of the dopant (boron, phosphorus, or arsenic) are extracted from a gas and separated using magnetic deflection. The ions are then accelerated by a high voltage (typically $50 \sim 150$ kV) and the resulting ion beam focused on the wafer surface. As a result, the dopant atoms penetrate into the wafer material to a depth which increases with increasing accelerating voltage. The distribution of charges versus depth usually has a bell-shaped (Gaussian) profile.

By using a mask, the implantation process can be performed selectively. Metal, photoresist, and polysilicon layers can all block ion penetration into the silicon below. A thin SiO_2 layer can transmit, at least partially, a high-energy ion beam. Thus, it is possible to use ion implantation through the gate oxide to modify the threshold voltage in the gate region.

The penetration depth of the dopant atoms is typically around 1 μm. If necessary, these atoms can be forced to diffuse deeper inside the silicon, by using a drive-in step similar to that discussed in connection with the diffusion operation.

Since the impact of the ion beam causes some damage to the single-crystal silicon structure, a subsequent annealing step is usually needed to restore the regularity of the lattice.

The equipment needed for ion implantation is much costlier than that needed for diffusion. However, it offers a more accurate control of both the total number of impurity atoms and their depth. This control is especially valuable for the creation of highly doped regions.

Using the six processing steps described above, a variety of devices can be fabricated. Historically, the first MOS integrated circuits were made in the mid-1960s, and used *p*-channel enhancement-mode devices on (111)-oriented wafers. (This orientation was used since the bipolar ICs then already prevalent were fabricated using it.) After a few years of development, the technology matured, and the device illustrated in Fig. 3.35 became an industrial standard.

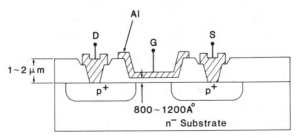

FIGURE 3.35. A *p*-channel enhancement-mode MOS transistor.

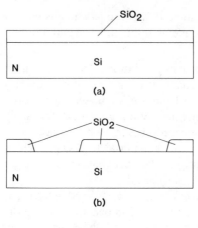

(a)

(b)

(c)

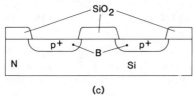

(d)

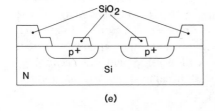

(e)

FIGURE 3.36.   The fabrication steps used to realize the PMOS transistor shown in Fig. 3.35.

It was fabricated in the following steps (Fig. 3.36):[15]

1. Oxidation was used to establish a thick (1 ~ 2 μm) field oxide layer over the whole wafer (Fig. 3.36a).
2. Photomasking and etching were used to remove the oxide over the source (Fig. 3.36b).
3. Boron was predeposited through the source and drain windows (Fig. 3.36c).

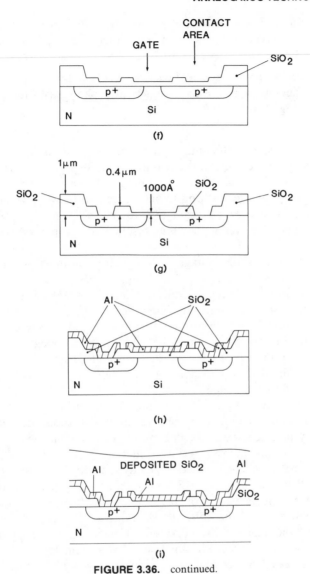

FIGURE 3.36.   continued.

4. The predeposited dopants were driven in and the windows covered by SiO₂ using oxidation (Fig. 3.36*d*).

5. Using a second photomask, new windows were opened through the oxide for the gates and the drain/source contacts (Fig. 3.36*e*).

6. A thin (~ 1000 Å) gate oxide was grown over the new windows (Fig. 3.36*f*).

7. The drain/source contact windows were opened again using a third photomask (Fig. 3.36*g*).

8. Aluminum was selectively deposited, with the aid of a fourth photomask, to create the gates, contacts, interconnects, and bonding pads (Fig. 3.36h).

9. A thick oxide was deposited over the whole wafer as a protective layer, and windows were opened (using a fifth photomask) for the bonding pads (Fig. 3.36i).

In this process, the aluminum gates were formed in Step 8 *after* the formation of the sources and gates by diffusion in Steps 3 and 4. Hence, it was necessary to extend the thin oxide and aluminum layers over the source/drain diffusion regions by at least 2 μm, to make sure that the gate metal covers the whole channel area between drain and source, even in the event of some alignment errors. This overlap, however, resulted in large parasitic capacitances $C_{gs_{ov}}$ and $C_{gd_{ov}}$.

The transistor threshold voltage V_T obtained using this technology, with a substrate doping of 10^{15} atoms/cm^3 and a gate oxide thickness of 1200 Å, was around -4 V. In the *field* (i.e., the inactive areas between devices) using a field oxide thickness of 1.5 μm, the threshold voltage $V_{T\,\text{field}}$ was about -37 V. Ideally, of course, the device threshold voltage should be low to allow operation with small bias voltages, while the field threshold voltage should be large to prevent the formation of parasitic channels under high-voltage lines, which may short circuit adjacent diffusions. While here $|V_{T\,\text{field}}|$ was sufficiently large, $|V_T|$ was not sufficiently small.

To reduce the magnitude $|V_T|$ of the device threshold voltage, several modifications were attempted to improve the basic process described above. The (111) crystal orientation was replaced by the (100) one, which has a lower charge density Q_{ss} at the oxide–silicon interface and hence makes a smaller $|V_T|$ possible.[2] This change resulted in greatly reduced threshold voltages, such as $V_T \simeq -1.9$ V and $V_{T\,\text{field}} \simeq -13$ V. The reduction of $|V_T|$ was an improvement; however, the reduction of $|V_{T\,\text{field}}|$ was detrimental. In addition, the effective mobility μ_p of the holes in the transistor channels is lower for the (100) orientation than for the (111) one. Hence, the "conductivity factor" $k' = \mu C_{\text{ox}}/2$ was also reduced, from about 2.75 to about 1.9 μA/V^2. Since the transconductance of the device is proportional to $k = k'(W/L)$, as can be seen from Table 3.3, a larger aspect ratio W/L was now needed to achieve the same g_m.

To improve this situation, the unit-area oxide capacitance C_{ox} can be increased. In addition to reducing the oxide thickness to the lowest practical value, the dielectric constant can also be increased. This was achieved by replacing the SiO$_2$ layer by a sandwich of SiO$_2$ and silicon nitride (Si$_3$N$_4$) layers. Si$_3$N$_4$ has a higher relative dielectric constant than SiO$_2$ ($K \simeq 7$ as opposed to $K = 3.9$ for silicon) so that values as high as $C_{\text{ox}} \simeq 6 \times 10^{-8}$ F/cm^2 can be obtained for a dielectric consisting of a 200-Å-thick layer of SiO$_2$ and an 800-Å-thick layer of Si$_3$N$_4$. (Note that Si$_3$N$_4$ should not be used to

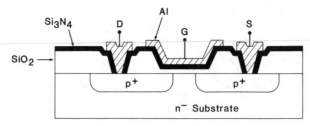

FIGURE 3.37. A p-channel NMOS transistor.

replace SiO_2 altogether, since the $Si–Si_3N_4$ interface results in charge-trapping effects and in a bias-dependent V_T.) A metal–nitride–oxide–silicon (MNOS) p-channel transistor is illustrated in Fig. 3.37. Using (111) substrate orientation, the values $V_T \simeq -2.4$ V, $V_{T \text{ field}} \simeq -42$ V, and $k' \simeq 5.7 \; \mu A/V^2$ were obtained for such a device.

Another technique for reducing $|V_T|$ is to reduce the substrate doping, which controls the charge in the depletion layer. This process, of course, reduces $|V_{T \text{ field}}|$ as well. To avoid the latter effect, only the channel regions are modified, as follows. Rather than try to remove locally the n-type doping, which is difficult, a small amount of p-type impurity (usually boron) is implanted in the n-doped channel region. This impurity contributes holes, which then recombine with some of the free electrons originating from the n-type substrate impurity atoms. Thus, the effective density of donors is reduced, and hence $|V_T|$ decreased. The implantation of the required boron ions can be performed through the thin gate oxide; the much thicker field oxide will prevent these ions from reaching the substrate in the field regions. Thus, the value of $|V_T|$ can be reduced without affecting $|V_{T \text{ field}}|$.

By applying a high dose of boron implant, it is possible to make the density of holes actually higher in the channel region than the density of electrons. Then, the substrate becomes p-type under the gate, and the transistor becomes a *depletion-mode device*. Figure 3.38 illustrates such a transistor. It requires an extra masking and processing step to make some transistors operate in the depletion mode, while keeping others in the enhancement mode of operation.

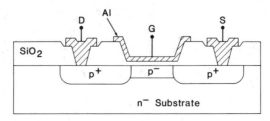

FIGURE 3.38. A p-channel depletion-mode MOS transistor.

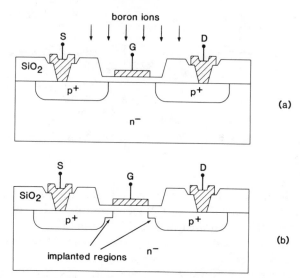

FIGURE 3.39. Self-aligned *p*-channel transistor fabricated using ion implantation.

Using ion implantation techniques, it is also possible to reduce the overlap of the gate metal and the source/drain diffusions, and thus to decrease the large overlap capacitances associated with the geometry of Fig. 3.35. The process is based on using *self-aligned* techniques, in which the gate acts as a mask for the definition of the source and drain regions. Thus, now the gate electrode is deposited *before* the boundaries of the source and drain are fully defined. To achieve this, after the source and drain have been diffused in, and the gate oxide grown, the gate metal is deposited in such a way that *gaps are left* intentionally between the region under the gate and the source and drain regions (Fig. 3.39*a*). Next, these gaps are implanted by boron ions while the gate protects the gate area from the implantation. The resulting structure is shown in Fig. 3.39*b*. Since a large dose of ions is required in the process, a high-temperature annealing must follow to restore the regularity of the crystal lattice in the implanted regions. This causes two problems. One is a slight lateral diffusion of the implanted dopants which introduces a small amount of overlap between the gate and the source and drain. The other is the damage to the gate metal (aluminum) which has a low melting point. The latter problem can be avoided by using a different gate material, such as molybdenum (a high melting-point metal) or polysilicon. The latter can be deposited in pure form, and doped in the same diffusion step in which the drain and source regions are established. Since the work function of polysilicon is different from that of aluminum, the threshold voltage value $|V_T|$ is also reduced by about 1 V,[14] which is a significant improvement. Using the (111)-oriented substrate to preserve the large value of $|V_{T \text{ field}}|$ and a 1000-Å gate oxide thickness, a threshold voltage of $V_T = -2.5$ V results. Figure 3.40 shows the structure of a

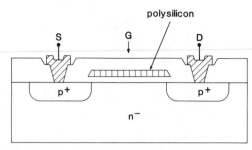

FIGURE 3.40. Self-aligned silicon-gate p-channel transistor.

silicon-gate p-channel device. This device can be fabricated as shown in Fig. 3.39. Alternatively, in the fabrication of such a transistor, first the gate oxide is grown, and then a thin layer of polysilicon is deposited over the entire wafer. This layer is next etched away everywhere, except where gates (and, in some cases, polysilicon lines or capacitor plates) are required. Next, windows are opened in the gate oxide where it is not covered by polysilicon. Through these windows, boron is diffused into the silicon, to create the source and drain regions. The gate thus again acts as a mask in defining the source and drain boundaries. Thus, the device is self-aligned, with only the lateral boron diffusion creating some overlap capacitances.

While historically p-channel MOSFETs preceded the n-channel ones, it was always recognized that NMOS devices have definite advantages over PMOS ones. As can be seen from Table 3.1, the mobility μ of electrons is about three times higher than that of holes. Since g_m is proportional to μ, much higher transconductance is thus achievable for an n-channel device with given sizes. Also, much lower diffusion resistance can be achieved for n^+-diffusions than for p^+-ones,[15] and the n^+-to-oxide interface has fewer defects than the p^+-to-oxide one.[15] A basic problem was, however, due to the surface charge Q_{ss} at the silicon-to-silicon-dioxide interface.* This charge increases the value of $|V_T|$ for a p-channel device, and reduces it for an n-channel one. The reduction is usually so large that the device operates in the depletion mode.

There are several techniques available for increasing the V_T of an NMOS device. One is to increase the p-doping of the substrate. This, however, reduces $V_{T \text{ field}}$ also, and also leads to reduced electron mobility. It also increases the body-effect coefficient γ (cf. Table 3.1). Another technique is to connect a negative bias voltage V_B to the substrate, and thus to increase the body-effect contribution to V_T. This, however, requires an additional bias supply. Another way of increasing V_T is to use the (100) rather than the (111) orientation. This reduces Q_{ss}, and thus increases V_T. Using polysilicon, rather than aluminum, as gate material also increases V_T somewhat. Finally, a local increase of the

*Caused by sodium (Na) contamination.

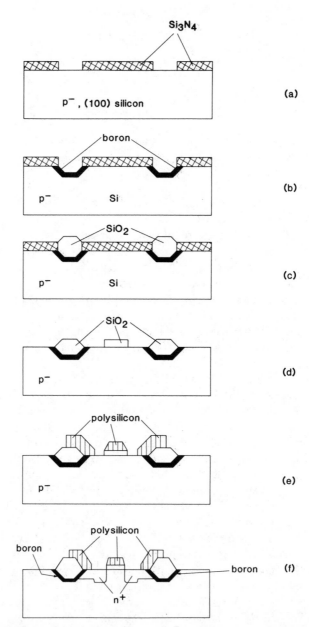

FIGURE 3.41. Processing steps in a self-aligned silicon-gate NMOS technology

doping in the channel regions (usually by ion implantation) can be used to obtain an effective and accurate adjustment of V_T. Hence, a self-aligned silicon-gate technique using ion implantation and (100)-oriented silicon is a very attractive process for fabricating n-channel devices.

To illustrate such a process, Fig. 3.41 shows a possible sequence of processing steps.[14] they are as follows:

1.   Silicon nitride (Si_3N_4, a dense dielectric with desirable chemical properties) is grown on the Si surface, and is selectively etched using a photomask. It is left only over the active areas (Fig. 3.41a).

2.   The silicon is etched to a depth of about 1 μm under the windows. Boron is deposited under these grooves to make the silicon p^+-type there. Both of these steps raise $V_{T \text{ field}}$, and hence tend to prevent any parasitic channel formation (Fig. 3.41b).

3.   A thick oxide $(1 \simeq 2\ \mu$m thick) is grown in the grooved areas. This oxide isolates adjacent devices (Fig. 3.41c).

4.   The silicon nitride is stripped off, and a thin oxide layer is grown over the entire wafer. If necessary, ion implantation is performed through the thin oxide to adjust V_T in the active areas. The thin oxide is then selectively etched away, using a photomask, in areas where polysilicon-to-diffusion contact will be needed (Fig. 3.41d).

5.   Polysilicon is deposited over the entire wafer, using CVD. With the help of a photomask, it is etched away in areas where it will not be used as a gate, an interconnect line, or a capacitor plate (Fig. 3.41e).

6.   The thin oxide is removed wherever it is not covered by polysilicon. An n^+-diffusion (arsenic or phosphorus) is introduced in each source or drain area, with the polysilicon–SiO_2 layers acting as the mask. During this step, the polysilicon is also doped to make it n^+-type (Fig. 3.41f).

The wafer is then covered by a SiO_2 layer, and the necessary metal contact windows are opened through this layer. Finally, aluminum is deposited and then selectively etched to provide all needed metal interconnections. These steps are not illustrated in Fig. 3.41.

The technology which provides the most flexibility to the circuit designer is the *complementary MOS* (CMOS) process. Here, both n- and p-channel transistors are fabricated on the same chip. The original motivation for developing this relatively complicated technology was the need for low-power and high-speed logic gates for digital circuits. In terms of the number of processing steps and complexity, CMOS is the most sophisticated MOS process. This is due to the required isolation of the two different device types, which is accomplished by the use of "wells," that is, large, low doping level, deep diffusions, which serve as the substrates for one of the two device types.* As an example, Fig. 3.42 shows part of a p-well CMOS chip, with a p-channel

---

*In the British literature, the well is often called a "tub."

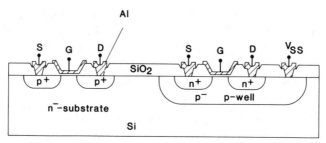

FIGURE 3.42. A *p*-well CMOS structure.

(left) and an *n*-channel (right) device. The processing step used to create the *p*-wells requires great precision, since it determines the threshold voltage of the *n*-channel devices in the *p*-wells, as well as field threshold in the well area.

As will be shown in Chapter 4, a CMOS circuit can be operated with a single power supply, and it can be used to realize high-speed, high-gain, low-power analog amplifier stages. An additional advantage is that for the devices in the well (in Fig. 3.42, the NMOS transistor), the source can be connected to the well, thereby eliminating the body effect and, if the device is used in an amplifier, increasing the gain of the circuit. This, however, results in a large stray capacitance between the source and the substrate, due to the large size of the well-to-body interface.

The fabrication of a CMOS wafer may require 30 or more processing steps. Hence, the yield tends to be lower than for PMOS or NMOS processes. Also, due to presence of the wells, the device density is lower than that achievable for single-channel technologies. However, the increased flexibility afforded to the circuit designer by the simultaneous availability of both transistor types usually outweighs these drawbacks. It is usually also possible to obtain depletion-mode devices and even bipolar transistors on a CMOS chip at little or no extra cost, increasing the potential of the technology even further.

It is possible to grow a thin (1 ~ 2 μm) layer of crystalline silicon (called an *epitaxial* layer) on a crystalline insulator such as sapphire. The resulting structure is called *silicon-on-sapphire* (SOS). Using the processing steps discussed earlier in this section, *p*- and *n*-channel MOSFETs can then be created on isolated silicon "islands" (Fig. 3.43). Thus, no wells are required to isolate the complementary devices, and all parasitic capacitances are greatly reduced due to the absence of junctions between the substrate and the bottom surfaces of the source and drain diffusions. The capacitance between the interconnect lines and substrate is also greatly reduced. In addition, the body effect can be eliminated in *all* devices without an excessive increase of the stray capacitances, and the leakage currents in the diffusion-to-substrate junctions is also greatly reduced. Thus, the speed and performance of CMOS circuits can be greatly enhanced by this technology.

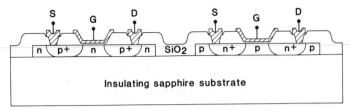

FIGURE 3.43.   A silicon-on-insulator (sapphire) MOS structure.

The obvious disadvantage of the SOS technology is its higher cost, which makes its use economically justified only under special circumstances, usually in very high-speed applications.

In recent years, MOS fabrication technology has gone through an evolution towards higher density and faster integrated circuits. To achieve this, the critical dimensional parameter, the *minimum linewidth*, has been gradually reduced. When this reduction results in source-to-drain spacings around 3 μm or less, some undesirable changes occur in the characteristics of the device. These are largely caused by the extension of the source and drain depletion regions into the gate area. This results in a reduction of the effective channel length, and hence of the threshold voltage V_T of the device. This phenomenon is called *short-channel effect*, and is a poorly controlled, undesirable property which is aggravated by a high drain voltage. A technique for avoiding it is to scale the vertical as well as the horizontal dimension of the device, and simultaneously to reduce the applied voltages and increase the substrate impurity concentrations, all of them proportionately.[17] This scaling process, combined with ion implantation techniques in the actual fabrication, results in improved device characteristics. In the fabrication process, a light substrate doping is used, but the surface layer in areas where the channels are formed is more heavily doped using ion implantation. The implantation depth is chosen to be about the same as the source/drain diffusion depth. This causes the depletion regions around these diffusions to extend deeply vertically into the lightly doped substrate (thus reducing the source and drain to substrate capacitances) but not so far into the channel area. Using this technique for a device with a channel length of about 1 μm, the structure illustrated schematically in Fig. 3.44 results.[17] The effective channel length is about twice as large as for a conventional structure, and the drain-to-substrate capacitance is much reduced.

If the channel length is reduced to below 1 μm, additional effects (carrier velocity saturation caused by high electric fields, inversion-layer thickness effects, ohmic and contact resistance effects, etc.) appear.[18] These also affect the device characteristics adversely.

Since in many recent systems digital and analog circuits share the area on the same chip, an ideal fabrication process satisfies the requirements of both. Digital circuits require high density, low power dissipation, and high reliability,

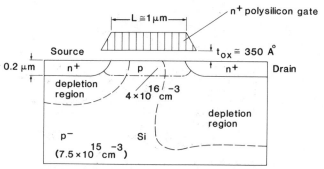

FIGURE 3.44.   A scaled-down NMOS device structure.

while devices used in analog circuits need high transconductance, low noise, small size, and the possibility of close matching for low offset error.

An up-to-date NMOS process has the following general features.[19] It has a scaled-down geometry with short (less than 3 μm) minimum channel lengths, self-aligned gate structure, and a thin ($t_{ox} \leq 500$ Å) gate oxide. Its substrate has high resistivity (i.e., low doping concentration) for reduced source/drain-to-substrate capacitances. Channel and field implants are used to control the device and field threshold voltages. The source and drain junctions are shallow (less than 1 μm deep) to control the short-channel effect. It has two polysilicon and one metal layers for interconnect lines, and a high-resistivity polysilicon layer as well as depletion devices for loads. Such a process is especially suitable for the fabrication of digital circuits; a functional density of 200 ~ 300 gates/ mm^2 can be achieved. The cross section of a device fabricated in this technology is schematically illustrated in Fig. 3.45.

The described process is suitable also for realizing analog components such as are needed for the realization of op-amps, voltage references, matched capacitor, and resistor arrays, and so on. However, scaling down the devices is

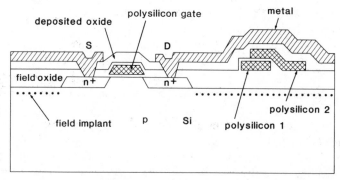

FIGURE 3.45.   Device structure fabricated using a high-performance NMOS process.

of limited usefulness in these applications. One reason is the decrease of the incremental drain resistance r_D with reduced channel length, due to channel-length modulation. For a high gain, and hence high drain resistance, usually channel lengths of 8 μm or more are needed, much longer than the minimum feature size. Also, the flicker $(1/f)$ noise is inversely proportional to the gate area. Hence, for low noise, the gate dimensions are often chosen to be much larger than the minimum size allowed by the technology. In addition, the matching accuracy of nominally identical devices decreases with decreasing size, due to random effects. This is true for capacitors, resistors, and transistors alike. Finally, the scaled-down devices have proportionately much higher leakage currents when they should be cut off, due to subthreshold conduction. This limits their use in such important applications as sample-and-hold stages.[17]

A major disadvantage of the NMOS process for analog applications is the lack of complimentary devices. This, as will be seen in Chapter 4, limits the gain achievable for the amplifier stages needed in op-amps. Also, if only NMOS switches can be used, then the signal level in the circuit is restricted to a value which is one threshold voltage below the clock voltage; otherwise, the switches cannot conduct. The NMOS process (in fact any MOS process) is in many ways inferior to the bipolar one: it leads to lower transconductance, higher noise (especially $1/f$ noise), and larger op-amp offset voltages. However, its higher density, superior charge storage capability, and the possibility to realize high-quality offset-free switches often outweigh these disadvantages.

As mentioned earlier, the CMOS process provides more flexibility to the circuit designer, and is hence usually preferable in spite of its complexity and reduced functional density. The cross section of a device structure fabricated in a high-performance p-well CMOS process is shown in Fig. 3.46. As mentioned earlier, for digital applications this technique gives a somewhat lower functional density (100 ~ 150 gates/mm^2) due to the extra area needed for the wells; however, it has much lower standby power dissipation. The latter is especially important for high density and fast circuits.

For analog applications, neither the functional density nor the standby power dissipation are important criteria; instead, the high-gain feature of

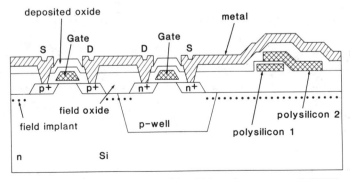

FIGURE 3.46. Device structure fabricated in a high-performance p-well CMOS process.

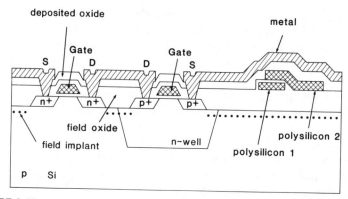

FIGURE 3.47. Device structure fabricated in a high-performance n-well CMOS process.

CMOS amplifier stages, and the availability of CMOS transmission gates (Fig. 3.30) to be used as switches make this technology attractive. When such transmission gates are used, the signal is no longer limited to a level which is a threshold voltage below that of the high clock signal, as is the case when single-channel switches are used. In addition, in CMOS chips a bipolar transistor can be fashioned from a source diffusion, the well, and the substrate. This can be used in an emitter–follower buffer stage, as will be described in Chapter 4, or in a bandgap voltage reference circuit, and so on.

It is also possible to use a high-resistivity p-type substrate for the n-channel devices, and diffused n-wells for the p-channel ones. The cross section of a device fabricated in such an n-well process is shown in Fig. 3.47. This process puts the faster n-channel devices in the high-resistivity p-type substrate which reduces the parasitic p-n junction capacitances and also the substrate body effect.

Recently, twin-well CMOS technology was also developed.[20] In this process, a lightly doped (or undoped) epitaxial silicon layer is deposited on an n^+-substrate. The p- and n-wells are then diffused into this layer, and the p- and n-channel devices independently optimized. Such technology allows a complete elimination of the body effect for all devices.

A potential reliability problem specifically associated with the CMOS circuit is *latchup*. This phenomenon is caused by the parasitic lateral p-n-p and n-p-n bipolar transistors created on the chip. The collectors of each transistor feed the other's base, and this creates an unstable device similar to a p-n-p-n thyristor.[16] This causes a sustained dc current which may cause the chip to stop functioning and may even destroy it. Latchup may be prevented using various techniques.[16] An effective one is to use guard rings to surround some critical transistors on the chip, and thereby interrupt the positive feedback loops causing latchup. Another strategy is to reduce the substrate resistance.[16] In this method the p- and n-channel transistors are formed in a lightly doped epitaxial layer that is grown on a low-resistivity substrate.

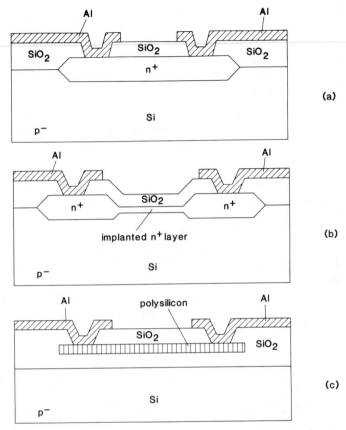

FIGURE 3.48.   Resistors realizable in an NMOS process.[13]

In addition to transistors, analog MOS circuits usually require on-chip capacitors, and sometimes also on-chip resistors. The realization and most important properties of MOS capacitors have already been discussed in Section 3.5. (See also Ref. 13 for a more detailed discussion and evaluation of the properties of the various structures.) Resistors can be created on an MOS chip using a diffused or implanted layer on the surface of the substrate, or a chemically deposited polycrystalline layer enclosed in SiO_2. For CMOS technology, the well diffusion is also available as a resistor. As an illustration, Fig. 3.48 shows some commonly used structures.[13] Since the sheet resistance of these resistive layers is relatively low (typically 20–50 Ω for a square layer), the size of the resistors obtainable on a reasonably small area is limited to about 100 kΩ. Again, the reader is referred to the excellent discussions in Ref. 13 for a detailed comparison of the properties of the different realizations. In Section 5.9, the realization of MOS resistors using transistors in their triode (i.e., nonsaturated) regions is also discussed.

PROBLEMS

3.1. A p-n junction diode is connected to an external voltage v in the forward direction (Fig. 3.3). Reversing the polarity of the voltage reduces the current by a factor 10^6. Assume that the diode satisfies Eq. (3.1) and is at room temperature. What is v?

3.2. For a p-n junction (Fig. 3.3), $N_A = N_D = 10^{16}$ ions/cm^3, $|v| = 5$ V, $A = 0.34$ mm^2, and the measured value of C is 27 pF. How much is x_d, the width of the depletion layer? How much is ϕ_i?

3.3. Using the definition $R = 1/(\partial i_D/\partial V_{DS})$, calculate the channel resistance of an NMOS transistor from (a) Eq. (3.6), (b) Eq. (3.7), and (c) Eq. (3.9).

3.4. For an NMOS transistor, $\mu_n = 10^3$ cm^2/V · s, the thickness of the gate oxide is 10^3 Å (1 Å = 10^{-8} cm), $W = 25$ μm, and L = 5 μm. The threshold voltage is 4 V. Calculate i_D for $v_s = v_B = 0$ V and $v_G = 6$ V, and (a) $v_D = 0.1$ V, (b) $v_D = 2$ V, and (c) $v_D = 4$ V.

3.5. Repeat the calculations of Problem 3.4, if $v_s = 0$, $v_B = -3$ V, and $\phi_p = 0.3$ V. What conclusions can you draw from your results regarding the body effect?

3.6. For an NMOS transistor, $k' = 2$ μA/V^2, $W = 30$ μm, $L = 10$ μm, $\phi_p = 0.3$ V, $\gamma = 1.5$ V$^{1/2}$, and $\lambda = 0.03$ V^{-1}. Find the incremental conductances g_m, g_d, and g_{mb} for $v_{SB} = 0$ V, $v_{DS}^0 = 5$ V, and $i_D^0 = 10$ μA. Repeat your calculations for $v_{SB} = 2$ V!

3.7. Prove Eq. (3.24). How large is C_j if $W = L = 20$ μm, $x_j = 2$ μm, and $C_s = C_a = 0.01$ pF/μm^2?

3.8. Prove Eq. (3.27). (*Hint*: If x and y are independent random variables and $z = x + y$, then $\sigma_z^2 = \sigma_x^2 + \sigma_y^2$.)

3.9. Show from Eq. (3.27) that for given $\sigma_L = \sigma_W$ and C, σ_C is minimized if $W = L$. What is $\sigma_{C \min}$?

3.10. Prove Eqs. (3.28) and (3.29) for capacitance-ratio errors.

3.11. Prove Eqs. (3.31) and (3.32) giving the conditions for undercut-insensitive capacitance ratios.

3.12. An NMOS switch transistor has a gate-to-source voltage $v_{GS} > V_T$. Its drain is open-circuited. How much is v_{DS}? Why?

3.13. In the circuit of Fig. 3.49, the switch S is opened at $t = 0$. (a) Is the transistor operating in its linear or saturation region? (b) Neglecting body effect and channel-length modulation, find $v(t)$ by solving the appropriate differential equation for the circuit.

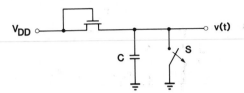

FIGURE 3.49.   Circuit for Problem 3.13.

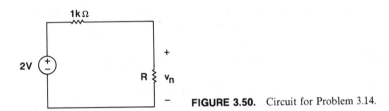

FIGURE 3.50.   Circuit for Problem 3.14.

3.14.   In the circuit of Fig. 3.50, a noise voltage v_n is generated due to thermal and shot noise effects. For what value of R will the two noise voltages v_{nT} and v_{nS} be equal?

3.15.   Calculate the incremental impedance $\partial v / \partial i$ seen at node Ⓐ of the circuits shown in Fig. 3.51.

3.16.   Show that the transconductance g_m in the saturation region is equal to the drain conductance in the triode region for a given device and a fixed V_G.

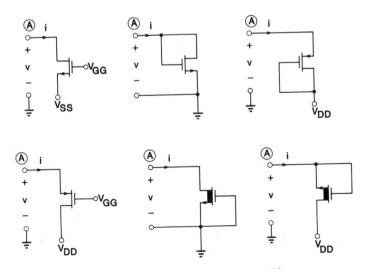

FIGURE 3.51.   Circuits for Problem 3.15.

REFERENCES

1. R. S. Muller and T. I. Kamins, *Device Electronics for Integrated Circuits*, Wiley, New York, 1977, Sec. 4.3.
2. A. S. Grove, *Physics and Technology of Semiconductor Devices*, Wiley, New York, 1967, Sec. 6.6.
3. A. S. Grove, Ref. 2, Sec. 9.2.
4. R. S. Muller and T. I. Kamins, Ref. 1, Sec. 7.4.
5. R. W. Brodersen, P. R. Gray, and D. A. Hodges, *Proc. IEEE*, **67**, 63 (1979).
6. J. L. McCreary, *IEEE J. Solid-State Circuits*, **SC-16**, 608 (1981).
7. J. L. McCreary and P. R. Gray, *IEEE J. Solid-State Circuits*, **SC-10**, 371 (1975).
8. D. J. Hamilton and W. G. Howard, *Basic Integrated Circuit Engineering*, McGraw-Hill, New York, 1975, Sec. 6-6.
9. M. B. Das and J. M. Moore, *IEEE Trans. Electron. Devices*, **ED-21**, 247–257 (1974).
10. P. R. Gray and R. G. Meyer, *Analysis and Design of Analog Integrated Circuits*, Wiley, New York, 1977, Sec 11.3.
11. P. R. Gray, D. A. Hodges, and R. W. Brodersen (Eds.), *Analog MOS Integrated Circuits*, IEEE Press, New York, 1980, p. 31.
12. H. Shichman and D. A. Hodges, *IEEE J. Solid-State Circuits*, **SC-3**, 285–289 (1968).
13. D. J. Allstot and W. C. Black, Jr., *IEEE Proc.*, **71**, 967–986, (1983).
14. O. J. McCarthy, *MOS Device and Circuit Design*, Wiley, Chichester, 1982.
15. J. McCreary and W. C. Black, Jr., unpublished lecture notes, UCLA.
16. S. M. Sze (Ed.), *VLSI Technology*, McGraw-Hill, New York, 1983.
17. R. H. Dennard, F. H. Gaensslen, H. N. Yu, V. L. Rideout, E. Bassows, and A. R. LeBlanc, *IEEE J. Solid-State Circuits*, **SC-9**, 256–268 (1974).
18. Y. A. El-Mansy, *Proc. IEEE Conf. Circuits and Components*, 457–460 (1980).
19. C. A. T. Salama, *IEEE J. Solid-State Circuits*, **SC-16**, 253–260 (1981).
20. L. C. Parrillo, R. S. Payne, R. E. Davis, G. W. Rentlinger, and R. L. Field, Digest IEEE International Electron Device Meeting, Washington, D.C., 752–755 (1980).
21. R. S. Payne, W. N. Grant, and W. J. Bertram, Ref. 20, 248–251 (1980).

MOS OPERATIONAL AMPLIFIERS

The MOS operational amplifier is the most intricate, and in many ways the most important, building block of switched-capacitor circuits. Its performance usually limits the high-frequency application and the dynamic range of the overall circuit. Without a thorough understanding of the operation and the basic limitations of these amplifiers, the circuit designer cannot determine or even predict the actual response of the overall system. Hence, this chapter gives a fairly detailed explanation of the usual configurations and performance limitations of operational amplifiers. A design example is also worked out to acquaint the reader with the problems and tradeoffs involved in amplifier design.

The technology, and hence the design techniques used for MOS amplifiers change rapidly. Therefore, the main purpose of the discussions is to illustrate the most important principles underlying the specific circuits and design procedures. Nevertheless, the treatment is detailed enough to enable the reader to design high-performance MOS operational amplifiers suitable for most switched-capacitor circuit applications.

4.1. OPERATIONAL AMPLIFIERS[1,2]

In switched-capacitor filters—in fact, in all active filters—the most commonly used active component is the *operational amplifier*, usually simply called *op-amp*. Ideally, the op-amp is a voltage-controlled voltage source (Fig. 4.1) with infinite voltage gain and with zero input admittance as well as zero output impedance. It is free of frequency and temperature dependence, distortion, and noise. Needless to say, practical op-amps can only approximate such an ideal

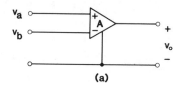

(a)

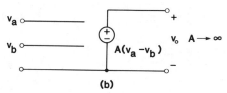

(b)

FIGURE 4.1. (a) Symbol for ideal op-amp. (b) Equivalent circuit.

device. The main differences between the ideal op-amp and the real device are the following:[2]

1.   *Finite Gain.*   For practical op-amps, the voltage gain is finite. Typical values for low frequencies and small signals are $A = 10^2 \sim 10^5$, corresponding to 40–100 dB gain.

2.   *Finite Linear Range.*   The linear relation $v_0 = A(v_a - v_b)$ between the input and output voltages is valid only for a limited range of v_0. Normally, the maximum value of v_0 for linear operation is somewhat smaller than the positive dc supply voltage; the minimum value of v_0 is somewhat positive with respect to the negative supply.

3.   *Offset Voltage.*   For an ideal op-amp, if $v_a = v_b$ (which is easily obtained by short circuiting the input terminals), then $v_0 = 0$. In real devices, this is not exactly true, and a voltage $v_{0,\text{off}} \neq 0$ will occur at the output for shorted inputs. Since $v_{0,\text{off}}$ is usually directly proportional to the gain, the effect can be more conveniently described in terms of the input offset voltage $v_{\text{in,off}}$, defined as the differential input voltage needed to restore $v_0 = 0$ in the real device. For MOS op-amps, $v_{\text{in,off}}$ is about 5–15 mV.

4.   *Common-Mode Rejection Ratio (CMRR).*   The common-mode input voltage is defined by

$$v_{\text{in},c} = \frac{v_a + v_b}{2} \tag{4.1}$$

as contrasted with the differential-mode input voltage

$$v_{\text{in},d} = v_a - v_b. \tag{4.2}$$

Accordingly, we can define the differential gain A_D (which is the same as the

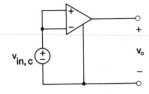

FIGURE 4.2. Op-amp with only common-mode input voltage.

gain A discussed earlier), and also the common-mode gain A_C which can be measured as shown in Fig. 4.2, where $A_C = v_0/v_{\text{in, }c}$.

The CMRR is now defined as A_D/A_C or (in logarithmic units) CMRR = $20\log_{10}(A_D/A_C)$ in dB. Typical CMRR values for MOS amplifiers are in the 60–80 dB range. The CMRR measures how much the op-amp can suppress common-mode signals at its inputs. These normally represent undesirable noise, and hence a large CMRR is an important requirement.

5. *Frequency Response.*   Because of stray capacitances, finite carrier mobilities, and so on, the gain A decreases at high frequencies. It is usual to describe this effect in terms of the unity-gain bandwidth, that is, the frequency f_0 at which $|A(f_0)| = 1$. For MOS op-amps, f_0 is usually in the range of 1–10 MHz. It can be measured with the op-amp connected in a voltage-follower configuration (Problem 4.29).

6. *Slew Rate.*   For a large input step voltage, some transistors in the op-amp may be driven out of their saturation regions or completely cut off. As a result, the output will follow the input at a slower finite rate. The maximum rate of change dv_0/dt is called the *slew rate*. It is *not* directly related to the frequency response. For typical MOS op-amps, slew rates of 1–20 V/μs can be obtained.

7. *Nonzero Output Resistance.*   For a real MOS op-amp, the open-loop output impedance is nonzero. It is usually resistive, and is of the order of 0.1–5 kΩ for op-amps with an output buffer; it can be much higher (~ 1 MΩ) for op-amps with unbuffered output. This affects the speed with which the op-amp can charge a capacitor connected to its output, and hence the highest signal frequency.

8. *Noise.*   As explained in Section 3.6, the MOS transistor generates noise, which can be described in terms of an equivalent current source in parallel with the channel of the device. The noisy transistors in an op-amp give rise to a noise voltage v_{on} at the output of the op-amp; this can be again modeled by an equivalent voltage source $v_n = v_{on}/A$ at the op-amp input. Unfortunately, the magnitude of this noise is relatively high, especially in the low-frequency band where the flicker noise of the input devices is high; it is about 10 times the noise occurring in an op-amp fabricated in bipolar technology. In a wide band (say, in the 10 Hz to 1 MHz range), the equivalent input noise source is usually of the order of 10–50 μV RMS, in contrast to the 3–5 μV achievable for low-noise bipolar op-amps.

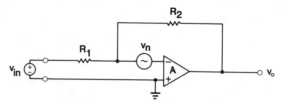

FIGURE 4.3. Noisy feedback amplifier.

9. *Dynamic Range.* Due to the limited linear range of the op-amp, there is a *maximum* input signal amplitude $v_{in,max}$ which the device can handle without generating an excessive amount of nonlinear distortion. If the power supply voltages of the op-amp are $\pm V_{cc}$, then an optimistic estimate is $v_{in,max} \approx V_{cc}/A$, where A is the open-loop gain of the op-amp. Due to spurious signals (noise, clock feedthrough, low-level distortion such as crossover distortion, etc.) there is also a *minimum* input signal $v_{in,min}$ which still does not drown in noise and distortion. Usually, $v_{in,min}$ is of the same order of magnitude as the equivalent input noise v_n of the op-amp. The dynamic range of the op-amp is then defined as $20\log_{10}(v_{in,max}/v_{in,min})$ measured in decibels. When the op-amp is in open-loop condition, $v_{in,max} \approx V_{cc}/A$ which is of the order of a millivolt, while $v_{in,min} \approx \sqrt{\overline{v_n^2}}$ which is around 30 μV. Thus, the *open-loop* dynamic range of the op-amp is only around 30–40 dB. However, the dynamic range of a circuit containing op-amps in negative feedback configuration can be much larger. As a simple illustration, consider the feedback amplifier shown in Fig. 4.3. It is easy to show (Problem 4.1) that the output due to the noise source v_n acting alone has the RMS value

$$v_{on} = \sqrt{\overline{v_n^2}} \Big/ \left(\frac{1}{A} + \frac{R_1}{R_1 + R_2} \right). \tag{4.3}$$

The voltage gain of the (noiseless) feedback circuit is

$$\frac{v_0}{v_{in}} = -1 \Big/ \left[\frac{1}{A} + \frac{R_1}{R_2}\left(1 + \frac{1}{A}\right) \right]. \tag{4.4}$$

The minimum input signal $v_{in,min}$ gives rise to an output voltage approximately equal to v_{on}. Hence

$$v_{in,min} \cong \sqrt{\overline{v_n^2}} \frac{1/A + (R_1/R_2)(1 + 1/A)}{1/A + R_1/(R_1 + R_2)} \tag{4.5}$$

and, for $v_{0,max} \approx V_{cc}$,

$$v_{in,max} \cong V_{cc}\left[\frac{1}{A} + \frac{R_1}{R_2}\left(1 + \frac{1}{A}\right) \right]. \tag{4.6}$$

Hence, the dynamic range is given by

$$20 \log_{10} \left[\frac{V_{cc}}{\sqrt{\overline{v_n^2}}} \left(\frac{1}{A} + \frac{R_1}{R_1 + R_2} \right) \right]$$

$$\approx 20 \log_{10} \left(\frac{V_{cc}}{\sqrt{\overline{v_n^2}}} \frac{R_1}{R_1 + R_2} \right), \tag{4.7}$$

where the indicated approximation is usually valid for $A \gg 1$. For typical values ($V_{cc} \sim 10$ V, $\sqrt{\overline{v_n^2}} \sim 30$ μV, $R_2/R_1 \sim 10$), a dynamic range of about 90 dB results for the overall circuit.

In switched-capacitor filters, dynamic range values around 80–90 dB are readily achievable. Even higher values (up to 100 dB) are possible if the large low-frequency noise ($1/f$ noise) is canceled using a differential circuit configuration and chopper stabilization.[3]

10. *Power-Supply Rejection Ratio (PSRR).* If a power-supply voltage contains an incremental component v due to noise, hum, and so on, then a corresponding voltage $A_p v$ will appear at the op-amp output. The PSRR is defined as A_D/A_p, where $A_D = A$ is the differential gain. It is common to express the PSRR in dB; then PSRR $= 20 \log_{10}(A_D/A_p)$. Usual PSRR values range from 60 to 80 dB for the op-amp alone; for the complete filter, 30–50 dB can be achieved.

11. *DC Power Dissipation.* Ideal op-amps require no dc power dissipated in the circuit; real ones do. Typical values for an MOS op-amp range from 0.25 to 10 mW dc power drain.

To obtain near-ideal performance for a practical op-amp, the general structure of Fig. 4.4 is usually employed.[1] The input differential amplifier (first block) is designed so that it provides a high input impedance, a large CMRR and PSRR, a low offset voltage, low noise, and high gain. Its output should preferably be single-ended, so that the rest of the op-amp need not contain symmetrical differential stages. Since the transistors in the input stage (and in subsequent stages) operate in their saturation regions (Fig. 3.10), there is an appreciable dc voltage difference between the input and output signals of the input stage.

The second block in the diagram of Fig. 4.4 may perform one or more of the following functions:

1. *Level Shifting.* This is needed to compensate for the dc voltage change occurring in the input stage, and thus to assure the appropriate dc bias for the following stages.

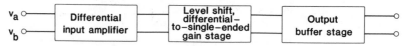

FIGURE 4.4. Block diagram for a practical op-amp.

2. *Added Gain.* In most cases, the gain provided by the input stage is not sufficient, and additional amplification is required.

3. *Differential-to-Single-Ended Conversion.* In some circuits, the input stage has a differential output, and the conversion to single-ended signals is performed in a subsequent stage.

The third block is the output buffer. It provides the low output impedance and larger output current needed to drive the load of the op-amp. It normally does not contribute to the voltage gain. If the op-amp is an internal component of a switched-capacitor filter, then the output load is a (usually small) capacitor, and the buffer need not provide a very large current or very low output impedance. However, if the op-amp is at the filter output, then it may have to drive a large capacitor and/or resistive load. This requires large current drive capability and very low output impedance, which can only be attained by using large output devices with appreciable dc bias currents. Thus, the dc power drain will be much higher for such output op-amps than for interior ones.

Next, some typical MOSFET circuits realizing the described functions will be discussed. The design often varies according to whether both PMOS and NMOS devices can be fabricated on the same chip using complementary MOS (CMOS) technology, or only one type (usually NMOS) is available. Hence, most circuits will be discussed twice; once for CMOS and once for NMOS realization.

Unless otherwise postulated, we shall assume that all devices are operated in the saturation region. Then i_D is to a good approximation independent of v_D, and is given by $i_D \cong k'(W/L)(v_{GS} - V_T)^2$. Here, due to body effect, V_T depends on the source-to-body voltage.

4.2. BIAS CIRCUITS IN MOS TECHNOLOGY

The amplifier stages which were discussed in general terms above, and which will be described in detail subsequently, need various dc bias voltages and currents for their operations. Since the op-amp normally has only two dc voltage supplies—$V_{DD} > 0$ and $V_{SS} \leq 0$—all other bias voltages and currents must be obtained from these voltages.

To obtain the dc bias voltages $V_{01}, V_{02}, \ldots, V_{0n}$, where $V_{SS} < V_{01} < V_{02} < \cdots < V_{0n} < V_{DD}$, voltage division can be used. Resistive dividers are seldom used in MOS technology, mainly because of the large silicon area required for resistors with suitable values. Instead, MOSFETs are used in a "totem-pole"

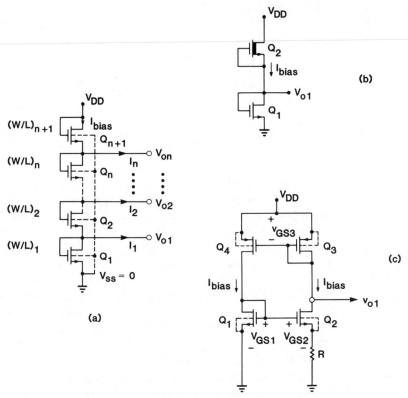

FIGURE 4.5. (*a*) NMOS voltage divider. (*b*) Supply-independent NMOS bias circuit. (*c*) Supply-independent CMOS bias circuit.

configuration; see Fig. 4.5*a*, where *n*-channel transistors are assumed and $V_{SS} = 0$ chosen. Since here $v_{GS} = v_{DS}$ for all devices, the condition for saturation

$$v_{DS} > v_{GS} - V_T \tag{4.8}$$

is satisfied. Hence, from Eq. (3.8), the common value of the drain currents is

$$I_{\text{bias}} = k'(W/L)_1 (V_{01} - V_{T1})^2$$

$$= k'(W/L)_2 (V_{02} - V_{01} - V_{T2})^2 \tag{4.9}$$

$$\vdots$$

$$= k'(W/L)_{n+1} (V_{DD} - V_{0n} - V_{T, n+1})^2,$$

where $I_i \ll I_{\text{bias}}$ is assumed for $i = 1, 2, \ldots, n$. Due to body effect, from (3.12), the threshold voltages $V_{T1}, V_{T2}, \ldots$ are different for different devices:

$$V_{T1} = V_T$$

$$V_{T2} = V_T + \gamma\left(\sqrt{2|\phi_p| + V_{01}} - \sqrt{2|\phi_p|}\right)$$

$$\vdots$$

$$V_{T, n+1} = V_T + \gamma\left(\sqrt{2|\phi_p| + V_{0n}} - \sqrt{2|\phi_p|}\right).$$

(4.10)

Here, the V_{0i} are usually specified, as are the currents I_i, $i = 1, 2, \ldots, n$ (Fig. 4.5). Then, $I_{\text{bias}} \gg I_i$ can be selected,* and Eqs. (4.9) and (4.10) used to find the W/L ratios for all devices

An undesirable feature of the circuit of Fig. 4.5a is that all voltages and currents depend on the supply voltages V_{DD} and V_{SS}. This makes the performance and dc power consumption of the overall circuit also a function of the supply voltages. To avoid this, the improved circuit of Fig. 4.5b (which provides only one bias voltage V_{01}) can be used. Here, the depletion-mode device Q_2 acts as current source, forcing Q_1 (which has a low input resistance at its common gate/drain terminal) to carry a drain current $I_{\text{bias}} = k_2'(W/L)_2 V_{TD}^2$, where V_{TD} is the threshold voltage of the depletion device. Then V_{01} is stabilized at $V_{T1} + |V_{TD}|\sqrt{[k_2'(W/L)_2]/[k_1'(W/L)_1]}$.

A CMOS circuit with (theoretically) perfect supply independence is shown in Fig. 4.5c. If Q_3 and Q_4 are matched transistors so that $(W/L)_3 = (W/L)_4$, they will carry equal currents $I_{\text{bias}} = (V_{GS_1} - V_{GS_2})/R$. Choosing $(W/L)_2 > (W/L)_1$, the bias current can then be calculated to be $I_{\text{bias}} = [1/(R^2 k')][(W/L)_1^{-1/2} - (W/L)_2^{-1/2}]^2$. It is thus independent of V_{DD}. Alternatively, $(W/L)_1 = (W/L)_2$ and $(W/L)_3 > (W/L)_4$ can be chosen. Then, $I_{\text{bias}} = [\sqrt{(W/L)_3/(W/L)_4} - 1]^2/[k'R^2(W/L)_1]$ results. It is again independent of V_{DD}, as is the bias voltage V_{01}.

Note that the above analysis neglected the effects of channel-length modulation. This is only permissible if the V_{Ti} and λ are not too large (cf. Problem 4.4).

As will be seen in Section 4.3, *constant-current sources* are important components in MOS differential amplifiers. The MOS current sources are quite similar to the bipolar ones (see Ref. 1, Sec. 9-2; see Ref. 2, Sec. 4.2). An NMOS realization of a "current mirror" is shown in Fig. 4.6. In this circuit, Q_1 is forced to carry a current I_1, since its input resistance at its shorted gate/drain terminals is low, and its gate potential adjusts accordingly. If Q_1 and Q_2 are in saturation, then their drain currents I_1 and I_2 are determined to a large extent by their v_{GS} values. Since $v_{GS1} = v_{GS2}$, the condition $I_1/I_2 \approx k_1/k_2$ will therefore hold. More accurately, from (3.9), if the transistors have the same λ,

---

*Usually, the V_{0i} are connected only to the gates of transistors, and the I_i are very small.

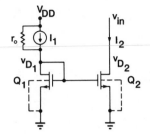

FIGURE 4.6. MOS current source.

k', and V_T, then

$$\frac{I_1}{I_2} = \frac{(W/L)_1}{(W/L)_2}\frac{1 + \lambda v_{D1}}{1 + \lambda v_{D2}}. \tag{4.11}$$

The current I_1 is thus "mirrored" in I_2.

For small-signal analysis, the equivalent circuit of Fig. 3.14 can be used to replace Q_1 and Q_2. The resulting circuit is shown in Fig. 4.7. Here, r_0 is the incremental resistance of the current source I_1, and v_{in} is the test voltage connected to the drain of Q_2 for measuring the output impedance of the circuit. Clearly, the small-signal output impedance is simply

$$r_{out} = \frac{v_{in}}{i_{in}} = r_{d2} = \frac{1}{g_{d2}} = \frac{1 + \lambda v_{D2}^0}{\lambda i_{D2}^0}, \tag{4.12}$$

where Eq. (3.20) was used.

The output impedance r_{out} can be increased, and thus the circuit made to perform more like an ideal current source, by adding one more device and modifying the connections slightly. The resulting circuit (Fig. 4.8) is the MOS equivalent of Wilson's current source (see Refs. 1 and 2, loc. cit.). In this circuit, if I_2 increases then v_1 becomes larger. This results in a drop of v_3 which then counteracts the increase of I_2. Thus, a negative feedback loop exists, which tries to hold I_2 constant. The small-signal equivalent circuit is shown in Fig. 4.9; a simplified circuit is shown in Fig. 4.10. The latter was obtained by combining r_0 and r_{d1} into $r_1 = (r_0^{-1} + r_{d1}^{-1})^{-1}$; replacing the self-controlled current source $g_{m2}v_2$ by a resistor $1/g_{m2}$; and neglecting r_{d2} which is now in parallel with the (usually much smaller) resistor $1/g_{m2}$.

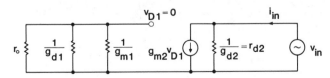

FIGURE 4.7. Small-signal equivalent circuit of the MOS current mirror.

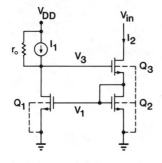

FIGURE 4.8.   MOS version of Wilson's current source.

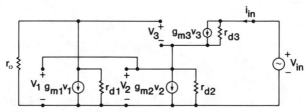

FIGURE 4.9.   Small-signal equivalent circuit of Wilson's current source.

Writing the loop equations (Fig. 4.10):

$$\text{Loop 1:}\quad -i_{\text{in}}/g_{m2} - v_3 - r_1 g_{m1}\left(i_{\text{in}}/g_{m2}\right) = 0,$$

$$\text{Loop 2:}\quad -v_{\text{in}} + \left(i_{\text{in}} - g_{m3}v_3\right)r_{d3} + i_{\text{in}}/g_{m2} = 0. \tag{4.13}$$

Solving for i_{in} gives

$$i_{\text{in}} = v_{\text{in}}\Big/\left\{r_{d3} + \left[1 + g_{m3}r_{d3}\left(1 + g_{m1}r_1\right)\right]\big/g_{m2}\right\}. \tag{4.14}$$

Typical values for g_m are around 1 mA/V, while r_d is of the order of hundreds of kΩ. Hence, $g_{m3}r_{d3} \gg 1$ and also (for reasonably large r_0) $g_{m1}r_1$

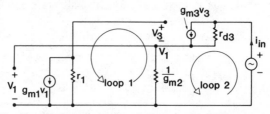

FIGURE 4.10.   Simplified small-signal equivalent circuit of Wilson's current source.

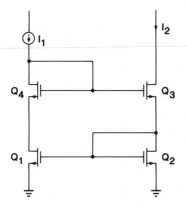

FIGURE 4.11.   Improved MOS Wilson's current source.

$\gg 1$. Then, from (4.14)

$$r_{out} = \frac{v_{in}}{i_{in}} \cong \frac{g_{m1}r_1 g_{m3} r_{d3}}{g_{m2}} = (g_{m1}r_1)\left(\frac{g_{m3}}{g_{m2}}\right) r_{d3}. \qquad (4.15)$$

Here, on the RHS the value of the first factor is around 100 while that of the second around 1. Hence, the output impedance r_{out} is two orders of magnitude larger than r_{d3}, or around 10 MΩ.

The circuit can be made symmetrical, and the drain/source voltage drops of Q_1 and Q_2 equalized, by adding another transistor Q_4 (Fig. 4.11). It can easily be shown that the output impedance of the resulting *improved MOS Wilson's current source* is also given by (4.15). The detailed analysis of the circuit is left to the reader (Problem 4.7a).

A slightly better version of the circuit of Fig. 4.11 (often called *cascode current source*) is shown in Fig. 4.12a. Its small-signal equivalent circuit is shown in Fig. 4.12b and a simplified form of it in Fig. 4.12c. This circuit also uses a feedback loop to maintain I_2 constant, and it also equalizes the drain potentials of Q_1 and Q_2. It can easily be shown [Problem 4.7b] that

$$r_{out} = \frac{v_{in}}{i_{in}} = r_{d2} + r_{d3} + g_{m3}r_{d3}r_{d2} \cong (g_{m3}r_{d2})r_{d3}. \qquad (4.16)$$

Hence, again there is an increase of about 100 over the MOSFET output resistance. In addition, now the internal impedance r_0 of the current source I_1 is in parallel with $1/g_{m1} + 1/g_{m4}$, a low input impedance, rather than with r_{d1}. Hence, its loading effect is reduced.

A common disadvantage of the circuits of Figs. 4.8–4.12 is that they need a higher dc bias voltage at their output terminals than that of Fig. 4.6. This reduces the available voltage swing of the stage(s) driven by the source.

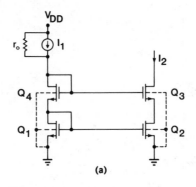

(a)

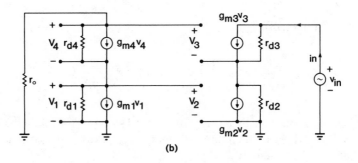

(b)

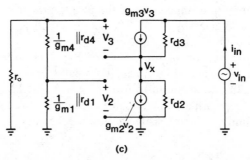

(c)

FIGURE 4.12. (*a*) Cascode current source. (*b*) Equivalent small-signal circuit of the cascode current source. (*c*) simplified small-signal circuit of the cascode current source.

The formulas and numerical estimates given for the current sources are somewhat optimistic, since they neglect the body effect of the floating devices in the circuits. Also, the voltage at the output terminal of the current source must be kept high enough in order to keep all devices in saturation so that r_d and $g_m r_d$ are high. This limits the operating range of the current source realized.

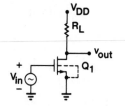

FIGURE 4.13. Resistive-load MOS gain stage.

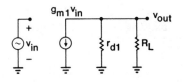

FIGURE 4.14. Small-signal low-frequency equivalent circuit of the resistive-load gain stage.

4.3. MOS GAIN STAGES[4-6]

A simple NMOS gain stage with resistive load is shown in Fig. 4.13. Q_1 is biased so that it operates in its saturation region. The small-signal equivalent circuit is shown in Fig. 4.14. The voltage gain of the stage is clearly

$$A_v = \frac{v_{out}}{v_{in}} = -g_{m1}\frac{R_L r_{d1}}{R_L + r_{d1}}. \tag{4.17}$$

In integrated-circuit realization, the resistor R_L is undesirable, and is usually replaced by a second MOSFET. If an NMOS enhancement-mode device is used as a load, then the circuit of Fig. 4.15 results. The drain and gate of Q_2 are shorted to ensure that $v_{ds} > v_{gs} - V_T$, and hence the device is in saturation. The small-signal equivalent circuit of the load device Q_2 alone is shown in Fig. 4.16. Here the voltage-controlled current source $g_m v_{ds}$ is across the voltage v_{ds}; hence, it behaves simply as a resistor of value $1/g_m$. Similarly, since $v_{sb} = v_{out}$, the source $g_{mb}v_{sb}$ corresponds to a resistor $1/|g_{mb}|$ (recall that, by Eq. (3.19), $g_{mb} < 0!$). In conclusion, Q_2 behaves like a resistor of value

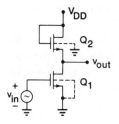

FIGURE 4.15. Enhancement-load NMOS gain stage.

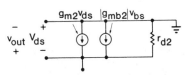

FIGURE 4.16. Small-signal equivalent circuit of the enhancement-load device Q_2.

$1/(g_{m2} + |g_{mb2}| + g_{d2})$. Replacing R_L in Fig. 4.14 by this resistor and neglecting g_{d1} and g_{d2} in comparison with $g_{m2} + |g_{mb2}|$ gives

$$A_v \cong \frac{-g_{m1}}{g_{m2} + |g_{mb2}|}. \tag{4.18}$$

If the body effect is small, so that $|g_{mb2}| \ll g_{m2}$, then using (3.18)

$$A_v \cong -\frac{g_{m1}}{g_{m2}} = -\frac{2\sqrt{k_1(1 + \lambda_1 v_{DS1}^0)i_{D1}^0}}{2\sqrt{k_2(1 + \lambda_2 v_{DS2}^0)i_{D2}^0}}$$

$$\cong -\sqrt{\frac{k_1}{k_2}} = -\sqrt{\frac{(W/L)_1}{(W/L)_2}} \tag{4.19}$$

results. Here, the channel-modulation terms $\lambda_i v_{DS_i}^0$ were also neglected, and the relations $i_{D1} = i_{D2}$, $k_1' = k_2'$ utilized.

The sad message conveyed by (4.19) is that a large gain can only be obtained if the "aspect ratio" W/L of Q_1 is many times that of Q_2. If, for example, a gain of 10 is required, then $(W/L)_1 = 100(W/L)_2$ must hold. This is only possible if a large silicon area is used. In addition, the body effect also reduces the gain significantly. Including body effect (but still neglecting channel-length modulation), using (3.12), (3.19), and (4.18),

$$A_v \cong \frac{-g_{m1}/g_{m2}}{1 + \gamma/\left(2\sqrt{2|\phi_p| + |v_{out}^0|}\right)} \tag{4.20}$$

results. For $|\phi_p| = 0.3$ V, $v_{out}^0 = 5$ V, and $\gamma = 1$, the denominator is 1.21; hence, the gain is reduced from 10 to 8.26.

In conclusion, the NMOS enhancement-load gain stage provides a low gain, needs a large silicon area, and is sensitive to the body effect. In addition, to keep the load device conducting, $v_{out} < V_{DD} - V_T$ must be satisfied. This reduces the signal-handling capability and hence the dynamic range of the stage.

To improve the performance, a depletion-mode device can be used as the load (Fig. 4.17). The small-signal model is shown in Fig. 4.18. In the equivalent circuit of the load, g_{m2} does not appear since $v_{GS2} = 0$. This increases the gain significantly. From Fig. 4.18, applying the current law to the output node

$$g_{m1}v_{in} + g_{d1}v_{out} + |g_{mb2}|v_{out} + g_{d2}v_{out} = 0 \tag{4.21}$$

results. Hence,

$$A_v = \frac{-g_{m1}}{g_{d1} + g_{d2} + |g_{mb2}|}. \tag{4.22}$$

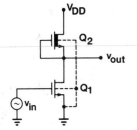

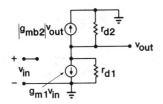

FIGURE 4.17. Depletion-load NMOS gain stage.

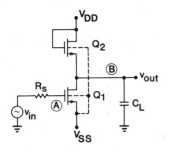

FIGURE 4.18. Small-signal equivalent circuit for a depletion-load gain stage.

The denominator is now much smaller than that in (4.18), and hence A_v is much larger. If the substrate of the load device is separated from that of Q_1, so that it can be connected to the output node, then the denominator becomes $g_{d1} + g_{d2}$ which is very small, of order $10^{-6} \ \Omega^{-1}$ or less, while g_{m1} is around $10^{-3} \ \Omega^{-1}$. Therefore, gains over 100 are (in the absence of other limitations) possible.

For high-frequency applications, all gain stages discussed so far have a common shortcoming. Consider the circuit of Fig. 4.19, which includes the source resistance R_S and the capacitive load C_L of the gain stage. Including the parasitic capacitances (introduced in Fig. 3.16) in the small-signal equivalent circuit, the diagram shown in Fig. 4.20a is obtained. Combining parallel-connected elements, the circuit of Fig. 4.20b results, where

$$G_{Leq} = g_{d1} + g_{d2} + g_{m2} + |g_{mb2}|,$$

$$(4.23)$$

$$C_{Leq} = C_{db1} + C_{gs2} + C_{sb2} + C_L.$$

FIGURE 4.19. Enhancement-load NMOS gain stage with capacitive load.

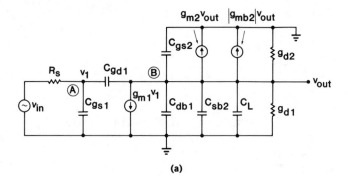

(a)

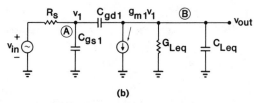

(b)

FIGURE 4.20. (*a*) Equivalent circuit of the MOS gain stage. (*b*) Simplified equivalent circuit of the MOS gain stage.

The node equations for nodes Ⓐ and Ⓑ are

$$(V_1 - V_{in})G_s + V_1 sC_{gs1} + (V_1 - V_{out})sC_{gd1} = 0,$$

$$(V_{out} - V_1)sC_{gd1} + g_{m1}V_1 + V_{out}(G_{Leq} + sC_{Leq}) = 0, \quad (4.24)$$

where all voltages are Laplace transformed functions.
Solving (4.24) gives

$$A_v(s) = \frac{V_{out}(s)}{V_{in}(s)}$$

$$= \frac{G_s(sC_{gd1} - g_{m1})}{[s(C_{gs1} + C_{gd1}) + G_s][G_{Leq} + s(C_{gd1} + C_{Leq})] - sC_{gd1}(sC_{gd1} - g_{m1})}. \quad (4.25)$$

To obtain the frequency response, s must be replaced by $j\omega$. For moderate frequencies, $g_{m1} \gg \omega C_{gd_1}$, $G_{Leq} \gg \omega(C_{gd1} + C_{Leq})$ hold. Then a good ap-

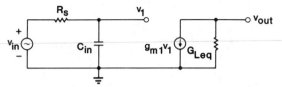

FIGURE 4.21. Approximate equivalent circuit of the MOS gain stage.

proximation is

$$A_v(j\omega) \approx \frac{-g_{m1}G_s}{G_sG_{Leq} + j\omega\left[G_{Leq}(C_{gs1} + C_{gd1}) + g_{m1}C_{gd1}\right]}$$

$$= \frac{-g_{m1}/G_{Leq}}{1 + j\omega R_s\left[C_{gs1} + C_{gd1}(1 + g_{m1}/G_{Leq})\right]}$$

$$= \frac{A_v^0}{1 + j\omega R_sC_{in}}. \tag{4.26}$$

Here, $A_v^0 = -g_{m1}/G_{Leq}$ is the dc value of $A_v(j\omega)$, and

$$C_{in} = C_{gs1} + C_{gd1}(1 + g_{m1}/G_{Leq}) = C_{gs1} + C_{gd1}(1 + |A_v^0|). \tag{4.27}$$

$A_v(j\omega)$ in (4.26) can be recognized as the transfer function of the circuit shown in Fig. 4.21. Thus, the capacitor C_{gd1}, which is connected between the input and output terminals of the gain stage (Fig. 4.20a), behaves like a capacitance $(1 + |A_v^0|)$ times its real size, loading the input terminal. This is the well-known *Miller effect*.[2] For $|A_v^0| \gg 1$, the high-frequency gain will be seriously affected, and the bandwidth considerably reduced by this phenomenon.

To prevent the Miller effect, the *cascode gain stage* of Fig. 4.22 can be used. Here, Q_2 is used to isolate the input and output nodes. It provides a low input

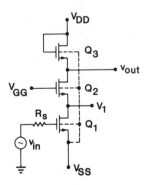

FIGURE 4.22. Cascode gain stage with enhancement load.

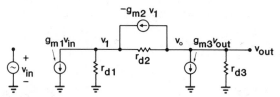

FIGURE 4.23.   Low-frequency small-signal circuit of cascode gain stage.

resistance $1/g_{m2}$ at its source, and a high one at its drain to drive Q_3. Ignoring the body effect, the low-frequency small-signal equivalent circuit is in the form shown in Fig. 4.23. Neglecting the small g_d admittances, clearly

$$g_{m1}v_{\text{in}} = -g_{m2}v_1 = -g_{m3}v_{\text{out}}. \tag{4.28}$$

Hence, for low frequencies

$$v_1 \cong -\frac{g_{m1}}{g_{m2}}v_{\text{in}},$$

$$\tag{4.29}$$

$$v_{\text{out}} \cong \frac{g_{m2}}{g_{m3}}v_1 \cong -\frac{g_{m1}}{g_{m3}}v_{\text{in}}.$$

The gate-to-drain gain of Q_1 is $-g_{m1}/g_{m2}$, and therefore the C_{gd1} of the driver transistor Q_1 is now multiplied by $(1 + g_{m1}/g_{m2})$. Choosing $g_{m1} = g_{m2}$, this factor is only 2. The overall voltage gain $-g_{m1}/g_{m3}$, however, can still be large, without introducing significant Miller effect, since there is no appreciable capacitance between the input and output terminals.

As before, the gain of the cascode gain stage can be increased by using a depletion-mode load device (Fig. 4.24). It can readily be shown (Problem 4.9)

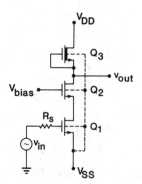

FIGURE 4.24.   Cascode gain stage with depletion load.

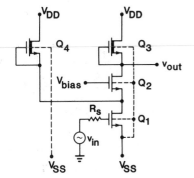

FIGURE 4.25. Improved cascode gain stage.

that the low-frequency voltage gain of this circuit is given by

$$A_v \cong -\frac{g_{m1}r_{d3}}{1 + |g_{mb3}|r_{d3}}. \tag{4.30}$$

To increase the gain further, g_{m1} should be increased. As (3.18) has shown, this can be achieved by increasing k and/or the dc bias current i_D^0. However, as (3.19) and (3.20) indicate, the values of $1/|g_{mb3}|$ and $r_{d3} = 1/g_{d3}$ will decrease faster with increasing $i_{D3}^0 = i_{D1}^0$, and $|A_v|$ will be reduced. To increase i_{D1}^0 without increasing i_{D3}^0, the improved cascode stage of Fig. 4.25 can be used.[7] The added device Q_4 acts as a current source, augmenting the current of Q_1 without affecting that of Q_3. It does not act to shunt the load of Q_1, since Q_2 creates a low-impedance node at the drain of Q_1.

The gain of the circuits of Figs. 4.24 and 4.25 is reduced by the body effect of Q_3. To decrease the body effect, Q_3 can be replaced by the composite load[8] shown in Fig. 4.26.

Physically, the circuit operates as follows. If v_{out} rises, $|v_{BS5}|$ increases and hence the body effect will act to reduce i_5. However, $i_5 = i_6$ is stabilized by Q_6, and hence v_G will rise more than v_{out} so as to keep i_5 stable. If Q_5 and Q_3 have the same γ and ϕ_p values, then the increase of v_G will also keep i_3 constant. It can be shown (Problem 4.10) that the current i_0 is given by

$$i_0 = i_6 + k_3\left(\sqrt{i_6/k_5} + V_{T50} - V_{T30}\right)^2. \tag{4.31}$$

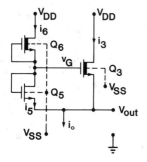

FIGURE 4.26. Improved constant-current source.

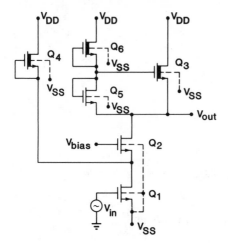

FIGURE 4.27. High-gain cascode NMOS stage.

Here, $k_3(k_5)$ is the constant factor in the $i_D - v_D$ relation, and $V_{T30}(V_{T50})$ is the threshold voltage for $v_{BS} = 0$ of $Q_3(Q_5)$. As (4.31) shows, as long as i_6 is kept constant (which can be achieved if i_6 is small so that $g_{d6} \approx 0$) i_0 will also be stabilized.

Figure 4.27 shows a high-gain NMOS stage using the techniques of Fig. 4.25 and 4.26. Very high gain (over 100) is practically achievable with this circuit. The voltage swing is somewhat reduced, however, due to the added load devices, as compared to a simple inverter.

All gain stages shown so far used exclusively NMOS devices. If CMOS technology is available, then the simple circuit of Fig. 4.28 can be used. Since both devices have their substrates and sources connected, no body effect is present, and g_{mb} does not appear in the small-signal circuit (Fig. 4.29). The node equation is

$$(g_{m1} + g_{m2})v_{in} + (g_{d1} + g_{d2})v_{out} = 0 \qquad (4.32)$$

so that

$$A_v = -\frac{g_{m1} + g_{m2}}{g_{d1} + g_{d2}}. \qquad (4.33)$$

Since g_m can be 100 times larger than g_d, the gain is high.

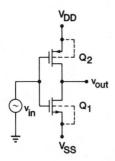

FIGURE 4.28. CMOS gain stage.

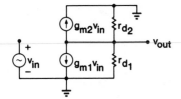

FIGURE 4.29. Small-signal equivalent circuit of the CMOS gain stage.

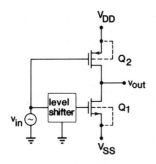

FIGURE 4.30. CMOS gain stage with level shifter.

To increase the linear range, both transistors must operate with their optimum dc bias voltages. This may make it necessary to use a dc-level shifter (Fig. 4.30) for large input signals.

4.4. MOS SOURCE FOLLOWERS AND OUTPUT BUFFERS[4-6]

MOS source followers are similar to bipolar emitter followers. They can be used as buffers or as dc-level shifters. The basic circuit is shown in Fig. 4.31; its small-signal low-frequency equivalent circuit in Fig. 4.32. The current equation for the output node is

$$(g_{d1} + g_{d2})v_{out} + |g_{mb1}|v_{out} - g_{m1}v_{gs1} = 0. \tag{4.34}$$

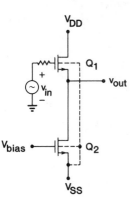

FIGURE 4.31. Basic structure of MOS source follower.

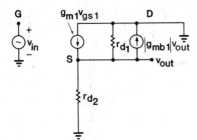

FIGURE 4.32. Small-signal low-frequency equivalent circuit of the source follower.

Substituting $v_{gs1} = v_{in} - v_{out}$ and solving (4.34)

$$A_v = \frac{v_{out}}{v_{in}} = \frac{g_{m1}/(g_{d1} + g_{d2} + |g_{mb1}|)}{g_{m1}/(g_{d1} + g_{d2} + |g_{mb1}|) + 1}. \qquad (4.35)$$

Hence, $A_v \simeq 1$ if $g_{m1} \gg g_{d1} + g_{d2} + |g_{mb1}|$.

The output impedance of the source follower can be calculated by applying a test source v_x at its output (Fig. 4.33). The current law gives

$$i_x = (g_{d1} + g_{d2})v_x + |g_{mb1}|v_x - g_{m1}v_{gs1}. \qquad (4.36)$$

Here, $v_{gs1} = -v_x$, and hence (4.36) gives

$$R_{out} = \frac{v_x}{i_x} = \frac{1}{g_{d_1} + g_{d_2} + g_{m1} + |g_{mb1}|} \approx \frac{1}{g_{m1}} \qquad (4.37)$$

since usually $g_{m1} \gg g_{d1}$, g_{d2}, and $|g_{mb1}|$. Thus, R_{out} has a relatively low value, of the order of 1 kΩ.

The dc bias current of the stage is determined by the current source Q_2, which drives Q_1 at its low-impedance source terminal. Thus, the dc drop V_{GS1} between the input and output terminals is determined by $V_{bias} - V_{SS}$ and the dimensions of Q_1 and Q_2; these parameters can be used to control the level shift provided by the stage.

The gate of the load device Q_2 may be connected to the drain to eliminate the gate bias voltage (Fig. 4.34). Analysis shows (Problem 4.11) that for

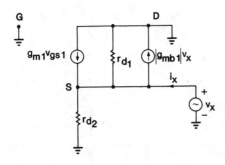

FIGURE 4.33. Equivalent circuit to calculate the output impedance of the source follower.

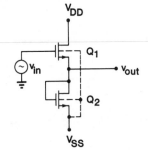

FIGURE 4.34.   Enhancement-load source follower.

$g_{m2} \gg g_{d1}, |g_{mb1}|$ the voltage gain is then $(1 + g_{m2}/g_{m1})^{-1}$. This can be close to 1 only if $g_{m1} \gg g_{m2}$, which, as discussed in connection with Eq. (4.19), requires a large area for the stage. Hence, this stage is rarely used. An improved version is shown in Fig. 4.35. This circuit uses a depletion-mode load device, and hence A_v satisfies (4.35) so that $A_v \simeq 1$ can be achieved without extravagant area requirements. Now the level shift can be controlled by choosing $(W/L)_1$ and $(W/L)_2$ appropriately.

The output stage of an op-amp (Fig. 4.4) must have a low output impedance and a large output voltage swing. The source follower has these attributes. The circuit of Fig. 4.31 has a gain close to 1, and a low output impedance, if g_{m1} is large. This can be achieved by choosing i_{D1}^0 and $(W/L)_1$ sufficiently large.

The large-signal operation of the circuit can be analyzed simply if the load device is regarded as a current source. Figure 4.36 shows the redrawn circuit; r_0 is the average large-signal output resistance of the current source and R_L is the load resistor. From Eq. (3.9), ignoring the body effect,

$$i_{D1} = I_0 + v_{\text{out}}(g_0 + G_L) = k_1(v_{\text{in}} - v_{\text{out}} - V_{T1})^2. \qquad (4.38)$$

If $(g_0 + G_L)/k_1 \ll 2|v_{\text{in}} - V_{T1}|$, then

$$v_{\text{out}} \cong v_{\text{in}} - V_{T1} - \sqrt{I_0/k_1}, \qquad (4.39)$$

so that the circuit operates as a linear buffer. To achieve this, $(W/L)_1$ must be sufficiently large.

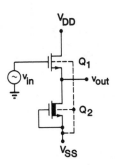

FIGURE 4.35.   Depletion-load source follower.

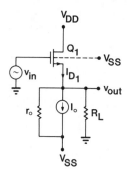

FIGURE 4.36. Source follower with a current source as load.

A major disadvantage of this stage is the following. If $v_{out} < 0$, then the load *supplies* current to the output stage. However, the latter can sink (absorb) an output current only if it is less than I_0. This represents a serious limitation. Also, for $v_{out} > 0$, Q_1 must supply the output current *plus* I_0. In addition, there is a voltage drop greater than V_{T1} between the input and output terminals. Thus, if v_{in} comes from a gain stage such as that shown in Fig. 4.15 where the output voltage must be less than $V_{DD} - V_{T2}$, then the maximum positive output voltage swing is $V_{DD} - 2V_T$. The negative swing is limited by the requirement that the device(s) in the current source must remain in saturation for the smallest output voltage.

By using negative feedback, better output stages can be designed. An example is shown in Fig. 4.37a and its small-signal equivalent circuit (drawn neglecting the body effect) is shown in Fig. 4.37b. Assuming small drain conductances ($g_{d1}, g_{d2}, g_{d3}, g_{d4} \approx 0$), the relations

$$g_{m2}(v_{out} - v_1) = g_{m1}v_{in} \qquad (4.40)$$

and

$$v_{out} = -g_{m3}v_1(1/g_{m4}). \qquad (4.41)$$

result. Solving for v_{out},

$$A_v = \frac{v_{out}}{v_{in}} = \frac{g_{m1}/g_{m2}}{1 + g_{m4}/g_{m3}}. \qquad (4.42)$$

For the proper choice of the W/L values, $A_v = 1$ can easily be obtained.

The output impedance can also readily be found (Problem 4.12). The result is

$$r_{out} \cong \frac{1}{g_{m3} + g_{m4}}. \qquad (4.43)$$

which can be made small by choosing the bias currents and W/L (aspect)

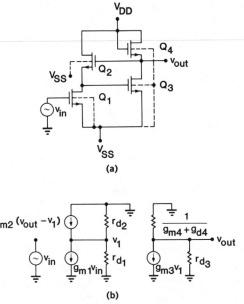

FIGURE 4.37. (a) Improved output stage. (b) Equivalent small-signal circuit of the improved output stage.

ratios of Q_3 and Q_4 large. Physically, if v_{out} drops due to loading, the drop is fed through Q_2 and Q_1 to the gate of Q_3. Then Q_3 conducts less, allowing the output voltage to rise, and thus counteracting the drop in v_{out}.

The output voltage can swing to $V_{DD} - v_{GS4}$ in the positive direction, and to $V_{SS} + v_{GS3} - V_T$ in the negative one.

The current driving capabilities of this circuit are superior to those discussed earlier. The variable v_{GS3} acts to cut off or open the current path through Q_3 so that it conducts only when it is needed to sink current from the load. Thus, if v_{in} goes negative, the gate voltage of Q_3 rises to turn its drain current on, and vice versa. The output current is thus limited only by the size of the output devices.

Class AB push–pull output stages (similar to the bipolar version) can also be realized in CMOS technology. An example is shown in Fig. 4.38. In the circuit, $Q_1 - Q_4$ forms a gain stage, while Q_5 and Q_6 drive the load R_L. Q_2 and Q_3 provide a voltage drop between the gates of Q_5 and Q_6, to reduce crossover distortion. The sizes of Q_2 and Q_3 are chosen such that the gate-to-source voltages of Q_5 and Q_6 are slightly larger than their threshold voltages.

Since the circuit does not use negative feedback, its output impedance is fairly high, comparable to that of the NMOS circuits of Figs. 4.34 and 4.35. The maximum output voltage for $R_L \to \infty$ is $V_{DD} - V_{T5}$ and the minimum $V_{SS} - |V_{T6}|$, where $V_{T5}(V_{T6})$ is the threshold voltage of $Q_5(Q_6)$. If R_L draws current from, say, Q_5 then the device must provide a drain current v_{out}/R_L,

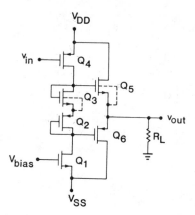

FIGURE 4.38. A CMOS class AB push-pull output stage.

and hence needs a gate-to-source voltage $v_{GS5} = V_{T5} + \sqrt{v_{out}/(k_5 R_L)}$. This increases rapidly with decreasing R_L, and hence represents an important limitation on the achievable positive output voltage swing. Similar considerations hold, of course, for negative swings due to the necessary v_{GS6}.

Due to its inherent symmetry, the CMOS output stage can provide an output signal with lower distortion than a similar NMOS circuit.

4.5. MOS DIFFERENTIAL AMPLIFIERS

As described in Section 4.1 in connection with Fig. 4.4, the input amplifier stage must provide a high input impedance, large common-mode rejection ratio (CMRR) and power-supply rejection ratio (PSRR), low dc offset voltage and noise, and much (or all) of the op-amp's voltage gain. The output signal of the input stage is much larger than the input one, and is hence no longer so sensitive to noise and offset voltage effects in the later stages. (Note that a large common-mode rejection is desirable even if the noninverting terminal is grounded in normal operation, to suppress noise in the ground line.)

The above requirements can often be met for NMOS circuits by using the source-coupled stage shown in Fig. 4.39. Since this circuit operates in a differential mode, it can provide high differential gain along with low common-mode gain, and hence a large CMRR. The differential configuration also helps in achieving a large PSRR, since variations of V_{DD} are, to a large extent, canceled in the differential output voltage $v_{01} - v_{02}$.

An approximate analysis of the amplifier can readily be performed. We assume that the current source I is ideal, that is, that its internal conductance g is zero. We also assume ideal symmetry between Q_1 and Q_2 and Q_3 and Q_4, and that all devices operate in saturation. Then, the incremental drain currents satisfy $i_{d_1} \simeq g_{mi}(v_{in1} - v)$, $i_{d_2} \simeq g_{mi}(v_{in2} - v)$, and $i_{d_1} + i_{d_2} \simeq 0$. This gives $v \simeq (v_{in1} + v_{in2})/2$ for the source voltages of Q_1 and Q_2, and $i_{d_1} \simeq -i_{d_2} \simeq g_{mi}(v_{in1} - v_{in2})/2$ for their drain currents. Hence, the output voltages are

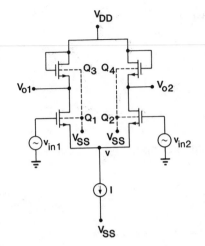

FIGURE 4.39. Source-coupled input stage with enhancement-mode load devices.

$v_{01} \simeq -v_{02} = -i_{d_1}/g_l = g_{mi}(v_{in1} - v_{in2})/(2g_l)$, where g_l is the load conductance. Defining the differential gain by $A_{dm} \triangleq (v_{01} - v_{02})/(v_{in1} - v_{in2})$, we obtain the simple result $A_{dm} \simeq -g_{mi}/g_l$. Thus, the differential gain is the same as for a simple inverter; however, the stage provides also a rejection of common-mode signals and of noise in the power supplies V_{DD} and V_{SS}, all of which are canceled (or, for actual circuits, reduced) by the differential operation of the stage. A more detailed analysis follows next.

The low-frequency small-signal equivalent circuit of the source-coupled stage is shown in Fig. 4.40. In the circuit, the body-effect transconductances of the input devices Q_1 and Q_2 are ignored to simplify the discussions. It will also be assumed that the circuit is perfectly symmetrical, so that the parameters of Q_1 and Q_2 are identical, as are those of Q_3 and Q_4. The load conductance g_l of Q_3 and Q_4 can be found as was done in connection with Fig. 4.16; the result is

$$g_l = g_{m3} + g_{d3} + |g_{mb3}| = g_{m4} + g_{d4} + |g_{mb4}|. \qquad (4.44)$$

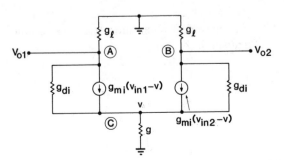

FIGURE 4.40. Small-signal equivalent circuit of the source-coupled pair.

Applying the current law at nodes Ⓐ and Ⓑ,

$$v_{01}g_l + (v_{01} - v)g_{di} + g_{mi}(v_{in1} - v) = 0$$

and   (4.45)

$$v_{02}g_l + (v_{02} - v)g_{di} + g_{mi}(v_{in2} - v) = 0$$

results. (Here, the subscripts i and l refer to the input and load devices, respectively.) The current law at node Ⓒ gives

$$(v - v_{01})g_{di} - g_{mi}(v_{in1} - v) + vg$$

$$+ (v - v_{02})g_{di} - (v_{in2} - v)g_{mi} = 0. \qquad (4.46)$$

Equations (4.45) and (4.46) represent three equations in the three unknowns v_{01}, v_{02}, and v. Solving them for v_{01} and v_{02}, we get

$$v_{01} = - \frac{g_l g_{mi}(g_{di} + g_{mi})(v_{in1} - v_{in2}) + g g_{mi}(g_l + g_{di})v_{in1}}{(g_l + g_{di})[2g_l(g_{di} + g_{mi}) + g(g_l + g_{di})]}$$

(4.47)

$$v_{02} = - \frac{g_l g_{mi}(g_{di} + g_{mi})(v_{in2} - v_{in1}) + g g_{mi}(g_l + g_{di})v_{in2}}{(g_l + g_{di})[2g_l(g_{di} + g_{mi}) + g(g_l + g_{di})]}.$$

The differential and common-mode input voltages are

$$v_{in, d} = v_{in1} - v_{in2}$$

and   (4.48)

$$v_{in, c} = \frac{v_{in1} + v_{in2}}{2}$$

(cf. (4.1) and (4.2) of Section 4.1). The differential and common-mode output voltages can be defined similarly:

$$v_{0, d} = v_{01} - v_{02}$$

and   (4.49)

$$v_{0, c} = \frac{v_{01} + v_{02}}{2}.$$

Then, the differential-mode gain can be obtained from (4.47):

$$A_{dm} = \frac{v_{0,d}}{v_{in,d}} = \frac{v_{01} - v_{02}}{v_{in1} - v_{in2}}$$

$$= -\frac{2g_l g_{mi}(g_{di} + g_{mi}) + gg_{mi}(g_l + g_{di})}{(g_l + g_{di})[2g_l(g_{di} + g_{mi}) + g(g_l + g_{di})]}.$$

(4.50)

For $g = 0$ and $g_{di} \ll g_l$, $A_{dm} \simeq -g_{mi}/g_l$ as predicted earlier. Similarly, the common-mode gain can be found:

$$A_{cm} = \frac{v_{0,c}}{v_{in,c}} = \frac{(v_{01} + v_{02})/2}{(v_{in1} + v_{in2})/2}$$

$$= -\frac{gg_{mi}}{2g_l(g_{di} + g_{mi}) + g(g_l + g_{di})}.$$

(4.51)

Hence, the common-mode rejection ratio is

$$\text{CMRR} = \left| \frac{A_{dm}}{A_{cm}} \right| = 1 + 2\frac{g_l}{g}\frac{g_{di} + g_{mi}}{g_{di} + g_l}.$$

(4.52)

Normally, $g, g_{di} \ll g_l, g_{mi}$, and the approximations

$$A_{dm} \simeq -\frac{g_{mi}}{g_l}$$

$$A_{cm} \simeq -\frac{g}{2g_l}$$

(4.53)

$$\text{CMRR} \simeq \frac{2g_{mi}}{g}$$

can be used. Clearly, to obtain a large CMRR, g must be small, that is, the current source should have a large output impedance. The circuits described earlier and shown in Figs. 4.6–4.11 are suitable to achieve this. All, however, require a dc voltage drop for operation which limits the achievable output voltage swing.

An alternative approach to increasing the CMRR is to use common-mode feedback (Fig. 4.41). In the circuit, the matched devices Q_5 and Q_6 sense and average v_{01} and v_{02}. The resulting current is replicated by the current mirror formed by Q_7 and Q_8 (cf. Fig. 4.6!). Thus an *increase* in the $v_{0,c}$ causes the current of Q_8 to increase. This increases the voltage drop across Q_3 and Q_4,

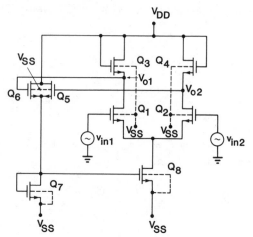

FIGURE 4.41.   Differential stage with common-mode feedback.

which in turn *reduces* $v_{0,c}$. The described negative feedback action tends to keep $v_{0,c}$ constant; this means that the *small-signal* (incremental) common-mode output is zero (or, in practice, very small).

As (4.53) indicates, with the described approximations (including the assumed absence of the body effect) the differential gain can be obtained from

$$A_{dm} \simeq - \sqrt{\frac{(W/L)_i}{(W/L)_l}}. \tag{4.54}$$

As discussed in Section 4.3, this value can only be made larger than $10 \sim 15$ at the cost of using a large silicon area for the stage. If larger A_{dm} is needed, therefore, it is preferable to use depletion-mode devices Q_3 and Q_4 as loads (Fig. 4.42). As shown in Section 4.3 (cf. Fig. 4.18), the load admittance presented by the depletion-mode device is only

$$g_l = g_{d3} + |g_{mb3}| = g_{d4} + |g_{mb4}|. \tag{4.55}$$

The $g_{m3} = g_{m4}$ term is absent, since $v_{gs} \equiv 0$ for Q_3 and Q_4. Equations (4.45) and (4.53) remain valid; however, both A_{dm} and A_{cm} are much larger for the depletion-mode load circuit since g_l is now much smaller.

The common-mode feedback arrangement shown in Fig. 4.41 can be applied for this circuit as well. Alternatively, the feedback configuration described by Tsividis et al.[18] can be used.

As mentioned earlier, the differential input stage is usually followed by a differential-to-single-ended converter. Such a stage[9] is shown in Fig. 4.43a; its small-signal equivalent circuit is shown in Fig. 4.43b. This circuit has two advantages. First, its output voltage swing is very close to the full differential

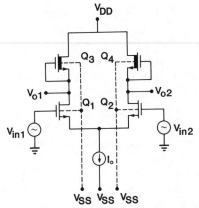

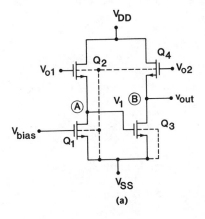

FIGURE 4.42. Differential stage with depletion-load devices.

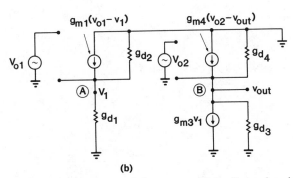

FIGURE 4.43. (a) Differential-to-single-ended converter. (b) Small-signal equivalent circuit of differential-to-single-ended converter.

output voltage swing of the preceding differential amplifier. Second, the converter itself can have a large CMRR, thus enhancing the CMRR of the input stage.

The two input signals v_{01} and v_{02} of the circuit of Fig. 4.43 (which were the output signals in Fig. 4.41 or 4.42) travel different paths to arrive at the output node Ⓑ. One (v_{01}) is transferred to node Ⓐ through the source follower $Q_2 - Q_1$, and is then amplified and inverted by the inverter $Q_3 - Q_4$. The other (v_{02}) arrives at Ⓑ via the source follower $Q_4 - Q_3$. Thus, the output is (for appropriate W/L values) proportional to $v_{02} - v_{01}$, and common-mode components cancel. The circuit also provides a level shift v_{GS4}.

The detailed analysis of the circuit can be based on the equivalent circuit of Fig. 4.43b. For simplicity, the body effect was neglected for all devices; that is, $g_{mb} = 0$ was used. Applying the current law at nodes Ⓐ and Ⓑ

$$g_{d1}v_1 + g_{d2}v_1 - g_{m1}(v_{01} - v_1) = 0$$

and   (4.56)

$$g_{d3}v_{out} + g_{d4}v_{out} + g_{m3}v_1 - g_{m4}(v_{02} - v_{out}) = 0$$

results. Solving the two equations in the two unknowns v_1 and v_{out} gives

$$v_{out} = \frac{1}{g_{d3} + g_{d4} + g_{m4}}\left(g_{m4}v_{02} - \frac{g_{m1}g_{m3}v_{01}}{g_{d1} + g_{d2} + g_{m1}}\right). \quad (4.57)$$

Here, the differential and common-mode input signals are

$$v_{in,d} = v_{01} - v_{02}$$

and   (4.58)

$$v_{in,c} = \frac{v_{01} + v_{02}}{2}.$$

In terms of these, (4.57) can be rewritten as

$$v_{out} = A_{dm}v_{in,d} + A_{cm}v_{in,c} \quad (4.59)$$

where

$$A_{dm} = -\frac{1}{2}\frac{g_{m1}(g_{m3} + g_{m4}) + g_{m4}(g_{d1} + g_{d2})}{(g_{d1} + g_{d2} + g_{m1})(g_{d3} + g_{d4} + g_{m4})}$$

and   (4.60)

$$A_{cm} = \frac{g_{m1}(g_{m3} - g_{m4}) + g_{m4}(g_{d1} + g_{d2})}{(g_{d1} + g_{d2} + g_{m1})(g_{d3} + g_{d4} + g_{m4})}.$$

For the usual case when $g_{m1}, g_{m4} \gg g_{di}$ $(i = 1, 2, 3, 4)$, we have

$$A_{dm} \simeq -\frac{1}{2}\left(\frac{g_{m3}}{g_{m4}} + 1\right)$$

$$A_{cm} \simeq \frac{1}{2}\left(\frac{g_{m3}}{g_{m4}} - 1\right) \tag{4.61}$$

$$\text{CMRR} = \left|\frac{A_{dm}}{A_{cm}}\right| \cong \frac{g_{m3} + g_{m4}}{g_{m3} - g_{m4}}.$$

For $g_{m3} = g_{m4}$, the ideal situation $A_{dm} \simeq -1$, $A_{cm} \simeq 0$, and CMRR $\to \infty$ is approximated. If $g_{m3} \gg g_{m4}$, then the circuit can be used as a gain stage,[10] and will not provide a large CMRR.

All the above differential and converter stages used NMOS devices only. If CMOS technology is available, then the circuit of Fig. 4.44a can be used. This

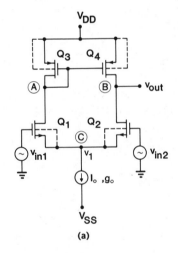

(a)

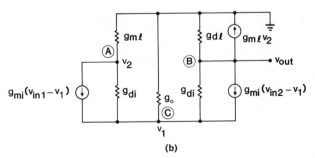

(b)

FIGURE 4.44. (*a*) CMOS differential stage with active load. (*b*) Small-signal equivalent circuit for CMOS differential stage.

circuit has differential input, but single-ended output. Hence, it performs as a combination of a differential gain stage and a differential-to-single-ended converter.

An approximate analysis of the circuit can be readily performed as follows. Assuming that the current source I_0 is ideal, the incremental drain currents of Q_1 and Q_2 must satisfy $i_{d1} + i_{d2} = 0$. Also, if both Q_1 and Q_2 are in saturation, then $i_{d1} \simeq g_{mi}(v_{in1} - v_1)$ and $i_{d2} \simeq g_{mi}(v_{in2} - v_1)$. Combining these equations, $v_1 \simeq (v_{in1} + v_{in2})/2$ results. Hence, $i_{d1} = -i_{d2} \simeq g_{mi}(v_{in1} - v_{in2})/2$. The current i_{d1} is easily imposed on Q_3 by Q_1, since the impedance at the common terminal of the gate and drain of Q_3 is only $1/g_{m3}$.

Transistors Q_3 and Q_4 form a current mirror similar to that shown in Fig. 4.6, and hence the current through Q_4 satisfies $i_{d4} = i_{d3} = i_{d1}$. Thus, *both* Q_2 and Q_4 send a current $i_{d1} = g_{mi}(v_{in1} - v_{in2})/2$ into the output terminal. Since the output is loaded by the drain resistances of Q_2 and Q_4, the output voltage is $v_{out} \simeq 2i_{d1}/(g_{di} + g_{dl}) \simeq g_{mi}(v_{in1} - v_{in2})/(g_{di} + g_{dl})$. The differential gain is thus $A_{dm} \triangleq v_{out}/(v_{in1} - v_{in2}) \simeq g_{mi}/(g_{di} + g_{dl})$. A more exact analysis follows next.

The small-signal equivalent circuit of the stage is shown in Fig. 4.44b. It was drawn under the assumption that both input devices Q_1 and Q_2 have the same conductances g_{mi} and g_{di}, and that both load devices have the parameters g_{ml} and g_{dl}; also that separate "wells" are provided for the NMOS devices. Then $v_{BS} = 0$ for all devices, and hence no body effect occurs. The output conductance of the current source is denoted by g_0.

Writing and solving the current law equations for nodes Ⓐ, Ⓑ, and Ⓒ (Problem 4.14), we obtain

$$v_{out} = \frac{g_{mi}g_{ml}}{D}\left\{2(g_{di} + g_{mi})(v_{in1} - v_{in2})\right.$$

$$\left. + g_0\left[v_{in1} - \left(\frac{g_{di}}{g_{ml}} + 1\right)v_{in2}\right]\right\}, \qquad (4.62)$$

$$D = (g_{di} + g_{mi})[g_{dl}g_{di} + 2g_{ml}(g_{dl} + g_{di})]$$

$$+ g_0(g_{di} + g_{ml})(g_{dl} + g_{di}).$$

We define, as before, the differential and common-mode input signals by (4.48). Then the differential gain A_{dm} and the common-mode gain A_{cm} can be defined by

$$v_{out} = A_{dm}v_{in, d} + A_{cm}v_{in, c}. \qquad (4.63)$$

From Eqs. (4.62) and (4.63),

$$A_{dm} = \frac{g_{mi}g_{ml}}{D}\left[2(g_{di} + g_{mi}) + g_0\left(1 + \frac{g_{di}}{2g_{ml}}\right)\right],$$

$$(4.64)$$

$$A_{cm} = -\frac{g_{mi}g_{di}g_0}{D},$$

and

$$\text{CMRR} = \left|\frac{A_{dm}}{A_{cm}}\right| = \frac{g_{ml}}{g_{di}g_0}[2(g_{di} + g_{mi}) + g_0] + \frac{1}{2}. \qquad (4.65)$$

For g_{mi}, $g_{ml} \gg g_0$, g_{di}, and g_{dl}, the approximations

$$A_{dm} \simeq \frac{g_{mi}}{g_{dl} + g_{di}},$$

$$(4.66)$$

$$A_{cm} \simeq \frac{-g_0 g_{di}}{2g_{ml}(g_{dl} + g_{di})},$$

and

$$\text{CMRR} \simeq 2\frac{g_{mi}g_{ml}}{g_0 g_{di}}$$

can be used. Note that A_{dm} is the same as that obtainable from a CMOS differential-input/differential-output stage (Problem 4.15). Thus, the single-ended output signal does not result in lower gain for the stage. By contrast, the CMRR is higher by a factor of g_{ml}/g_{di} which (for usual values) is much greater than one.

To calculate the small-signal output impedance of the CMOS stage, a test source i_0 can be applied at the output of the small-signal equivalent circuit and the input voltages v_{in1} and v_{in2} set to zero (Fig. 4.45). Analysis shows (Problem

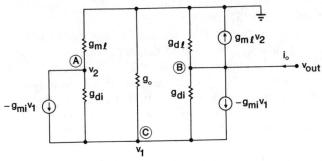

FIGURE 4.45. Equivalent circuit for the calculation of the output impedance of the CMOS differential stage.

4.16) that

$$r_{out} = \frac{v_{out}}{i_0} = \frac{1}{D}\{(g_{di} + g_{ml})[2(g_{di} + g_{mi}) + g_0]$$

$$- g_{di}(g_{di} + g_{mi})\} \qquad (4.67)$$

where D is defined in (4.62).

For $g_{mi}, g_{ml} \gg g_0, g_{di}, g_{dl}$, the approximation

$$r_{out} \simeq \frac{1}{g_{dl} + g_{di}} \qquad (4.68)$$

can be used.

4.6.   THE FREQUENCY RESPONSE OF MOS AMPLIFIER STAGES

In the previous sections, the linearized (small-signal) performance of MOS amplifier stages was analyzed at low frequencies. Thus, the parasitic capacitances illustrated in the equivalent circuit of Fig. 3.16 were ignored. For high-frequency signals, however, the admittances of these branches are no longer negligible, and hence neither are the currents which they conduct. Then the gains and the input and output impedances of the various circuits all become functions of the signal frequency ω. These effects will be analyzed next.

Consider again the NMOS single-ended amplifier with enhancement-mode load device (Fig. 4.19), discussed in Section 4.3. Using the equivalent circuit of Fig. 3.16, the high-frequency small-signal equivalent circuit of Fig. 4.20a resulted. In the circuit, R_s is the output impedance of the signal source, and C_L is the load capacitance. This circuit was then simplified to that of Fig. 4.20b, which (in the Laplace-transform domain) was shown to have the frequency response given in Eq. (4.25). With the approximations $g_{ml} \gg \omega C_{gd1}, G_{Leq} \gg \omega(C_{gd1} + C_{Leq})$, the frequency response of Eq. (4.26) resulted. It corresponded to the simplified equivalent circuit of Fig. 4.21, which was used to introduce the Miller effect.

The accuracy of the simplified circuit can be improved by restoring the two capacitances C_{Leq} and C_{gd1} which load the output node in the exact circuit of Fig. 4.20b. Also, in the numerator of Eq. (4.25), the term sC_{gd1} was neglected in comparison with g_{ml}. At higher frequencies, this is no longer justified. To restore the sC_{gd1} term, the gain of the controlled source g_{ml} can be changed to $g_{ml} - sC_{gd1}$ in the equivalent circuit. The resulting circuit is shown in Fig. 4.46. The corresponding transfer function is

$$A_v(s) \cong \frac{G_s(sC_{gd1} - g_{ml})}{(sC_{in} + G_s)[s(C_{Leq} + C_{gd1}) + G_{Leq}]}, \qquad (4.69)$$

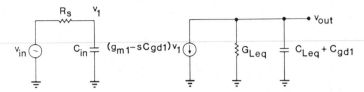

FIGURE 4.46.   Simplified equivalent circuit of MOS gain stage using Miller's thoerem.

where C_{in} is given by Eq. (4.27). This function has a right-half-plane (positive) real zero at

$$s_z = g_{m1}/C_{gd1} \qquad (4.70)$$

and two left-half-plane (negative) poles at

$$s_{p1} = -G_s/C_{in}$$

and $\qquad (4.71)$

$$s_{p2} = -G_{Leq}/(C_{Leq} + C_{gd1}).$$

Normally, C_{gd1} is small. Hence, $s_z \gg |s_{p1}|$ and, if C_{Leq} is also small, then $|s_{p2}| \gg |s_{p1}|$. Then, s_{p1} is closest to the $j\omega$-axis and is therefore the dominant pole of the circuit.

The frequency response $A_v(j\omega)$ can be obtained simply by replacing s by $j\omega$ in (4.69). It can be arranged in the form

$$A_v(j\omega) = \frac{G_s C_{gd1}}{C_{in}(C_{Leq} + C_{gd1})} \frac{j\omega - s_z}{(j\omega - s_{p1})(j\omega - s_{p2})}. \qquad (4.72)$$

If $|s_{p1}| \ll |s_{p2}|$ and s_z, then the 3 dB frequency (that is, the frequency where $|A_v(j\omega)|$ is $1/\sqrt{2}$ times its dc value) is

$$\omega_{3\ dB} \cong |s_{p1}| = G_s/C_{in}. \qquad (4.73)$$

If a high dc gain is required, then by (4.19) the ratio $(W/L)_2$ must be small. Since the minimum possible width is limited by the geometrical resolution of the fabrication process, this necessitates a long load device Q_2, with a large gate area. The latter, in turn, causes C_{gs2} and hence C_{Leq} to be large. Then, $|s_{p2}|$ will be lowered and the frequency response degraded. This problem can be alleviated (at the cost of reducing the positive signal swing) by using a split load (Fig. 4.47). It is readily shown that for given g_{m2} the load of Fig. 4.47 requires half the gate area of a single load device (Problem 4.17).

It should be pointed out that the discussions resulting in Eqs. (4.69)–(4.72) were based on somewhat heuristic approximations. More exact calculations

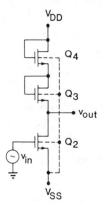

FIGURE 4.47. Split-load gain stage.

can be based on Eq. (4.25). However, the resulting formulas for s_{p1} and s_{p2} are complicated and hence do not provide such easy physical insight as (4.71).

A depletion-load gain stage with capacitive loading is shown in Fig. 4.48a; its high-frequency small-signal equivalent circuit is shown in Fig. 4.48b. Both the exact and the approximating relations obtained from the analysis of the circuit of Fig. 4.19 remain valid, with minor modifications, for this stage. The modifications are evident from a comparison of Fig. 4.20a and 4.48b; in the latter,

$$G_{Leq} = |g_{mb2}| + g_{d1} + g_{d2}$$

and

$$(4.74)$$

$$C_{Leq} = C_{db1} + C_{gd2} + C_{sb2} + C_L$$

replace G_{Leq} and C_{Leq} given by (4.23). As a result, the dc gain is higher, and G_{Leq} is smaller.

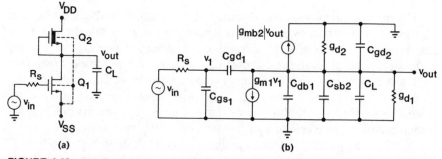

FIGURE 4.48. (a) Depletion-load gain stage with capacitive loading. (b) Equivalent circuit of depletion-load gain stage.

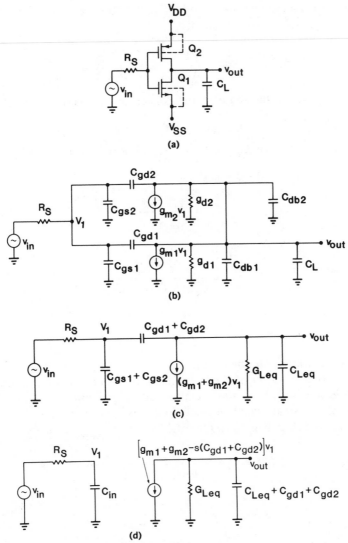

FIGURE 4.49. (*a*) CMOS gain stage with capacitive loading. (*b*) Equivalent circuit of the CMOS gain stage. (*c*) Simplified equivalent circuit of the CMOS gain stage. (*d*) Simplified equivalent circuit of CMOS gain stage using Miller's theorem.

A CMOS gain stage with capacitive loading is shown in Fig. 4.49*a*; the corresponding high-frequency linear equivalent circuit in Fig. 4.49*b*. Defining

$$G_{Leq} = g_{d1} + g_{d2}$$

and

$$C_{Leq} = C_{db1} + C_{db2} + C_L,$$

(4.75)

the simplified circuit of Fig. 4.49*c* results.

Using the approximations yielding Fig. 4.46 for the NMOS gain stage, the circuit of Fig. 4.49d can be obtained. Here,

$$C_{in} = C_{gs1} + C_{gs2} + \left(1 + \frac{g_{m1} + g_{m2}}{G_{Leq}}\right)(C_{gd1} + C_{gd2}). \qquad (4.76)$$

From Fig. 4.49d, the approximate transfer function is (Problem 4.18)

$$A_v(s) \simeq \frac{G_s[s(C_{gd1} + C_{gd2}) - (g_{m1} + g_{m2})]}{[s(C_{gd1} + C_{gd2} + C_{Leq}) + G_{Leq}](sC_{in} + G_s)}. \qquad (4.76)$$

Hence, the zero and the poles are

$$s_z = \frac{g_{m1} + g_{m2}}{C_{gd1} + C_{gd2}} \qquad (4.77)$$

and

$$s_{p2} = \frac{-G_{Leq}}{C_{gd1} + C_{gd2} + C_{Leq}},$$

$$s_{p1} = -\frac{G_s}{C_{in}}. \qquad (4.78)$$

The dominant pole is normally s_{p1}.

The analysis of the NMOS enhancement-load source follower is straightforward. Figure 4.50a shows the actual circuit; Figs. 4.50b and 4.50c show the high-frequency small-signal equivalent circuits. The transfer function can be readily derived (Problem 4.19); the result is

$$A_v(s) = \frac{sC_{gs1} + g_{m1}}{s(C_{gs1} + C_{Leq}) + (g_{m1} + G_{Leq})}, \qquad (4.79)$$

where

$$C_{Leq} = C_L + C_{sb1} + C_{db2} + C_{gd2}$$

and $\qquad (4.80)$

$$G_{Leq} = g_{d1} + g_{d2}.$$

The zero and the pole are hence

$$s_z = -g_{m1}/C_{gs1}$$

and $\qquad (4.81)$

$$s_p = -\frac{g_{m1} + G_{Leq}}{C_{gs1} + C_{Leq}}.$$

Choosing $G_{Leq}/g_{m1} \cong C_{Leq}/C_{gs1}$, $s_z \cong s_p$ can be achieved. Then $A_v(s) \simeq$

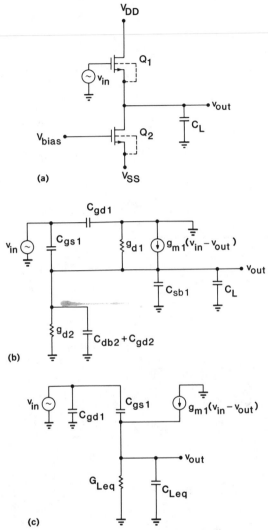

FIGURE 4.50. (*a*) NMOS source follower with capacitive loading. (*b*) Equivalent circuit of the NMOS source follower. (*c*) Simplified equivalent circuit of the NMOS source follower.

$C_{gs1}/(C_{gs1} + C_{Leq})$, and hence the gain is constant up to very high frequencies, where higher-order effects cause it to drop. In the actual implementation, in order to meet the condition on C_{Leq}/C_{gs1} it may be necessary to connect a capacitor C in parallel with C_{gs1}, that is, between the input and output terminals. Then C_{gs1} should be replaced by $C_{gs1} + C$ in the above relations.

Figure 4.51*a* shows an NMOS cascode gain stage with enhancement-mode load.* C_L represents the capacitive load of the stage. Figures 4.51*b* and 4.51*c*

*With separate wells for Q_2 and Q_3, so no body effect occurs in the stage.

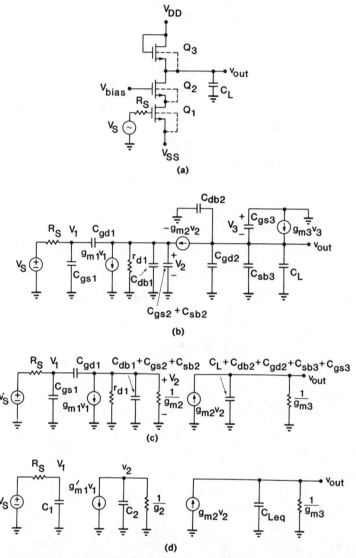

FIGURE 4.51. (*a*) Enhancement-load cascode stage with capacitive loading. (*b*) High-frequency equivalent circuit of the cascode stage. (*c*) Simplified high-frequency equivalent circuit of the cascode stage. (*d*) Simplified equivalent circuit of the cascode stage obtained using Miller's theorem.

show the detailed and simplified high-frequency linearized equivalent circuits, respectively. Using Miller's theorem in the circuit of Fig. 4.51c, the approximate equivalent circuit of Fig. 4.51d results, where

$$g_2 = g_{m2} + \frac{1}{r_{d1}},$$

$$C_1 = C_{gs1} + (1 + g_{m1}/g_{m2})C_{gd1},$$

$$C_2 = C_{gd1} + C_{db1} + C_{gs2} + C_{sb2},$$

$$C_{Leq} = C_L + C_{gd2} + C_{db2} + C_{sb3} + C_{gs3},$$ (4.82)

$$g'_{m1} = g_{m1} - sC_{gd1}.$$

The circuit of Fig. 4.51d can be easily analyzed (thanks to the Miller approximation which neatly partitioned it into buffered sections). The result is

$$A_v(s) = \frac{G_s g_{m2}(sC_{gd1} - g_{m1})}{(sC_1 + G_s)(sC_2 + g_2)(sC_{Leq} + g_{m3})}$$ (4.83)

from which the zero and the poles can be directly recognized:

$$s_z = g_{m1}/C_{gd1},$$

$$s_{p1} = -G_s/C_1,$$

$$s_{p2} = -g_2/C_2,$$ (4.84)

$$s_{p3} = -g_{m3}/C_{Leq}.$$

For practical values, often $|s_{p1}| \ll s_z,\ |s_{p2}|,\ |s_{p3}|$. Then s_{p1} is the dominant pole, and the 3 dB frequency is given by

$$f_{3\ dB} \simeq \frac{|s_{p1}|}{2\pi} = \frac{G_s}{2\pi C_1}.$$ (4.85)

Typically, $g_{m1} = g_{m2}$; then $C_1 = C_{gs1} + 2C_{gd1}$ and $f_{3dB} \simeq G_s/[2\pi(C_{gs1} + 2C_{gd1})]$. By contrast, for the simple inverter stage of Fig. 4.19, the corresponding value is $G_s/\{2\pi[C_{gs1} + (1 + g_{m1}/G_{Leq})C_{gd1}]\}$ as Eqs. (4.69) and (4.27) show. Since g_{m1}/G_{Leq} is the magnitude of the dc gain of the stage, it is usually large. Hence, the dominant pole (and thus the 3 dB frequency) is much smaller for the simple gain stage than for the cascode circuit. This confirms the effectiveness of the latter for high-frequency amplification.

The small-signal analysis of the differential amplifier stages of Fig. 4.39–4.44 can, in principle, be performed similarly. Thus, in the small-signal equivalent circuit of each transistor the stray capacitances can be included, and a nodal analysis performed in the Laplace-transform domain. The process becomes quite complicated, however, since the numbers of both nodes and branches are high.

A much more efficient process can be based on the *half-circuit* concept.[2] Consider the stage of Fig. 4.39, and the corresponding equivalent circuit of Fig. 4.40. From Eq. (4.48), the input voltages can be expressed as

$$v_{in1} = v_{in,c} + v_{in,d}/2$$

and   (4.86)

$$v_{in2} = v_{in,c} - v_{in,d}/2.$$

Since the circuit operates in its linear region, superposition may be used in the analysis. Accordingly, let $v_{in1} = v_{in,d}/2$ and $v_{in2} = 0$; the corresponding value of v (Fig. 4.40) will be denoted by v_d. Next, let $v_{in1} = 0$, while $v_{in2} = -v_{in,d}/2$. It follows from the symmetry of the circuit that for $v_{in2} = +v_{in,d}/2$, v would again be v_d; by linearity then, for $v_{in2} = -v_{in,d}/2$, $v = -v_d$.

Assume, next, purely differential-mode inputs $v_{in1} = v_{in,d}/2 = -v_{in2}$. By superposition, $v = v_d - v_d \equiv 0$. Hence, in the analysis the equivalent circuit of Fig. 4.52a may be used. Clearly, the two symmetric halves of the circuit operate independently. Thus the differential-mode analysis of the circuit of Fig. 4.39 is the same as for the circuit of Fig. 4.52b, which contains two isolated half-circuits. Each half-circuit is a simple gain stage, similar to that shown in Fig. 4.15.

Consider now the analysis for the common-mode input voltages. The situation is illustrated in Fig. 4.53a, where the current source I has been split into two symmetric sources. Because of the inherent symmetry of the circuit, the lead connecting the two symmetric nodes marked Ⓑ and Ⓒ carries no current. Hence, none of the currents or voltages in the circuit will change if this wire is cut at the point Ⓐ. Then the circuit is again split into two independent half-circuits, each a simple gain stage with a current source in the source branch. Thus, for low frequencies, the equivalent circuit of Fig. 4.53b is applicable.

Having found all voltages and currents for both differential- and common-mode inputs using the process of half-circuits, the same quantities for general inputs given by Eq. (4.86) can be found simply by superposition, that is, by adding the partial results.

The process was so far discussed for low-frequency conditions. Clearly, however, the same argument holds also for high frequencies as long as the stage remains symmetric. Thus, for example, the differential-mode analysis of the circuit of Fig. 4.39 for high frequencies reduces to the analyses of two half-circuits of the form of Fig. 4.46. For common-mode analysis, each half-circuit is of the form shown in Fig. 4.54, where C_L is the load capacitance

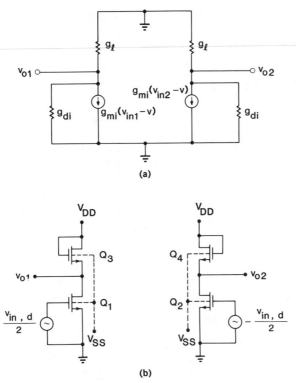

FIGURE 4.52. (*a*) Low-frequency small-signal equivalent circuit of the differential stage of Fig. 4.39 for differential-mode input voltages. (*b*) Differential-mode equivalent circuit of the stage of Fig. 4.39.

at each output terminal, while g and C are the output conductance and capacitance of the current source I, respectively. The subscript "i" refers to an input device (Q_1 or Q_2), while "l" to a load device (Q_3 or Q_4). The output voltage can then be obtained using nodal analysis (Problem 4.21).

The procedure is quite complicated even with the half-circuit method. Furthermore, this method assumes a perfect symmetry in the circuit, which may or may not hold. Consider, for example, the CMOS differential stage of Fig. 4.55. If only Q_1 has an input voltage, while the output voltage is used only at node Ⓑ, then the load capacitances usually satisfy $C_{LB} \gg C_{LA}$. Furthermore, since Q_3 is driven with its gate and drain shorted, by Fig. 3.14 it presents a large load conductance $g_{m3} + g_{d3} \simeq g_{m3}$. By contrast, the conductance connected to node Ⓑ is $g_{d2} + g_{d4}$, a small value. Hence, the time constant of the admittance connected to node Ⓑ, $t_B = C_{LB}/(g_{d2} + g_{d4})$ is likely to be several orders of magnitude larger than that at node Ⓐ, $t_A \cong C_{LA}/g_{m3}$. The time constant at node Ⓒ is also small, since Q_1 and Q_2 load this node with the large conductance $g_{m1} + g_{m2}$.

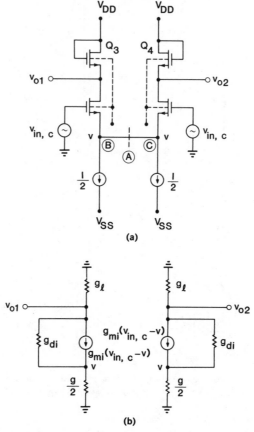

FIGURE 4.53. (*a*) The common-mode equivalent circuit of the differential stage shown in Fig. 4.39. (*b*) Low-frequency linear equivalent circuit of the circuit of Fig. 4.53*a*.

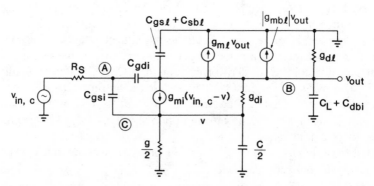

FIGURE 4.54. High-frequency half-circuit for the common-mode analysis of the differential stage shown in Fig. 4.39.

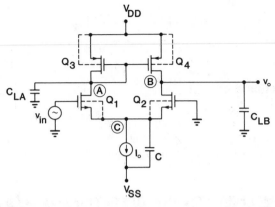

FIGURE 4.55.   CMOS differential stage with capacitive loading.

Clearly, in a situation like that represented by the circuit of Fig. 4.55, the half-circuit method cannot be used, while the general nodal analysis is very complicated. Thus, either a computer program (such as SPICE) which can perform the frequency analysis of linearized MOS circuits should be used, or some simplifying assumptions used in the theoretical analysis. For our example (Fig. 4.55), it was verified above that the dominant pole is that corresponding to the largest time constant t_B; its value is $s_{p1} \simeq (g_{d2} + g_{d4})/C_{LB}$. Therefore, for example, the differential-mode voltage gain can be approximated by

$$A_{dm}(s) \simeq \frac{A_{dm}(0)}{1 + s/s_{p1}} \simeq \frac{g_{mi}}{sC_{LB} + (g_{di} + g_{dl})}. \qquad (4.87)$$

Here, it was assumed that Q_1 and Q_2 as well as Q_3 and Q_4 are matched devices, and (4.66) was used. The 3 dB frequency can also be (approximately) predicted from (4.87) as $(g_{di} + g_{dl})/(2\pi C_{LB})$.

The same approximation can be used to find the common-mode voltage gain:

$$A_{cm}(s) \simeq \frac{A_{cm}(0)}{1 + s/s_{p1}} = -\frac{g_{di}}{2g_{mi}} \frac{g}{sC_{LB} + (g_{di} + g_{dl})}. \qquad (4.88)$$

Here, g is the output conductance of the current source. In a CMOS op-amp, the common source of the n-channel devices is tied to a p-well. There is a large stray capacitance C between the p-well and the V_{ss} lead, which reduces the impedance between node Ⓒ and ground at high frequencies. The effect of C can be incorporated in (4.88) simply by replacing g by $g + sC$. Then

$$A_{cm}(s) \simeq -\frac{g_{di}}{2g_{mi}} \frac{g + sC}{sC_{LB} + (g_{di} + g_{dl})} \qquad (4.89)$$

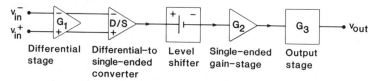

FIGURE 4.56. Basic building blocks of an operational amplifier.

results. The zero at $s_z = -g/C$ will cause $|A_{cm}/A_{dm}|$ to increase by 20 dB/decade at high frequencies, thus causing a reduced CMRR.

4.7. UNCOMPENSATED CMOS OPERATIONAL AMPLIFIERS

Using the stages discussed in Sections 4.2 to 4.5 as building blocks, it is possible to design either NMOS or CMOS op-amps. Since CMOS circuits tend to be simpler than their single-channel counterparts, CMOS op-amps will be discussed first. A practical block diagram of an MOS op-amp is shown in Fig. 4.56; it represents a more detailed version of that shown in Fig. 4.4. The required voltage gain is obtained in the differential (G_1) and single-ended (G_2) gain stages. The output stage G_3 is usually a wide-band, unity-gain, low-output impedance buffer, capable of driving large capacitive and/or resistive loads. In op-amps used in the internal (as opposed to output) stages of a switched-capacitor filter, the load may be only a small capacitor. In such a situation the output buffer G_3 may be omitted, and the load may then be directly connected to the output of G_2.

A simple CMOS implementation of the scheme of Fig. 4.56 is shown in Fig. 4.57. In this circuit, Q_5 acts as a simple current source, and $Q_1 - Q_5$ form a differential stage (cf. Fig. 4.44a) with a single-ended output. Q_6 (acting as the driver device) and Q_7 (acting as the load) form the second gain stage which also acts as a level shifter. Finally, the output buffer is realized by the source follower consisting of Q_8 as driver and Q_9 as load.

The low-frequency differential-mode gain of the input stage can be obtained from (4.66):

$$A_{v1} \simeq \frac{g_{mi}}{g_{di} + g_{dl}}. \tag{4.90}$$

Here, it is assumed that Q_1 is matched to Q_2, and Q_3 to Q_4. The low-frequency gain of the inverter formed by Q_6 and Q_7 is clearly

$$A_{v2} = \frac{-g_{m6}}{g_{d6} + g_{d7}}. \tag{4.91}$$

The overall voltage gain A_v is $A_{v1}A_{v2}$. For typical biasing conditions and device geometries, $A_v = 10{,}000 \sim 20{,}000$ can be achieved.

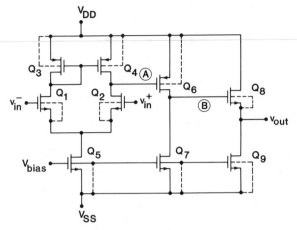

FIGURE 4.57.   Uncompensated CMOS operational amplifier.

The output terminals (A) and (B) of both stages are high-impedance nodes; the low-frequency output impedance of the input stage driving node (A) is

$$R_{01} \simeq \frac{1}{g_{dl} + g_{di}};$$ (4.92)

that of the second stage (Q_6, Q_7) is

$$R_{02} \simeq \frac{1}{g_{d6} + g_{d7}}.$$ (4.93)

An equivalent circuit showing these impedances and also the parasitic capacitances C_A, C_B loading nodes (A) and (B), respectively, is shown in Fig. 4.58. It is evident from the figure that the transfer function of the amplifier $A_v(s) = V_{out}(s)/[V_{in}^+(s) - V_{in}^-(s)]$ will contain the factors

$$\frac{1/sC_A}{R_{01} + 1/sC_A} \frac{1/sC_B}{R_{02} + 1/sC_B} = \frac{1}{(1 - s/s_A)(1 - s/s_B)},$$ (4.94)

where the poles are $s_A = -1/R_{01}C_A$ and $s_B = -1/R_{02}C_B$. Since R_{01} and R_{02}

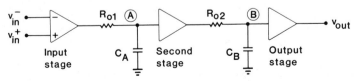

FIGURE 4.58.   Block diagram showing the origin of the dominant poles.

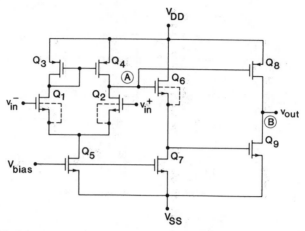

FIGURE 4.59. Improved uncompensated CMOS operational amplifier.

are large, s_A and s_B will be close to the $j\omega$-axis in the s-plane. Hence, they will be the dominant poles of the amplifier. The effects of other poles will only be noticeable at very high frequencies.

If the op-amp is required to drive small internal capacitive loads only, then the output source follower (Q_8, Q_9) may be eliminated, and the output taken directly from node (B). However, then the maximum output current which can be sunk is determined by the current source Q_7.

An improved CMOS op-amp[11] with increased output range is shown in Fig. 4.59. The output stage is now in the form of Fig. 4.30, with Q_6 and Q_7 acting as the level shifter, while Q_8 and Q_9 act as the class-B push–pull output stage (Q_1 and Q_2 in Fig. 4.30). The dc biasing is designed so that Q_8 and Q_9 have equal-valued small gate-to-source dc biases. This maximizes the linear v_{out} range. The low-frequency small-signal differential gain can easily be found from Eqs. (4.66) and (4.33):

$$A_v \cong \frac{-g_{mi}}{g_{dl} + g_{di}} \frac{g_{m8} + g_{m9}}{g_{d8} + g_{d9}}. \tag{4.95}$$

Thus $|A_v|$ can be as high as 20,000. However, if the circuit has to drive a resistive load G_L, then $g_{d8} + g_{d9}$ is replaced by $g_{d8} + g_{d9} + G_L$ which normally reduces the gain significantly. Also, for a large load capacitance C_L, the pole of the compensated op-amp in a feedback arrangement resulting from the time constant $C_L/(g_{d8} + g_{d9})$ may move so close to the $j\omega$-axis that instability occurs. Hence, this op-amp is again only suited for driving small-to-moderate-sized internal capacitive loads.

If the circuit of Fig. 4.59 is to be used in the output buffer of the chip, then an output stage must be added. This may be simply a source follower, similar

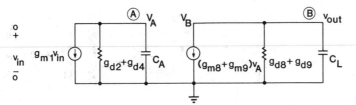

FIGURE 4.60. Two-stage representation of the CMOS operational amplifier of Fig. 4.59.

to the $Q_8 - Q_9$ stage in Fig. 4.57. However, better output current sourcing and lower output impedance can be obtained using a bipolar-MOS emitter-follower output, in which the bipolar driver transistor is formed by the parasitic *npn* transistor realized in most *n*-substrate CMOS processes.[12] More elaborate CMOS output stages, operating in the class-B mode, can also be utilized.[13]

Consider now the high-frequency behavior of the circuit of Fig. 4.59. As before, nodes (A) and (B) are at a high-impedance level, and are responsible for the dominant poles. The approximate equivalent circuit is shown in Fig. 4.60. Here, the input stage is represented by a simple Norton equivalent, obtained using eqs. (4.66) and (4.92). Similarly, the Norton equivalent of the output stage can be found from Eqs. (4.33) and (4.93). As before, the transfer function contains a factor similar to that given in (4.94), where now $s_A = -(g_{d2} + g_{d4})/C_A$ and $s_B = -(g_{d8} + g_{d9})/C_L$. The overall transfer function is therefore

$$A_v(s) = \frac{V_{out}(s)}{V_{in}^+ - V_{in}^-} \approx \frac{g_{m1}}{g_{d2} + g_{d4}} \frac{g_{m8} + g_{m9}}{g_{d8} + g_{d9}} \frac{1}{(1 - s/s_A)(1 - s/s_B)}$$

$$= \frac{A_v(0)}{(1 - s/s_A)(1 - s/s_B)}. \tag{4.96}$$

The frequency response is obtained by replacing s by $j\omega$. For low frequencies ($\omega \ll |s_A|, |s_B|$),

$$A_v(j\omega) \approx A_v(0) = \frac{g_{m1}}{g_{d2} + g_{d4}} \frac{g_{m8} + g_{m9}}{g_{d8} + g_{d9}}. \tag{4.97}$$

For high frequencies ($\omega \gg |s_A|, |s_B|$)

$$A_v(j\omega) \approx \frac{A_v(0)}{-\omega^2/s_A s_B} = -\frac{g_{m1}(g_{m8} + g_{m9})}{\omega^2 C_A C_L}. \tag{4.98}$$

Hence, for high frequencies, the amplifier inverts the input voltage. As will be shown later, in switched-capacitor applications the op-amp always has a

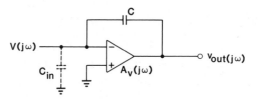

FIGURE 4.61. Operational amplifier with feedback capacitor C and input capacitor C_{in}.

feedback capacitor C connected between its output and its inverting input terminals. A typical circuit is shown in Fig. 4.61. A sine-wave signal $V(j\omega)$ appearing at the inverting input terminal will thus be amplified by $-A_v(j\omega)$, and fed back to the input via the capacitive divider C and C_{in}; here, C_{in} represents the overall capacitance of the input circuit driving the op-amp, including stray capacitances, and so on. For a low-frequency signal $V(j\omega)$, the feedback signal will be 180° out of phase with $V(j\omega)$; it will be

$$V_{fb} \simeq -A_v(0)\frac{C}{C + C_{in}}V(j\omega). \qquad (4.99)$$

Hence, the feedback is indeed negative. However, for high frequencies, $V_{fb}(j\omega)$ will be in phase with $V(j\omega)$, since from (4.98)

$$V_{fb} = -A_v(j\omega)\frac{C}{C + C_{in}}V(j\omega) \cong \frac{A_v(0)s_As_BC}{\omega^2(C + C_{in})}V(j\omega). \qquad (4.100)$$

This corresponds to a positive feedback. Thus, if the loop gain $|A_v(j\omega)C/(C + C_{in})|$ is 1 at any high frequency, the circuit may oscillate. In theory, for our two-pole model, $A_v(j\omega)$ becomes negative real only for $\omega \to \infty$; however, for large loop gain the circuit is only marginally stable for high frequencies so that any additional small phase shift due to the high-frequency poles neglected in Fig. 4.60 may cause oscillation. Even if stability is retained, the transient response contains a lightly damped oscillation which is unacceptable in most applications.

To prevent oscillation in feedback amplifiers, and to ensure a good transient response, an additional design step (called *compensation*) is needed. It is based on the stability theory of feedback systems, and is discussed briefly in the next section.

4.8. STABILITY THEORY AND COMPENSATION OF CMOS AMPLIFIERS

In the previous section, it was shown that the CMOS op-amp of Fig. 4.59 is only marginally stable when used in a feedback circuit. In this section, the analysis of stability, and the design steps required to ensure stable feedback op-amps, will be discussed.

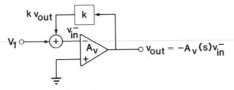

FIGURE 4.62. Operational amplifier with negative feedback.

A systematic investigation of stability can be based on the general block diagram of Fig. 4.62, which shows an op-amp in a negative feedback configuration. The voltage at the inverting input terminal is

$$V_{in}^- = V_1 + kV_{out} \tag{4.101}$$

and the output voltage is

$$V_{out} = -A_v(s)V_{in}^-. \tag{4.102}$$

Hence, the voltage gain is

$$A_{vf}(s) = \frac{V_{out}(s)}{V_1(s)} = \frac{-A_v(s)}{kA_v(s) + 1}. \tag{4.103}$$

A_{vf} is often called the *closed-loop gain*, while A_v the *open-loop gain* of the system; kA_v is the *loop gain*.

We shall next assume that all poles s_i of $A_v(s)$ are due to stray capacitances to ground in an otherwise resistive circuit. (This is an acceptable approximation if the capacitances loading the high-impedance nodes are connected between voltages which are in phase or 180° out of phase of each other.) Then all s_i are negative real numbers, and $A_v(s)$ is in the form

$$A_v(s) = \frac{K}{(s - s_1)(s - s_2) \cdots (s - s_n)}. \tag{4.104}$$

For $s = j\omega$, $A_v(j\omega)$ gives the frequency response of the op-amp. Its magnitude is

$$|A_v(j\omega)| = \frac{|K|}{\prod_{i=1}^{n}(\omega^2 + |s_i|^2)^{1/2}} \tag{4.105}$$

and its phase is given by

$$\angle A_v(j\omega) = \angle K - \sum_{i=1}^{n} \tan^{-1}(\omega/|s_i|). \tag{4.106}$$

Note that both $|A_v(j\omega)|$ and $\angle A_v(j\omega)$ are monotone decreasing functions of ω.

The natural frequencies of the overall feedback system are the poles s_p of $A_{vf}(s)$, which by (4.103) satisfy the relation

$$kA_v(s_p) + 1 = 0. \tag{4.107}$$

For stability, all s_p must be in the negative half of the s-plane; that is, the real parts of all poles must be negative. Now assume that $\text{Re}[kA_v(j\omega)] > -1$ for all real values of ω. Then $kA_v(j\omega) \neq -1$, and hence no s_p can occur on the $j\omega$-axis; furthermore, it can easily be proven that if $A_v(s)$ has only poles with negative real parts, then under the stated assumption so will $A_{vf}(s)$. The proof is implied in Problem 4.23. Thus, the condition

$$\text{Re}[kA_v(j\omega)] > -1 \tag{4.108}$$

is *sufficient* to ensure stability. It is not, however, a *necessary* condition. Two other sufficient conditions for stability can also readily be stated. Let ω_{180} be the frequency at which the monotone decreasing phase of $kA_v(j\omega)$ reaches $-180°$; that is,

$$\angle kA_v(j\omega_{180}) = -180°. \tag{4.109}$$

If now $|kA_v(j\omega_{180})| < 1$, then (4.107) cannot hold on the $j\omega$-axis, and hence the circuit is stable. A measure of its stability is the *gain margin* defined as

$$\text{gain margin (in dB)} = 20\log|kA_v(j\omega_{180})|. \tag{4.110}$$

The gain margin must be negative for stability; the more negative it is, the larger the margin of stability of the circuit. Normally, a margin of at least 20 dB is desirable.*

Next, let $|kA_v(j\omega)|$, which also decreases monotonically with ω, reach the value 1 (i.e., 0 dB) at the *unit-gain frequency* ω_0. Then, if the phase at ω_0 satisfies $\angle kA_v(j\omega_0) > -180°$, the system will be stable. The *phase margin*, defined as $\angle kA_v(j\omega_0) + 180°$, is a measure of its stability; the larger the phase margin, the more stable the circuit. Usually, at least a 60° (and preferably larger) margin is required. This will also give a desirable (i.e., nonringing) step response for the closed-loop amplifier.

All the above stability conditions can readily be visualized and checked using the *Bode plots*. These show $|kA_v(j\omega)|$ (in dB) and $\angle kA_v(j\omega)$ (in degrees) as functions of ω on a logarithmic scale. Typical plots are shown in Fig. 4.63 (broken curves) for the three-pole loop gain

$$kA_v(j\omega) = \frac{A_0}{(1 - j\omega/s_1)(1 - j\omega/s_2)(1 - j\omega/s_3)}, \tag{4.111}$$

---

*The gain margin is harder to control, and hence much less often used, than the phase margin described next.

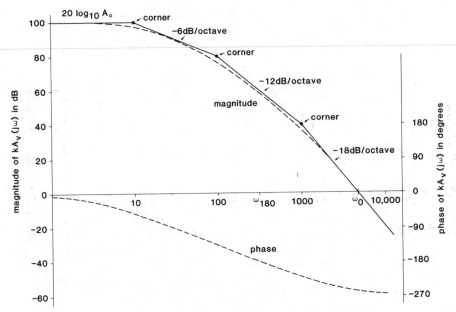

FIGURE 4.63. An example of a Bode plot for three real poles.

with $A_0 = 10^5$, $s_1 = -10$ rad/s, $s_2 = -10^2$ rad/s, and $s_3 = -10^3$ rad/s. Drawing the magnitude plot is simplified by using an asymptotic approximation to the logarithmic magnitude of the general term $a_i(j\omega) = 1/(1 - j\omega/s_i)$:

$$20 \log_{10}|a_i| = -20 \log_{10}|1 - j\omega/s_i|$$

$$= -10 \log_{10}(1 + \omega^2/|s_i|^2). \qquad (4.112)$$

Clearly, $20 \log_{10}|a_i| \approx 0$ for $|\omega| \ll |s_i|$, and $20 \log_{10}|a_i| \approx -20(\log_{10}\omega - \log_{10}|s_i|)$ for $|\omega| \gg |s_i|$. Figure 4.64 illustrates the approximation of $|a_i|$, and also the phase $\measuredangle a_i(j\omega)$ of $a_i(j\omega)$. An important conclusion which can be drawn from the figure is that, for $|\omega| \gg |s_i|$, $20 \log_{10}|a_i|$ approaches a straight line with a slope of -6 dB/octave (i.e., it decreases by 6 dB for each doubling of ω), while $\measuredangle a_i$ approaches $-90°$ in this same region. In particular, $|\measuredangle a_i| \approx 90°$ for $\omega > 5|s_i|$. Also, $|\measuredangle a_i| \leq 30°$ for $\omega < 0.5|s_i|$; this fact will be used later.

The logarithmic form of the loop gain satisfies

$$20 \log_{10}\left[kA_v(j\omega)\right] = 20 \log_{10}A_0$$

$$+ \sum_{i=1}^{n} 20 \log_{10}\left[a_i(j\omega)\right]. \qquad (4.113)$$

Therefore, at the unity-gain frequency ω_0 the slope of the logarithmic loop gain

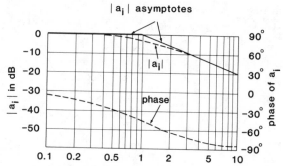

FIGURE 4.64. Gain and phase responses for a factor $a_i(j\omega) = (1 - j\omega/s_i)^{-1}$.

versus logarithmic frequency is approximately $-6m$ dB/octave, while its angle is around $-90m$ degrees. Here, m is the number of those poles whose magnitude $|s_i|$ is less than ω_0. Clearly, for a substantial positive phase margin (say 60°, so that $\angle kA_v(j\omega_0) > -120°$), m should be less than 2. Ideally, m is one (i.e., there is only one pole satisfying $|s_i| < \omega_0$), and the other poles have much larger magnitudes than ω_0. Then the phase margin is close to 90°. (For $A_0 \gg 1$, $m = 0$ is impossible.)

Returning to the example of Fig. 4.63, the broken lines show the asymptotic approximation to the logarithmic magnitude of $kA_v(j\omega)$. The curves indicate that at the unity-gain frequency $\omega_0 \approx 4$ krad/s, the phase of $kA_v(j\omega)$ is about $-270°$. Hence, the phase margin is negative, and the feedback system is potentially unstable.

The modification of $kA_v(s)$ which changes an unstable feedback system into a stable one is called *frequency compensation*. Its purpose is usually to achieve the ideal situation described above; thus, we aim to realize a loop gain which contains exactly one pole smaller in magnitude than ω_0, while all others are much larger. Since the feedback factor k can be anywhere in the $0 < k \le 1$ range, and $k = 1$ represents the worst case (i.e., the largest ω_0 and hence the smallest phase margin), this will be assumed from here on. Note that $k \cong 1$ corresponds to $C \gg C_{in}$ in Fig. 4.61; $k = 1$ represents a short circuit between the output and the inverting input of the amplifier. Such a circuit is shown in Fig. 4.131 in connection with Problem 4.29.

It will next be shown how to carry out the compensation for the op-amp of Fig. 4.59. Referring to its equivalent small-signal representation (Fig. 4.60), we will first attempt to achieve compensation by connecting a compensating capacitor C_c between the high-impedance nodes Ⓐ and Ⓑ, as shown in Fig. 4.65. It is well known (see Ref. 2, Sec. 9.4) that for bipolar op-amps the addition of such capacitor moves the pole associated with node Ⓐ to a much lower frequency, while that corresponding to node Ⓑ becomes much larger. It is therefore often called a *pole-splitting capacitor*, and (for bipolar op-amps) accomplishes the desired compensation. The situation is less favorable for

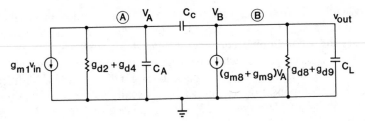

FIGURE 4.65. Two-stage representation of a CMOS operational amplifier with a pole-splitting capacitor C_c.

MOS op-amps, as will be shown next. The node equations for nodes (A) and (B) in Fig. 4.65 are

$$g_{m1}V_{in} + (g_{d2} + g_{d4} + sC_A)V_A + sC_c(V_A - V_{out}) = 0$$

and

$$(4.114)$$

$$sC_c(V_{out} - V_A) + (g_{m8} + g_{m9})V_A + (g_{d8} + g_{d9} + sC_L)V_{out} = 0.$$

Solving for V_{out}, the voltage gain

$$A_V(s) = \frac{A_0(1 - s/s_z)}{(1 - s/s_{p1})(1 - s/s_{p2})}$$

$$(4.115)$$

results. Here, the dc gain is

$$A_0 = \frac{g_{m1}(g_{m8} + g_{m9})}{(g_{d2} + g_{d4})(g_{d8} + g_{d9})}$$

$$(4.116)$$

and the zero is

$$s_z = \frac{g_{m8} + g_{m9}}{C_c}.$$

$$(4.117)$$

The calculation of the poles is simplified if it is a priori assumed that $|s_{p2}| \gg |s_{p1}|$, and that $g_{m8} + g_{m9} \gg g_{d2} + g_{d4}$ or $g_{d8} + g_{d9}$. Then, after some calculation (see Ref. 2, p. 519 and Problem 4.24)

$$s_{p1} \approx \frac{(g_{d2} + g_{d4})(g_{d8} + g_{d9})}{(g_{m8} + g_{m9})C_c} = -\frac{g_{m1}}{A_0 C_c}$$

and

$$(4.118)$$

$$s_{p2} \approx \frac{-(g_{m8} + g_{m9})C_c}{C_A C_L + (C_A + C_L)C_c} = \frac{-(g_{m8} + g_{m9})/C_A C_L}{1/C_c + 1/C_A + 1/C_L},$$

where A_0 is the dc gain given in (4.116).

Physically, C_c (multiplied by the Miller effect) is added in parallel to C_A, thus reducing $|s_{p1}|$ by a very large ($\sim 10^3$) factor, while at the same time it increases the second pole frequency $|s_{p2}|$ via shunt feedback.

Clearly, $|s_{p1}|$ decreases, while $|s_{p2}|$ increases with increasing values of C_c. Thus, C_c indeed splits the poles apart, as originally intended. Unfortunately, the desired compensation is nevertheless usually not achieved, due to the positive (right-half-plane) zero s_z. For the usual case of $1/C_c \ll 1/C_A + 1/C_L$, the inequalities

$$|s_{p2}| \approx \frac{g_{m8} + g_{m9}}{C_A + C_L} \gg s_z > |s_{p1}| \qquad (4.119)$$

hold. The logarithmic magnitude of the factor $(1 - j\omega/s_z)$ is near zero for $|\omega| \ll s_z$, while it *increases* by about 6 dB/octave for $\omega \gg s_z$. The phase of the factor is $-\tan^{-1}(\omega/s_z)$; it *decreases* from 0 to $-90°$ as ω grows from zero to infinity. As a result, the plots shown in Fig. 4.66 are obtained. Clearly, at the unit-gain frequency ω_0 the phase is less than $-180°$. Hence, in a feedback configuration the amplifier can become unstable. Note that if $g_{m8} + g_{m9}$ would be increased, then $s_z/|s_{p1}|A_0$ would, by Eqs. (4.116) to (4.118), increase proportionally. It is clear from Fig. 4.66 that if $s_z/|s_{p1}|$ (in octaves) is greater than A_0 (in dB)/6, then the unit-gain frequency ω_0 is less than s_z, and the phase margin is positive. Thus, for sufficiently high g_m's (such as are afforded by bipolar transistors), the inclusion of C_c accomplishes the desired stabilization. Unfortunately, the transconductance of MOSFETs is normally not high enough for the purpose, and other arrangements must be found to eliminate s_z.

One scheme for getting rid of s_z is to shift it to infinite frequency. Physically, the zero is due to the existence of *two* paths through which the signal can propagate from node Ⓐ to node Ⓑ. The first is through C_c, while

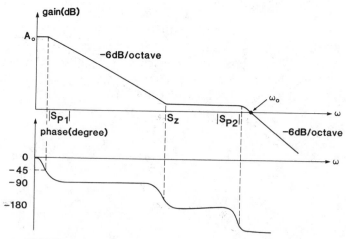

FIGURE 4.66. Amplitude and phase plots of the CMOS op-amp.

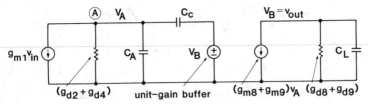

FIGURE 4.67. Unit-gain buffer arrangement used to eliminate the right-plane zero.

the second is by way of the controlled source $(g_{m8} + g_{m9})V_A$. For $s = s_z$, the two signals from these paths cancel, and a transmission zero occurs. The zero can be shifted to infinite frequency by eliminating the feed-forward path through C_c, at the cost of an extra unit-gain buffer (Fig. 4.67). A detailed analysis shows (Problem 4.25) that the numerator of $A_v(s)$ is now simply A_0, while the denominator remains nearly the same as in Eq. (4.115). A circuit implementing this scheme is shown in Fig. 4.68, where Q_{10}/Q_{11} form the buffer.

An alternative (and simpler) scheme[11] can also be used. Consider the circuit shown in Fig. 4.69. Nodal analysis shows (Problem 4.26) that its transfer function is

$$A_v(s) = \frac{A_0(1 - s/s_z)}{(1 - s/s_{p1})(1 - s/s_{p2})(1 - s/s_{p3})}. \qquad (4.120)$$

Here, A_0, s_{p1}, and s_{p2} are (as before) given by Eqs. (4.116) and (4.118), while now

$$s_z = -\frac{1}{[R_c - 1/(g_{m8} + g_{m9})]C_c} \qquad (4.121)$$

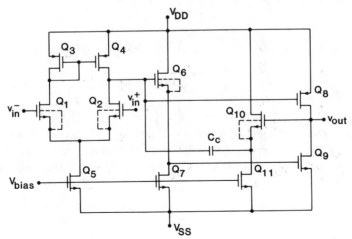

FIGURE 4.68. Internally compensated CMOS op-amp with unit-gain buffer used to avoid the right-half-plane zero.

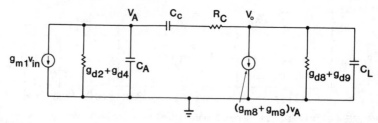

FIGURE 4.69. Small-signal equivalent circuit of CMOS op-amp with nulling resistor for compensation.

and

$$s_{p3} \approx -\frac{1}{R_c}\left(\frac{1}{C_c} + \frac{1}{C_A} + \frac{1}{C_L}\right).$$ (4.122)

As Eq. (4.121) shows, it is again possible for this circuit to shift s_z to infinity, if $R_c = 1/(g_{m8} + g_{m9})$ is chosen. Then, the poles can be split by choosing a sufficiently large value for C_c. To quantify this, it is reasonable to require that $|s_{p2}| > \omega_0$, the unit-gain frequency. For this choice, since in the frequency region between $|s_{p1}|$ and $|s_{p2}|$

$$A_v(j\omega) \cong \frac{A_0}{j\omega/|s_{p1}|}$$ (4.123)

holds, the approximation $\omega_0 \approx A_0|s_{p1}|$ can be used. Thus, $|s_{p2}| > A_0|s_{p1}|$ may be specified. From (4.118), with $1/C_A \gg 1/C_c + 1/C_L$, this requires

$$\frac{g_{m8} + g_{m9}}{C_L} > \frac{g_{m1}}{C_c},$$ (4.124)

so that a feedback capacitor satisfying

$$C_c > \frac{g_{m1}C_L}{g_{m8} + g_{m9}}$$ (4.125)

is needed. Since experience indicates that normally $C_c \sim C_L$ is a good choice, we require $g_{m1} < g_{m8} + g_{m9}$. Another way of eliminating s_z for the circuit of Fig. 4.69 is by pole/zero cancellation. Choosing $s_z = s_{p2}$, from Eqs. (4.118) and (4.121)

$$R_c \cong \frac{1 + (C_A + C_L)/C_c}{g_{m8} + g_{m9}}$$ (4.126)

is obtained. The resulting cancellation leaves the op-amp with a two-pole

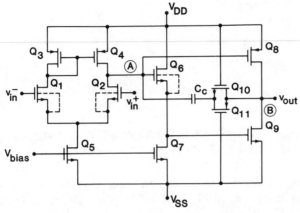

FIGURE 4.70. Improved internally compensated CMOS operational amplifier.

response. Compensation now requires $|s_{p3}| > A_0|s_{p1}|$. Using (4.118), (4.122), and (4.126), this condition can be rewritten in the form

$$C_c > C_c^0 \frac{1/C_c + 1/(C_A + C_L)}{1/C_c + 1/C_L + 1/C_A}. \tag{4.127}$$

Here, C_c^0 is the bound given for C_c on the RHS of Eq. (4.125). The factor multiplying C_c^0 in (4.127) is usually much smaller than 1; hence, now a smaller C_c can be used. Its value can be obtained from the bound*

$$C_c > \frac{1}{2} \frac{g_{m1}/(g_{m8} + g_{m9}) - 1}{1/C_A + 1/C_L}$$

$$+ \left[\frac{1}{4} \left(\frac{g_{m1}/(g_{m8} + g_{m9}) - 1}{1/C_A + 1/C_L} \right)^2 + \frac{g_{m1} C_A C_L}{g_{m8} + g_{m9}} \right]^{1/2}. \tag{4.128}$$

The actual implementation of the scheme of Fig. 4.69 in the CMOS op-amp of Fig. 4.59 is shown in Fig. 4.70. The parallel-connected channels of the complementary transistors Q_{10} and Q_{11} form R_c. This push–pull arrangement helps to suppress even harmonics and thus improves the linearity of the resistor R_c.

Note that the condition $R_c = 1/(g_{m8} + g_{m9})$ is easily obtained by matching Q_{10} and Q_{11} to Q_8 and Q_9. Satisfying Eq. (4.126) is somewhat harder; however, the accuracy is not critical and, as explained above, this choice for R_c results in a smaller value for C_c.

---

*In practice, it is usual to choose $C_c \simeq C_L$. This choice satisfies the constraints of Eqs. (4.127) and (4.128) with a large margin.

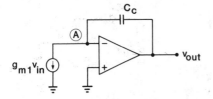

FIGURE 4.71.   Small-signal model of the CMOS op-amp used to calculate its frequency response.

4.9. FREQUENCY RESPONSE, TRANSIENT RESPONSE, AND SLEW RATE OF THE COMPENSATED CMOS OP-AMP[2, 14]

Next, an approximating frequency- and time-domain analysis of the compensated CMOS op-amp of Fig. 4.68 will be given. (It is also applicable for the circuit of Fig. 4.70 if $R_c \approx 0$ is used.) For small input signals v_{in} the transistors will operate in their saturation regions, and their small-signal models can be used. Then, for moderate frequencies (i.e., for $|s_{p1}| \ll \omega \ll |s_{p2}|$) the input stage $Q_1 - Q_5$ can be replaced by a frequency-independent voltage-controlled current source, while the subsequent stages $Q_6 - Q_{11}$ can be replaced by a frequency-independent amplifier with the feedback capacitor C_c connected between its input and output terminals (Fig. 4.71). The model is valid as long as the signal frequencies are much larger than $|s_{p1}|$, but are negligibly small compared to the magnitude of the high-frequency pole s_{p2}. From Fig. 4.71, $V_{out}(s) = g_{mi}V_{in}(s)/sC_c$, so that the high-frequency gain is given by $A_v(j\omega) = V_{out}(j\omega)/V_{in}(j\omega) = g_{mi}/j\omega C_c$. *The unity-gain frequency is thus* $\omega_0 = g_{mi}/C_c$. For $|s_{p2}| \gg \omega_0$, the phase of A_v at ω_0 will thus be close to 90°. This can be obtained by choosing C_c sufficiently large.

Consider next the voltage inverter shown in Fig. 4.72a. Assume again that the amplifier is compensated, so that its voltage gain can be approximated by

$$A_v(s) = \frac{A_0}{1 - s/s_{p1}}. \tag{4.129}$$

Hence, for an input step $v_{in}(t) = V_1 u(t)$, the output voltage is in the form

$$v_{out}(t) = -V_1 u(t)\frac{A_0}{A_0 + 2}[1 - e^{-(A_0/2 + 1)|s_{p1}|t}] \tag{4.130}$$

(Problem 4.27). Thus, for a square input voltage (Fig. 4.72b) the exponentially varying waveform of Fig. 4.72c should occur at the output.* If the amplitude V_1 is small (say, much less than a volt), this is in fact what happens. If, however, the input voltage is large (e.g., $V_1 = 5$ V), then the experimentally observable output voltage is of the form shown in Fig. 4.72d. The nearly linear (rather than exponential) rise and fall of $v_{out}(t)$ is called *slewing*, and the nearly constant slope dv_{out}/dt of the curve is called the *slew rate*. Slewing is a

---

*The time constant is $t_0 \simeq 2/(A_0|s_{p1}|) \simeq 2/\omega_0$.

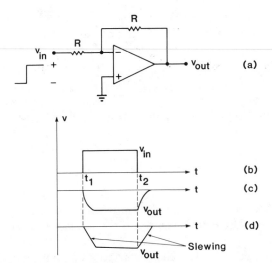

FIGURE 4.72. Slewing response of the CMOS op-amp connected in the inverting mode: (a) circuit; (b) input signal; (c) small-signal output waveform; (d) large-signal output waveform.

nonlinear (large-signal) phenomenon, and hence it must be analyzed in terms of the large-signal model of the op-amp shown in Fig. 4.73. Prior to the arrival of the input step, $v_{in} = 0$ and the currents in Q_1 and Q_2 are both equal to $I_0/2$. After the large step occurs at the input, Q_1 conducts more current and cuts off Q_2. Hence, the current conducted by Q_1 and Q_3 is now I_0 (Fig. 4.73). Since Q_3 and Q_4 form a current mirror, the current in Q_4 (which charges C_c) is also I_0. Hence, assuming the output stage A_2 can sink the current I_0, the slew rate is

$$S_r = \left| \frac{dV_{out}}{dt} \right| = \left| -\frac{1}{C_c} \frac{dQ_c}{dt} \right| = \frac{I_0}{C_c}, \qquad (4.131)$$

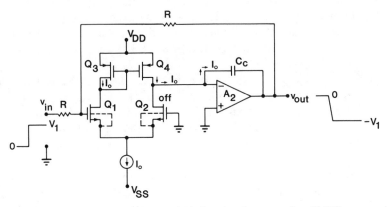

FIGURE 4.73. Large-signal model for calculating the slew rate of a CMOS op-amp in the inverting mode.

where Q_c is the charge in C_c. Here, $C_c = g_{mi}/\omega_0$ where [from Eq. (3.18) of Chapter 3] the transconductance of the input stage is

$$g_{mi} = 2\sqrt{\frac{I_0}{2}k'\frac{W}{L}} \qquad (4.132)$$

and ω_0 is the unity-gain frequency of the op-amp. Combining these relations, we obtain

$$S_r = \frac{I_0\omega_0}{g_{mi}} = \omega_0\sqrt{\frac{I_0}{2k'W/L}}. \qquad (4.133)$$

Thus, the slew rate can be increased by *increasing* the unity-gain bandwidth and the bias current of the input stage, and by *decreasing* the W/L ratio of the input transistors.

It should be noted that the transconductance of MOSFETs is much lower than that of bipolar devices. This is ordinarily a major disadvantage; however, it results in significantly higher slew rates for MOS op-amps than for bipolar ones for a given unity-gain bandwidth since C_c can be smaller.

The negative slewing of v_{out} continues until it reaches $-V_1$. At that time, the gate voltage of Q_1 [which, due to the two resistors R, equals $(v_{in} + v_{out})/2$] reaches zero voltage. Hence, at that time the quiescent bias conditions are restored, and $Q_1 - Q_4$ all carry a current $I_0/2$. Therefore, the charging of C_c and the decrease of the output voltage cease.

The complementary process takes place when v_{in} drops back to zero at $t = t_2$. Now Q_1 cuts off, since $v_{out} = -V_1$ still holds and hence its gate voltage drops to $-V_1/2$. Thus, Q_3 and Q_4 cut off, and C_c is discharged through Q_2 with a current I_0, provided A_2 can source at least the same current. The slew rate of v_{out} is hence again I_0/C_c. The process stops when v_{out} (and hence the gate voltage of Q_1) reaches zero voltage.

In Fig. 4.73, the op-amp operates in the inverting mode. Figure 4.74*a* illustrates the use of the op-amp as a unity-gain voltage follower. Figure 4.74*b* shows an input pulse waveform; Fig. 4.74*c* shows the corresponding output response under large-signal conditions. As the diagram shows, the rising edge contains a positive step followed by a fast slewing rise, while the falling edge is a relatively slow linear slope.

The behavior of the rising edge can be understood by considering the equivalent circuit shown in Fig. 4.75. In the circuit, the stray capacitance C_w across the input-stage current source I_0 is included. Note that C_w is quite large in CMOS op-amps where the common sources of the input devices Q_1 and Q_2 are connected to the p-well, since this creates a large capacitance between the source and the substrate.

A large input signal $v_{in}(t) = V_1u(t)$ turns Q_2 fully on. Therefore its source voltage v_w rises and hence Q_1 and Q_3 are turned off. Thus, Q_2 carries the full current $I_0 + i_w$, where $i_w(t)$ is the current through C_w. Since normally the combined impedance of C_w and the current source I_0 is much larger than the

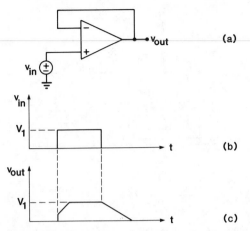

FIGURE 4.74. Slewing in a voltage follower: (a) op-amp used as a voltage follower; (b) large input signal; (c) output response.

driving impedance $(1/g_{m2})$ of Q_2, the incremental source voltage is $v_w(t) \simeq v_{\text{in}}(t)$. Hence,

$$i_w(t) = C_w \frac{dv_w(t)}{dt} \simeq C_w \frac{dv_{\text{in}}(t)}{dt}, \qquad (4.134)$$

which is the impulse function $V_1 C_w \delta(t)$. The output voltage satisfies

$$v_{\text{out}}(t) = \frac{1}{C_c} \int_0^t (I_0 + i_w)\, dt$$

$$= \frac{I_0}{C_c} t + \frac{C_w}{C_c} \int_0^t \frac{dv_{\text{in}}}{dt}\, dt = \frac{I_0}{C_c} t + \frac{C_w}{C_c} V_1 u(t). \qquad (4.135)$$

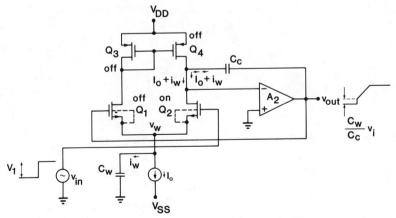

FIGURE 4.75. Equivalent circuit of the voltage follower used to calculate the large-signal behavior for positive inputs.

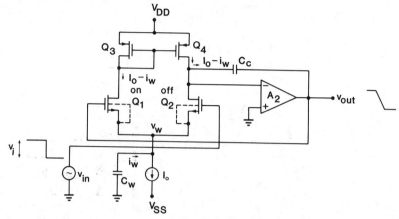

FIGURE 4.76. Equivalent circuit of the voltage follower used to calculate the large-signal behavior for negative inputs.

The first term represents the linear rise, with a slew rate I_0/C_c, while the second represents the small pedestal seen at the beginning of the rising edge.

For a negative step, the equivalent circuit of Fig. 4.76 applies. Now the input signal turns Q_2 off, and Q_1, Q_3, and Q_4 all carry the current $I_0 - i_w$. Considering next the two capacitors C_c and C_w, we note that C_c is connected between $v_{out}(t)$ and (virtual) ground, while C_w is connected between v_w and (true) ground. Now $v_w(t)$ follows the gate voltage $v_{out}(t)$ of Q_1, and hence $v_w \simeq v_{out}$ so that

$$\frac{dv_{out}}{dt} = -\frac{I_0 - i_w}{C_c} = -\frac{i_w}{C_w}. \tag{4.136}$$

Therefore, $i_w = I_0 C_w/(C_c + C_w)$ and

$$\frac{dv_{out}}{dt} = -\frac{I_0}{C_c + C_w}. \tag{4.137}$$

Thus, the negative slew rate is reduced by the presense of C_w, from I_0/C_c to $I_0/(C_c + C_w)$; that is, by a factor $1 + C_w/C_c$.

4.10. NOISE PERFORMANCE OF CMOS AMPLIFIERS

Noise represents a fundamental limitation of the performance of MOS op-amps: the equivalent noise voltage may be 10 times higher than for a comparable bipolar amplifier. Hence, it is important to analyze the causes of noise, and the possible measures that can reduce it.

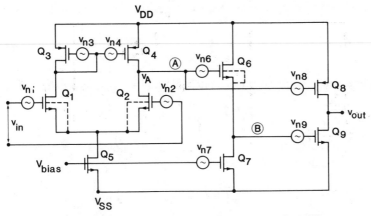

FIGURE 4.77. Noise sources in a CMOS operational amplifier.

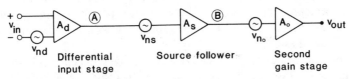

FIGURE 4.78. Block diagram of a three-stage CMOS operational amplifier with noise sources.

Figure 4.77 shows the uncompensated CMOS op-amp of Fig. 4.57, with the noise generated by each device Q_i represented symbolically by an equivalent voltage source v_{ni} connected to its gate.* (The calculation of the gate-referred noise voltages v_{ni} was briefly described in Section 3.7.) We can next combine the noise sources $v_{n1}-v_{n4}$ in the differential input stage into a single equivalent source v_{nd} connected to the input of an otherwise noiseless input stage, as shown in Fig. 4.78. (Note that the noise of Q_5 is a common-mode signal, and is hence suppressed by the CMRR of the op-amp; it is therefore omitted in Fig. 4.77.) The voltage gain from the noise sources v_{n1} and v_{n2} to the output node Ⓐ of the input stage can be calculated using its low-frequency equivalent circuit. This gives

$$A_d = \frac{v_A}{v_{n1}} = \frac{v_A}{v_{n2}} = \frac{g_{m1}}{g_{d2} + g_{d4}}. \qquad (4.138)$$

This is the same as the differential signal gain of the stage. Similarly, the gain between sources v_{n3} and v_{n4} and node Ⓐ can be calculated. Physically, the noise source v_{n3} introduces a noise current $g_{m3}v_{n3}$ into Q_3, which is mirrored in Q_4. Hence, v_{n3} causes currents of Q_3 and Q_4 to change by $g_{m3}v_{n3}$, and thus

---

*Such a source indicates that a noise current $g_{mi}v_{ni}$ flows in Q_i.

v_A by $g_{m3}v_{n3}/(g_{d2} + g_{d4})$. The effect of v_{n4} is similar. The gain is therefore

$$A_v = \frac{v_A}{v_{n3}} = \frac{v_A}{v_{n4}} = \frac{g_{m3}}{g_{d2} + g_{d4}}. \tag{4.139}$$

Since these sources are all uncorrelated, they result in a mean-square voltage

$$\overline{v_A^2} = A_d^2\left(\overline{v_{n1}^2} + \overline{v_{n2}^2}\right) + A_v^2\left(\overline{v_{n3}^2} + \overline{v_{n4}^2}\right) \tag{4.140}$$

at node Ⓐ. Hence, the equivalent input noise voltage $v_{nd} = v_A/A_d$ has the mean-square value

$$\overline{v_{nd}^2} = \overline{v_{n1}^2} + \overline{v_{n2}^2} + \left(g_{m4}/g_{m1}\right)^2\left(\overline{v_{n3}^2} + \overline{v_{n4}^2}\right). \tag{4.141}$$

Hence, to minimize $\overline{v_{nd}^2}$, clearly v_{n1} and v_{n2} should be small and $g_{m4} \ll g_{m1}$. The former, by the discussions of Section 3.7, requires that the area ($W \cdot L$) and transconductance (g_m) of Q_1 and Q_2 be large. To obtain large g_m, the bias current and W/L ratio should be large—this, however, requires large devices and high power dissipation.

The noise contribution of the load devices can be reduced, as (4.141) shows, by making their transconductances as *small* as their biasing conditions permit. This can be achieved by increasing their lengths L. Thus, assuming that the areas of the input and load devices are given, the W/L ratios of the input devices Q_1 and Q_2 should be as *large*, while those of the load devices Q_3 and Q_4 as *small* as other considerations permit.* Also, it has been found experimentally[15] that the RMS equivalent $1/f$ noise voltage v_n is about three times larger for an n-channel device than for a p-channel one. Since in (4.141) $(g_{m4}/g_{m1})^2 \ll 1$, it is hence advantageous to use p-channel input devices with n-channel loads, rather than the other way around, as shown in Fig. 4.77. Applying all these principles, the equivalent input noise voltage v_{nd} can be reduced appreciably.[15]

Similarly, the noise sources of the source follower (Q_6, Q_7) can be replaced by an equivalent source v_{ns} (Fig. 4.78). From the low-frequency small-signal equivalent circuit,

$$\overline{v_{ns}^2} = \overline{v_{n6}^2} + \left(g_{m7}/g_{m6}\right)^2 \overline{v_{n7}^2}. \tag{4.142}$$

Referring $\overline{v_{ns}^2}$ back to the input of the op-amp, the total equivalent input noise voltage becomes

$$\overline{v_n^2} \cong \overline{v_{nd}^2} + \overline{v_{ns}^2}/A_d^2 = \overline{v_{n1}^2} + \overline{v_{n2}^2} + \left(g_{m4}/g_{m1}\right)^2\left(\overline{v_{n3}^2} + \overline{v_{n4}^2}\right)$$

$$+ \left[\overline{v_{n6}^2} + \left(g_{m7}/g_{m6}\right)^2 \overline{v_{n7}^2}\right]/A_d^2. \tag{4.143}$$

---

*Thus, as (4.89) shows, the CMRR decreases if g_{ml} is reduced.

For low frequencies where $A_d^2 \gg 1$, the effect of v_{ns} is negligible; however, at higher frequencies this will no longer be true. Since Q_6 and Q_7 are used as a level shifter, the gate/source voltage drop of Q_6 must be large. By Eq. (3.9), this will be achieved for a given i_{D6} if $k_6 = k'(W/L)_6$ is small. Hence, Q_6 is a long thin device, and $(g_{m7}/g_{m6})^2 \gg 1$. At frequencies where $|A_d(\omega)| \approx g_{m7}/g_{m6}$, the effect of v_{n7} is comparable to that of v_{n1} and v_{n2}. Hence, care must be taken in the design of Q_7 to make it a low-noise device.

The effect of the noise sources v_{n8} and v_{n9} can be analyzed similarly and can be represented by an equivalent source v_{no}. However, they usually do not affect the total equivalent input noise voltage significantly.

Normally, all v_{ni} contain a $1/f$ noise component which dominates it at low frequencies. Hence, the equivalent input noise voltage is greatest at low frequencies (below 1 kHz) where $|A_d(\omega)| \gg 1$. Thus, the input devices Q_1 and Q_2 tend to be the dominant noise sources, and their optimization is the key to low-noise design.

Using chopper-stabilized differential configuration,[16] the low-frequency $1/f$ noise of the op-amp can be canceled, and a large (over 100 dB) dynamic range obtained for an integrated MOS low-pass filter.

For wide-band operational amplifiers and a low clock frequency, aliasing can increase the effect of the high-frequency noise to the point where it overwhelms the $1/f$ noise. (This problem will be discussed in Chapter 7.) Hence, the unity-gain frequency ω_0 should be kept as low as is permitted by the application at hand.

4.11. AN ENHANCEMENT-MODE NMOS OP-AMP[10]

The CMOS amplifiers discussed in earlier sections provide an excellent op-amp realization for high-performance low-power filter applications. There may be, however, situations when single-channel (usually NMOS) op-amps should be used. An obvious one is when only NMOS technology is available. However, even if both technologies are present, sometimes NMOS offers advantages. If, for example, analog and digital functions must be combined on the same chip, then an NMOS realization (which can provide very densely packed logic circuits) may result in a much smaller overall area on the chip.

It was shown in Section 4.3 that an NMOS circuit using depletion loads can provide a much higher gain for comparable dimensions than an all-enhancement-mode stage. Nevertheless, all-enhancement NMOS op-amps can be designed[10] which provide an open-loop gain of 60 dB and a performance comparable to that of a depletion-load amplifier. Such an op-amp can be used in some communication circuits, where only moderate gain is required and where the op-amp has to drive only capacitive loads. An important advantage of the all-enhancement amplifier over the enhancement–depletion one is that it

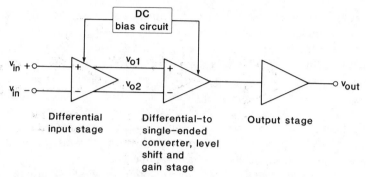

FIGURE 4.79. Block diagram of an all-enhancement NMOS op-amp.

is less process dependent. This is because the threshold voltage of the enhancement-mode transistors is generally better defined (and hence less sensitive to process variations) than that of the depletion-mode MOSFETs.

The block diagram of an all-enhancement op-amp is shown in Fig. 4.79. The differential input stage can be simply a source-coupled pair (Fig. 4.39, reproduced for convenience in Fig. 4.80). In the circuit of Fig. 4.80, Q_4 is biased by a bias string, and acts as a current source.

The linear range of the input stage for common-mode voltages can readily be calculated. Let $v_{in+} = v_{in-} = V_c$. As V_c is increased, the currents (ideally) remain unchanged, with $i_1 = i_2 = i/2$. Hence, so do $v_{01} = v_{02}$. However, the source potential v of Q_5 and Q_6 follows V_c, with a voltage drop $V_{GS5} = V_{GS6}$ necessary to support the current $i/2$ in Q_5 and Q_6. Thus, V_{DS5} and V_{DS6} are reduced. The *maximum* permissible positive common-mode voltage $V_{C\,max}$ is the largest value for which the input transistors Q_5 and Q_6 are still in

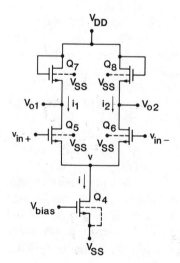

FIGURE 4.80. Source-coupled pair with enhancement loads, used as a differential input stage.

saturation. The drain-to-source voltage of Q_5 is then

$$V_{DS5} = V_{DD} - V_{GS7} - (V_{C\,\text{max}} - V_{GS5}). \qquad (4.144)$$

Here, V_{GS5} and V_{GS7} are the gate-to-source voltages of Q_5 and Q_7, respectively, and V_{DS5} the drain-to-source voltage of Q_5. For Q_5 to remain in saturation, we must have

$$V_{DS5} \geq V_{GS5} - V_{T5} \qquad (4.145)$$

where V_{T5} is the threshold voltage of Q_5. From (4.144) and (4.145)

$$V_{C\,\text{max}} = V_{DD} + V_{T5} - V_{GS7}. \qquad (4.146)$$

Here, V_{GS7} can be calculated from the drain current i_1 (which is half the bias current i) and source-to-bulk potential $(V_{01} - V_{SS})$ of Q_7.

The *minimum* permissible common-mode voltage $V_{C\,\text{min}}$ is the lowest value for which Q_4 still remains in saturation. Thus, it satisfies

$$V_{DS4} = V_{C\,\text{min}} - V_{GS5} - V_{SS} \geq V_{GS4} - V_{T4}. \qquad (4.147)$$

From (4.147),

$$V_{C\,\text{min}} = V_{GS4} + V_{GS5} - V_{T4} + V_{SS}. \qquad (4.148)$$

The common-mode input range of the input stage is hence

$$V_{C\,\text{max}} - V_{C\,\text{min}} = V_{DD} - V_{SS} - V_{GS4} - V_{GS5} - V_{GS7}$$

$$+ V_{T4} + V_{T5}. \qquad (4.149)$$

This range can thus be increased by decreasing the dc gate-to-source voltage drops of all (input, load, and current–source) transistors. This can be achieved by reducing the bias current, or by increasing the W/L ratios of all devices. The former measure will, however, reduce the slew rate, while the latter will increase the size of the circuit.

In most switched-capacitor filter applications, the op-amp has its noninverting input terminal grounded. Then a wide common-mode range is not necessary.

The differential gain of the source-coupled pair was derived in Section 4.5 and given by Eq. (4.47). Assuming that g, g_{dl}, and g_{di} are much smaller than g_{mi}, g_{ml}, and $|g_{mbl}|$,

$$v_{01} = -v_{02} \cong \frac{-g_{mi}/2}{g_{ml} + |g_{mbl}|}(v_{\text{in}+} - v_{\text{in}-}) \qquad (4.150)$$

results. Here, the subscript i refers to the input devices Q_5 and Q_6, while l

refers to the load devices Q_7 and Q_8. Hence, the differential-mode gain is given by

$$A_{dm} = \frac{v_{01} - v_{02}}{v_{in+} - v_{in-}} = \frac{-g_{mi}/g_{ml}}{1 + |g_{mbl}|/g_{ml}}. \tag{4.151}$$

Using Eqs. (3.18) and (3.19) with $\lambda \simeq 0$,

$$A_{dm} \simeq -\frac{\sqrt{(W/L)_i/(W/L)_l}}{1 + (\gamma/2)/\sqrt{2|\phi_p| + v_{BSl}^0}} \tag{4.152}$$

is obtained. Typically, $|A_{dm}| = 10 \sim 15$ can be achieved with this circuit.

The common-mode gain of the source-coupled pair was given in Eqs. (4.44) and (4.53); for $g_{di} \approx 0$, these relations give

$$A_{cm} \simeq \frac{-g_{d4}/2}{g_{ml} + |g_{mbl}|}. \tag{4.153}$$

From (4.151) and (4.153), the common-mode rejection ratio can be found:

$$\text{CMRR} \simeq \frac{2g_{mi}}{g_{d4}}. \tag{4.154}$$

This can be as large as $50 \sim 60$ dB; it can be enhanced by common-mode feedback, as will be shown later.

A possible realization[10] of the second stage (level shifter, etc.) is shown in Fig. 4.81. This circuit is the same as that given in Fig. 4.43 (with an added feedback capacitor C_c which only affects the high-frequency response, and will be discussed later). The low-frequency response is hence given by (4.57). For

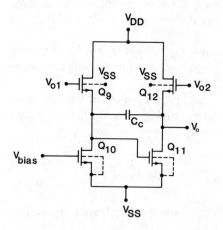

FIGURE 4.81. Level-shifter-and-gain stage.

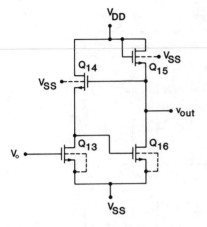

FIGURE 4.82. Output stage.

the usual case where the transconductances are much larger than the drain conductances, the approximating relation

$$v_0 \simeq v_{02} - \frac{g_{m11}}{g_{m12}} v_{01} \tag{4.155}$$

results. Combining Eqs. (4.150) and (4.155), the low-frequency voltage gain of the first two stages is obtained:

$$A_V^{1,2} = \frac{v_0}{v_{in+} - v_{in-}} \simeq \frac{g_{m5}/2}{g_{m7} + |g_{mb7}|} \left(1 + \frac{g_{m11}}{g_{m12}}\right). \tag{4.156}$$

For practical transconductance values, $A_V^{1,2} \approx 300$ can be achieved.

A possible output circuit for an all-enhancement NMOS op-amp was shown in Fig. 4.37; it is reproduced in Fig. 4.82. It uses negative feedback to reduce the output impedance. For this buffer circuit, the low-frequency voltage gain was given in (4.42); changing subscripts appropriately, for the circuit of Fig. 4.82

$$A_V^3 = \frac{v_{out}}{v_0} = \frac{g_{m13}/g_{m14}}{1 + g_{m15}/g_{m16}} \tag{4.157}$$

results. $A_V^3 = 7 \sim 10$ is feasible.

The input capacitance C_{in} of the output circuit loads the output node of the level-shifter stage. The output impedance of that stage is readily found (from Fig. 4.43b) to be

$$R_0 \simeq \frac{1}{g_{m12}}. \tag{4.158}$$

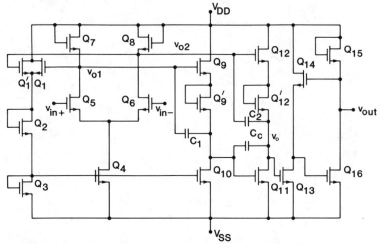

FIGURE 4.83. The complete circuit for an NMOS all-enhancement operational amplifier. The bodies of all transistors are connected to V_{SS}.

Hence, the gate area of Q_{13} must be chosen sufficiently small, so that the pole $1/R_0 C_{in}$ due to C_{in} is much larger than the unity-gain frequency ω_0.

The total low-frequency small-signal voltage gain of the op-amp can now be obtained from Eqs. (4.156) and (4.157):

$$A_V = \frac{v_{out}}{v_{in+} - v_{in-}} = A_V^{1,2} A_V^3 \simeq \frac{g_{m13}}{g_{m14}} \frac{g_{m5}/2}{g_{m7} + |g_{mb7}|} \frac{1 + g_{m11}/g_{m12}}{1 + g_{m15}/g_{m16}}. \quad (4.159)$$

$A_V > 2000$ is feasible.

The complete circuit of a successfully fabricated all-enhancement NMOS op-amp[10] based on the described principles is shown in Fig. 4.83. In the circuit, Q_1, Q_1', Q_2 and Q_3 form the bias chain. Q_1 and Q_1' provide common-mode feedback for enhanced CMRR, as described in Section 4.5 (cf. Fig. 4.41). The level shifter differs from that shown in Fig. 4.81, in that split loads (Q_9 and Q_9', Q_{12} and Q_{12}') rather than single-transistor loads are used. As described earlier in Section 4.6, in connection with Fig. 4.47, this increases the bandwidth while making high gain possible.

The added capacitors C_c, C_1, and C_2 are used to obtain compensation and to improve the frequency response of the op-amp. Consider the high-frequency input–output transfer characteristics of the level-shifter-and-gain stage (Fig. 4.81). It can be written in the Laplace-transform domain as

$$V_0(s) = A_1(s)V_{01}(s) + A_2(s)V_{02}(s), \quad (4.160)$$

where V_0, V_{01}, and V_{02} denote the Laplace transforms of v_0, v_{01}, and v_{02}, respectively. The path from v_{02} to v_0 is simply through the source follower $Q_{11} - Q_{12}$ (Fig. 4.81); hence, $A_2(s)$ is a transfer function in the form

$$A_2(s) = \frac{V_0(s)}{V_{02}(s)}\bigg|_{V_{01}(s)=0} = A_2(0)\frac{1 + s/s_{z1}}{1 + s/s_{p1}} \tag{4.161}$$

as can be seen from Eqs. (4.79)–(4.81) of Section 4.6. Here, s_{z1} and s_{p1} are given by (4.81), and are determined by g_{m12}, C_{gs12} and by the total load capacitance connected to the source of Q_{12}.

By contrast, the path from v_{01} to v_0 is through the source follower $Q_9 - Q_{10}$ *and* the gain stage $Q_{11} - Q_{12}$ (Fig. 4.81). Hence,

$$A_1(s) = \frac{V_0(s)}{V_{01}(s)}\bigg|_{V_{02}(s)=0} = A_1(0)\frac{1 + s/s_{z2}}{1 + s/s_{p2}}\frac{1 - s/s_{z3}}{1 + s/s_{p3}} \tag{4.162}$$

as can be seen from Eqs. (4.79)–(4.81) and (4.69)–(4.71) of Section 4.6.* Here, s_{z2} and s_{p2} are due to the source follower $Q_9 - Q_{10}$, and hence their values are determined by g_{m9}, C_{gs9} and by the total load capacitance C_L at the source of Q_9. If the capacitor C_c is connected as shown in Fig. 4.81, then its value (as seen from the source of Q_9) becomes multiplied by the gain of the $Q_{11} - Q_{12}$ stage due to Miller effect; thus, the load capacitance is

$$C_L = C_c(1 + g_{m11}/g_{m12}) + C_{in}, \tag{4.163}$$

where C_{in} is the total junction and gate capacitance connected to the source of Q_9. The output impedance of the $Q_9 - Q_{10}$ source follower is approximately $1/g_{m9}$; hence, the pole frequency associated with C_L is

$$s_{p2} \simeq \frac{g_{m9}}{C_{in} + C_c(1 + g_{m11}/g_{m12})}. \tag{4.164}$$

Since $C_c(1 + g_{m11}/g_{m12})$ can be made large, s_{p2} is chosen as the dominant pole of the op-amp. The dc gain $A_1(0)$ is negative.

The second pole/zero pair (s_{p3} and s_{z3}) corresponds to the gain stage $Q_{11} - Q_{12}$: s_{p3} is the product of the output resistance of the stage multiplied by the capacitance loading its output terminal, while the right-half-plane zero is given by $s_{z3} = g_{m11}/C_c$. Since C_c is large and g_{m11} is relatively small, s_{z3} is at a low frequency.

To find the overall transfer function of the cascaded input and level-shifter stages, we note from (4.148) that $V_{01} = -V_{02} = A_i(s)(V_{in+} - V_{in-})$. Here, $A_i(s)$ is the transfer function of the input stage which (for this symmetric

---

*Note that Eq. (4.69) contains *two* poles for the transfer function of the gain stage. However, the pole $-G_s/C_{in}$ is already included in $A_1(s)$ as s_{p2}.

circuit) is the same for $V_{01}(s)$ and $-V_{02}(s)$. Hence, using (4.160),

$$V_0 = [A_1(s) - A_2(s)]V_{01}(s)$$

$$= A_i(s)[A_1(s) - A_2(s)](V_{\text{in}+} - V_{\text{in}-}). \qquad (4.165)$$

Therefore, the transfer function is by (4.161) and (4.162)

$$A(s) = \frac{V_0(s)}{V_{\text{in}-}(s) - V_{\text{in}+}(s)} = A_i(s)[A_1(s) - A_2(s)]$$

$$= A_i(s)\left(A_1(0)\frac{1 + s/s_{z2}}{1 + s/s_{p2}}\frac{1 - s/s_{z3}}{1 + s/s_{p3}} - A_2(0)\frac{1 + s/s_{z1}}{1 + s/s_{p1}} \right)$$

$$= A_i(s)[A_1(0) - A_2(0)]\frac{(1 + s/s'_{z1})(1 + s/s'_{z2})(1 + s/s'_{z3})}{(1 + s/s_{p1})(1 + s/s_{p2})(1 + s/s_{p3})}. \qquad (4.166)$$

As explained earlier, the dominant pole is s_{p2}. The other two poles can be canceled by two of the new zeros, say s'_{z1} and s'_{z2}. To achieve this, the bypass capacitors C_1 and C_2 (Fig. 4.83) can be used. Their capacitances add to C_{gs9} and C_{gs12}, respectively, thus lowering s_{z1}, s_{z2}, s_{p1} and s_{p2}, and hence also lowering s'_{z1} and s'_{z2}. These can then cancel s_{p1} and s_{p3}. In the design of this circuit, care must be taken to avoid any mismatched pole zero pair below the unity-gain frequency ω_0. Such pair leads to a slow settling time for the op-amp.[17]

An op-amp based on the circuit of Fig. 4.83 has been fabricated using NMOS metal gate process.[10] It achieved a dc open-loop gain of 60 dB, and a 3 MHz unit-gain bandwidth. This is adequate for many noncritical applications.

4.12.   NMOS OP-AMPS WITH DEPLETION LOADS

If the gain achievable with all-enhancement-mode NMOS amplifier stages is insufficient, depletion-mode loads can be used. As the analysis of Section 4.3 showed, these can provide inherently more gain than enhancement-mode loads. An NMOS differential stage with depletion loads was shown in Fig. 4.42. Its differential-mode gain can be found from Eqs. (4.53) and (4.55):

$$A_{dm} \simeq \frac{-g_{mi}}{g_{dl} + |g_{mbl}|} \simeq -\frac{g_{mi}}{|g_{mbl}|}. \qquad (4.167)$$

Here, g_{mi} is the transconductance of the input devices Q_1 and Q_2, g_{dl} is the (small) drain conductance of the load devices Q_3 and Q_4, and g_{mbl} is their body-effect transconductance.

Using Eqs. (3.18) and (3.19) in (4.167), we get

$$A_{dm} \simeq -\sqrt{\frac{(W/L)_i}{(W/L)_l}} \; \frac{2\sqrt{2|\phi_p| + v_{SBl}^0}}{\gamma}. \tag{4.168}$$

Here, $v_{SBl}^0 > 0$.

A comparison with (4.152) shows that for equal dimensions and equal constants γ, ϕ_p, and v_{SBl}^0 the gain is now higher by a factor, which for usual values is around 10; including the effects of g_{di}, gains of the order of $50 \sim 100$ are thus achievable with a depletion-load input stage. As (4.168) shows, $|A_{dm}|$ can be increased by increasing $(W/L)_i$, reducing $(W/L)_l$, increasing v_{SBl}^0, and reducing γ. The latter can be achieved by reducing the substrate doping density N_{imp}, as (3.13) shows.

As mentioned earlier, the price paid for the higher gain of the depletion-load input stage is its sensitivity to the poorly controlled threshold voltage of the depletion-mode devices. This is illustrated[18] in Fig. 4.84. Figure 4.84a shows

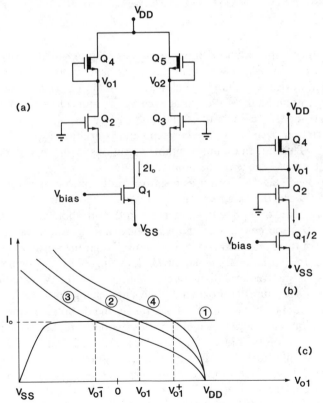

FIGURE 4.84. (*a*) Depletion-load input stage with grounded input terminals. (*b*) Half-circuit of Fig. 4.84a. (*c*) Voltage–current curve for the input devices (curve ①), and load lines (curves ② – ④) for the load devices with different threshold voltages.

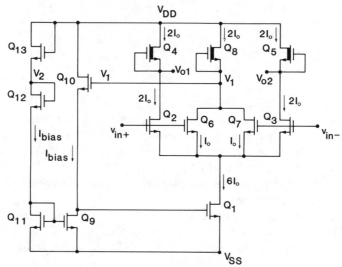

FIGURE 4.85. Improved source-coupled depletion-load NMOS input stage.

the input stage with grounded input terminals; Fig. 4.84b the corresponding half-circuit. Finally, Fig. 4.84c illustrates the voltage–current characteristics of the cascaded devices Q_1 and Q_2 combined (curve ①), as well as the load lines of Q_4 for its nominal threshold voltage (curve ②) for a low threshold voltage (curve ③) and for a high threshold voltage (curve ④). For a threshold voltage variation of ± 0.3 V, the corresponding variation $V_{01}^+ - V_{01}^-$ can be as high as 5 V. This can cause unacceptable variations in the gain and common-mode input voltage range of the op-amp.

An improved stage[18] which uses negative feedback to reduce the sensitivity of V_{01} and V_{02} to threshold voltage variations is shown in Fig. 4.85. In this circuit, an increase of V_1 due to an increase in the threshold voltage results (via the Q_9 and Q_{10} source follower) in an increase of V_{GS1}, and hence an increase of I_0. This, in turn, reduces V_1. It can be shown[18] that the sensitivity of V_{01}, V_1, and V_{02} to the variations of the threshold voltages of the depletion-mode MOSFETs is thus reduced to about 1; that is a ± 0.3 V variation in the threshold voltage causes approximately the same change in V_{01}, V_1, and V_{02}.

For maximum common-mode voltage range, the dc voltage drop across the three depletion-load devices should be just enough to keep them in their saturation regions for a range of V_{DD} values. Hence, V_1 and the dc values of v_{01} and v_{02} should track V_{DD}. This is achieved approximately for the circuit of Fig. 4.85 as follows. Due to the current mirror $Q_9 - Q_{11}$, the currents in Q_{10} and Q_{12} match. Hence, if Q_{10} and Q_{12} are matched devices, then $V_{GS10} \simeq V_{GS12}$. If we also choose $(W/L)_9/(W/L)_1 = I_9/I_1 = I_{\text{bias}}/6I_0$, then $V_{GS9} = V_{GS11} \simeq V_{GS1}$. Hence, $V_1 = V_{GS10} + V_{GS1} \simeq V_{GS12} + V_{GS11} = V_2$. Thus, V_1 (and hence also V_{01} and V_{02}) track V_2. V_2, in turn, tracks V_{DD} since it is obtained

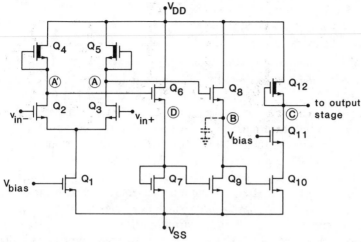

FIGURE 4.86.   NMOS operational amplifier with depletion loads.

from V_{DD} by voltage division via Q_{11}, Q_{12}, and Q_{13}. Hence, the desired tracking between V_1, V_{01} and V_{02} and V_{DD} is obtained.

The differential voltage gain $(v_{01} - v_{02})/(v_{in+} - v_{in-})$ in a source-coupled pair is twice as large as the individual gains $v_{01}/(v_{in+} - v_{in-})$ or $v_{02}/(v_{in+} - v_{in-})$. Hence, it is advantageous to utilize the full differential output swing $(v_{01} - v_{02})$ by using a differential-to-single-ended converter. As in the all-enhancement NMOS op-amp, it is also possible to achieve additional voltage gain in this stage. Usually, however, this gain is low, comparable to that of an all-enhancement gain stage. Hence, to obtain a high-performance op-amp, the converter must be followed by a high-gain depletion-load second gain stage. An example of such a circuit[4] is shown in Fig. 4.86; the gain stage here uses the cascode circuit discussed in Section 4.6 (see Fig. 4.51). The converter stage (Q_6, Q_7, Q_8, Q_9) uses only enhancement-mode devices.

The compensation of this op-amp can be achieved by pole splitting. Ideally, the only high-impedance nodes in the circuit are Ⓐ, Ⓐ', and Ⓒ (Fig. 4.86). Hence, a negative feedback via a capacitor connected between nodes Ⓐ and Ⓒ will result in compensation. If, however, V_{DD} and V_{SS} are large, then a large dc voltage drop is needed across the level-shifter transistors Q_6 and Q_8 for correct biasing of the output stage. This requires small W/L ratios and hence small g_ms for these devices. Since the output impedance at node Ⓑ is approximately $1/g_{m8}$, the resulting pole at $-g_{m8}/C_B$ is at a fairly low frequency, and hence will effect the transfer function of the forward path from node Ⓐ to the output node Ⓒ. Therefore, when the pole-splitting capacitor is added between nodes Ⓐ and Ⓒ, the feedback circuit will have an extra pair of complex poles. Depending on the element values, these may introduce peaking or even damped oscillations into the time response.[14]

To reduce the high-frequency output impedance at node Ⓑ, a small capacitor C_8 may be connected between nodes Ⓐ and Ⓑ. Similarly, the effects of the extra pole introduced by the high impedance at node Ⓓ may be reduced by connecting a small capacitor C_6 between nodes Ⓐ' and Ⓓ. As discussed earlier in Section 4.6, in connection with Fig. 4.50c and Eqs. (4.79)–(4.81), by appropriate choice of C_6 and C_8 an approximate pole/zero cancellation can be achieved, and the extra poles eliminated. When this technique is used, care must be taken to avoid any pole/zero pair below the unity-gain frequency; such a pair degrades the transient response.[17]

An alternative approach, which avoids altogether the additional poles introduced by the extra high-impedance nodes in the converter, is to use a low-impedance level shifter. A simplified schematic of such an op-amp[4,6] is shown in Fig. 4.87. The level-shifter stage now consists of two equal floating voltage sources $V_1 = V_2 = V$ and a current mirror ($Q_6 - Q_7$). It is followed by a second gain stage ($Q_{10} - I_0$) with high output impedance. Hence, an additional output stage is needed in most applications.

The incremental impedance of Q_6 at node Ⓒ is approximately $1/g_{m6}$, which is low since $(W/L)_6 \gg 1$. Hence, the only two high-impedance nodes are now Ⓐ and Ⓑ, and a pole-splitting capacitor C can be connected between these nodes. The second gain stage ($Q_{10} - I_0$) then acts as an integrator, and it almost solely determines the gain and phase responses of the entire op-amp.

Let $v_{in+} = v_{in-}$; then clearly for a symmetrical input stage $I_2 = I_3 = I_1/2$, $I_4 = I_5$ and $I_6 = I_7 = I = I_4 - I_1/2$. The value of I determines the gate-to-source voltage V_{03} of Q_6, which is by symmetry for matched Q_6 and Q_7 the same as V_{04}, the quiescent output voltage of the level-shifter stage. Any symmetrical variation of ΔV of V_1 and V_2 will result in proportional changes

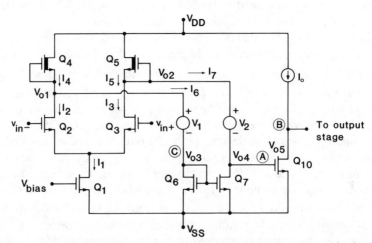

FIGURE 4.87. Simplified circuit schematic of an NMOS operational amplifier with a low-impedance level shifter.

ΔI in I, and ΔV_{03} in V_{03} and V_{04}. The output impedance of the $Q_2 - Q_4$ stage at the drain of Q_2 is $r_{01} = 1/(g_{d2} + g_{d4} + |g_{mb4}|)$; the impedance of Q_6 at node Ⓒ is $1/g_{m6}$. Hence, when V_1 changes, I changes by

$$\Delta I = -\Delta V/(r_{01} + 1/g_{m6}) \qquad (4.169)$$

and V_{03} and V_{04} change by

$$\Delta V_{03} = \Delta V_{04} = \Delta I/g_{m6} = -\Delta V/(r_{01}g_{m6} + 1). \qquad (4.170)$$

Since typically $r_{01} \approx 1/|g_{mb4}| \gg 1/g_{m6}$, a good approximation is to use $\Delta I \approx -|g_{mb4}|\,\Delta V$ and $\Delta V_{04} \approx -(|g_{mb4}|/g_{m6})\,\Delta V$. To reduce the sensitivities of I and V_{04} to changes in V, $|g_{mb4}|$ should hence be small, while g_{m6} should be large. Since variations of V_{04} change the dc offset voltage, this is an important consideration.

It should be noted that the high output impedance of the input stage is loaded by the low-impedance level shifter, and hence the input stage provides a *current*, rather than *voltage*, output. Hence, it operates as a transconductance amplifier. The differential gain $\Delta V_{04}/V_d$ can be found (cf. Problem 4.44) to be $g_{mi}r_{d7}$.

A circuit which can be used to realize the floating voltage source V_1 is shown in Fig. 4.88. In this circuit, V_{GS8} is determined by I and V_{GS9} by I_b, and the floating voltage is $V_1 = V_{GS9} + V_{GS8}$. For a change ΔI in the current I, V_{GS8} (and hence also V_1) changes by $\Delta V_1 = \Delta V_{GS8} = \Delta I/g_{m8}$. Hence, the larger g_{m8} is, the less sensitive will be V_1 to variations of I. Therefore, $(W/L)_8 \gg 1$ is desirable. Then $V_{GS8} \ll V_{GS9} \approx V_1$. Since V_{GS9} is independent of I, the circuit will behave as an ideal voltage source. For large $V_1 \approx V_{GS9}$, $(W/L)_9 \ll 1$ may be needed.

The gate potential V_{01} of Q_9 should be $V_{DD} - |V_{TD}|$; that is, one depletion threshold below V_{DD}, since then Q_4 will be just barely in saturation. Similarly, $V_{02} = V_{DD} - |V_{TD}|$ is desirable. As explained earlier in this section (in connec-

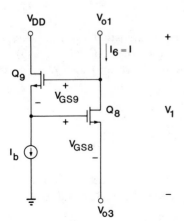

FIGURE 4.88. Low-impedance floating voltage source.

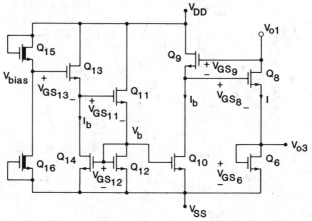

FIGURE 4.89. Circuit schematic for the level shifter and its bias circuit.

tion with Fig. 4.85), this condition allows maximum dynamic range for the input stage. This optimum dc bias condition on V_{01} can be achieved by using the circuit of Fig. 4.89, which includes the floating voltage source (Q_8 and Q_9) of Fig. 4.88. In the circuit, Q_{13} is matched to Q_9, and Q_{14} to Q_{10}. Hence, the dc currents of Q_{10} and Q_{14} (which have the same gate-to-source voltages) are equal; both have the value I_b. Therefore, the currents of Q_9 and Q_{13} are equal; hence, $V_{GS9} = V_{GS13}$. Since $(W/L)_9 = (W/L)_{13} \ll 1$ (typically, 0.05), V_{GS9} and V_{GS13} are large, as much as 5 ~ 8 V above the threshold voltage. By contrast, devices Q_6, Q_8, Q_{11}, and Q_{12} all have large aspect ratios ($W/L = 10$ ~ 20); hence their gate-to-source voltages are all within a few hundreds of millivolts of the threshold voltage. Therefore, $V_{GS6} \approx V_{GS8} \approx V_{GS11} \approx V_{GS12}$ and

$$V_{bias} = V_{GS13} + V_{GS11} + V_{GS12} \simeq V_{01}$$

$$= V_{GS9} + V_{GS8} + V_{GS6}. \qquad (4.171)$$

Thus, by setting V_{bias} to $V_{DD} - |V_{TD}|$, it can be assured that V_{01} will have the same value, within a few tens of millivolts. It can be shown (Problem 4.45) that the voltage divider formed by Q_{15} (in its triode region) and Q_{16} (in saturation) sets V_{bias} to

$$V_{bias} = V_{DD} - |V_{T15}| \left[-1 + \sqrt{1 + \left(\frac{V_{T16}}{V_{T15}}\right)^2 \frac{(W/L)_{16}}{(W/L)_{15}}} \right], \qquad (4.172)$$

where V_{T15} is affected by the body effect. By choosing the aspect ratios of Q_{15} and Q_{16} appropriately, the quantity in the brackets can be made equal to $+1$.

Then, $V_{bias} = V_{DD} - |V_{T15}|$, which is close to the desired value. The circuit is called, for obvious reasons, the *replica bias* circuit.[4]

Note that the depletion-mode transistors (Q_4, Q_5, Q_{15}, and Q_{16}) are connected differently and thus have different body biases. Hence, the values achieved for V_{bias} and V_{01} will not be exact, and they will also be somewhat process dependent. However, the resulting input-referred offset voltage is typically only about 2 mV, much less than that due to random mismatch.[4]

Naturally, a similar circuit can be used to set the dc value of v_{02} to $V_{DD} - |V_{TD}|$. However, the basic bias-replicating circuit (Q_{11} through Q_{16}) can be shared between the two floating voltage sources. The complete circuit of the differential input stage, level-shifter, and differential-to-single-ended converter[4] is shown in Fig. 4.90.

Most of the voltage gain of the op-amp discussed is provided by the second gain stage ($Q_{10} - I_0$ in Fig. 4.87). Thus, this stage must have high gain over a wide frequency band. The cascode gain stages discussed earlier (Figs. 4.22–4.25) can be used here. Figure 4.91 illustrates a possible circuit. Devices $Q_{20} - Q_{23}$ form the amplifier;[4,6] it is similar to that shown in Fig. 4.25 but the load contains only depletion-mode devices. As described in connection with Fig. 4.25, Q_{20}, Q_{22}, and Q_{23} operate as a basic cascode gain stage, while Q_{21} injects dc current into Q_{20} to enhance its bias current and hence to increase g_{m20} without decreasing the load impedance. This enhancement is especially important for that part of the output voltage (v_{05}) swing which is near V_{DD}. In the absence of gain enhancement, the gain greatly decreases in that region since the load device enters its linear (triode) region while the g_m of the driver is low since its drain current is minimum. The latter effect is eliminated by the current injection from Q_{21}.

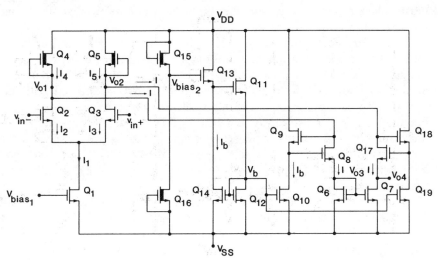

FIGURE 4.90. Differential stage, level-shifter, and differential-to-single-ended converter of an NMOS op-amp.

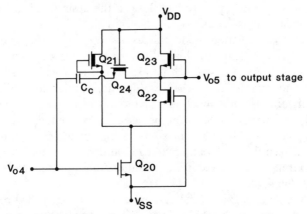

FIGURE 4.91. Cascode gain stage with compensation.

The gain of the stage can be increased by *increasing* the aspect ratio and drain current of Q_{20}, thus enhancing g_{m20}, and/or *reducing* the current and aspect ratio of Q_{23}, thus increasing the load impedance. However, the drain current of Q_{23} should be sufficient to charge and discharge the compensating capacitor C_c fast enough for an acceptable slew rate. Consider the simplified schematic of Fig. 4.87, with C_c connected between nodes Ⓐ and Ⓑ. In a typical design, the dc bias currents $I_4 = I_5 = 2I = I_1$ may be chosen. The maximum current i_c in C_c occurs if v_{in-} drops suddenly, cutting Q_2 off. Then $I_6 \to I_4 = I_1$, $I_3 = I_5 \to I_1$, and hence $I_7 \to 0$. However, the current mirror $Q_6 - Q_7$ forces Q_7 to absorb a current I_1 from C_c. Hence, under these conditions $i_c = I_1$. To make this possible, the drain currents of Q_{22} and Q_{23} must be larger than I_1; otherwise, the current i_c cannot be diverted into C_c and the slew rate will be reduced.

A detailed Laplace-transform analysis of the circuit of Fig. 4.91, including the stray capacitances at its input and output nodes can be carried out using the same techniques (node analysis, Miller approximation) as were used in Sections 4.6 and 4.8. This analysis reveals that *without* the series resistor provided by the channel of Q_{24}, a right-half-plane zero exists in the transfer function $A_v(s)$ of the stage. As explained in Section 4.8, this zero reduces the phase margin, and makes compensation difficult. *With Q_{24} in the circuit*, this zero shifts to the left half-plane, where it can be used to cancel the second most dominant pole of $A_v(s)$. This improves the phase margin and broadens the frequency response.

Since the output impedance of the second gain stage is high, it is usually followed by a wide-band unity-gain buffer stage. The class-A negative-feedback circuit of Fig. 4.37 can be used as this buffer. If a class-AB output stage is required, then the augmented circuit[4,6] of Fig. 4.92 can be used. Here, Q_{26} and Q_{28}, which are matched wide devices ($W/L \sim 8$), form a current mirror, thus keeping the currents in the matched load devices Q_{25} and Q_{27} ($W/L \sim 0.1$)

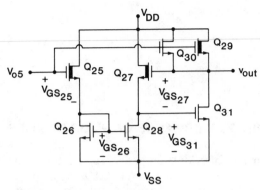

FIGURE 4.92. Circuit schematic for the output stage.

equal. Hence, the feedback through devices $Q_{29} - Q_{31}$ will cause $V_{GS25} = V_{GS27}$. Hence, from Fig. 4.92,

$$v_{05} - v_{out} = v_{GS29} = v_{GS30} \simeq v_{GS26} - v_{GS31}. \tag{4.173}$$

Let first $v_{05} = 0$. Then by design the feedback loop will force $v_{out} \approx v_{05}$. Hence, v_{GS30} will be less than the enhancement-mode threshold voltage V_{TE}, and Q_{30} will be off. Thus, it does not dissipate power in the idling mode.

If now v_{05} is driven *positive*, then v_{GS26} (due to the source-follower action of Q_{25} and Q_{26}) will *increase*. The voltage inverter $Q_{27} - Q_{28}$ then causes v_{GS31} to *decrease*. Thus, by Eq. (4.173), v_{GS30} will *increase*. If the increase is greater than V_{TE}, Q_{30} turns on.* Now Q_{30} will help Q_{29} greatly in driving the load: since Q_{30} does not carry idling current (as shown above), it can be made large without increasing the idle power dissipated by the op-amp. At the same time, the decreasing v_{GS31} will decrease (or even cut off) the current of Q_{31}, which will thus no longer deflect the currents of Q_{29} and Q_{30} from the load.

For a *negative*-going v_{05}, the opposite process will take place. Thus, v_{GS29} and v_{GS30} will go negative, Q_{30} will cut off, and Q_{29} will carry a reduced current. Q_{31} will turn on heavily, and hence can sink a large load current.

Typical dimensions[4] for the output devices are $(W/L)_{29} \approx 10$, $(W/L)_{30} = 2$, and $(W/L)_{31} = 25$.

In addition to providing bias current for $v_{05} < 0$, Q_{29} also helps to increase the maximum positive output swing. Since v_{05} originates from the gain stage of Fig. 4.91, it can approach (but cannot reach) V_{DD}. However, Q_{30} requires $v_{GS30} = v_{05} - v_{out} > V_{TE}$ for conduction. Hence, it cannot supply current to the load if $v_{05} > V_{DD} - V_{TE}$. Q_{29}, on the other hand, can since it operates in the depletion mode.

As the above discussion shows, $v_{05} - v_{out} > 0$ for $v_{05} > 0$, and $v_{05} - v_{out} < 0$ for $v_{05} < 0$. Thus, the voltage gain of the stage is slightly less than one. The

---

*Here, V_{TE} is the enhancement-mode threshold voltage.

output resistance can be shown to be

$$R_0 \simeq \frac{1}{g_{m29} + g_{m30} + g_{m31}} \tag{4.174}$$

(Problem 4.46).

4.13.   THE NOISE PERFORMANCE OF NMOS OP-AMPS

The noise performance of NMOS amplifiers can be discussed using arguments similar to those applied in Section 4.10 for CMOS op-amps. Thus, consider the all-enhancement differential input stage of Fig. 4.31. An equivalent circuit in which the noise sources are symbolically included is shown in Fig. 4.93. A similar representation for the depletion-load differential stage of Fig. 4.42 is shown in Fig. 4.94. As was the case of the noisy CMOS stage, analysis gives, for the mean-squared value of the equivalent input noise v_{nd}, the value

$$\overline{v_{nd}^2} = \overline{v_{n1}^2} + \overline{v_{n2}^2} + \left(g_{m4}/g_{m1}\right)^2 \left(\overline{v_{n3}^2} + \overline{v_{n4}^2}\right) \tag{4.175}$$

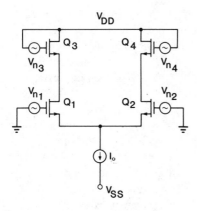

FIGURE 4.93.   Noise-equivalent circuit of the all-enhancement differential stage.

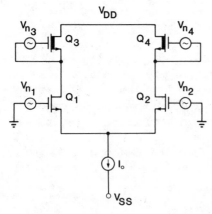

FIGURE 4.94.   Noise-equivalent circuit of the depletion-load differential stage.

which is the same as that given in (4.141). Equation (4.175) holds for both of the circuits in Figs. 4.93 and 4.94.

Neglecting the body effect, the differential signal voltage gain of the stage of Fig. 4.93 is approximately $(-g_{m1}/g_{m4})$, as can be seen from Eqs. (4.53) and (4.44). Hence, even for a modest gain of 10, $(g_{m4}/g_{m1})^2 \approx 0.01$ and hence the last term on the RHS of (4.175) is negligible. For the depletion-load stage, the W/L ratio of Q_3 and Q_4 is determined by the dc bias current carried by the stage. This normally gives $(W/L)_3 = (W/L)_4 < 1$; hence again $(g_{m4}/g_{m1})^2 \ll 1$, and the noise originating from the load devices is normally negligible.

The situation is different if the low-impedance level shifter of Figs. 4.87–4.90 is used in conjunction with the depletion-load circuit. The corresponding equivalent circuit is then that shown in Fig. 4.95. Analysis gives for the input-referred noise voltage (Problem 4.32)

$$\overline{v_{nd}^2} = \overline{v_{n1}^2} + \overline{v_{n2}^2} + \left(g_{m4}/g_{m1}\right)^2 \left(\overline{v_{n3}^2} + \overline{v_{n4}^2}\right)$$

$$+ \left(g_{m6}/g_{m1}\right)^2 \left(\overline{v_{n5}^2} + \overline{v_{n6}^2}\right). \tag{4.176}$$

To minimize the noise contribution of Q_5 and Q_6, $g_{m5} = g_{m6}$ should thus be made much smaller than $g_{m1} = g_{m2}$; that is, the W/L ratio of Q_5 and Q_6 should be as low as permitted by other considerations. The noise generated in the devices realizing the floating voltage sources can usually be neglected.

From the results of Section 4.10 and of this section, we can conclude that if the circuit is carefully designed to minimize the noise contribution of the load devices, level shifters, and so on, then (at least for low and medium frequencies) *the noise generated in the input devices will dominate the overall op-amp noise*. The amplifier noise will therefore be determined mostly by the fabrica-

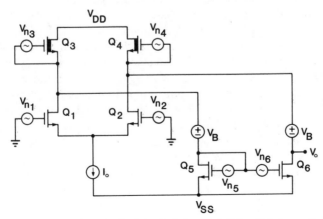

FIGURE 4.95. Noise-equivalent circuit of the depletion-load differential stage with a low-impedance level-shift circuit.

tion process as it affects the input devices, and by the dimensions of these devices: the larger their areas are the less noise will be generated.

At high frequencies, the voltage gain of the input stage decreases due to the compensation capacitor C_c. Then the noise contribution of the other stages can also become significant. In switched-capacitor circuits, aliasing often enhances the high-frequency op-amp noise, and may make it the dominant source of filter noise. Then the effects of noise generated by devices other than the input ones must also be considered and optimized.

4.14.   PRACTICAL CONSIDERATIONS IN MOS OP-AMP DESIGN

In Section 4.1, several nonideal effects which can degrade the performance of practical op-amps were listed. The minimization of these effects is an important aspect of op-amp design. The corresponding considerations are briefly discussed next for the most important nonideal effects.

1.   *Finite Gain.*   In the preceding sections we have already discussed the available gain for various MOS gain stages, and also the special circuits which may be utilized (cascode devices, composite loads, etc.) to enhance the voltage gain without reducing the bandwidth.

2.   *Finite Linear Range.*   This question was also briefly discussed, and circuits were introduced (see, e.g., Figs. 4.37, 4.38, and 4.89) for maximizing the allowable signal swing.

3.   *Offset Voltage.*   As defined in Section 4.1 the input-referred offset voltage $v_{in,off}$ is the differential input voltage needed to restore the output voltage v_{out} to zero. It contains two components: a *systematic offset* which is due to improper dimensions and/or bias conditions, and a *random offset* which is due to the random errors in the fabrication process resulting, for example, in the mismatch of ideally symmetrical devices.

To illustrate the generation of systematic offset (and ways to avoid it) consider the simple two-stage CMOS op-amp shown in Fig. 4.96. The first stage $(Q_1 - Q_5)$ is the differential-input/single-ended-output input stage originally introduced in Fig. 4.44; the output stage is a single-ended gain stage with a driver Q_6 and a current–source load Q_7. Clearly if the circuit has no systematic offset, then grounding both input terminals (as shown) results in $v_{out} = 0$. Then, if the output terminal is also grounded (as shown in Fig. 4.96) the current I_g in the grounding lead will also be zero. Thus, the condition for zero offset is equivalent to requiring $I_g = 0$ for grounded input and output terminals; this in turn requires $I_6 = I_7$.

Assuming symmetry in the input stage, $(W/L)_1 = (W/L)_2$ and $(W/L)_3 = (W/L)_4$. Then all currents and voltages will also be symmetrical, and hence $V_{DS3} = V_{DS4}$. Then also $V_{GS3} = V_{GS6}$. If this value of V_{GS6} results in I_6 being equal to the source current I_7 when $V_{DS6} = 0 - V_{DD} = -V_{DD}$, then $I_g = 0$ as required. If this is *not* the case, then $I_g \neq 0$ and systematic offset exists.

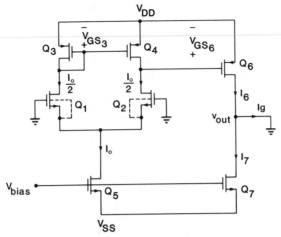

FIGURE 4.96.   Two-stage CMOS op-amp.

Specifically, let $\hat{V}_{GS6}$ denote the value of V_{GS6} needed to make I_6 equal to I_7. Then, the *input* offset voltage is clearly

$$v_{\text{in, off}} = \frac{V_{GS6} - \hat{V}_{GS6}}{A_d} = \frac{V_{GS3} - \hat{V}_{GS6}}{A_d}, \qquad (4.177)$$

where A_d is the voltage gain of the *input* stage. Thus, for example, an error of 0.1 V in the bias voltage of Q_6 will result in a 1 mV input offset if $A_d = 100$.

Assuming that all devices are in saturation, and neglecting channel-length modulation effects, the voltages of Q_3 and Q_4 can be expressed as

$$V_{GS3} = V_{DS3} = V_{GS4} = V_{DS4} = V_{Tp} + \sqrt{\frac{I_0/2}{k'_p(W/L)_3}}. \qquad (4.178)$$

Here, V_{Tp} is the threshold voltage of the *p*-channel devices Q_3 and Q_4, and k'_p is the constant "transconductance" factor $\mu_p C_{ox}/2$ of the drain–current equation for PMOS devices. Similarly, for Q_6

$$V_{GS6} = V_{Tp} + \sqrt{\frac{I_6}{k'_p(W/L)_6}}. \qquad (4.179)$$

Substituting $V_{GS6} = V_{GS3}$ and the required condition $I_6 = I_7$ into (4.179),

$$V_{GS3} = V_{Tp} + \sqrt{\frac{I_7}{k'_p(W/L)_6}} \qquad (4.180)$$

results. From (4.178) and (4.180), the condition

$$\frac{(W/L)_3}{(W/L)_6} = \frac{I_0/2}{I_7}$$

is obtained for zero offset.

Turning to Q_5 and Q_7, since they have equal gate-to-source voltages, neglecting channel-length modulation we obtain

$$\frac{(W/L)_5}{(W/L)_7} = \frac{I_0}{I_7}. \tag{4.181}$$

Combining our equations, the design relations are

$$\frac{(W/L)_3}{(W/L)_6} = \frac{(W/L)_4}{(W/L)_6} = \frac{1}{2}\frac{(W/L)_5}{(W/L)_7} = \frac{I_0}{2I_7}. \tag{4.182}$$

Physically, if Eq. (4.182) is satisfied then the current I_7 induced in Q_7 by its gate-to-source voltage V_{GS7} and the current I_6 induced in Q_6 by the gate-to-source voltage $V_{GS6} = V_{DS4}$ are the same and hence $v_{out} = 0$ is possible when Q_6 and Q_7 are in saturation. If the gate-to-source voltages are *not* compatible, and the output terminal is open-circuited, then v_{out} assumes a nonzero value such that the drain voltages of Q_6 and Q_7 will compensate for the discrepancies of the gate voltages. This may also result in Q_6 or Q_7 operating out of saturation. It usually represents a major systematic offset voltage effect, and may reduce the gain and the bandwidth of the op-amp.

To minimize the effects of *random* process-induced channel-length variations on the matching of the devices, and thus the *random* offset, the channel lengths of Q_3, Q_4, and Q_6 should be chosen equal. Then the current density I_d/W is the same for these devices when (4.182) is satisfied, and the required current ratios are determined by the ratios of the widths. If ratios as large as (or larger than) two are required, then the wider transistor can be realized by the parallel connection of two (or more) "unit transistors" of the size of the narrower one. Note, however, that this process is in conflict with the guidelines for the minimization of noise established in Sections 4.10 and 4.13: according to those rules the transconductances of the load devices Q_3 and Q_4 should be low while that of Q_6 should be high for high gain and good high-frequency response. The channel lengths for all three devices Q_3, Q_4, and Q_6 should be long for high output impedance and high gain. While this helps to reduce the noise of Q_3 and Q_4 and increase the gain of the differential stage, it decreases the transconductance of Q_6 and hence reduces the gain of the second stage. Clearly, the optimum tradeoff among these conflicting requirements will vary from application to application.

The *random* offset voltage will be affected by several factors, including mismatch between the (ideally symmetrical) input devices Q_1 and Q_2 and/or between the load devices Q_3 and Q_4. This can be caused either by geometrical mismatch, or by a process gradient causing different threshold voltages. Assume first that the current mirror $(Q_3 - Q_4)$ is imperfect, so that

$$I_3 = \tfrac{1}{2}(1 - \varepsilon_1)I_0 \neq I_4 = \tfrac{1}{2}(1 + \varepsilon_1)I_0. \tag{4.183}$$

The differential voltage $v_{G1} - v_{G2}$ needed at the input terminal to restore symmetry is clearly

$$v_{\text{off1}} = \frac{\varepsilon_1 I_0}{g_{mi}}. \tag{4.184}$$

Thus, v_{off1} can be reduced by increasing the transconductance g_{mi} of the input devices or by reducing the bias current I_0.

Assume next that the dimensions and the threshold voltages of the input devices are mismatched while the load devices are symmetrical. Thus let

$$(W/L)_1 = (1 - \varepsilon_2)(W/L)_2 \tag{4.185}$$

and

$$V_{T1} = V_{T2} - \Delta V_T. \tag{4.186}$$

Clearly, it requires an input offset voltage $v_{\text{off2}} = \Delta V_T$ to cancel the effect of the threshold voltage mismatch. The geometric mismatch causes a current imbalance $\Delta I_1 \cong -\varepsilon_2 I_1 \cong -\varepsilon_2 k_1 (V_{GS1} - V_{T1})^2$. This can be balanced by a change v_{off3} in v_{G1} such that

$$g_{mi} v_{\text{off3}} \cong 2k_1 (V_{GS1} - V_{T1}) v_{\text{off3}} = -\Delta I_1$$

$$\cong \varepsilon_2 k_1 (V_{GS1} - V_{T1})^2. \tag{4.187}$$

Here, we used Eq. (3.18) of Chapter 3 to express g_{mi}, with $v_{SB}^o = 0$ and $\lambda \cong 0$. From (4.187),

$$v_{\text{off3}} = \frac{\varepsilon_2}{2}(V_{GS1} - V_{T1}) = \frac{\varepsilon_2}{2}\sqrt{\frac{I_0/2}{k'(W/L)_1}}. \tag{4.188}$$

Thus, v_{off3} can be reduced (as could v_{off1}) by increasing $(W/L)_1$—and thus g_{mi}—or by reducing I_0. Both will reduce $V_{GS1} - V_{T1}$.

The variation of the threshold voltage ΔV_T is independent of I_0 or W/L; it only depends on process uniformity. It can be reduced by building Q_1 and Q_2 from unit transistors arranged in a common-centroid structure.[2,17]

4. *Common-Mode Rejection Ratio (CMRR)*. As defined in Section 4.1, CMRR $= A_D/A_C$, where A_D is the differential gain, while A_C is the common-

mode one. For the op-amp of Fig. 4.96, the common-mode rejection is provided by the input stage. The value of the CMRR for this stage was found earlier, and was given by Eq. (4.66) as

$$\text{CMRR} \cong 2\frac{g_{mi}g_{ml}}{g_0 g_{di}}. \tag{4.189}$$

As explained in Section 3.4 of Chapter 3, both g_{mi} and g_{ml} are proportional to $\sqrt{I_0}$, while g_0 and g_{di} are proportional to I_0. Thus, the rejection ratio is inversely proportional to I_0. Values of 10^3–10^4 can readily be achieved, as (4.189) shows.

If a mismatch exists between Q_1 and Q_2 such that $g_{ml} = (1 + \varepsilon)g_{m2}$, then a common-mode voltage v_c will cause as much differential output voltage as a differential input voltage $(g_{di}/g_{mi})\varepsilon v_c$ (Problem 4.33). Thus, now we have CMRR $\cong g_{mi}/(\varepsilon g_{di})$. This again illustrates the importance of making the input devices symmetrical, using common-centroid geometry if necessary: a 1% mismatch may lower the CMRR to 60 dB!

5. *Frequency Response, Slew Rate, Biasing, Power Dissipation.* The requirements on the speed (i.e., high-frequency gain and slew rate) of an MOS amplifier depend very much on its application. In switched-capacitor circuits, most op-amps drive capacitive loads only. The speed requirement is then that the op-amp must be able to charge the load capacitance C_L and settle to within a specified accuracy (usually 0.1% of the final voltage) in a specified time interval.

Figure 4.97 shows the simple op-amp of Fig. 4.96, supplemented by a feedback branch (Q_8, C_c) for compensation and driving a capacitive load C_L. As explained in Section 4.9, the unity-gain bandwidth of the stage is given by $\omega_0 = g_{mi}/C_c$. A detailed analysis of the performance of a compensated op-amp used to charge or discharge a capacitor reveals[23,24] that (in linear operation)

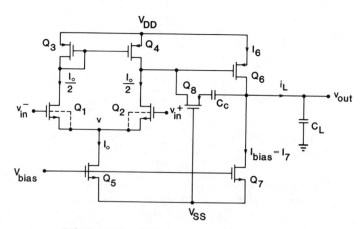

FIGURE 4.97. CMOS op-amp with capacitive load.

the condition upon the unity-gain bandwidth,

$$\omega_0 = \frac{g_{mi}}{C_c} \geq \frac{15}{T_{ch}}, \tag{4.190}$$

is usually sufficient to guarantee adequate speed. Here, T_{ch} is the time available for recharging C_L; for a two-phase circuit, usually $T_{ch} \cong 1/2f_c$ where f_c is the clock frequency.

The minimum value of g_{mi} is usually determined by the required dc gain and by noise considerations. In addition, there is also an upper bound on g_{mi}/C_c, based on the requirement that the second pole frequency $|s_{p2}|$ must be considerably higher than ω_0. Often, $|s_{p2}| = 3\omega_0$ is chosen. This, as derived in (4.120)–(4.125), requires for our circuit that

$$C_c \sim 3C_L \frac{g_{mi}}{g_{m6}}. \tag{4.191}$$

A rule of thumb, which usually results in a good compromise among all requirements, is to choose $C_c = C_L$. Then (4.190) gives a lower bound for g_{mi}, and (4.191) a lower bound for g_{m6}.

For large-signal operation, the slew rate S_r of the input stage must also be considered. From Eq. (4.131), $S_r = I_0/C_c$. For $C_c = C_L$, the minimum value of I_0 (the bias current of the input stage) is thus determined. In addition, for the circuit of Fig. 4.97, the slew rate of the output also needs attention. For *positive-going* v_{out}, the output current i_L is supplied by Q_6. The magnitude of i_L is limited only by the size of Q_6, and that of v_{out} only by the v_{DS6} needed to keep Q_6 in saturation. For *negative* v_{out}, by contrast, the output stage must *sink* the load current i_L. This is performed by reducing i_6 below I_{bias} so that $i_6 = I_{bias} - |i_L|$.

The maximum value $|i_L| = I_{bias}$ is obtained when $i_6 = 0$, that is, when Q_6 is cut off. Thus, the negative-going slew rate due to the output stage is

$$S_{r0} = \left| \frac{dv_{out}}{dt} \right| = \frac{|i_L|}{C_L} = \frac{I_{bias}}{C_L}. \tag{4.192}$$

This establishes the minimum value of I_{bias}. Note that by Eqs. (4.119) and (4.124), the transconductances of the output devices (here Q_6 and Q_7) must be large for stability. A large I_{bias} will help to satisfy this with moderate-sized Q_6 and Q_7.

The dc power dissipated by the circuit of Fig. 4.97 in the quiescent state is thus

$$(V_{DD} - V_{SS})(I_0 + I_{bias})$$

$$\geq (V_{DD} - V_{SS})(S_r C_c + S_{r0} C_L) = (V_{DD} - V_{SS})(S_r + S_{r0})C_L, \tag{4.193}$$

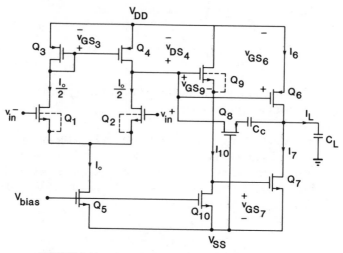

FIGURE 4.98. CMOS op-amp with class-AB output stage.

where $C_c = C_L$ has been set. Thus, the higher the slew rates and the larger C_L, the more dc standby power is needed by the stage.

The standby current of the output stage can often be reduced by using a class-AB stage. The resulting circuit has been discussed earlier (cf. Fig. 4.70), and is reproduced in Fig. 4.98. The level-shifter stage (Q_9 and Q_{10}) should be dimensioned such that the voltages v_{GS6} and v_{GS7}, and hence the quiescent drain currents of Q_6 and Q_7, are not too large. Since the gates of both Q_6 and Q_7 are driven by signal voltages, the load current can now be much larger than the bias current for either positive- or negative-going v_{out}. The minimum value of the output-dc bias current is thus determined only by the requirement on the transconductances g_{m6} and g_{m7} needed for a good phase margin.

The value of the standby current I_{10} will determine the locations of the zero and the pole of the source follower $Q_9 - Q_{10}$, as given by Eq. (4.81). Obviously, the pole (which is at a lower frequency) will result in a positive phase shift, while the higher-frequency left-half-plane zero will cause a negative phase, resulting in a dip in the phase characteristic. The location of the dip is determined by the frequency of the pole, and its depth and width by the separation between the pole and the zero. The minimum value of I_{10} can be thus determined such that the dip moves to a sufficiently high frequency where it affects only slightly the phase shift of the stage at ω_0. This requirement gives the minimum value of g_{m9}, and thus of the bias current I_{10}. The depth of the dip can be reduced by forcing the zero as close to the pole as possible. It is worth mentioning that the source follower's time response, due to the presence of the pole/zero pair, will contain an exponential term. In this term, the time constant is determined by the pole frequency, while the amplitude (residue) is determined by the difference between the pole and zero frequencies. Increasing

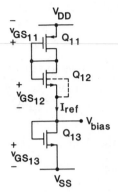

FIGURE 4.99. Bias circuit for the op-amp of Fig. 4.98.

the pole frequency and reducing the distance between the pole and zero therefore will improve the time response as well.

The voltage V_{bias} for the op-amps of Figs. 4.97 and 4.98 can be obtained using the circuits of Section 4.2. In particular, for the op-amp of Fig. 4.97 a supply-independent bias is desirable. This will keep I_0 and I_{bias} independent of the supply voltages, and hence the parameters which affect the stability unchanged. (The dc power dissipation will, of course, vary with the supply voltages.) Thus V_{bias} may be obtained from the circuit of Fig. 4.5c.

For the circuit of Fig. 4.98, by contrast, supply-independent biasing can cause problems. In particular, if I_0 remains constant with supply voltage variation then V_{GS3} does also; if I_{10} remains constant, then V_{GS9} does too. Hence, in the expression for the gate-to-source voltage of Q_7

$$V_{GS7} = V_{DD} + V_{DS4} - V_{GS9} - V_{SS} \cong (V_{DD} - V_{SS}) + (V_{GS3} - V_{GS9}) \quad (4.194)$$

$(V_{GS3} - V_{GS9})$ is invariant of V_{DD} and V_{SS}. Therefore, all power-supply changes appear directly in V_{GS7}, and Q_7 can cut off if $V_{DD} - V_{SS}$ drops significantly.

A suitable bias circuit for the op-amp of Fig. 4.98 is shown in Fig. 4.99. When this circuit is used, the dimensions of the NMOS devices Q_5, Q_{10}, and Q_{13} are obviously related by

$$\frac{(W/L)_5}{(W/L)_{13}} = \frac{I_0}{I_{\text{ref}}}, \qquad \frac{(W/L)_{10}}{(W/L)_{13}} = \frac{I_{10}}{I_{\text{ref}}}. \quad (4.195)$$

In addition, the PMOS transistors Q_3, Q_4, and Q_{11} can be dimensioned such that

$$\frac{(W/L)_3}{(W/L)_{11}} = \frac{(W/L)_4}{(W/L)_{11}} = \frac{I_0/2}{I_{\text{ref}}} \quad (4.196)$$

holds. Then, we have

$$V_{GS3} = V_{DS3} = V_{GS4} = V_{DS4} = V_{GS11}. \tag{4.197}$$

In addition, we can dimension the NMOS devices Q_9 and Q_{12} so as to satisfy

$$\frac{(W/L)_9}{(W/L)_{12}} = \frac{I_{10}}{I_{\text{ref}}} = \frac{(W/L)_{10}}{(W/L)_{13}}. \tag{4.198}$$

This will cause $V_{GS9} \cong V_{GS12}$, and using (4.194) and (4.197) we find

$$V_{GS7} = V_{DD} + V_{DS4} - V_{GS9} - V_{SS} = V_{DD} + V_{GS11} - V_{GS12} - V_{SS}$$

$$= V_{\text{bias}} - V_{SS} = V_{GS13} = V_{GS5} = V_{GS10}. \tag{4.199}$$

This matching of voltages is independent of V_{DD} and V_{SS}.

Thus, if I_7 is the desired value of the output bias current, then we must choose the dimensions of the NMOS devices Q_7 and Q_{13} to satisfy

$$\frac{(W/L)_7}{(W/L)_{13}} = \frac{I_7}{I_{\text{ref}}}. \tag{4.200}$$

Finally, since clearly

$$V_{GS6} = V_{DS4} \cong V_{DS3} = V_{GS3} \tag{4.201}$$

we should also choose

$$\frac{(W/L)_6}{(W/L)_{11}} = \frac{I_7}{I_{\text{ref}}}. \tag{4.202}$$

Note that this choice of dimensions will establish the desired bias currents without introducing any systematic offset voltage. Assume now that the supply voltages vary in the circuit. Then I_{ref} will change, and so will v_{GS11}, v_{GS12}, and v_{GS13}. However, Q_{11}, Q_{12}, and Q_{13} will certainly continue to conduct; in fact, they will all also remain in saturation since their gates and drains are connected together. But since the currents and gate-to-source voltages of Q_5, Q_{10}, and Q_7 mirror those of Q_{13}, they too will conduct and remain in saturation. Also, since the currents and gate-to-source voltages of Q_3, Q_4, and Q_6 mirror those of Q_{11}, these transistors will conduct and remain saturated. Finally, the conduction and saturation conditions of Q_9 follow from those of Q_{12}. Thus, all transistors have stabilized dc bias conditions, and the operation of the circuit will be insensitive to process variations.

For NMOS op-amps, similar biasing considerations can be derived. In general, NMOS op-amps are more complex and hence require more dc power and more elaborate bias circuitry than their CMOS counterparts.

For applications where only very low dc power dissipation is permitted, the "dynamic" op-amps[25-27] briefly described in Section 4.16 can be used. These circuits utilize time-varying bias currents, and hence draw full dc power only during a small fraction of each clock cycle.

6. *Nonzero Output Resistance.* This is usually important only for the output amplifier of the overall filter, which may have to drive a large capacitive and/or resistive load. The low-frequency output resistance R_{out} of the op-amp *without* negative feedback (i.e., open loop) is of the order of $r_d/2$ for an unbuffered circuit, such as shown in Fig. 4.97 or 4.98. Here r_d is the drain resistance of the output devices, of order $0.1 \sim 1$ MΩ. For a buffered circuit (such as shown in Fig. 4.83), the output impedance is around $1/g_m$, where g_m is the transconductance of the output device; hence $R_{out} \approx 1$ kΩ. In closed-loop operation, the effective output impedance is $R_{out}(1 - A_C)/A$, where A is the open-loop gain and A_C the closed-loop one (Problem 4.31). Since usually $A > 1000$, the effective closed-loop output impedance is around 1 kΩ for unbuffered op-amps, and very low (of the order of a few ohms) for buffered ones. This value is sufficiently low for most applications.

7. *Noise and Dynamic Range.* These subjects were discussed in some detail earlier in Section 4.10 for CMOS op-amps and in Section 4.13 for NMOS ones. Hence, they will not be analyzed here.

8. *Power-Supply Rejection.* This is one of the most important nonideal effects in MOS analog integrated circuits, for several reasons. First, several circuits (some analog, some digital) may operate off the same power supply. Hence, a number of analog and digital signal currents can enter the supply lines. Since these lines have nonzero impedances, digital and analog voltage noise will be superimposed on the dc voltage provided by the supply. If the op-amp circuit does not reject this noise, then the noise will appear at its output, reducing the signal/noise ratio and the dynamic range. Second, if switching regulators or dc-voltage multipliers are used, then a substantial amount of high-frequency switching noise will be present on the supply line. Finally, the clock signals of the various circuits fed from the same line will also usually appear superimposed on the supply voltage with a reduced but nonzero amplitude.

In a sampled-data circuit (such as a switched-capacitor filter), all signals are periodically sampled. This results in a frequency mixing, and as a result the high-frequency noise will be "aliased" into the frequency band of the signal (cf. the discussions of Section 2.5). Hence *any* noise in the overall frequency range is detrimental if it can enter the signal path.

The most likely path through which supply noise can be coupled to the signal is via the op-amps. Thus, a high value of the power-supply rejection ratio (PSRR), defined in Section 4.1 as the ratio of the open-loop differential gain A_D and the noise gain A_p from the supply to the op-amp output, is of great importance.

At low frequencies, the supply noise is coupled into the op-amp mostly through the bias circuits, and also can enter due to the asymmetries in the differential input stage. At high frequencies, on the other hand, the noise gain

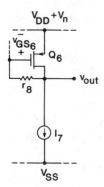

FIGURE 4.100.   High-frequency model of the op-amp output stage.

is mostly determined by the capacitive branches. Consider the circuits of Figs. 4.97 and 4.98. At high frequencies, the compensation capacitor C_c becomes nearly a short circuit, and (since the drain/source resistance r_8 of Q_8 is small in the linear region), the gate and drain voltages of Q_6 are nearly equal. As shown in Fig. 4.100, therefore, the *incremental* output voltage due to the supply noise v_n is

$$\Delta v_{\text{out}} = v_n + \Delta v_{GS6} = v_n$$

since $v_{GS6} = I_7/g_{m7}$ is constant. Thus, $A_p = 1$, independent of frequency, for this stage. The output impedance of Q_6 driving C_L with this noise voltage is low, around $1/g_{m6}$. Since the open-loop gain A_D decreases with increasing frequency at a rate of -6 dB/octave while A_p stays constant, the PSRR due to noise in V_{DD} decreases at the same rate as A_D, reaching 0 dB near the unity-gain frequency.

The situation is more favorable with respect to noise on the V_{SS} line. The gain for noise entering via Q_5 is the low common-mode gain. Any noise entering via Q_{10} and Q_7 is added to the signal and attenuated at the same rate (-6 dB/octave) as the signal by the load capacitor C_L since now the noise output impedance is high.

An effective technique[19] for increasing the PSRR for noise coming from the positive supply is illustrated in Fig. 4.101. In contrast to the circuits of Figs. 4.97 and 4.98 where Q_8 operated as a linear resistor for both feedback and feedforward signals, Q_8 now is biased in its saturation region. Hence, in the feedback direction (from v_{out} through C_c and Q_8 to the gate of Q_6) the resistance is $1/g_{m8}$, while viewed from node Ⓐ the drain of Q_8 shows a high impedance r_{d8}. Thus, while the feedback (and thus also the compensation) remains functional, the feedforward path for noise is interrupted. Specifically, if V_{DD} changes, the source voltage of Q_6 does too and (as explained earlier) the gate voltage of v_{G6} of Q_6 must follow. Now, however, the output terminal is not shorted to v_{G6}, and hence v_{out} need not follow v_{G6}. Thus, A_p is considerably reduced at high frequencies.

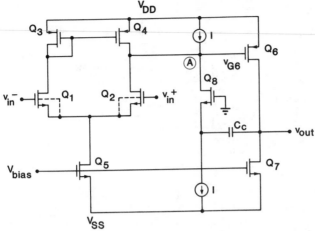

FIGURE 4.101. Compensation scheme for improving the positive-supply PSRR.

The currents of the two sources I (needed to keep Q_8 in its saturated region) must be carefully matched. This is possible using the strategy explained in connection with Fig. 4.99. Any mismatch will introduce a systematic offset voltage. Also, the impedance at node Ⓐ is somewhat reduced and therefore so is the gain of the input stage. Finally, due to the added devices, the internally generated noise of the op-amp is increased; however, the effect of increased PSRR usually outweighs this and the overall noise at the op-amp output is reduced.

Another path for power supply noise injection is provided by the stray capacitances.[20,21] Consider the circuit of Fig. 4.102. It illustrates a typical switched-capacitor integrator (to be discussed in the next chapter) in one of its switched states. The two parallel-connected transistors Q_1 and Q_2 constitute the switch. The parasitic junction capacitances coupling the drain of Q_1 and the source of Q_2 to the substrate, as well as the stray capacitances between the lines connected to the inverting input (node Ⓐ) and the substrate and power lines are illustrated as C_{DD} and C_{SS}. (As shown, these capacitors contain both linear and nonlinear components.) Consider the effect of v_{nD}. Since node Ⓐ

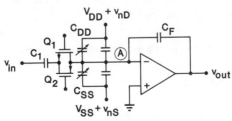

FIGURE 4.102. Parasitic capacitances affecting the summing node Ⓐ of an integrator.

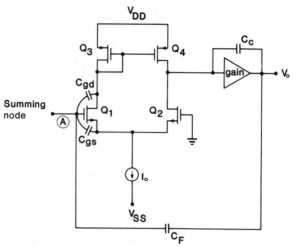

FIGURE 4.103. Equivalent circuit of the op-amp connected as an integrator.

is a virtual ground, the noise charge entering is $C_{DD}v_{nD}$. This charge flows into C_F, and causes an output noise voltage $-(C_{DD}/C_F)v_{nD}$. Thus, the noise gain is $-C_{DD}/C_F$. Similarly, the noise gain for v_{nS} is $-C_{SS}/C_F$. Since the signal gain is $v_{out}/v_{in} = -C_1/C_F$, the PSRR of the integrator is $C_1/(C_{SS} + C_{DD})$.* Thus, the PSRR can be increased by minimizing the stray capacitances and by choosing the values of the capacitors C_1 and C_F sufficiently large—this latter, of course, increases the overall area occupied on the chip by the stage. To reduce the stray capacitances and thereby the influence of the substrate noise, the dimensions of the switches should be chosen as small as possible, and whenever feasible, all lines connected to the input nodes of the op-amps should be shielded by grounded polysilicon or diffusion planes placed between the lines and the substrate.

Parasitic capacitances inside the op-amp also contribute to the power-supply noise gain.[20] Consider the equivalent circuit of the op-amp with an external feedback capacitor C_F, as shown in Fig. 4.103. The figure also illustrates the stray capacitances C_{gd} and C_{gs} of the input device Q_1. Noise in the positive line will appear at the drains of Q_3 and Q_4 and, from the former, will be coupled to the input node Ⓐ via C_{gd} and from there to the output via C_F. In addition, variations of the bias current I_0 due to the noise in V_{DD} and V_{SS} will change the gate-to-source voltages v_{gs} of Q_1 and Q_2 by $\Delta v_{gs} \cong (\Delta I_0/2)/g_{mi}$. The corresponding change in the source voltages will be coupled to node Ⓐ by C_{gs}. Similarly, changes in V_{SS} will alter the threshold voltage V_{Tn} of Q_1 and Q_2, unless these devices are placed in an isolated p-well. The resulting change

---

*This calculation assumes the worst-case condition that v_{nD} and v_{nS} are fully correlated.

in the source voltages will also be coupled to the input node by C_{gs} and to the output via C_F. Overall, it can be shown[20] that the relations

$$\frac{\partial v_{\text{out}}}{\partial V_{ss}} \cong \frac{C_{gs}}{C_F}\left(\frac{\partial I_0}{\partial V_{ss}}\frac{1}{2g_{mi}} + \frac{\partial V_{Tn}}{\partial V_{ss}}\right) + \frac{C_{gd}}{C_F}\frac{1}{2g_{ml}}\frac{\partial I_0}{\partial V_{ss}}$$

$$\frac{\partial v_{\text{out}}}{\partial V_{DD}} \cong -\frac{C_{gd}}{C_F}\left(1 - \frac{\partial I_0}{\partial V_{DD}}\frac{1}{2g_{ml}}\right) + \frac{C_{gs}}{C_F}\frac{1}{2g_{mi}}\frac{\partial I_0}{\partial V_{DD}} \qquad (4.203)$$

give the power-supply noise gains of the circuit. The terms containing $\partial I_0/\partial V_{ss}$ and $\partial I_0/\partial V_{DD}$ can be eliminated by using supply-independent biasing for the current source I_0. The $\partial V_{Tn}/\partial V_{ss}$ term can be eliminated *in CMOS amplifiers* by using a *p*-well for Q_1 and Q_2, connected to their sources. *In NMOS amplifiers*, the dependence of V_{Tn} on the supply voltage can be reduced by using a very lightly doped substrate.

If all the above steps are taken to reduce the noise gain, then the remaining gain is $\partial v_{\text{out}}/\partial V_{DD} \cong -C_{gd}/C_F$. This can, in principle, be reduced by making Q_1 and Q_2 small and/or C_F large. The former measure, however results in the increase of internally generated noise, while the latter increases the chip area needed by the stage. A technique which eliminates the problem, at the cost of a slightly reduced common-mode input range, is to use cascode circuitry (Fig. 4.104) in the input stage. The added devices Q_5 and Q_6 then buffer the drains of Q_1 and Q_2 from the variations of V_{DD} provided V_{bias} is independent of the supply voltages.

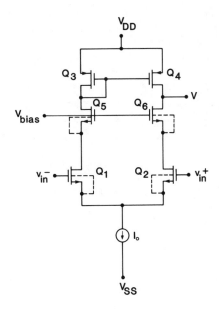

FIGURE 4.104. Cascode CMOS differential input stage.

4.15. OP-AMP DESIGN TECHNIQUES AND EXAMPLES

The design of MOS op-amps is not an exact scientific process. Typically, the circuit must satisfy many requirements, often including conflicting ones. The op-amp performance parameters most often specified are collected in Table 4.1. Other important design criteria include the noise level, dynamic range, output impedance and the area occupied on the chip. The specific steps followed in the design depend on the application, the circuit chosen, and the relative importance of the various criteria.

To illustrate the process, the devices in the circuit of Fig. 4.97 will be dimensioned such that the following specifications are satisfied:

Low-frequency gain	$A_0 \geq 70$ dB
Unity-gain frequency	$f_0 \geq 2$ MHz
Slew rate	$S_r \geq 4$ V/μs
Common-mode rejection	CMRR ≥ 80 dB
Phase margin	$\phi M > 60°$
Load impedance	$C_L = 10$ pF
dc supply voltages	$V_{DD} = -V_{SS} = 5$ V

It will be assumed that the transconductance factor $k' \triangleq \mu C_{\text{ox}}/2$ is 30 μA/V^2 for the NMOS devices and 12 μA/V^2 for the PMOS ones. Also the threshold voltages are assumed to be $V_{Tn} = 1.2$ V and $V_{Tp} = -1$ V.

TABLE 4.1. Op-Amp Performance Parameters

Design Parameter	Symbol	Relation to Other Parameters[a]	Typical Values
Low-frequency open-loop gain	A_0	$\dfrac{g_{mi}(g_{m6} + g_{m7})}{(g_{dl} + g_{di})(g_{d6} + g_{d7})}$	$10^3 \sim 10^4$
Unity-gain frequency	f_0	$g_{mi}/2\pi C_c$	$f_0 = 1 \sim 10$ MHz
Slew rate	S_r	I_0/C_c	$2 \sim 20$ V/μs
Common-mode rejection ratio	CMRR	$2\dfrac{g_{mi}g_{ml}}{g_{d5}g_{di}}$	$60 \sim 80$ dB
dc Power drain	P_{dc}	$(V_{DD} - V_{SS})(I_0 + I_7 + I_{10})$	$0.5 \sim 10$ mW
Phase margin (open loop)	ϕM	$\phi M > 60°$ for $\|s_{p_2}\| \geq 3\omega_0$	$45° \sim 90°$
Load impedance	R_L, C_L	None	$1 \sim 100$ kΩ $1 \sim 100$ pF

[a] The formulas in this column are given for the circuit of Fig. 4.98. Set g_{m7} and I_{10} equal to zero to obtain the relations for the amplifier of Fig. 4.97.

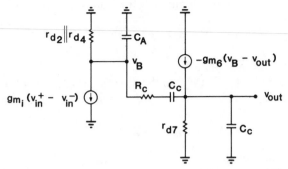

FIGURE 4.105.   Small-signal equivalent circuit of the op-amp of Fig. 4.97.

As mentioned already in Section 4.14, it is usual to choose the value of the compensation capacitor C_c equal to C_L. Hence, we set

$$C_c = C_L = 10 \text{ pF}. \tag{4.204}$$

For an adequate phase margin ϕM, the frequency of the second pole s_{p2} of the open-loop gain should be sufficiently higher than ω_0, the unity-gain frequency. As Fig. 4.64 shows, for $|s_{p2}| \sim 2\omega_0$ the contribution of the factor $j\omega - s_{p2}$ to the phase at $\omega = \omega_0$ is about 30°, and hence the phase margin is 60°. Thus, $|s_{p2}| = 3\omega_0$ gives a margin greater than 60°. The value of s_{p2} can be found from the small-signal equivalent circuit of the op-amp of Fig. 4.97, shown in Fig. 4.105. Alternatively, we can use the formula (4.118) derived for the circuit of Fig. 4.70; here, however, we must replace g_{m8} by g_{m6} (since Q_6 is now the output driver) and omit g_{m9} (since Q_7, the lower output device, is used merely as a current source). Thus, assuming $C_A \ll C_L = C_c$ and $g_{m6} \gg g_{d7}$,

$$s_{p2} = \frac{-g_{m6}C_c}{C_A(C_L + C_c) + C_LC_c} \cong -\frac{g_{m6}}{C_L} \tag{4.205}$$

results. Therefore, as already given in (4.191),

$$|s_{p2}| = g_{m6}/C_L = 3\omega_0 = 3g_{mi}/C_c \tag{4.206}$$

is the design equation. This gives

$$g_{m6} = 3g_{mi} = 3\omega_0 C_L = 3 \times 2\pi \times 2 \times 10^6 \times 10^{-11} \tag{4.207}$$

and hence

$$g_{m6} \cong 377 \times 10^{-6} \text{ A/V} = 377\,\mu\text{A/V}$$

$$g_{mi} \cong 125.7\,\mu\text{A/V}. \tag{4.208}$$

The specified slew rate requires that the bias current of the input stage satisfy

$$I_0 = S_r C_c \geq 4 \times 10^6 \times 10^{-11} = 40 \ \mu\text{A}. \qquad (4.209)$$

We can thus choose $I_0 = 40 \ \mu\text{A}$.

As explained in Section 4.14—see the discussions preceding Eq. (4.192)—the negative-going slew-rate limitation due to the use of Q_7 as a current source is

$$S_{r0} \leq I_{\text{bias}}/C_L. \qquad (4.210)$$

To make this effect small, we can set $S_{r0} = 2.5 S_r = 10 \ \text{V}/\mu\text{s}$. Then

$$I_{\text{bias}} = C_L S_{r0} = 10^{-11} \times 10^7 = 100 \ \mu\text{A}. \qquad (4.211)$$

Such large current also enables the realization of the required large g_{m6} without an excessive aspect ratio $(W/L)_6$.

From Eq. (4.182), to avoid systematic offset voltage, the condition

$$\frac{(W/L)_3}{(W/L)_6} = \frac{(W/L)_4}{(W/L)_6} = \frac{I_0/2}{I_{\text{bias}}} = \frac{1}{5} \qquad (4.212)$$

must hold. Since, by Eq. (3.18) of Section 3.4, g_m is proportional to $\sqrt{(W/L)i_{D0}}$, Eqs. (4.209)–(4.212) give for the transconductance of the loads

$$g_{ml} = g_{m3} = g_{m4} = \sqrt{\frac{(W/L)_3 I_0/2}{(W/L)_6 I_{\text{bias}}}} \ g_{m6} = \frac{I_0/2}{I_{\text{bias}}} g_{m6} = \frac{g_{m6}}{5}$$

$$\cong 75.4 \ \mu\text{A}/\text{V}. \qquad (4.213)$$

At this point, an estimate of the low-frequency gain A_0 and the common-mode rejection ratio can be found. From Eq. (3.20) of Section 3.4 in Chapter 3, the drain conductance of a MOSFET is approximately $g_d \approx \lambda i_D^0$, where λ is the channel-length modulation constant* ($\lambda \cong 0.03 \ \text{V}^{-1}$ for $L \approx 10 \ \mu\text{m}$) and i_D^0 the dc drain current. Hence, from Fig. 4.97 or Table 4.1,

$$A_0 = \frac{g_{mi}g_{m6}}{(g_{dl} + g_{di})(g_{d6} + g_{d7})} \cong \frac{g_{mi}g_{m6}}{(\lambda I_0)(2\lambda I_{\text{bias}})}$$

$$\cong \frac{125.7 \times 10^{-6} \times 377 \times 10^{-6}}{(0.03 \times 40 \times 10^{-6})(0.06 \times 100 \times 10^{-6})} \cong 6582 \qquad (4.214)$$

which corresponds to over 76 dB gain. Similarly, from (4.66), the common-mode

*Since in fact n- and p-channel devices have slightly different λ values, this calculation gives only a rough estimate of A_0 and the CMRR.

rejection can be approximated as follows:

$$\text{CMRR} = 2\frac{g_{mi}g_{ml}}{g_{d5}g_{di}} \cong \frac{2g_{mi}g_{ml}}{(\lambda I_0)(\lambda I_0/2)}$$

$$= \frac{2 \times 125.7 \times 10^{-6} \times 75.4 \times 10^{-6}}{(0.03 \times 40 \times 10^{-6})(0.03 \times 20 \times 10^{-6})} \cong 26{,}327 \quad (4.215)$$

which corresponds to about 88 dB common-mode rejection ratio.

Both values exceed the specifications. If this would not have been the case, the specifications would have been inconsistent. This can be seen by using Eqs. (4.205)–(4.213) to express the parameters entering A_0 and CMRR:

$$g_{mi} = C_c\omega_0,$$

$$I_0 = C_cS_r,$$

$$g_{m6} = 3\omega_0C_L, \quad (4.216)$$

$$I_{\text{bias}} = S_{r0}C_L,$$

$$g_{ml} = I_0g_{m6}/(2I_{\text{bias}}) = 3C_cS_r\omega_0/2S_{r0}.$$

Hence,

$$A_0 \cong \frac{3\omega_0^2}{2\lambda^2S_rS_{r0}} \quad (4.217)$$

and

$$\text{CMRR} \cong \frac{6\omega_0^2}{\lambda^2S_rS_{r0}} \cong 4A_0. \quad (4.218)$$

Thus, both A_0 and CMRR are fully determined by ω_0 and the slew rates.* They can be increased by choosing g_{mi} and g_{m6} (and hence the realized values of ω_0 and $|s_{p2}|$) larger.

Next, the channel resistance r_8 of Q_8 is found so as to place the zero s_z of $A_v(s)$ at a desirable location. Simple analysis based on Fig. 4.105 shows that for the circuit of Fig. 4.97 the zero is at

$$s_z = \frac{-1}{(R_c - 1/g_{m6})C_c}, \quad (4.219)$$

---

*In single-ended switched-capacitor circuits where the noninverting input is grounded, the CMRR as such is not very important.

and hence the required resistance is related to the desired zero location s_z by the formula

$$R_c = \frac{1}{|s_z| C_c} + \frac{1}{g_{m6}}. \tag{4.220}$$

As discussed in Section 4.8, there exist different strategies for choosing s_z. One possibility is to use $s_z = s_{p2}$; another is to shift s_z to ∞. For the former choice, using (4.205) and $C_c = C_L$, we get

$$R_c = \frac{1}{|s_{p2}| C_c} + \frac{1}{g_{m6}} \cong \frac{2}{g_{m6}} \cong 5.3 \text{ k}\Omega. \tag{4.221}$$

For the latter case, $R_c = 1/g_{m6} \cong 2.65 \text{ k}\Omega$. (Note that an even larger phase margin can be obtained by choosing $|s_z|$ only slightly above ω_0; $|s_z| \sim 1.2\omega_0$ is usually a reasonable value.*) Here, we choose the value given in (4.221). Note that Q_8 is clearly in its linear region, since its gate is at V_{SS} while its dc drain-to-source voltage is zero. Hence, by Eq. (3.6) of Section 3.2,

$$\frac{1}{R_c} = \left| \frac{\partial i_D}{\partial v_D} \right| = \mu C_{ox} \left(\frac{W}{L} \right)_8 |v_{GS8} - V_T|$$

$$= 2k_8(|V_{SS} - v_{D8}| - |V_T|). \tag{4.222}$$

Next, the design of the current sources Q_5 and Q_7 will be discussed. The aspect ratios W/L of these transistors should not be too small since otherwise for the given currents (I_0, I_{bias}) the required "excess" gate-to-source voltage ($v_{GS} - V_T$) will be large. This is inconvenient, since the voltages v and v_{out} (Fig. 4.97) are not allowed to drop below $V_{SS} + v_{GS} - V_T$ if Q_5 and Q_7 are to stay in saturation. Hence, a large $v_{GS} - V_T$ for Q_5 and Q_7 restricts the voltage swing and thus the dynamic range of the op-amp.

On the other hand, the areas of Q_5 and Q_7 should not be too large, either. One reason is that, of course, real estate is very expensive on the chip; the other, that a large area for Q_5 increases the stray capacitance C_W across the current source. This capacitance consists of two parallel-connected reverse-biased junction capacitances: the drain-to-substrate capacitance of Q_5, and the p-well-to-substrate capacitance of Q_1 and Q_2. C_w causes a decrease of the CMRR at high frequencies, since then g_{d5} in (4.215) is replaced by $g_{d5} + j\omega C_w$. Also, as explained in connection with Figs. 4.74–4.76, C_w causes a distortion in the step response of the op-amp. A large stray capacitance across Q_7, caused by a large drain diffusion, will increase C_L and hence reduce the phase margin.

*Professor K. Martin, personal communication.

Thus, a compromise should be found when Q_5 and Q_7 are dimensioned. From Eq. (3.8) of Chapter 3, the excess gate-to-source voltages are

$$v_{GS5} - V_T \cong \sqrt{\frac{I_0}{k'_n(W/L)_5}}$$

and   (4.223)

$$v_{GS7} - V_T \cong \sqrt{\frac{I_{\text{bias}}}{k'_n(W/L)_7}} \, .$$

We assumed $k'_n = 30 \ \mu\text{A}/\text{V}^2$ for the NMOS devices and $k'_p = 12 \ \mu\text{A}/\text{V}^2$ for the PMOS ones; hence allowing 0.5 V excess voltage for both Q_5 and Q_7 we get

$$(W/L)_5 = \frac{I_0}{k'_n(v_{GS5} - V_T)^2} = \frac{40}{30 \times (0.5)^2} \cong 5.33$$

and   (4.224)

$$(W/L)_7 = \frac{I_{\text{bias}}}{k'_n(v_{GS7} - V_T)^2} = \frac{100}{30 \times (0.5)^2} \cong 13.33.$$

To avoid short-channel effects which occur for $L < 10 \ \mu\text{m}$ and which would increase the drain conductance g_d, we choose $L_5 = L_7 = 10 \ \mu\text{m}$. Then $W_5 = 54 \ \mu\text{m}$ and $W_7 = 133 \ \mu\text{m}$ can be used.

We can next calculate the aspect ratios of $Q_1 - Q_4$ and Q_6 from their transconductances.

From Eq. (3.18) of Chapter 3, assuming $|\lambda v_{Ds}| \ll 1$, the transconductance is given by

$$g_m \cong 2\sqrt{k'(W/L)i_D^0} \, .$$ (4.225)

Hence, the aspect ratios can be found from

$$(W/L)_1 = (W/L)_2 \cong \frac{g_{mi}^2}{4k'_n I_0/2} = \frac{(125.7)^2}{4 \times 30 \times 20} \cong 6.58$$

and   (4.226)

$$(W/L)_3 = (W/L)_4 \cong \frac{g_{m1}^2}{4k'_p I_0/2} = \frac{(75.4)^2}{4 \times 12 \times 20} \cong 5.92,$$

and, by (4.212),

$$(W/L)_6 = 5(W/L)_3 \cong 29.6. \tag{4.227}$$

Choosing again $L = 10 \ \mu m$ for all transistors, $W_1 = W_2 = 66 \ \mu m$, $W_3 = W_4 = 60 \ \mu m$, and $W_6 = 300 \ \mu m$ result. (Note that often noise considerations require that the width of the input devices be chosen much larger, say 200 μm or more!)

Next, we can estimate the (common) dc bias voltages at the drains of $Q_1 - Q_4$. Since they all carry a dc current $I_0/2$, we have

$$i_{D3} = I_0/2 = k_p'(W/L)_3 \big(|v_{GS3}| - |V_{Tp}| \big)^2. \tag{4.228}$$

Since the PMOS threshold voltage is $V_{Tp} = -1 \ V$,

$$|v_{GS3}| = |V_{Tp}| + \sqrt{\frac{I_0/2}{k_p'(W/L)_3}} = 1 + \sqrt{\frac{20}{12 \times 6}}$$

$$\cong 1.527 \ V. \tag{4.229}$$

Hence, the drains of $Q_1 - Q_4$ are at a dc bias voltage $V_{DD} - |v_{Gs3}| \cong 5 - 1.527 = 3.473 \ V$. This is also the drain-and-source bias voltage v_{D8} of Q_8, and hence from (4.221) and (4.222)

$$2k_p'(W/L)_8(|-5 - 3.473| - 1) = 1/R_c \tag{4.230}$$

$$(W/L)_8 = \frac{1}{2 \times 12 \times 10^{-6} \times 5300 \times 7.473} \cong 1.052.$$

Hence, $W_8 = L_8 = 10 \ \mu m$ can be used.

At this point, the dimensions of all devices have been (tentatively) determined, and we also know the values of all currents. The drain voltages of Q_1 and Q_2 have been found; their sources are (for $v_{in}^- = v_{in}^+ = 0$) at a voltage v such that

$$k_n'(W/L)_1(-v - V_{Tn})^2 = I_0/2 \tag{4.231}$$

which gives, for $V_{Tn} \cong 1.2 \ V$,

$$-v = V_{Tn} + \sqrt{\frac{I_0}{2k_n'(W/L)_1}}$$

$$-v = 1.2 + \sqrt{\frac{40}{2 \times 30 \times 6.6}} = 1.518 \ V \tag{4.232}$$

so $v \cong -1.52 \ V$.

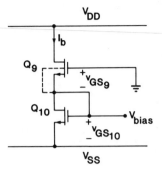

FIGURE 4.106.   Bias circuit for the op-amp of Fig. 4.97.

The only remaining task is to design a bias chain which provides V_{bias}. In (4.224) Q_5 and Q_7 have been dimensioned such that $v_{GS5} = V_{Tn} + 0.5$ V $= 1.7$ V. Thus, $V_{\text{bias}} = V_{SS} + v_{GS5} = -3.3$ V. This can be achieved by the simple circuit shown in Fig. 4.106. Choosing the current of the bias chain $I_b = 20 \ \mu$A, the aspect ratios of Q_9 and Q_{10} are easily found: since $v_{GS9} = 0 - V_{\text{bias}} = 3.3$ V and $v_{GS10} = V_{\text{bias}} - V_{SS} = 1.7$ V are known

$$(W/L)_9 = \frac{I_b}{k'_n(v_{GS9} - V_{Tn})^2}$$

$$= \frac{20}{30 \times 2.1^2} \cong 0.1512 \qquad (4.233)$$

and

$$(W/L)_{10} = \frac{I_b}{k'_n(v_{GS10} - V_{Tn})^2} \cong 2.667. \qquad (4.234)$$

Hence, $W_9 = 10 \ \mu$m, $L_9 = 66 \ \mu$m and $W_{10} = 27 \ \mu$m, $L_{10} = 10 \ \mu$m can be used.

It should be noted that V_{bias} and I_b are insensitive to variations of V_{DD}, but not to changes in V_{SS}. If $V_{SS} \to V'_{SS} = V_{SS} + \Delta V_{SS}$, then v_{GS9} and v_{GS10} also change, such that

$$v'_{GS9} + v'_{GS10} = -V_{SS} - \Delta V_{SS} = |V'_{SS}|; \qquad (4.235)$$

and I_b changes to

$$I'_b = k_9 \left(v'_{GS9} - V_{Tn}\right)^2 = k_{10}\left(v'_{GS10} - V_{Tn}\right)^2. \qquad (4.236)$$

Here, the prime denotes changed values.
    From (4.236)

$$v'_{GS9} = \sqrt{k_{10}/k_9}\left(v'_{GS10} - V_{Tn}\right) + V_{Tn}. \qquad (4.237)$$

Combining with (4.235), and solving for v'_{GS9},

$$v'_{GS10} = \frac{V_{Tn}\left(\sqrt{k_{10}/k_9} - 1\right) - V_{SS} - \Delta V_{SS}}{\sqrt{k_{10}/k_9} + 1}.$$ (4.238)

Thus, a positive change of $+0.1$ V in V_{SS} will change v_{GS10} by

$$\Delta v_{GS10} = \frac{-\Delta V_{SS}}{\sqrt{k_{10}/k_9} + 1} \cong -0.01915 \text{ V}.$$ (4.239)

Hence, $V_{bias} = V_{SS} + v_{GS10}$ changes by $0.1 - 0.01915 \cong 0.081$ V. The corresponding changes in I_0 and I_{bias} can be found approximately from

$$\Delta I_0 \cong g_{m5} \Delta v_{GS5} = g_{m5} \Delta v_{GS10}$$

$$\cong 2\sqrt{k_5 I_0} \, \Delta v_{GS10} = -2\sqrt{30 \times 5.4 \times 40} \times 0.01915$$

$$\cong -3.1 \, \mu\text{A}$$ (4.240)

and

$$\Delta I_{bias} \cong g_{m7} \Delta v_{GS10} \cong -2\sqrt{30 \times 13.3 \times 100} \times 0.01915$$

$$= -7.7 \, \mu\text{A}.$$ (4.241)

If these changes are not acceptable, then the bias-independent circuit of Fig. 4.5c can be used to provide $v_{01} = V_{bias}$.

To verify the accuracy of the design, the overall circuit was analyzed using the popular program SPICE2.[30] Figure 4.107 shows the computed gain and phase responses of the circuit under open-circuit conditions. The unity-gain bandwidth is about 2.5 MHz, the dc gain 77.1 dB, and the phase margin about 90°. Thus, these specifications are met. The open-loop slew-rate (S_r) performance is illustrated in Figs. 4.108 (for a negative voltage input step) and 4.109 (for a positive input step).* The maximum slope gives S_r; it is over 5 V/μs in both directions. Hence, this requirement is also satisfied. Figure 4.110 illustrates the common-mode gain response. The required 80 dB CMRR is clearly obtained across the full dc-to-unity-gain frequency range:

The systematic input offset voltage can be estimated from the output voltage v_{out}^0 for $v_{in} = 0$ (Fig. 4.108) as $v_{out}^0/A_0 \cong 1.3$ V/$10^{77.1/20} \cong 0.2$ mV. This is very low, and likely to be negligible compared to the random offset.

All computed responses have been obtained using the simple square-law MOSFET model described by the drain/current relation (3.6)–(3.11) of

---

*In these figures, the input waveform is shown by crosses, the output by stars.

```
****      AC ANALYSIS                        TEMPERATURE =   27.000 DEG C

LEGEND:

*: VDB(8)
+: VP(8)

    FREQ      VDB(8)

(*)------------- -5.000D+01    0.000D+00    5.000D+01   1.000D+02 1.500D+02

                 - - - - - - - - - - - - - - - - - - - - - - - -

(+)------------- -1.500D+02   -1.000D+02   -5.000D+01   0.000D+00 5.000D+01

                 - - - - - - - - - - - - - - - - - - - - - - - -
  1.000D+01    7.711D+01 .                    .            .       *      +        .
  1.259D+01    7.711D+01 .                    .            .       *    +.
  1.585D+01    7.710D+01 .                    .            .       *    +.
  1.995D+01    7.710D+01 .                    .            .       *    +.
  2.512D+01    7.709D+01 .                    .            .       *    +.
  3.162D+01    7.707D+01 .                    .            .       *   + .
  3.981D+01    7.705D+01 .                    .            .       *   + .
  5.012D+01    7.701D+01 .                    .            .       *   + .
  6.310D+01    7.696D+01 .                    .            .       *  +  .
  7.943D+01    7.687D+01 .                    .            .       * +   .
  1.000D+02    7.673D+01 .                    .            .       * +
  1.259D+02    7.652D+01 .                    .            .       *+
  1.585D+02    7.620D+01 .                    .            .       X
  1.995D+02    7.575D+01 .                    .            .    +  *
  2.512D+02    7.511D+01 .                    .            .   +   *
  3.162D+02    7.426D+01 .                    .            .  +    *
  3.981D+02    7.318D+01 .                    .            . +     *
  5.012D+02    7.188D+01 .                    .            +      .*
  6.310D+02    7.039D+01 .                    .         +  .       *
  7.943D+02    6.874D+01 .                    .       +    .       *
  1.000D+03    6.698D+01 .                    .      +     .       *
  1.259D+03    6.514D+01 .                    .     +      .      *
  1.585D+03    6.325D+01 .                    .    +       .     *
  1.995D+03    6.131D+01 .                    .   +        .    *
  2.512D+03    5.935D+01 .                    .   +        .   *
  3.162D+03    5.738D+01 .                    .  +         .  *
  3.981D+03    5.540D+01 .                    .  +         . *
  5.012D+03    5.341D+01 .                    .  +         .*
  6.310D+03    5.142D+01 .                    .  +         *
  7.943D+03    4.942D+01 .                    . +        *.
  1.000D+04    4.742D+01 .                    . +       *.
  1.259D+04    4.542D+01 .                    . +       *.
  1.585D+04    4.343D+01 .                    . +      *.
  1.995D+04    4.143D+01 .                    . +      *.
  2.512D+04    3.943D+01 .                    . +     *
  3.162D+04    3.743D+01 .                    . +    *
  3.981D+04    3.543D+01 .                    . +    *
  5.012D+04    3.343D+01 .                    . +   *
  6.310D+04    3.143D+01 .                    . +   *
  7.943D+04    2.943D+01 .                    . +  *
  1.000D+05    2.743D+01 .                    . +  *
  1.259D+05    2.543D+01 .                   ., +  *
  1.585D+05    2.343D+01 .                    . + *
  1.995D+05    2.143D+01 .                    . +  *
  2.512D+05    1.944D+01 .                    . + *
  3.162D+05    1.744D+01 .                    .+  *
  3.981D+05    1.545D+01 .                    +*
  5.012D+05    1.346D+01 .                    +*
  6.310D+05    1.148D+01 .                    *+
  7.943D+05    9.507D+00 .                    *+
  1.000D+06    7.552D+00 .                  . *  +
  1.259D+06    5.621D+00 .                  . *  +
  1.585D+06    3.725D+00 .                 .*    +
  1.995D+06    1.880D+00 .                 .*    +
  2.512D+06    9.964D-02 .               *      +     .
  3.162D+06   -1.602D+00 .               *      +     .
```

FIGURE 4.107. Computed gain and phase responses for the op-amp example.

```
****      TRANSIENT ANALYSIS                TEMPERATURE =    27.000 DEG C

LEGEND:

*: V(8)
+: V(6)

    TIME       V(8)

(*)------------- 1.000D+00     2.000D+00       3.000D+00      4.000D+00  5.000D+00

                 - - - - - - - - - - - - - - - - - - - - - - - - - - - - - - -

(+)------------- -1.500D+00    -1.000D+00      -5.000D-01     0.000D+00  5.000D-01

                 - - - - - - - - - - - - - - - - - - - - - - - - - - - - - - -
 0.000D+00  1.293D+00 .    *                    .               .        +        .
 2.000D-08  1.293D+00 .    *                    .               .        +        .
 4.000D-08  1.293D+00 .    *                    .               .        +        .
 6.000D-08  1.293D+00 .    *                    .               .        +        .
 8.000D-08  1.293D+00 .    *                    .               .        +        .
 1.000D-07  1.293D+00 .    *                    .               .        +        .
 1.200D-07  1.342D+00 .      *                  .          +                       .
 1.400D-07  1.457D+00 .        *                .       +                          .
 1.600D-07  1.580D+00 .          *          +                                      .
 1.800D-07  1.688D+00 .       *        +                                           .
 2.000D-07  1.790D+00 .       *  +                                                 .
 2.200D-07  1.891D+00 .       * +                                                  .
 2.400D-07  1.991D+00 .       X                                                    .
 2.600D-07  2.090D+00 .      +*                                                    .
 2.800D-07  2.190D+00 .      +   *                                                 .
 3.000D-07  2.290D+00 .      +     *                                               .
 3.200D-07  2.390D+00 .      +       *                                             .
 3.400D-07  2.489D+00 .      +          *                                          .
 3.600D-07  2.589D+00 .      +            *                                        .
 3.800D-07  2.689D+00 .      +                                                     .
 4.000D-07  2.789D+00 .      +              *                                      .
 4.200D-07  2.888D+00 .      +                *                                    .
 4.400D-07  2.988D+00 .      +                                                     .
 4.600D-07  3.088D+00 .      +                  *                                  .
 4.800D-07  3.187D+00 .      +                    *                                .
 5.000D-07  3.287D+00 .      +                      *                              .
 5.200D-07  3.387D+00 .      +                                                     .
 5.400D-07  3.487D+00 .      +                         *                           .
 5.600D-07  3.586D+00 .      +                           *                         .
 5.800D-07  3.686D+00 .      +                             *                       .
 6.000D-07  3.786D+00 .      +                              *                      .
 6.200D-07  3.885D+00 .      +                                *                    .
 6.400D-07  3.985D+00 .      +                                                     .
 6.600D-07  4.085D+00 .      +                                   *                 .
 6.800D-07  4.183D+00 .      +                                     *               .
 7.000D-07  4.280D+00 .      +                                       *             .
 7.200D-07  4.370D+00 .      +                                                     .
 7.400D-07  4.447D+00 .      +                                         *           .
 7.600D-07  4.517D+00 .      +                                                     .
 7.800D-07  4.575D+00 .      +                                          *          .
 8.000D-07  4.623D+00 .      +                                             *       .
 8.200D-07  4.664D+00 .      +,                                                    .
 8.400D-07  4.698D+00 .      +                                              *      .
 8.600D-07  4.726D+00 .      +                                                     .
 8.800D-07  4.751D+00 .      +                                               *     .
 9.000D-07  4.771D+00 .      +                                                *    .
 9.200D-07  4.789D+00 .      +                                                *    .
 9.400D-07  4.804D+00 .      +                                                *    .
 9.600D-07  4.817D+00 .      +                                                *    .
 9.800D-07  4.829D+00 .      +                                                 *   .
 1.000D-06  4.839D+00 .      +                                                 *   .
 1.020D-06  4.849D+00 .      +                                                 *   .
 1.040D-06  4.857D+00 .      +                                                 *   .
 1.060D-06  4.864D+00 .      +                                                 *   .
 1.080D-06  4.871D+00 .      +                                                 *   .
 1.100D-06  4.877D+00 .      +                                                 *   .
```

FIGURE 4.108. Step response of the op-amp example for a negative input step.

```
****      TRANSIENT ANALYSIS                  TEMPERATURE =   27.000 DEG C

LEGEND:

*: V(8)
+: V(6)

    TIME       V(8)

(*)------------ -6.000D+00    -4.000D+00     -2.000D+00      0.000D+00  2.000D+00

                      - - - - - - - - - - - - - - - - - - - - - - - - - - - - -

(+)------------ -5.000D-01     0.000D+00      5.000D-01      1.000D+00  1.500D+00

                      - - - - - - - - - - - - - - - - - - - - - - - - - - - - -
 0.000D+00  1.293D+00 .             +             .             .        *    .
 2.000D-08  1.293D+00 .             +             .             .        *    .
 4.000D-08  1.293D+00 .             +             .             .        *    .
 6.000D-08  1.293D+00 .             +             .             .        *    .
 8.000D-08  1.293D+00 .             +             .             .        *    .
 1.000D-07  1.293D+00 .             +             .             .        *    .
 1.200D-07  1.256D+00 .             .     +       .             .        *    .
 1.400D-07  1.179D+00 .             .          +  .             .       *     .
 1.600D-07  1.089D+00 .             .             .  +          .       *     .
 1.800D-07  9.956D-01 .             .             .       +     .      *      .
 2.000D-07  8.990D-01 .             .             .             +     *       .
 2.200D-07  8.007D-01 .             .             .             +     *       .
 2.400D-07  7.016D-01 .             .             .             +    *        .
 2.600D-07  6.020D-01 .             .             .             +   *         .
 2.800D-07  5.021D-01 .             .             .             +   *         .
 3.000D-07  4.020D-01 .             .             .             +  *          .
 3.200D-07  3.018D-01 .             .             .             + *           .
 3.400D-07  2.015D-01 .             .             .             +*            .
 3.600D-07  1.011D-01 .             .             .             +*            .
 3.800D-07  7.462D-04 .             .             .             X             .
 4.000D-07 -9.965D-02 .             .             .            *+             .
 4.200D-07 -2.000D-01 .             .             .            *+             .
 4.400D-07 -3.004D-01 .             .             .           * +             .
 4.600D-07 -4.008D-01 .             .             .          *  +             .
 4.800D-07 -5.012D-01 .             .             .         *   +             .
 5.000D-07 -6.016D-01 .             .             .         *   +             .
 5.200D-07 -7.020D-01 .             .             .        *    +             .
 5.400D-07 -8.023D-01 .             .             .       *     +             .
 5.600D-07 -9.027D-01 .             .             .       *     +             .
 5.800D-07 -1.003D+00 .             .             .      *      +             .
 6.000D-07 -1.103D+00 .             .             .     *       +             .
 6.200D-07 -1.204D+00 .             .             .     *       +             .
 6.400D-07 -1.304D+00 .             .             .    *        +             .
 6.600D-07 -1.404D+00 .             .             .   *         +             .
 6.800D-07 -1.505D+00 .             .             .  *          +             .
 7.000D-07 -1.605D+00 .             .             .  *          +             .
 7.200D-07 -1.705D+00 .             .             . *           +             .
 7.400D-07 -1.805D+00 .             .             .*            +             .
 7.600D-07 -1.906D+00 .             .             .*            +             .
 7.800D-07 -2.006D+00 .             .             *             +             .
 8.000D-07 -2.106D+00 .             .            *.             +             .
 8.200D-07 -2.206D+00 .             .            *.             +             .
 8.400D-07 -2.307D+00 .             .           * .             +             .
 8.600D-07 -2.407D+00 .             .          *  .             +             .
 8.800D-07 -2.507D+00 .             .         *   .             +             .
 9.000D-07 -2.607D+00 .             .         *   .             +             .
 9.200D-07 -2.707D+00 .             .        *    .             +             .
 9.400D-07 -2.807D+00 .             .       *     .             +             .
 9.600D-07 -2.907D+00 .             .       *     .             +             .
 9.800D-07 -3.008D+00 .             .      *      .             +             .
 1.000D-06 -3.108D+00 .             .     *       .             +             .
 1.020D-06 -3.208D+00 .             .     *       .             +             .
 1.040D-06 -3.308D+00 .             .    *        .             +             .
 1.060D-06 -3.408D+00 .             .   *         .             +             .
 1.080D-06 -3.508D+00 .             .  *          .             +             .
 1.100D-06 -3.608D+00 .             .  *          .             +             .
```

FIGURE 4.109. Step response of the op-amp example for a positive input step.

```
1.120D-06 -3.708D+00  .              .  *       .       .              +       .
1.140D-06 -3.808D+00  .               .*      .       .              +       .
1.160D-06 -3.908D+00  .               .*      .       .              +       .
1.180D-06 -4.008D+00  .               .*      .       .              +       .
1.200D-06 -4.108D+00  .             *.      .       .              +       .
1.220D-06 -4.207D+00  .             *.      .       .              +       .
1.240D-06 -4.307D+00  .            *.       .       .              +       .
1.260D-06 -4.407D+00  .           *.       .       .              +       .
1.280D-06 -4.504D+00  .         *         .       .              +       .
1.300D-06 -4.596D+00  .        *          .       .              +       .
1.320D-06 -4.676D+00  .       *           .       .              +       .
1.340D-06 -4.737D+00  .      *            .       .              +       .
1.360D-06 -4.787D+00  .    *              .       .              +       .
1.380D-06 -4.820D+00  .    *              .       .              +       .
1.400D-06 -4.840D+00  .    *              .       .              +       .
1.420D-06 -4.853D+00  .    *              .       .              +       .
1.440D-06 -4.859D+00  .    *              .       .              +       .
1.460D-06 -4.865D+00  .    *              .       .              +       .
1.480D-06 -4.870D+00  .    *              .       .              +       .
1.500D-06 -4.877D+00  . - -*- - - - - - - . - - - - . - - - - + - - - - -
```

FIGURE 4.109. continued.

Section 3.2. Using a more sophisticated and realistic model, very similar results are obtained. The main difference is that the g_{DS} parameters of the realistic model are larger than the value λi_D^0 predicted by the simpler model. Hence, the low-frequency gain drops to 72.4 dB, and the low-frequency CMRR to about 80 dB.

For illustration, the gains for noise entering via the positive (V_{DD}) supply (Fig. 4.111) as well as the negative (V_{SS}) one (Fig. 4.112) are also shown. As predicted in Section 4.14, the supply rejection becomes a problem at higher frequencies; at ω_0 the PSRR is near 0 dB. These curves were computed using the more realistic MOSFET model, and hence the behavior of the fabricated circuit will be similar.

Next, we will repeat the design for the same specifications, but using the circuit of Fig. 4.98 with its Class-AB output stage. Again, we select $C_c = C_L = 10$ pF, and since $\omega_0 = g_{mi}/C_c$, g_{mi} is given by (4.208), as before. Also, I (determined by the slew rate and C_c) remains the same, as given by (4.209). By the argument leading earlier to (4.224), $(W/L)_5 = 5.4$ can again be used. To obtain the specified CMRR, from Eqs. (4.66) and (3.20)

$$g_{ml} \geq (\text{CMRR}) \frac{g_{d5} g_{di}}{2 g_{mi}} \cong 10^4 \frac{(\lambda I_0)(\lambda I_0/2)}{2 g_{mi}}$$

$$\cong 10^4 \frac{(0.03)^2 (40 \times 10^{-6})^2}{2 \times 2 \times 125.7 \times 10^{-6}} \cong 28.6 \ \mu\text{A/V}. \qquad (4.242)$$

Hence, the values given in (4.213) remain suitable. In conclusion, the dimensions of the input stage can remain unchanged for the new circuit, since they are determined by the (unchanged) requirements on ω_0, S_r, C_c, and CMRR.

```
****    AC ANALYSIS                    TEMPERATURE =   27.000 DEG C

      FREQ      VDB(8)

                  -1.500D+02     -1.000D+02     -5.000D+01      0.000D+00  5.000D+01
                  - - - - - - - - - - - - - - - - - - - - - - - - - - - - - - - - -
1.000D+01 -5.815D+00 .             .              .              *  .              .
1.259D+01 -5.818D+00 .             .              .              *  .              .
1.585D+01 -5.822D+00 .             .              .              *  .              .
1.995D+01 -5.827D+00 .             .              .              *  .              .
2.512D+01 -5.837D+00 .             .              .              *  .              .
3.162D+01 -5.852D+00 .             .              .              *  .              .
3.981D+01 -5.875D+00 .             .              .              *  .              .
5.012D+01 -5.912D+00 .             .              .              *  .              .
6.310D+01 -5.969D+00 .             .              .              *  .              .
7.943D+01 -6.059D+00 .             .              .              *  .              .
1.000D+02 -6.197D+00 .             .              .              *  .              .
1.259D+02 -6.407D+00 .             .              .              *  .              .
1.585D+02 -6.721D+00 .             .              .              *  .              .
1.995D+02 -7.177D+00 .             .              .              *  .              .
2.512D+02 -7.813D+00 .             .              .              *  .              .
3.162D+02 -8.662D+00 .             .              .             *   .              .
3.981D+02 -9.739D+00 .             .              .            *    .              .
5.012D+02 -1.104D+01 .             .              .           *     .              .
6.310D+02 -1.253D+01 .             .              .          *      .              .
7.943D+02 -1.417D+01 .             .              .         *       .              .
1.000D+03 -1.593D+01 .             .              .        *        .              .
1.259D+03 -1.777D+01 .             .              .       *         .              .
1.585D+03 -1.967D+01 .             .              .      *          .              .
1.995D+03 -2.160D+01 .             .              .     *           .              .
2.512D+03 -2.356D+01 .             .              .    *            .              .
3.162D+03 -2.553D+01 .             .              .   *             .              .
3.981D+03 -2.752D+01 .             .              .  *              .              .
5.012D+03 -2.950D+01 .             .              . *               .              .
6.310D+03 -3.150D+01 .             .              .*                .              .
7.943D+03 -3.349D+01 .             .             .*                 .              .
1.000D+04 -3.549D+01 .             .            * .                 .              .
1.259D+04 -3.749D+01 .             .           *  .                 .              .
1.585D+04 -3.949D+01 .             .          *   .                 .              .
1.995D+04 -4.149D+01 .             .         *    .                 .              .
2.512D+04 -4.349D+01 .             .        *     .                 .              .
3.162D+04 -4.549D+01 .             .       *      .                 .              .
3.981D+04 -4.749D+01 .             .      *       .                 .              .
5.012D+04 -4.948D+01 .             .     *        .                 .              .
6.310D+04 -5.148D+01 .             .    *         .                 .              .
7.943D+04 -5.348D+01 .             .   *.         .                 .              .
1.000D+05 -5.548D+01 .             .  * .         .                 .              .
1.259D+05 -5.748D+01 .             . *  .         .                 .              .
1.585D+05 -5.947D+01 .             .*   .         .                 .              .
1.995D+05 -6.147D+01 .            .*    .         .                 .              .
2.512D+05 -6.346D+01 .           .*     .         .                 .              .
3.162D+05 -6.544D+01 .          .*      .         .                 .              .
3.981D+05 -6.742D+01 .         .*       .         .                 .              .
5.012D+05 -6.938D+01 .         *        .         .                 .              .
6.310D+05 -7.132D+01 .        *.        .         .                 .              .
7.943D+05 -7.322D+01 .       *.         .         .                 .              .
1.000D+06 -7.508D+01 .      *.          .         .                 .              .
1.259D+06 -7.686D+01 .     *.           .         .                 .              .
1.585D+06 -7.855D+01 .    *.            .         .                 .              .
1.995D+06 -8.011D+01 .    *             .         .                 .              .
2.512D+06 -8.153D+01 .   *.             .         .                 .              .
3.162D+06 -8.282D+01 .  *.              .         .                 .              .
3.981D+06 -8.399D+01 . *.               .         .                 .              .
5.012D+06 -8.512D+01 . *.               .         .                 .              .
                  - - - - - - - - - - - - - - - - - - - - - - - - - - - - - - - - -
```

FIGURE 4.110. Common-mode gain frequency response.

```
 ****      AC ANALYSIS                      TEMPERATURE =   27.000 DEG C

        FREQ       VDB(8)

                     -1.500D+02     -1.000D+02     -5.000D+01     0.000D+00  5.000D+01
                      - - - - - - - - - - - - - - - - - - - - - - - - - - - - - - - -
 1.000D+01 -8.357D+01  .                     .         *         .            .
 1.259D+01 -8.356D+01  .                     .         *         .            .
 1.585D+01 -8.354D+01  .                     .         *         .            .
 1.995D+01 -8.352D+01  .                     .         *         .            .
 2.512D+01 -8.348D+01  .                     .         *         .            .
 3.162D+01 -8.342D+01  .                     .         *         .            .
 3.981D+01 -8.332D+01  .                     .         *         .            .
 5.012D+01 -8.317D+01  .                     .         *         .            .
 6.310D+01 -8.294D+01  .                     .         *         .            .
 7.943D+01 -8.261D+01  .                     .         *         .            .
 1.000D+02 -8.213D+01  .                     .         *         .            .
 1.259D+02 -8.145D+01  .                     .         *         .            .
 1.585D+02 -8.057D+01  .                     .         *         .            .
 1.995D+02 -7.945D+01  .                     .           *       .            .
 2.512D+02 -7.812D+01  .                     .            *      .            .
 3.162D+02 -7.660D+01  .                     .             *     .            .
 3.981D+02 -7.493D+01  .                     .              *    .            .
 5.012D+02 -7.315D+01  .                     .               *   .            .
 6.310D+02 -7.130D+01  .                     .                *  .            .
 7.943D+02 -6.940D+01  .                     .                 * .            .
 1.000D+03 -6.746D+01  .                     .                  *.            .
 1.259D+03 -6.550D+01  .                     .                   *            .
 1.585D+03 -6.352D+01  .                     .                    *           .
 1.995D+03 -6.154D+01  .                     .                     *          .
 2.512D+03 -5.955D+01  .                     .                      *         .
 3.162D+03 -5.755D+01  .                     .                       *        .
 3.981D+03 -5.556D+01  .                     .                *                .
 5.012D+03 -5.356D+01  .                     .                 *               .
 6.310D+03 -5.156D+01  .                     .                  *              .
 7.943D+03 -4.956D+01  .                     .                   *             .
 1.000D+04 -4.756D+01  .                     .                    *            .
 1.259D+04 -4.557D+01  .                     .                     *           .
 1.585D+04 -4.357D+01  .                     .                      *          .
 1.995D+04 -4.157D+01  .                     .                       *         .
 2.512D+04 -3.957D+01  .                     .                        *        .
 3.162D+04 -3.757D+01  .                     .                         *       .
 3.981D+04 -3.557D+01  .                     .                          *      .
 5.012D+04 -3.357D+01  .                     .                           *     .
 6.310D+04 -3.157D+01  .                     .                            *    .
 7.943D+04 -2.957D+01  .                     .                             *   .
 1.000D+05 -2.758D+01  .                     .                              *  .
 1.259D+05 -2.559D+01  .                     .                               * .
 1.585D+05 -2.360D+01  .                     .                                *.
 1.995D+05 -2.162D+01  .                     .                                 *
 2.512D+05 -1.965D+01  .                     .                                 *
 3.162D+05 -1.769D+01  .                     .                                  *
 3.981D+05 -1.577D+01  .                     .                                   *
 5.012D+05 -1.388D+01  .                     .                                    *
 6.310D+05 -1.206D+01  .                     .                                     *
 7.943D+05 -1.032D+01  .                     .                                      *
 1.000D+06 -8.711D+00  .                     .                                       *
 1.259D+06 -7.272D+00  .                     .                                        *
 1.585D+06 -6.046D+00  .                     .                                         *
 1.995D+06 -5.067D+00  .                     .                                         *
 2.512D+06 -4.354D+00  .                     .                                          *
 3.162D+06 -3.907D+00  .                     .                                          *
 3.981D+06 -3.721D+00  .                     .                                          *
 5.012D+06 -3.789D+00  .                     .                                          *
                      - - - - - - - - - - - - - - - - - - - - - - - - - - - - - - - -
```

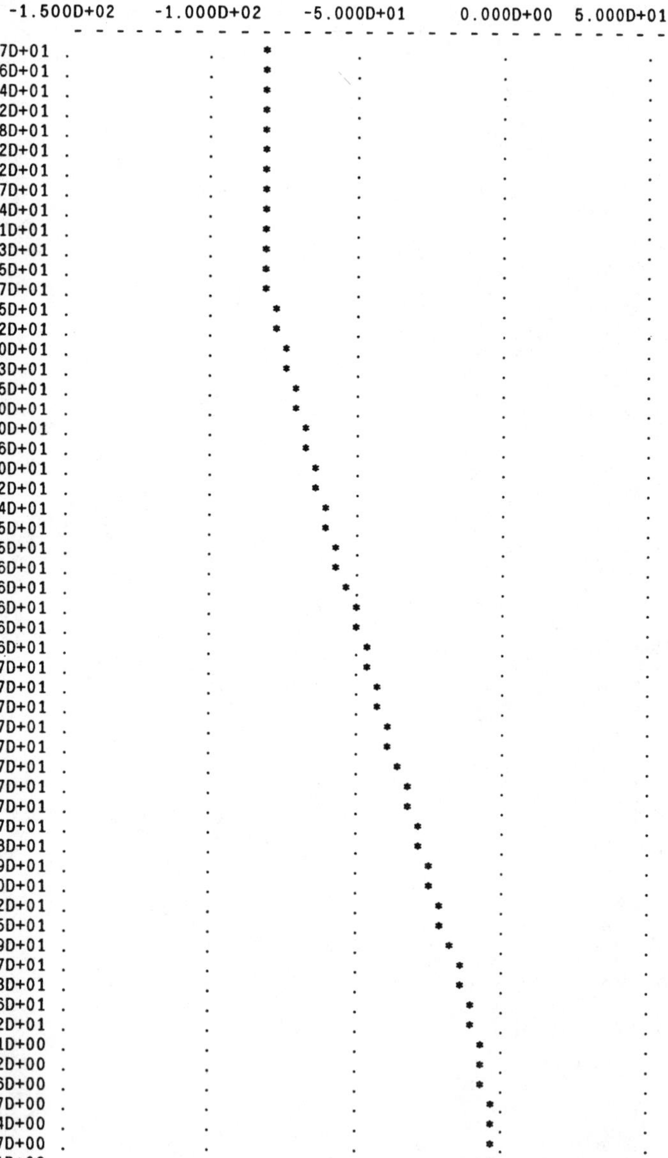

FIGURE 4.111. V_{DD} noise gain response.

```
****    AC ANALYSIS                    TEMPERATURE =   27.000 DEG C

    FREQ      VDB(8)

                  -8.000D+01    -6.000D+01    -4.000D+01    -2.000D+01  0.000D+00
                  - - - - - - - - - - - - - - - - - - - - - - - - - - - - - -
1.000D+01 -7.858D+01 .*            .             .             .             .
1.259D+01 -7.858D+01 .*            .             .             .             .
1.585D+01 -7.858D+01 .*            .             .             .             .
1.995D+01 -7.858D+01 .*            .             .             .             .
2.512D+01 -7.858D+01 .*            .             .             .             .
3.162D+01 -7.858D+01 .*            .             .             .             .
3.981D+01 -7.857D+01 .*            .             .             .             .
5.012D+01 -7.857D+01 .*            .             .             .             .
6.310D+01 -7.857D+01 .*            .             .             .             .
7.943D+01 -7.856D+01 .*            .             .             .             .
1.000D+02 -7.855D+01 .*            .             .             .             .
1.259D+02 -7.854D+01 .*            .             .             .             .
1.585D+02 -7.851D+01 .*            .             .             .             .
1.995D+02 -7.847D+01 .*            .             .             .             .
2.512D+02 -7.841D+01 .*            .             .             .             .
3.162D+02 -7.831D+01 .*            .             .             .             .
3.981D+02 -7.816D+01 .*            .             .             .             .
5.012D+02 -7.793D+01 .*            .             .             .             .
6.310D+02 -7.759D+01 . *           .             .             .             .
7.943D+02 -7.710D+01 . *           .             .             .             .
1.000D+03 -7.642D+01 .  *          .             .             .             .
1.259D+03 -7.553D+01 .  *          .             .             .             .
1.585D+03 -7.441D+01 .   *         .             .             .             .
1.995D+03 -7.307D+01 .    *        .             .             .             .
2.512D+03 -7.154D+01 .     *       .             .             .             .
3.162D+03 -6.987D+01 .      *      .             .             .             .
3.981D+03 -6.809D+01 .       *     .             .             .             .
5.012D+03 -6.624D+01 .        *    .             .             .             .
6.310D+03 -6.433D+01 .         *   .             .             .             .
7.943D+03 -6.239D+01 .          *  .             .             .             .
1.000D+04 -6.043D+01 .           * .             .             .             .
1.259D+04 -5.846D+01 .            *.             .             .             .
1.585D+04 -5.647D+01 .             *             .             .             .
1.995D+04 -5.448D+01 .             . *           .             .             .
2.512D+04 -5.249D+01 .             .  *          .             .             .
3.162D+04 -5.049D+01 .             .   *         .             .             .
3.981D+04 -4.850D+01 .             .    *        .             .             .
5.012D+04 -4.650D+01 .             .     *       .             .             .
6.310D+04 -4.460D+01 .             .      *      .             .             .
7.943D+04 -4.251D+01 .             .       *     .             .             .
1.000D+05 -4.051D+01 .             .        *    .             .             .
1.259D+05 -3.852D+01 .             .         *   .             .             .
1.585D+05 -3.653D+01 .             .          *  .             .             .
1.995D+05 -3.455D+01 .             .           * .             .             .
2.512D+05 -3.258D+01 .             .            *.             .             .
3.162D+05 -3.063D+01 .             .             .*            .             .
3.981D+05 -2.870D+01 .             .             . *           .             .
5.012D+05 -2.681D+01 .             .             .   *         .             .
6.310D+05 -2.499D+01 .             .             .     *       .             .
7.943D+05 -2.325D+01 .             .             .       *     .             .
1.000D+06 -2.164D+01 .             .             .         *   .             .
1.259D+06 -2.020D+01 .             .             .           * .             .
1.585D+06 -1.898D+01 .             .             .            *.             .
1.995D+06 -1.800D+01 .             .             .             .*            .
2.512D+06 -1.729D+01 .             .             .             . *           .
3.162D+06 -1.684D+01 .             .             .             . *           .
3.981D+06 -1.665D+01 .             .             .             .*            .
5.012D+06 -1.672D+01 .             .             .             .*            .
                  - - - - - - - - - - - - - - - - - - - - - - - - - - - - - -
```

FIGURE 4.112. V_{SS} noise gain response.

Using Eq. (4.119) and changing subscripts appropriately, the second pole s_{p2} can be found from

$$s_{p2} \cong -\frac{g_{m6} + g_{m7}}{C_L}. \tag{4.243}$$

Hence, for $|s_{p2}| = 3\omega_0$, now the relation

$$g_{m6} + g_{m7} = 3g_{mi} \cong 377\,\mu\text{A}/\text{V} \tag{4.244}$$

must be satisfied. To determine g_{m6} and g_{m7} individually, we note that $i_6 = i_7$, and hence the bias voltages must satisfy

$$k_6\big(|v_{GS6}| - |V_{Tp}|\big)^2 = k_7\big(v_{GS7} - V_{Tn}\big)^2. \tag{4.245}$$

Furthermore,

$$v_{GS6} = v_{DS4} = v_{GS3} \cong -1.527 \tag{4.246}$$

as given earlier in (4.229), since the input stage remained unchanged. Also, as suggested in Eq. (4.199) of Section 4.14, the bias voltages and currents can be made insensitive to process variations if $v_{GS7} = v_{GS5} = v_{GS10} = V_{\text{bias}} - V_{SS} = -3.3 + 5 = 1.7$ V is chosen. Hence, using (3.18) and (4.245),

$$\frac{g_{m6}}{g_{m7}} \cong \frac{2\sqrt{k_6 i_6^0}}{2\sqrt{k_7 i_7^0}} = \sqrt{\frac{k_6}{k_7}} = \frac{v_{GS7} - V_{Tn}}{|v_{GS6}| - |V_{Tp}|}$$

$$= \frac{1.7 - 1.2}{1.527 - 1} \cong 0.9488. \tag{4.247}$$

Combining Eqs. (4.244) and (4.247), we obtain

$$g_{m6} = 183.5\,\mu\text{A}/\text{V}$$

and $\tag{4.248}$

$$g_{m7} = 193.5\,\mu\text{A}/\text{V}.$$

To avoid systematic offset, as discussed in connection with Eq. (4.182), we must have

$$\frac{(W/L)_3}{(W/L)_6} = \frac{k_3}{k_6} = \frac{i_3^0}{i_6^0}. \tag{4.249}$$

Hence also

$$\frac{g_{m3}}{g_{m6}} = \frac{2\sqrt{k_3 i_3^0}}{2\sqrt{k_6 i_6^0}} = \frac{k_3}{k_6} = \frac{i_3^0}{i_6^0} = \frac{I_0/2}{i_6^0}. \tag{4.250}$$

Since, from (4.213) and (4.248)

$$\frac{g_{m3}}{g_{m6}} \cong \frac{75.4}{183.5} \cong 0.4108 \tag{4.251}$$

and $I_0/2 = 20\ \mu A$, we find $i_6^0 = 49\ \mu A$. Also, from $(W/L)_3 = 6$, $(W/L)_6 \cong$ 14.6. Hence, $L_6 = 10\ \mu m$ and $W_6 = 146\ \mu m$ can be used.
   Next, since $i_7^0 = i_6^0 = 49\ \mu A$

$$(W/L)_7 = \frac{g_{m7}^2}{4k_7' i_7^0} = \frac{(193.5)^2}{4 \times 30 \times 49} \cong 6.37. \tag{4.252}$$

Therefore, we can choose $L_7 = 10\ \mu m$ and $W_7 = 64\ \mu m$.
   Finally, the transistors Q_9 and Q_{10} of the level shifter will be dimensioned. As before, $i_9^0 = i_{10}^0$ leads to

$$k_9 (v_{GS9} - V_{Tn})^2 = k_{10} (v_{GS10} - V_{Tn})^2. \tag{4.253}$$

Here, as Fig. 4.98 shows, $v_{GS9} = V_{DD} - V_{SS} + v_{DS4} - v_{GS7} = 10 - 1.527 - 1.7 \cong 6.77$ and $v_{GS10} = v_{GS7} = 1.7\ V$. Hence,

$$\frac{k_9}{k_{10}} = \left(\frac{v_{GS10} - V_{Tn}}{v_{GS9} - V_{Tn}}\right)^2 \cong 8.058 \times 10^{-3}. \tag{4.254}$$

The transconductance g_{m9} of Q_9 can be found from the phase shift introduced by the pole/zero pair due to the stray capacitances loading the source terminal of Q_9. Since these are driven by the source follower $Q_9 - Q_{10}$, the pole and zero are located near $s_{p3} \cong -g_{m9}/C_p$. Estimating (pessimistically) $C_p = 0.5\ pF$, and requiring $|s_{p3}| = 3\omega_0$ to make the contribution of this pole to the phase at ω_0 small, we obtain

$$g_{m9} = 3\omega_0 C_p = 3 \times 2\pi \times 2 \times 10^6 \times 0.5 \times 10^{-12} \tag{4.255}$$

$$g_{m9} \cong 19\ \mu A/V.$$

From

$$g_{m9} \cong 2\sqrt{k_9 i_9^0} = 2k_n'(W/L)_9(v_{GS9} - V_{Tn}), \qquad (4.256)$$

we find

$$(W/L)_9 = \frac{g_{m9}}{2k_n'(v_{GS9} - V_{Tn})} = \frac{19}{2 \times 30(6.77 - 1.2)}$$

$$\cong 0.05685 \qquad (4.257)$$

Hence, $W_9 = 10 \ \mu$m and $L_9 = 176 \ \mu$m can be used. From (4.254),

$$(W/L)_{10} = (W/L)_9/(k_9/k_{10}) \cong 7.06. \qquad (4.258)$$

Hence, $L_{10} = 10 \ \mu$m and $W_{10} = 71 \ \mu$m can be chosen. The common current of Q_9 and Q_{10} is then

$$i_9^0 = i_{10}^0 = k_n'(W/L)_{10}(v_{GS10} - V_{Tn})^2$$

$$\cong 53 \ \mu\text{A}. \qquad (4.259)$$

As explained in Section 4.14, this circuit can be biased by the bias chain circuit of Fig. 4.99. The design formulas for the aspect ratios have been derived in Section 4.14, and given by Eqs. (4.195)–(4.201). Choosing $I_{ref} = I_0/2 = 20$ μA, we get

$$(W/L)_{13} = (W/L)_5(I_{ref}/I_0) = (W/L)_5/2 = 2.7,$$

$$(W/L)_{11} = (W/L)_3(2I_{ref}/I_0) = (W/L)_3 = 6,$$

$$(W/L)_{12} = (W/L)_9(I_{ref}/I_{10}) \cong 0.05685 \times 20/53 \qquad (4.260)$$

$$\cong 0.02145.$$

Hence, we can use $L_{11} = 10$, $W_{11} = 60$, $L_{12} = 466$, $W_{12} = 10$, and $L_{13} = 10$, $W_{13} = 27$, all in μm.

The compensation branch of the circuit remains unchanged if we choose again $s_z = s_{p2}$. This is because s_{p2} remained at $-3\omega_0$, and the dc potential of the drain of Q_8 also remained the same. Hence, we can once again use $W_8 = L_8 = 10 \ \mu$m.

SPICE2 analysis of the class-AB amplifier with the aspect ratios calculated above indicates that the dc bias voltage of the output terminal is unsuitable for proper operation. Its value is too low (about -4.5 V) to allow Q_7 to saturate.

This occurs only under open-circuit conditions, and is a consequence of the simplifying assumptions, primarily the neglect of the channel-length modulation factor $(1 + \lambda v_{Ds})$, made in the calculations. The problem is an artificial one, since the circuit never functions without a dc load and/or feedback. Adding a 1-MΩ load resistor between the output terminal and ground, or a feedback resistor of (say) 100 MΩ between the inverting input terminal (v_{in}^-) and the output terminal of the op-amp, the output voltage returns to a value sufficient to keep both Q_6 and Q_7 in saturation.

As an exercise in design, however, as well as a way to show how to reduce systematic offset, we are next going to redesign the output stage so as to obtain a satisfactory dc bias value for v_{out} even under open-circuit conditions. The SPICE2 analysis for the circuit gave $v_{GS6} = -1.572$ V, $v_{GS7} = 1.821$ V, and $i_6^0 = i_7^0 = 74$ μA. To achieve $v_{out} = 0$ V, while retaining the values of v_{GS6}, v_{GS7}, and $i_6^0 = i_7^0$, we must have by Eq. (3.11)

$$k_p'(W/L)_6\left(v_{GS6} - V_{Tp}\right)^2(1 - \lambda V_{SS})$$

$$= k_n'(W/L)_7\left(v_{GS7} - V_{Tn}\right)^2(1 + \lambda V_{DD}) = i_6^0. \qquad (4.261)$$

This relation now includes the channel-length modulation effect, and is hence more accurate.

Substituting $\lambda = 0.03$ V^{-1}, as well as the given values, $(W/L)_6 \cong 16.4$ and $(W/L)_7 \cong 50/9$ results. Thus, $W_6 = 164$ μm, $L_6 = 10$ μm, $W_7 = 50$ μm, and $L_7 = 9$ μm can be used. The resulting output bias voltage is only 0.04 V.

Figure 4.113 shows the gain and phase responses of the redesigned circuit under open-circuited output conditions. The unity-gain bandwidth is again near 2.5 MHz, while the dc gain is over 80 dB. The phase margin is over 90°; this is more than adequate, and indicates that the bias current of Q_6, Q_7, Q_9, and Q_{10} (and thus their transconductances) can be reduced, and still adequate stability maintained. This was not attempted. The common-mode gain response is shown in Fig. 4.114; the CMRR is over 80 dB across the full $0 - \omega_0$ frequency range. The systematic input-referred offset voltage is negligible.

The slew-rate performance of the op-amp for positive and negative voltage input steps was computed, and was found to be around 5 V/μs for both polarities.

Using the more realistic SPICE2 MOSFET model, the gain and phase response changed only slightly. The dc gain dropped to 76 dB, while the unity-gain bandwidth ω_0 and the phase margin remained unchanged. The CMRR remained near 80 dB at low frequencies, but dropped gradually, and was only about 70 dB near ω_0. The slew rate remained the same for both positive and negative signals.

The gain response for V_{DD} noise is illustrated in Fig. 4.115; for V_{SS} noise it is illustrated in Fig. 4.116. As before, the PSRR is near 0 dB around ω_0.

```
****      AC ANALYSIS                          TEMPERATURE =   27.000 DEG C
LEGEND:
*: VDB(6)   +: VP(6)
    FREQ      VDB(6)

*)------------- -5.000D+01    0.000D+00    5.000D+01    1.000D+02  1.500D+02
                                   - - -      - - -      - - -      - - -
+)------------  5.000D+01    1.000D+02    1.500D+02    2.000D+02  2.500D+02
                                   - - -      - - -      - - -      - - -
 1.000D+01  8.050D+01 .            .            .          +*     .
 1.259D+01  8.050D+01 .            .            .        +  *     .
 1.585D+01  8.049D+01 .            .            .        +  *     .
 1.995D+01  8.048D+01 .            .            .        +  *     .
 2.512D+01  8.046D+01 .            .            .        +  *     .
 3.162D+01  8.042D+01 .            .            .       +   *     .
 3.981D+01  8.037D+01 .            .            .       +   *     .
 5.012D+01  8.029D+01 .            .            .      +    *     .
 6.310D+01  8.017D+01 .            .            .    +     *      .
 7.943D+01  7.998D+01 .            .            .   +     *       .
 1.000D+02  7.969D+01 .            .            .  +     *        .
 1.259D+02  7.928D+01 .            .            . +    *          .
 1.585D+02  7.869D+01 .            .            +    *            .
 1.995D+02  7.790D+01 .            .           +   *              .
 2.512D+02  7.688D+01 .            .         +    *               .
 3.162D+02  7.564D+01 .            .       +     *                .
 3.981D+02  7.419D+01 .            .     +    *                   .
 5.012D+02  7.258D+01 .            .    +   *                     .
 6.310D+02  7.085D+01 .            .   +   *                      .
 7.943D+02  6.903D+01 .            . +   *                        .
 1.000D+03  6.714D+01 .            .+   *                         .
 1.259D+03  6.521D+01 .          +    *                           .
 1.585D+03  6.326D+01 .        +.    *                            .
 1.995D+03  6.129D+01 .       +.    *                             .
 2.512D+03  5.931D+01 .       +.   *                              .
 3.162D+03  5.732D+01 .     +.    *                               .
 3.981D+03  5.533D+01 .     + .  *                                .
 5.012D+03  5.334D+01 .     + . *                                 .
 6.310D+03  5.134D+01 .     + .*                                  .
 7.943D+03  4.934D+01 .     + .*                                  .
 1.000D+04  4.734D+01 .     + . *                                 .
 1.259D+04  4.534D+01 .    +       *                              .
 1.585D+04  4.334D+01 .    +      *                               .
 1.995D+04  4.134D+01 .    +      *                               .
 2.512D+04  3.934D+01 .    +    *                                 .
 3.162D+04  3.734D+01 .    +   *                                  .
 3.981D+04  3.535D+01 .    +   *                                  .
 5.012D+04  3.335D+01 .    +   *                                  .
 6.310D+04  3.135D+01 .    +   *                                  .
 7.943D+04  2.935D+01 .    +     *                                .
 1.000D+05  2.735D+01 .    +     *                                .
 1.259D+05  2.535D+01 .    +   *                                  .
 1.585D+05  2.335D+01 .    +     *                                .
 1.995D+05  2.135D+01 .    +    *                                 .
 2.512D+05  1.935D+01 .   +    *                                  .
 3.162D+05  1.736D+01 .   +   *                                   .
 3.981D+05  1.537D+01 .   +   *                                   .
 5.012D+05  1.338D+01 .   +   *                                   .
 6.310D+05  1.140D+01 .   +  *                                    .
 7.943D+05  9.425D+00 .   + . *                                   .
 1.000D+06  7.470D+00 .  +. *                                     .
 1.259D+06  5.540D+00 .  +. *                                     .
 1.585D+06  3.646D+00 .  +.*                                      .
 1.995D+06  1.802D+00 .  +*                                       .
 2.512D+06  2.364D-02 .  X                                        .
 3.162D+06 -1.675D+00 .  X                                        .
 3.981D+06 -3.292D+00 .  *.+                                      .
 5.012D+06 -4.843D+00 .  *.+                                      .
```

FIGURE 4.113. Gain and phase response of the op-amp of Fig. 4.98.

```
****     AC ANALYSIS                      TEMPERATURE =    27.000 DEG C
     FREQ        VDB(6)
                   -1.500D+02     -1.000D+02     -5.000D+01      0.000D+00  5.000D+01
                 - - - - - - - - - - - - - - - - - - - - - - - - - - - - - - - - -
 1.000D+01 -2.703D+00  .                .              .             *.
 1.259D+01 -2.709D+00  .                .              .             *.
 1.585D+01 -2.717D+00  .                .              .             *.
 1.995D+01 -2.730D+00  .                .              .             *.
 2.512D+01 -2.751D+00  .                .              .             *.
 3.162D+01 -2.784D+00  .                .              .             *.
 3.981D+01 -2.835D+00  .                .              .             *.
 5.012D+01 -2.915D+00  .                .              .             *.
 6.310D+01 -3.039D+00  .                .              .             *.
 7.943D+01 -3.229D+00  .                .              .             *.
 1.000D+02 -3.513D+00  .                .              .             *.
 1.259D+02 -3.929D+00  .                .              .             *.
 1.585D+02 -4.516D+00  .                .              .            * .
 1.995D+02 -5.309D+00  .                .              .           *  .
 2.512D+02 -6.328D+00  .                .              .          *   .
 3.162D+02 -7.572D+00  .                .              .        *     .
 3.981D+02 -9.016D+00  .                .              .       *      .
 5.012D+02 -1.063D+01  .                .              .      *       .
 6.310D+02 -1.236D+01  .                .              .     *        .
 7.943D+02 -1.418D+01  .                .              .    *         .
 1.000D+03 -1.607D+01  .                .              .   *          .
 1.259D+03 -1.799D+01  .                .              .  *           .
 1.585D+03 -1.995D+01  .                .              . *            .
 1.995D+03 -2.192D+01  .                .              .*             .
 2.512D+03 -2.390D+01  .                .             *.              .
 3.162D+03 -2.588D+01  .                .            * .              .
 3.981D+03 -2.788D+01  .                .           *  .              .
 5.012D+03 -2.987D+01  .                .          *   .              .
 6.310D+03 -3.187D+01  .                .         *    .              .
 7.943D+03 -3.387D+01  .                .        *     .              .
 1.000D+04 -3.586D+01  .                .       *      .              .
 1.259D+04 -3.786D+01  .                .      *       .              .
 1.585D+04 -3.986D+01  .                .     *        .              .
 1.995D+04 -4.186D+01  .                .    *         .              .
 2.512D+04 -4.386D+01  .                .   *          .              .
 3.162D+04 -4.586D+01  .                .  *           .              .
 3.981D+04 -4.786D+01  .                . *            .              .
 5.012D+04 -4.986D+01  .                .*             .              .
 6.310D+04 -5.186D+01  .               *.              .              .
 7.943D+04 -5.386D+01  .              * .              .              .
 1.000D+05 -5.586D+01  .             *  .              .              .
 1.259D+05 -5.786D+01  .            *   .              .              .
 1.585D+05 -5.986D+01  .           *    .              .              .
 1.995D+05 -6.186D+01  .          *     .              .              .
 2.512D+05 -6.385D+01  .         *      .              .              .
 3.162D+05 -6.585D+01  .        *       .              .              .
 3.981D+05 -6.784D+01  .       *        .              .              .
 5.012D+05 -6.983D+01  .      *         .              .              .
 6.310D+05 -7.181D+01  .      *         .              .              .
 7.943D+05 -7.378D+01  .     *          .              .              .
 1.000D+06 -7.574D+01  .    *           .              .              .
 1.259D+06 -7.767D+01  .   *            .              .              .
 1.585D+06 -7.956D+01  .  *             .              .              .
 1.995D+06 -8.141D+01  . *              .              .              .
 2.512D+06 -8.318D+01  . *              .              .              .
 3.162D+06 -8.488D+01  .*               .              .              .
 3.981D+06 -8.650D+01  *                .              .              .
 5.012D+06 -8.805D+01  .*               .              .              .
                 - - - - - - - - - - - - - - - - - - - - - - - - - - - - - - - - -
```

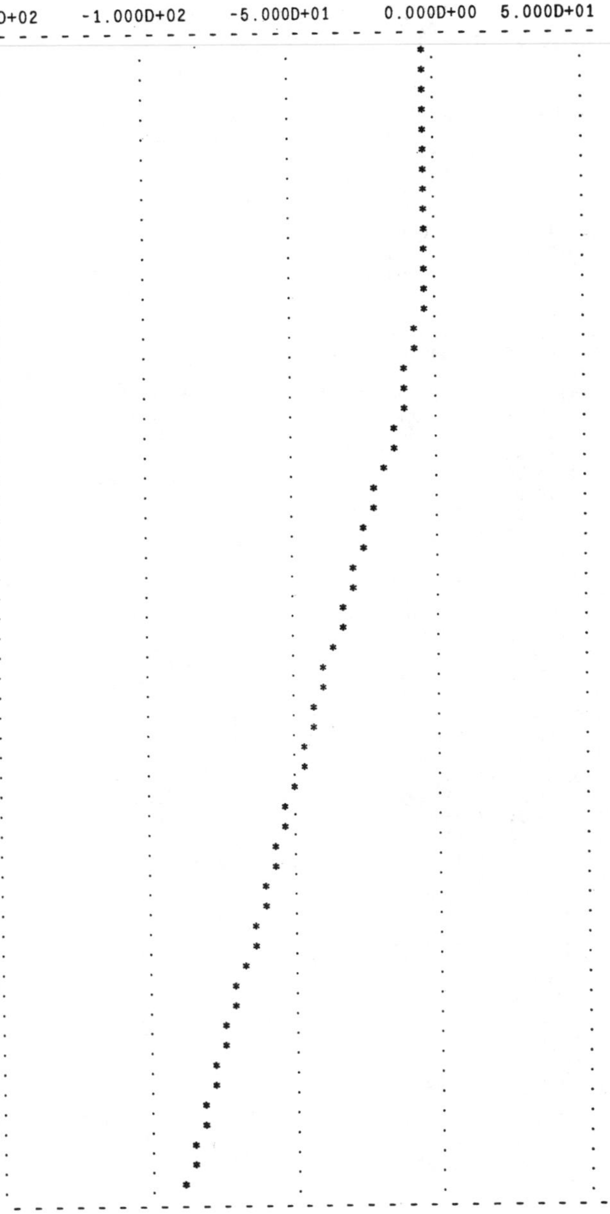

FIGURE 4.114. Common-mode gain response of the op-amp of Fig. 4.98.

```
        FREQ      VDB(6)

                 -3.000D+01    -2.000D+01    -1.000D+01    0.000D+00   1.000D+01
1.000D+01 -2.778D+01  .    *         .             .            .             .
1.259D+01 -2.693D+01  .     *        .             .            .             .
1.585D+01 -2.584D+01  .        *     .             .            .             .
1.995D+01 -2.454D+01  .          *   .             .            .             .
2.512D+01 -2.306D+01  .            * .             .            .             .
3.162D+01 -2.142D+01  .              .*            .            .             .
3.981D+01 -1.967D+01  .              .   *         .            .             .
5.012D+01 -1.786D+01  .              .      *      .            .             .
6.310D+01 -1.600D+01  .              .         *   .            .             .
7.943D+01 -1.414D+01  .              .            *.            .             .
1.000D+02 -1.229D+01  .              .             .  *         .             .
1.259D+02 -1.048D+01  .              .             .     *      .             .
1.585D+02 -8.754D+00  .              .             .        *   .             .
1.995D+02 -7.137D+00  .              .             .           *.             .
2.512D+02 -5.675D+00  .              .             .            .  *          .
3.162D+02 -4.408D+00  .              .             .            .     *       .
3.981D+02 -3.362D+00  .              .             .            .       *     .
5.012D+02 -2.542D+00  .              .             .            .         *   .
6.310D+02 -1.931D+00  .              .             .            .           * .
7.943D+02 -1.496D+00  .              .             .            .            *.
1.000D+03 -1.197D+00  .              .             .            .             *
1.259D+03 -9.973D-01  .              .             .            .             .*
1.585D+03 -8.664D-01  .              .             .            .             .*
1.995D+03 -7.816D-01  .              .             .            .             .*
2.512D+03 -7.273D-01  .              .             .            .             .*
3.162D+03 -6.927D-01  .              .             .            .             .*
3.981D+03 -6.707D-01  .              .             .            .             .*
5.012D+03 -6.568D-01  .              .             .            .             .*
6.310D+03 -6.479D-01  .              .             .            .             .*
7.943D+03 -6.424D-01  .              .             .            .             .*
1.000D+04 -6.388D-01  .              .             .            .             .*
1.259D+04 -6.366D-01  .              .             .            .             .*
1.585D+04 -6.351D-01  .              .             .            .             .*
1.995D+04 -6.341D-01  .              .             .            .             .*
2.512D+04 -6.334D-01  .              .             .            .             .*
3.162D+04 -6.329D-01  .              .             .            .             .*
3.981D+04 -6.323D-01  .              .             .            .             .*
5.012D+04 -6.316D-01  .              .             .            .             .*
6.310D+04 -6.307D-01  .              .             .            .             .*
7.943D+04 -6.293D-01  .              .             .            .             .*
1.000D+05 -6.271D-01  .              .             .            .             .*
1.259D+05 -6.237D-01  .              .             .            .             .*
1.585D+05 -6.183D-01  .              .             .            .             .*
1.995D+05 -6.098D-01  .              .             .            .             .*
2.512D+05 -5.964D-01  .              .             .            .             .*
3.162D+05 -5.754D-01  .              .             .            .             .*
3.981D+05 -5.424D-01  .              .             .            .             .*
5.012D+05 -4.909D-01  .              .             .            .             .*
6.310D+05 -4.115D-01  .              .             .            .             .*
7.943D+05 -2.908D-01  .              .             .            .            *.
1.000D+06 -1.113D-01  .              .             .            .           * .
1.259D+06  1.471D-01  .              .             .            .          *  .
1.585D+06  5.017D-01  .              .             .            .        *    .
1.995D+06  9.571D-01  .              .             .            .       *     .
2.512D+06  1.491D+00  .              .             .            .     *       .
3.162D+06  2.040D+00  .              .             .            .   *         .
3.981D+06  2.497D+00  .              .             .            .    *        .
5.012D+06  2.720D+00  .              .             .            .   *         .
```

FIGURE 4.115. V_{DD} noise gain response.

```
       FREQ       VDB(6)

                       -2.500D+01    -2.000D+01    -1.500D+01    -1.000D+01  -5.000D+00
                       - - - - - - - - - - - - - - - - - - - - - - - - - - - - - - - -
   1.000D+01 -1.713D+01 .                            *                .                .
   1.259D+01 -1.713D+01 .                            *                .                .
   1.585D+01 -1.714D+01 .                            *                .                .
   1.995D+01 -1.714D+01 .                            *                .                .
   2.512D+01 -1.715D+01 .                            *                .                .
   3.162D+01 -1.716D+01 .                            *                .                .
   3.981D+01 -1.717D+01 .                            *                .                .
   5.012D+01 -1.719D+01 .                            *                .                .
   6.310D+01 -1.723D+01 .                           *                 .                .
   7.943D+01 -1.728D+01 .                           *                 .                .
   1.000D+02 -1.737D+01 .                          *                  .                .
   1.259D+02 -1.749D+01 .                         *                   .                .
   1.585D+02 -1.769D+01 .                        *                    .                .
   1.995D+02 -1.796D+01 .                      *                      .                .
   2.512D+02 -1.835D+01 .                    *                        .                .
   3.162D+02 -1.886D+01 .                 *                           .                .
   3.981D+02 -1.948D+01 .              *                              .                .
   5.012D+02 -2.018D+01 .          *                                  .                .
   6.310D+02 -2.091D+01 .       *                                     .                .
   7.943D+02 -2.160D+01 .     *                                       .                .
   1.000D+03 -2.219D+01 .   *                                         .                .
   1.259D+03 -2.267D+01 .  *                                          .                .
   1.585D+03 -2.303D+01 . *                                           .                .
   1.995D+03 -2.328D+01 .*                                            .                .
   2.512D+03 -2.345D+01 *                                             .                .
   3.162D+03 -2.357D+01 *                                             .                .
   3.981D+03 -2.365D+01 *                                             .                .
   5.012D+03 -2.370D+01 *                                             .                .
   6.310D+03 -2.373D+01 *                                             .                .
   7.943D+03 -2.375D+01 *                                             .                .
   1.000D+04 -2.376D+01 *                                             .                .
   1.259D+04 -2.376D+01 *                                             .                .
   1.585D+04 -2.376D+01 *                                             .                .
   1.995D+04 -2.376D+01 *                                             .                .
   2.512D+04 -2.376D+01 *.                                            .                .
   3.162D+04 -2.375D+01 *                                             .                .
   3.981D+04 -2.373D+01 *                                             .                .
   5.012D+04 -2.370D+01 *                                             .                .
   6.310D+04 -2.365D+01 *                                             .                .
   7.943D+04 -2.358D+01 *                                             .                .
   1.000D+05 -2.347D+01 *                                             .                .
   1.259D+05 -2.330D+01 .*                                            .                .
   1.585D+05 -2.304D+01 . *                                           .                .
   1.995D+05 -2.265D+01 .   *                                         .                .
   2.512D+05 -2.211D+01 .     *                                       .                .
   3.162D+05 -2.137D+01 .       *                                     .                .
   3.981D+05 -2.041D+01 .          *.                                 .                .
   5.012D+05 -1.923D+01 .             *                               .                .
   6.310D+05 -1.785D+01 .               *                             .                .
   7.943D+05 -1.632D+01 .                  *                          .                .
   1.000D+06 -1.468D+01 .                      *                      .                .
   1.259D+06 -1.298D+01 .                          *                  .                .
   1.585D+06 -1.128D+01 .                              *              .                .
   1.995D+06 -9.628D+00 .                                  *          .                .
   2.512D+06 -8.094D+00 .                                       *     .                .
   3.162D+06 -6.757D+00 .                                           * .                .
   3.981D+06 -5.717D+00 .                                             .  *             .
   5.012D+06 -5.089D+00 .                                             .     *           .
                       - - - - - - - - - - - - - - - - - - - - - - - - - - - - - - - -
```

FIGURE 4.116. V_{SS} noise gain response.

4.16.   HIGH-PERFORMANCE CMOS OP-AMPS[19]

The simple op-amps described in the preceding sections perform adequately in most SC filters, and also in many nonfiltering applications. In some circuits (such as high-frequency filters, high-accuracy/high-speed D/A or A/D converters, instrumentation amplifiers, etc.), however, their gains or settling rates may not be adequate. The gain of the basic two-stage op-amp, discussed earlier, is equal to the overall gain of two cascaded common-source gain stages, and is hence normally no more than about 80 dB. If this is not sufficient, additional common-source stages are needed. However, this results in op-amp configurations with three or more high-impedance nodes. These in turn introduce three or more poles at relatively low frequencies, and hence the circuit will be very difficult to compensate. To avoid this, the structure of the stages can be altered instead, so as to enhance their gain. One possibility is to use composite loads which present an increased impedance to the driver transistor. In Figs. 4.24–4.27, we have already introduced such high-gain stages using NMOS devices. In this section, high-performance CMOS amplifier circuits will be discussed.

The simplest way of enhancing the load impedance is to use two series-connected common-gate devices (Fig. 4.117a). Using the small-signal equivalent circuit illustrated in Fig. 4.117b, it can readily be shown that the incremental load impedance of the circuit is

$$R_0 = \frac{v}{i} = (g_{m2}r_{d2})r_{d1} \tag{4.262}$$

(Problem 4.34). Since normally $g_{m2}r_{d2} \gg 1$, the impedance R_0 is much greater than that of a single load device, that is, r_{d1}. A disadvantage of this circuit is that (since both Q_1 and Q_2 must remain in saturation), the load voltage v can

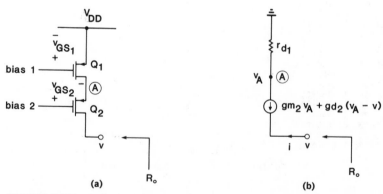

FIGURE 4.117.   Series load devices: (a) circuit; (b) small-signal equivalent circuit.

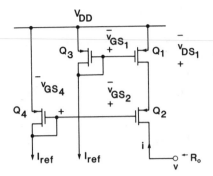

FIGURE 4.118. High-impedance (R_0) load circuit.

rise at most to $V_{DD} - (|v_{GS1}| - |V_{Tp}| + |v_{GS2}| - |V_{Tp}|)$. For identical transistors, $V_{DD} - v \geq 2(|v_{GS}| - |V_{Tp}|)$, and the equal sign can only be obtained if Q_1 is biased exactly at the verge of saturation. Since, in practice, MOS transistors display an indistinct transition from the ohmic (triode) region to the saturation region, to obtain a high incremental drain impedance r_{d1}, Q_1 must be biased by a few hundred millivolts into the saturation region.

A circuit[19] which can bias Q_1 and Q_2 effectively is shown in Fig. 4.118. Choosing $(W/L)_1 = (W/L)_2 = (W/L)_3 = 4(W/L)_4$, from the saturated drain–current relations of $Q_1 - Q_4$ we have

$$|v_{GS1}| = |v_{GS2}| = |v_{GS3}| = |V_{Tp}| + \sqrt{\frac{I_{\text{ref}}}{k'_p(W/L)_1}} \qquad (4.263)$$

and

$$|v_{GS4}| = |V_{Tp}| + \sqrt{\frac{4I_{\text{ref}}}{k'_p(W/L)_1}}. \qquad (4.264)$$

As Fig. 4.118 shows,

$$|v_{DS1}| = |v_{GS4}| - |v_{GS2}| = \sqrt{\frac{I_{\text{ref}}}{k'_p(W/L)_1}}$$

$$= |v_{GS1}| - |V_{Tp}|. \qquad (4.265)$$

Thus, Q_1 is at the verge of saturation, regardless of the exact values of V_{DD} and I_{ref} (as long as the two I_{ref} sources remain matched). As discussed above, it is better to make $|v_{DS1}|$ somewhat larger (by, say, 0.2 V) than the saturation limit $|v_{GS1}| - |V_{Tp}|$. This can be achieved by making $(W/L)_4$ smaller than $(W/L)_1/4$ (Problem 4.35).

Another way of creating a high-impedance load is to use one of the current sources described in Section 4.2. Consider, for example, the cascode current

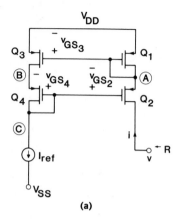

(a)

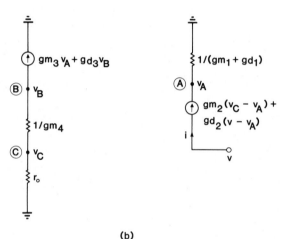

(b)

FIGURE 4.119. Current source as high-impedance load: (*a*) circuit; (*b*) small-signal equivalent circuit.

source shown in Fig. 4.12. Replacing all NMOS devices by PMOS ones, and turning the circuit upside down, the circuit of Fig. 4.119*a* results. Using the equivalent circuit of Fig. 4.119*b*, it can be shown that the impedance R_0 seen at the open-circuited terminal is

$$R_0 \cong \left(r_{d2} g_{m2} \right) \frac{g_{m3}/g_{m1}}{g_{d3} + g_0} \tag{4.266}$$

(Problem 4.36), where $r_0 = 1/g_0$ is the output impedance of the current source. Since $r_{d2} g_{m2} \gg 1$, $R_0 \gg 1/(g_{d3} + g_0)$ can be achieved.

The maximum value $v_{\max}$ which v may have so as to keep Q_2 still in saturation is $V_{DD} - (|v_{GS3}| + |v_{GS2}| - |V_{Tp}|)$. If $(W/L)_1 = (W/L)_2$, then $v_{\max} \cong V_{DD} - 2|v_{GS3}| + |V_{Tp}|$.

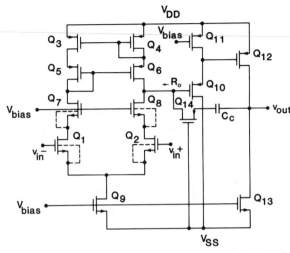

FIGURE 4.120. CMOS op-amp with a cascode load in the differential stage.

The use of the load circuit of Fig. 4.119a in an op-amp[19] is illustrated in Fig. 4.120. Here, Q_7 and Q_8 form a Wilson's current source which increases the output impedance and reduce the Miller effect, while $Q_3 - Q_6$ form the composite load of Fig. 4.119a. The output impedance of the stage is, for symmetrical dimensions $(W/L)_1 = (W/L)_2$, $(W/L)_3 = (W/L)_4$, $(W/L)_5 = (W/L)_6$, and $(W/L)_7 = (W/L)_8$,

$$R_0 = \frac{1}{g_{d4}/g_{m6}r_{d6} + g_{d2}/g_{m8}r_{d8}}. \tag{4.267}$$

(Problem 4.37). Since $g_{m6}r_{d6}$ and $g_{m8}r_{d8}$ are normally much greater than 1, $R_0 \gg 1/(g_{d4} + g_{d2})$ which would be the output impedance for the differential input stage with single-device loads. The voltage gain of the stage is simply $-g_{m1}R_0$.

The second stage $(Q_{10} - Q_{11})$ is a source follower which drives a common-source output gain stage with a current/source load Q_{13}.

The common-mode swing of the input voltages of this circuit is small because there are many devices connected in series in the input stage. However, the output voltage v_{out} can swing to within a voltage difference $(v_{GS} - V_T)$ of both supply voltages.

To obtain even higher gain, the cascode load of Fig. 4.117a can be used in the output stage, and the biasing scheme of Fig. 4.118 applied to keep the output voltage swing large.

The circuit of Fig. 4.120 has two high-impedance nodes, and hence the usual series RC compensation branch (Q_{14} and C_c) can be used to improve the phase margin.

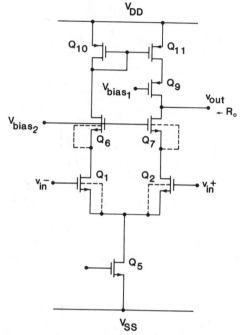

FIGURE 4.121. CMOS differential amplifier with cascode load.

A disadvantage of composite-load circuits is that they require additional devices, all of which contribute to the noise at the output. Their mismatches also contribute to the random offset voltage of the op-amp.

Another high-gain differential amplifier with a composite cascode load is shown in Fig. 4.121. In this circuit, the common-gate devices Q_6 and Q_7 increase R_0 and also reduce the Miller effect for the input transistors Q_1 and Q_2. Devices Q_{10} and Q_{11} form a current mirror, and Q_9 increases the output impedance R_0. It can be shown (Problem 4.38) that

$$R_0 = \frac{1}{g_{d10}/g_{m9}r_{d9} + g_{d2}/g_{m7}r_{d7}}. \qquad (4.268)$$

Again, $R_0 \gg 1/(g_{d10} + g_{d2})$ which would be the output impedance without the cascode devices Q_6, Q_7, and Q_9. The voltage gain is $-g_{m1}R_0$ and is thus also enhanced.

Normally, the differential stage of Fig 4.120 or 4.121 must be followed by a level shifter (usually a source follower) and often also by an output amplifier stage. It needs then to be compensated by a pole-splitting capacitor and a resistor or source follower, as discussed in the preceding sections. The resulting circuit has several nondominant poles due to the level shifter and the load

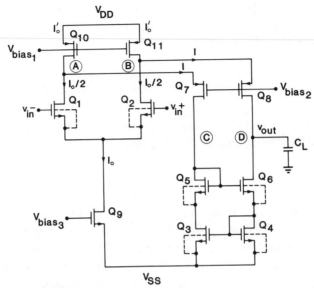

FIGURE 4.122. A wide-band "folded-cascode" op-amp consisting of a cascade of common source and common-gate amplifiers.

capacitor C_L. For a large load capacitor, the bandwidth will be limited by the corresponding nondominant pole. Also, for $C_c = C_L$, the slew rate g_{m1}/C_c will be limited by the size of C_L. Finally, as discussed in Section 4.14, the high-frequency PSRR will be reduced by the pole-splitting compensation capacitor, unless special circuits (such as that shown in Fig. 4.101) are used. These, however, complicate the circuit considerably.

All these problems can be eliminated, and the circuit made faster as well as simpler, by using the "folded cascode" configuration.[19] Consider the circuit of Fig. 4.120. Let the top terminals (i.e., the sources of Q_3 and Q_4) of the composite load $Q_3 - Q_8$ be disconnected from V_{DD}, folded down, and connected to V_{SS} instead. To assure proper dc bias currents, all PMOS devices must be replaced by NMOS types and vice versa in the composite loads; also, two additional current sources (Q_{10} and Q_{11}) must be added between V_{DD} and the drains of Q_1 and Q_2 to supply bias currents to these input devices. The resulting op-amp is shown in Fig. 4.122. The basic operation of the circuit is the following. The dc current I_0 of the current source Q_9 is shared equally by Q_1 and Q_2. Also, the matched sources Q_{10} and Q_{11} provide equal bias currents I'_0 to nodes Ⓐ and Ⓑ. Hence Q_7 and Q_8 carry equal dc bias currents $I = I'_0 - I_0/2$. A differential input voltage $\Delta v_{in}^+ = \Delta v_{in}/2$, $\Delta v_{in}^- = -\Delta v_{in}/2$ applied to the gates of Q_1 and Q_2 will offset their drain currents by $\pm \Delta I_0 = \pm g_{m1} \Delta v_{in}/2$. Since the currents I'_0 of Q_{10} and Q_{11} remain unchanged, the currents I of Q_7 and Q_8 (which are driven at their low-impedance source terminals) will change by $\pm \Delta I_0$. The current mirror $Q_3 - Q_6$ transfers the current change in Q_3 and Q_5 to Q_4 and Q_6. Hence, the output voltage increment is $g_{m1} R_0 \Delta v_{in}$, where R_0 is the output impedance at node Ⓓ. It can be shown (Problem 4.39) that

$$R_0 \cong \frac{1}{g_{d3}/g_{m6} r_{d6} + (g_{d2} + g_{d11})/g_{m8} r_{d8}}. \qquad (4.269)$$

The incremental gain is, of course, $-g_{m1} R_0$.

The dominant pole of the circuit is due to the load capacitance C_L in parallel with the output impedance R_0; hence, its value is

$$s_{p1} = -\frac{1}{R_0 C_L}. \qquad (4.270)$$

Higher-frequency nondominant poles are contributed by the stray capacitances loading the low-impedance nodes Ⓐ, Ⓑ, and Ⓒ. The impedance at node Ⓐ is approximately $1/g_{m7}$; at node Ⓑ, it is approximately $1/g_{m8}$; and at node Ⓒ, it is approximately $1/g_{m6} + 1/g_{m4}$. Since $1/g_m$ is of order of 1 kΩ, the corresponding poles s_{p2}, s_{p3}, and s_{p4} are at much higher frequencies than s_{p1}. Depending on the actual values of the s_{p1} and the dc gain A_0, the op-amp may or may not be stable under unity-gain closed-loop conditions. Figure 4.123 illustrates the gain and phase responses of the op-amp for two different

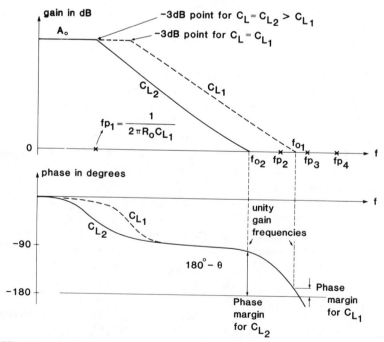

FIGURE 4.123. Loss and phase responses of the op-amp of Fig. 4.122 for two different values of the load capacitance C_L.

values of C_L. As the figure illustrates, the *larger* C_L, the *greater* the phase margin of the op-amp. This is the opposite of the conditions of the two- or three-stage op-amp where C_L contributes to a nondominant high-frequency pole. There, increasing C_L reduces the distance between the dominant and nondominant poles, and thus decreases the phase margin. Thus, the folded-cascode op-amp of Fig. 4.122 is particularly suitable for achieving wide and stable closed-loop bandwidths with large capacitive load, such as required in high-frequency SC filters.

In addition, the compensation in this circuit is achieved without coupling high-frequency noise from the power supplies to the output, as for the multistage op-amps. Hence, the high-frequency PSRR can be high.

A disadvantage is the reduced output voltage swing due to the many cascaded devices. The swing can be increased if the bias circuit of 4.118 is used to establish the gate voltages of Q_7, Q_8, Q_{10}, and Q_{11} such that the drain-to-source voltages of Q_{10} and Q_{11} are only a few hundred millivolts beyond $V_{d\,sat}$. In addition, as for the other composite-load amplifier stages, the added devices contribute to the output noise and hence reduce the dynamic range.

The circuit of Fig. 4.121 can also be modified into a folded-cascode op-amp. The resulting configuration, including some of the bias circuitry, is shown in Fig. 4.124. Using the design principles discussed in Section 4.14 and earlier in

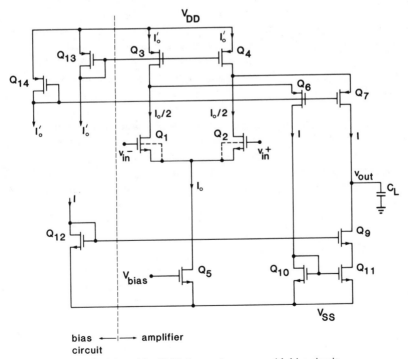

FIGURE 4.124. Folded-cascode op-amp with bias circuit.

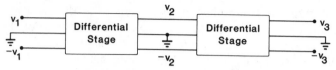

FIGURE 4.125. Fully differential circuit.

this section, the design equations are

$$(W/L)_{13} = (W/L)_3 = (W/L)_4,$$

$$(W/L)_{14} = (W/L)_{13}/4,$$

$$(W/L)_6 = (W/L)_7 = (I/I_0')(W/L)_4,$$

$$(W/L)_9 = (W/L)_{10} = (W/L)_{11} = 4(W/L)_{12}.$$

(4.271)

The advantages and disadvantages of this circuit are identical to those discussed in connection with Fig. 4.122.

In cases when power-supply rejection is an important consideration, the use of fully differential (balanced) signal paths may be advantageous. In such circuits, all signals appear with both positive and negative polarities with respect to ground (Fig. 4.125). This makes the circuit more complicated; the chip area required is 50–100% larger than for the single-ended realization of the same network. However, there are compensating advantages in terms of noise immunity. In a fully differential op-amp, power-supply noise appears as a common-mode signal, and is hence rejected by the circuit. In addition, as will be shown, the effective output voltage swing is doubled by the balanced op-amp configuration, while the input circuit (and hence most of the noise) remains the same as for the single-ended-output op-amps. Thus, the dynamic range is greater by nearly 6 dB than for single-ended op-amps.

Additional advantages also exist. Figure 4.126 shows the circuit of a fully differential switched-capacitor (SC) integrator. As will be discussed in Chapters 5 and 7, the switches schematically illustrated in the figure introduce a clock-feedthrough noise into the circuit. This can be minimized by the differential configuration, since (just as the power-supply noise) it will appear as a

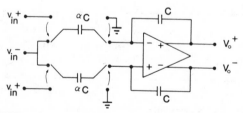

FIGURE 4.126. A differential switched-capacitor integrator.

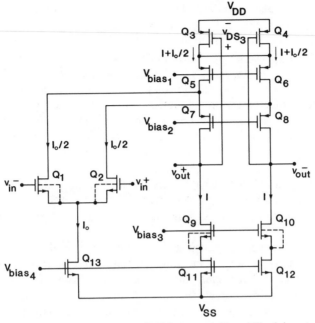

FIGURE 4.127. A fully differential CMOS op-amp with stabilized dc output level.

common-mode signal. The symmetry of the circuit should be fully preserved in the physical layout to obtain good rejection of common-mode signals even in the presence of stray elements and nonidealities.

The differential configuration also eliminates systematic offset voltages. Finally, as will be discussed in Chapter 7, fully differential op-amps can be used in a chopper-stabilized configuration, in which low-frequency noise and offset are also rejected.

The circuit diagram of a differential-output op-amp[19,22] is shown in Fig. 4.127. The circuit is a fully differential folded-cascode op-amp, using the split-load arrangement of Fig. 4.117a to achieve a high output impedance. The bias voltage V_{bias4} establishes the bias current I_0 of the input stage, and also a bias current I in the output devices $Q_7 - Q_{12}$. The bias current in the devices $Q_3 - Q_6$ is hence $I + I_0/2$ in equilibrium.

An important task, accomplished by the feedback from the output terminals to the gates of Q_3 and Q_4, is common-mode feedback. This is necessary, since in the fully differential op-amp the common-mode output voltage must be internally forced to ground, or some other reference potential. By contrast, in a single-ended op-amp one of the input terminals is usually grounded and the other becomes virtual ground due to an externally applied negative feedback. This stabilizes the common-mode voltages at both input and output terminals. The common-mode feedback operates the following way. Since the gate voltages of Q_5 and Q_6 are fixed at V_{bias1}, and their currents are $I + I_0/2$, their source voltages are also stabilized. This fixes the drain-to-source voltages v_{DS3} of Q_3 and Q_4. The value of V_{bias1} is chosen such that $|v_{DS3}| \ll V_{d\,sat}$, so that both Q_3 and Q_4 operate in their linear (ohmic) regions. Their aspect ratios $(W/L)_3 = (W/L)_4$ are chosen such that in equilibrium the common-mode output voltage $v_{out,\,c} = (v_{out}^+ + v_{out}^-)/2$ has some desired value (usually ground potential). If now the common-mode voltage $v_{out,\,c}$ would (say) rise due to any reason, the resistance of Q_3 and Q_4 increases. This reduces $|v_{GS5}|$ and $|v_{GS6}|$, and since the current in Q_5 and Q_6 remains unchanged, it forces $|v_{DS5}|$ and $|V_{DS6}|$ to increase. Thus, the drain voltages of Q_5 and Q_6 drop. This, by the argument just presented, reduces $|v_{GS7}|$ and $|v_{GS8}|$, and thus drops *their* drain voltages which are v_{out}^+ and v_{out}^-. The common-mode voltage $v_{out,\,c}$ is thus reduced. The gain of the negative feedback loop is readily seen to be $g_{m3}r_{d3}g_{m5}r_{d5}g_{m7}r_{d7}$, which can be very high. This feedback loop also stabilizes $v_{out,\,c}$ against transistor parameter variations arising from fabrication imperfections.

Since Q_3 and Q_4 operate in their linear (ohmic) regions, their drain currents are linear functions of their gate voltages. Thus, it can readily be shown (Problem 4.40) that a *differential* voltage $\pm v$ at the output terminals does not affect the overall drain/source resistance of the parallel combination of Q_3 and Q_4. Thus, the common-mode output voltage does *not* change if a differential input signal is applied; this is, of course, a desirable feature.

Using small-signal analysis, it can be shown (Problem 4.41) that the differential gain of the stage is

$$A_D = 2g_{m1}R_0, \tag{4.272}$$

where R_0 is the output impedance at either output node:

$$R_0 = R_0^+ = \frac{1}{(g_{d1} + g_{d5})/g_{m7}r_{d7} + g_{d11}/g_{m9}r_{d9}}$$

$$= R_0^- = \frac{1}{(g_{d2} + g_{d6})/g_{m8}r_{d8} + g_{d12}/g_{m10}r_{d10}}. \tag{4.273}$$

Since the circuit is a folded cascode, it does not require a level shifter. Also, since the desired output is a differential signal, no differential-to-single-ended conversion is required. Thus, the nondominant poles introduced by these stages do not appear. The only high-impedance nodes are the output terminals, and the corresponding dominant poles are those due to the time constants $R_0^+ C_L^+$ and $R_0^- C_L^-$, where C_L^+ and C_L^- are the load capacitances at the output terminals. To achieve compensation, the dominant poles can hence be shifted to lower frequencies by increasing $C_L^{\pm}$. Since no internal compensation is required, the op-amp can have a fast settling time, and is hence well suited for the implementation of high-frequency switched-capacitor filters.

Since the output impedances of the circuit can be made very high, the dc differential gain A_D can be large, comparable to that of a basic two-stage op-amp.

A possible bias chain circuit for the op-amp of Fig. 4.127 is shown in Fig. 4.128. The relations for the aspect ratios can be found from the design

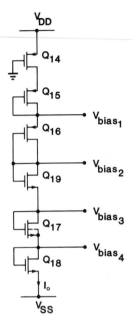

FIGURE 4.128. Bias chain for the fully differential CMOS op-amp of Fig. 4.127.

principles discussed in Section 4.14 (Problem 4.42). Choosing $I = I_0/2$ the resulting values are

$$(W/L)_{11} = (W/L)_{12} = (W/L)_{13}/2 = (W/L)_{18}/2,$$

$$(W/L)_3 = (W/L)_4 = (W/L)_{14},$$

$$(W/L)_5 = (W/L)_6 = (W/L)_{15}, \qquad (4.274)$$

$$(W/L)_7 = (W/L)_8 = (W/L)_{16}/2,$$

$$(W/L)_9 = (W/L)_{10} = (W/L)_{17}/2.$$

$(W/L)_{19}$ should be chosen (in conjunction with the other aspect ratios in the bias chain) to set I_0 to its desired value. For this bias circuit with the aspect ratios given above, the dc currents of Q_3, Q_4, and Q_{14} are equal. Also, their dc drain voltages are approximately the same. Hence, their gate-to-source dc bias voltages satisfy

$$v_{GS3} = v_{GS4} \cong v_{GS14} = -V_{DD}. \qquad (4.275)$$

This shows that $v_{G3} = v_{G4} \cong v_{G14} = 0$. Thus the output common-mode voltage is zero when this bias circuit is designed using Eq. (4.274).

NMOS technology can also be used to realize fully differential op-amps. Such amplifiers are described in References 3 and 29.

For special applications, where a very low dc power drain is imperative, the *dynamic op-amp* configuration[25-27] can be used. One type of these circuits, which is especially suited for switched-capacitor applications, uses time-varying clock-controlled biasing. The circuit diagram of a CMOS differential stage[26] with dynamic biasing is schematically illustrated in Fig. 4.129. As ϕ_1 turns high and ϕ_2 low, C_0 is discharged, while the stage current I_0 is zero. When next ϕ_1 turns off and ϕ_2 on, C_0 will recharge with a large initial current I_0, and transistor Q_1 and Q_2 conduct heavily in their strong inversion regions, since the potential v_A of node Ⓐ is close to V_{SS}. As C_0 gets charged, v_A increases and hence the gate-to-source voltages of Q_1 and Q_2 decrease. Therefore, all transistors enter their weak inversion regions and eventually cut off. During these transients, the gain and unity-gain bandwidth of the stage also vary. The open-loop dc gain is given by $A_0 = g_{m1}/(g_{d2} + g_{d4})$, and the unity-gain bandwidth by $\omega_0 = g_{m1}/C_L$, where C_L is the load capacitance. Now, from Eq. (3.18) of Section 3.4, $g_{m1} \propto \sqrt{I_0}$; by Eq. (3.20), $g_d \propto I_0$ in strong inversion. In weak inversion,[26] by contrast, g_m and g_d are both proportional to I_0. Therefore, for large I_0 and in strong inversion, $A_0 \propto 1/\sqrt{I_0}$ and is hence small but $\omega_0 \propto \sqrt{I_0}$ and is large. For small I_0 and in the weak inversion region, the gain will be large and independent of I_0, while $\omega_0 \propto I_0$ and hence low. Thus, if the circuit is used to charge the load capacitor C_L, at the beginning of the clock phase when ϕ_2 is high, the bias current I_0 is large, ω_0 is high, and hence the initial current charging C_L can increase rapidly. As I_0 is reduced, the gain is increased and the final voltage across C_L corresponds to a very high dc op-amp gain.

Figure 4.130 illustrates a two-stage op-amp with dynamic biasing. In this circuit, instead of a switched bias capacitor C_0 acting as the source impedance, a current–source device Q_5 is used. Q_5 and the current source Q_7 of the output stage are turned on and off by a dynamic bias chain. The current in the bias chain (containing Q_9, Q_{10}, and the two switched capacitors) is high when

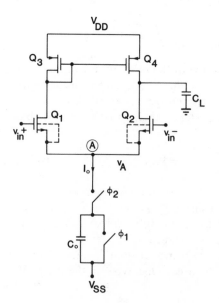

FIGURE 4.129. CMOS differential stage with dynamic biasing.

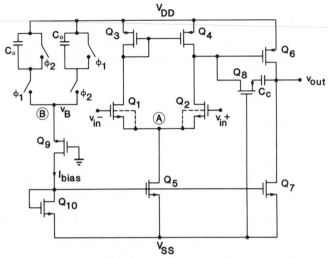

FIGURE 4.130.   CMOS operational amplifier with dynamic biasing.

a discharged capacitor C_0 is just connected into the bias chain; afterwards, it exponentially decreases. These bias current variations are mirrored by Q_5 and Q_7. With both capacitors present as illustrated, this op-amp can be used in both clock phases. If it is needed only in one phase, one of the capacitors and its associated switches may be eliminated. Also, a single dynamic bias chain can be used to bias several dynamic op-amps on the same chip.

The operation of the circuit of Fig. 4.130 closely parallels that of Fig. 4.129. As ϕ_1 turns high, the left-side capacitor (earlier discharged) is connected in the bias chain. Hence, node Ⓑ is initially at a potential V_{DD}, and Q_9 conducts a large current $I_{\text{bias}} = k_9(V_{DD} - V_{Tp})^2$. This is mirrored in Q_7 and Q_5, and hence both amplifier stages conduct heavily. As C_0 gets charged, the voltage v_B at node Ⓑ drops towards zero, Q_9 cuts off, and all currents in the circuit turn off. Thus, the power dissipation is initially large but is then reduced to zero. The initial bandwidth and slew rate are both high; in the later part of each cycle, the bandwidth decreases, but the dc gain becomes high, and hence the load capacitor will be fully charged at the end of the clock phase.

PROBLEMS

4.1.   Prove Eqs. (4.3)–(4.7) for the circuit of Fig. 4.3. How much is the dynamic range for (a) the op-amp alone and (b) the feedback amplifier if $V_{cc} = 10$ V, $A = 10^3$, $\sqrt{v_m^2} = 50$ μV, and $R_2 = 10R_1$?

4.2.   For the circuit of Fig. 4.5a, $n = 3$, $\gamma = 2$ V$^{1/2}$, $|\phi_p| = 0.3$ V, $V_T = 2$ V, $V_{DD} = 10$ V, and $V_{0i} = 2.5i$, $i = 1, 2, 3,$. Find the W/L values using (4.9) and (4.10).

4.3.   Derive the formulas for the W/L ratios in Fig. 4.5a, if the currents drawn at the nodes $V_{01}, V_{02}, \ldots, V_{0n}$ are *not* negligible.

4.4.   Derive the equivalents of Eqs. (4.9) and (4.10) with the channel-length modulation included in the analysis. Repeat the solution of Problem 4.2 using the new relations and $\lambda = 0.02$ V^{-1}.

4.5.   Prove Eq. (4.13) for the circuit of Fig. 4.10.

4.6.   In Figs. 4.9 and 4.10, assume that $r_0 \gg r_{d1}$ and $g_{m2} = g_{m3}$. Show that the output resistance is increased by the open-circuit voltage gain of Q_1.

4.7.   (a) Prove that Eq. (4.15) holds for the circuit of Fig. 4.11. (b) Analyze the circuit of Fig. 4.12. How much is r_{out}? Show that the output resistance is that of Q_2, magnified by the voltage gain of Q_3.

4.8.   Calculate the gain of the circuit of Fig. 4.23 without neglecting the r_{di}.

4.9.   Prove that the low-frequency small-signal gain of the circuit of Fig. 4.24 is given by Eq. (4.30).

4.10.   Prove Eq. (4.31) for the composite-load device of Fig. 4.26. [*Hint:* use Eqs. (3.8) and (3.12) for Q_3 and Q_5. Assume that Q_3 and Q_5 have the same γ and ϕ_p values.]

4.11.   Analyze the circuits of Figs. 4.34 and 4.35. Show that for equal geometries the circuit of Fig. 4.35 has a much higher gain than that of Fig. 4.34!

4.12.   Find the output impedance of the circuit of Fig. 4.38. (*Hint:* Set $v_{in} = 0$, and connect the output terminal to a test source. Calculate the current through the source.)

4.13.   (a) Derive the relations for $v_{0,d}$ and $v_{0,c}$ of the source-coupled stage (Fig. 4.39) if the circuit is not exactly symmetrical. (b) Rewrite your relations in the form

$$v_{0,d} = A_{dd}v_{in,d} + A_{dc}v_{in,c},$$

$$v_{0,c} = A_{cd}v_{in,d} + A_{cc}v_{in,c}.$$

What are A_{dd}, A_{dc}, A_{cd}, and A_{cc}? (c) Let the maximum difference between symmetrically located elements in the small-signal equivalent circuit be 1%. How much are the maximum values of $|A_{cd}|$ and $|A_{dc}|$?

4.14.   Derive Eq. (5.62) for the circuit of Fig. 4.44b. (*Hint:* Write and solve the current equations for nodes Ⓐ, Ⓑ, and Ⓒ.)

4.15.   Modify the CMOS differential stage of Fig. 4.44a so that it has differential output signals. Compare the differential gain with that of the original circuit!

4.16. Prove that the small-signal output impedance of the circuit of Fig. 4.45 is given by Eq. (5.67). (*Hint:* Write and solve the current law for nodes Ⓐ, Ⓑ, and Ⓒ.)

4.17. Show that the split load ($Q_3 - Q_4$) in the circuit of Fig. 4.47 requires half the gate area of Q_2 in a single-load circuit (Fig. 4.15) if the load admittances and dc currents are the same. (*Hint:* Ignore the body effect in both circuits; assume that Q_3 and Q_4 have the same dimensions.)

4.18. Analyze the CMOS gain stage of Fig. 4.49 in the Laplace-transform domain: (a) find the exact transfer function $A_v(s)$ from Fig. 4.49c; (b) use Miller-effect approximation to derive the simplified circuit of Fig. 4.49d; (c) analyze the simplified circuit to verify Eq. (4.76).

4.19. Analyze the NMOS source follower of Fig. 4.50; verify Eq. (4.79).

4.20. Analyze the NMOS cascode gain stage of Fig. 4.51 in the Laplace-transform domain: (a) verify the equivalent circuits of Figs. 4.51b–4.51d; (b) show that Eq. (4.83) holds for the circuit of Fig. 4.51d.

4.21. Analyze the circuit of Fig. 4.54. (*Hint:* Work in the s-domain; write the current laws at nodes Ⓐ, Ⓑ, and Ⓒ.)

4.22. Show that the load conductance represented by Q_3 in Fig. 4.55 is $g_l = g_{m3} + g_{d3}$.

4.23. Prove that if the following conditions are satisfied, that is, (a) $A_v(s)$ is a rational function with all its poles having negative real parts and (b) $Re[kA_v(j\omega)] > -1$ for all ω, then the s values satisfying $kA_v(s) = -1$ all have negative real parts. [*Hint:* By the maximum-modulus theorem of complex functions, if $kA_v(s)$ has no poles in the right half of the s plane, then its real part in the same region has its minimum value on the $j\omega$-axis.]

4.24. Prove Eq. (4.118) for the poles of the circuit of Fig. 4.65. [*Hints:* Calculate $A_v(s)$ from (4.114). Write its denominator as

$$(1 - s/s_{p1})(1 - s/s_{p2}) \simeq 1 - s/s_{p1} + s^2/s_{p1}s_{p2}.$$

Find s_{p1}, s_{p2}; use $(g_{m8} + g_{m9})/(g_{d2} + g_{d4}) \gg 1$ and $(g_{m8} + g_{m9})/(g_{d8} + g_{d9}) \gg 1$.]

4.25. Prove that the zeros of $A_v(s)$ for the circuit of Fig. 4.67 are at $s \to \infty$, while its poles are the same as for the circuit of Fig. 4.65.

4.26. Derive Eqs. (4.120)–(4.122) for the circuit of Fig. 4.69. Why is $A_v(s)$ now a third-order function, while for the circuit of Fig. 4.65 it was only second order?

4.27. Show that Eq. (4.130) gives the small-signal output voltage of the circuit of Fig. 4.72a, if $v_{in} = V_1 u(t)$ and the op-amp transfer function is given by Eq. (4.129).

4.28. Using the low-frequency small-signal models for devices $Q_1 - Q_4$, show that the voltage-gain relations of Eqs. (4.138) and (4.139) hold for the noisy input stage shown in Fig. 4.77.

4.29. The circuit of Fig. 4.131 can be used to measure the unity-gain bandwidth ω_0 of an op-amp. Show that ω_0 is the frequency at which $V_{out}(\omega) = V_{in}(\omega)/\sqrt{2}$, that is, the voltage gain is 3 dB below its dc value.

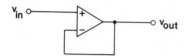

FIGURE 4.131. Op-amp in unity-gain configuration (Problem 4.29).

4.30. For the circuit of Problem 4.29, let the open-loop gain of the op-amp have a phase margin of 60° at the unity-gain bandwidth ω_0. How much is the phase shift between V_{out} and V_{in} at ω_0?

4.31. Show that in the circuit of Fig. 4.132, the effective output impedance is $R_{out}(-A_c + 1)/A$, where $A_c \cong -Z_2/Z_1$ is the closed-loop gain of the stage. Assume that $A \gg 1$ and $R_{out}/A \ll |Z_1|$ and $|Z_2|$.

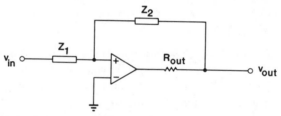

FIGURE 4.132. Op-amp with non-zero output impedance (Problem 4.31).

4.32. Prove Eq. (4.176) for the circuit of Fig. 4.95.

4.33. In the input stage of the circuit of Fig. 4.97 assume that Q_1 and Q_2 are imperfectly matched, such that $g_{m1} = (1 + \varepsilon)g_{m2}$. Show that if a common-mode voltage v_c exists at the input, then the corresponding CMRR is $(1/\varepsilon)g_{m1}/g_{di}$.

4.34. Show that the small-signal impedance of the cascode load of Fig. 4.117a is $g_{m2}r_{d2}r_{d1}$. (*Hints:* Use the equivalent circuit of Fig. 4.117b. You may also use the approximations $g_{m2} \gg g_{d1}, g_{d2}$, but only in the final formula!)

4.35. What should $(W/L)_4$ be chosen in Fig. 4.118 if $|v_{DS1}| = |v_{GS1}| - |V_{Tp}| + 0.2$ V is desired? Assume that $(W/L)_1 = (W/L)_2 = (W/L)_3$.

4.36. Show that the small-signal output impedance of the composite load of Fig. 4.119a is given by Eq. (4.266). (*Hints:* Use the equivalent circuit of Fig. 4.119b. Use also the approximations $g_{mi}r_{dk} \gg 1$ for any i, k, but *only* in the final formula!)

4.37. Prove Eq. (4.267) for the input stage of the op-amp shown in Fig. 4.120.

4.38. Show that the output impedance R_0 of the circuit of Fig. 4.121 is given by Eq. (4.268), and its voltage gain by $-g_{m1}R_0$. Assume that Q_1 and Q_2, Q_6 and Q_7, and Q_{10} and Q_{11} are pairwise matched devices, and all g_m's are much larger than the g_d's.

4.39. Show that the output impedance R_0 of the circuit of Fig. 4.122 is given by Eq. (4.269), and its voltage gain by $-g_{m1}R_0$. Assume that Q_1 and Q_2, Q_3 and Q_4, Q_5 and Q_6, Q_7 and Q_8, and Q_9 and Q_{10} are pairwise matched devices, and that all g_m's are much larger than the g_d's.

4.40. Let the output voltages of the differential op-amp of Fig. 4.127 change by a differential amount, so $v_{out}^+ \rightarrow v_{out}^+ + v$ and $v_{out}^- \rightarrow v_{out}^- - v$. Show that the sum of the drain currents of Q_3 and Q_4 remains unchanged. [*Hint:* assume that Q_3 and Q_4 are in their linear regions, and hence you can use Eq. (3.7).]

4.41. Prove the relations (4.272) and (4.273) giving the gain and output impedances, respectively, of the differential op-amp of Fig. 4.127.

4.42. Show that the aspect ratios of the bias circuit of Fig. 4.128 are given by Eq. (4.274) if we choose $I = I_0/2$ in Fig. 4.127, $I_{bias} = I_0$ as the bias chain current, and zero dc common-mode output voltage.

4.43. Show that (if necessary) a differential output op-amp can be constructed from two single-ended-output op-amps using the circuit[19] of Fig. 4.133.

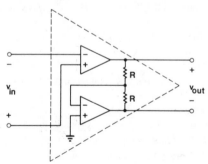

FIGURE 4.133. Simplified equivalent circuit of a differential output op-amp (Problem 4.43).

4.44. Show that the differential gain of the input stage of Fig. 4.87 is

$$A_d = \frac{\Delta V_{04}}{V_d} = -g_{mi}r_{d7}.$$

4.45. Prove Eq. (4.172) for the circuit of Fig. 4.89.

4.46. Show that the output resistance R_0 of the buffer of Fig. 4.92 is given by Eq. (4.174).

REFERENCES

1. D. J. Hamilton and W. G. Howard, *Basic Integrated Circuit Engineering*, McGraw-Hill, New York, 1975, Chap. 10.

2. P. R. Gray and R. G. Meyer, *Analysis and Design of Analog Integrated Circuits*, Wiley, New York, 1977, Chap. 6.

3. K. C. Hsieh, P. R. Gray, D. Senderowicz, and D. G. Messerschmitt, *IEEE J. Solid-State Circuits*, **SC-16**, 708–715 (1981).

4. P. R. Gray, *Basic MOS Operational Amplifier Design—An Overview*, in *Analog MOS Integrated Circuits*, IEEE Press, New York, 1980, Pt. II.

5. Y. Tsividis, *IEEE J. Solid-State Circuits*, **SC-13**, 383–391 (1978).

6. D. Senderowicz, D. A. Hodges, and P. R. Gray, *IEEE J. Solid-State Circuits*, **SC-13**, 760–766 (1978).

7. B. J. Hosticka, R. W. Brodersen, and P. R. Gray, *IEEE J. Solid-State Circuits*, **SC-12**, 600–608 (1977).

8. E. Toy, *Digest of Tech. Papers*, International Solid-State Circuits Conference, 134–135 (1979).

9. Y. P. Tsividis and P. R. Gray, *IEEE J. Solid-State Circuits*, **SC-11**, 748–754 (1976).

10. I. A. Young, *IEEE J. Solid-State Circuits*, **SC-14**, 1070–1077 (1979).

11. R. Gregorian and W. E. Nicholson, Jr., *IEEE J. Solid-State Circuits*, **SC-14**, 970–980 (1979).

12. Y. A. Haque et al., *IEEE J. Solid-State Circuits*, **SC-14**, 961–969 (1979).

13. P. R. Gray, Ref. 4, Fig. 19.

14. J. E. Solomon, *IEEE J. Solid-State Circuits*, **SC-9**, 314–332 (1974).

15. J.-C. Bertails, *IEEE J. Solid-State Circuits*, **SC-14**, 773–775 (1979).

16. K.-C. Hsieh and P. R. Gray, *Digest of Tech. Papers*, Internat. Solid-State Circuits Conference, 128–129 (1981).

17. O. H. Schade, Jr., *IEEE J. Solid-State Circuits*, **SC-13**, 791–798 (1978).

18. Y. P. Tsividis, D. L. Fraser, and J. E. Dziak, *IEEE J. Solid-State Circuits*, **SC-15**, 921–928 (1980).

19. P R. Gray and R. G. Meyer, *IEEE J. Solid-State Circuits*, **SC-17**, 969–982 (1982).

20. W. C. Black, D. J. Allstot, and R. A. Reed, *IEEE J. Solid-State Circuits*, **SC-15**, 929–938 (1980).

21. H. Ohara, P. R. Gray, W. M. Baxter, C. F. Rahim, and J. L. McCreary, *IEEE J. Solid-State Circuits*, **SC-15**, 1005–1013 (1980).

22. P. R. Gray, R. W. Brodersen, D. A. Hodges, T. C. Choi, R. Kaneshiro, and K. C. Hsieh, Proceedings of the International Symposium on Circuits and Systems, 419–422 (1982).

23. G. C. Temes, *IEEE J. Solid-State Circuits*, **SC-15**, 358–361 (1980).

24. Ken Martin and A. S. Sedra, *IEEE Trans. Circuits and Systems*, **CAS-28**, 822–829 (1981).

25. M. A. Copeland and J. M. Rabaey, *Electron. Lett.*, **15**, 301–302 (1979).

26. B. J. Hosticka, *IEEE J. Solid-State Circuits*, **SC-15**, 887–894 (1980).

27. M. G. Degrauwe, J. Rijmenants, E. A. Vittoz, and H. J. De Man, *IEEE J. Solid-State Circuits*, **SC-17**, 522–528 (1982).

28. E. Vittoz and J. Fellrath, *IEEE J. Solid-State Circuits*, **SC-12**, 224–231 (1977).

29. D. Senderowicz, S. F. Dreyer, J. M. Huggins, C. F. Rahim, and C. A. Laber, *IEEE J. Solid-State Circuits*, **SC-17**, 1014–1023 (1982).

30. L. W. Nagel, *SPICE 2: A Computer Program to Simulate Semiconductor Circuits*, University of California, Berkeley, ERL, Memo., ERL-M520, May 1975.

Chapter Five _____

SWITCHED-CAPACITOR FILTERS

The most common, and most difficult, application of switched-capacitor circuits is to perform the frequency-domain filtering of signals. This chapter deals with the design principles, building blocks, and actual circuits of such filters. The relative advantages and disadvantages of the two most useful filter realizations—cascade and ladder-type circuits—are described and illustrated by a typical design example. The important scaling process is also discussed in detail.

In special applications, very narrow-band switched-capacitor bandpass filters are required. In such a case, SC N-path filters may provide the only feasible realization. The design of these filters is hence also discussed. Finally, integrated MOS active-RC filters are described. These circuits are continuous-time filters, and hence do not require antialiasing or smoothing peripheral circuits; in fact, they can themselves be utilized for such tasks. In addition, they do not require clock signals and switches, and hence are free of some of the inherent shortcomings (clock feedthrough noise, noise aliasing, etc.) of switched-capacitor filters.

5.1. INTEGRATED FILTERS[1,2]

Probably the most important reason for the development of analog MOS integrated circuits was the need to fabricate fully integrated high-quality analog filters. Historically, such filters were first realized as passive circuits, built from resistors (R), inductors (L), and capacitors (C). However, inductors

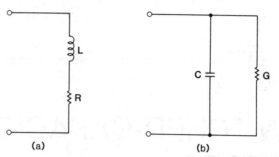

FIGURE 5.1. Lossy reactive elements: (*a*) lossy inductor; (*b*) lossy capacitor.

have several drawbacks in such applications:[1]

1. Inductors are usually very lossy. At low frequencies, their losses can be represented by the equivalent circuit of Fig. 5.1*a*. Defining the *quality factor* Q_L of the inductor by the relation

$$Q_L = \frac{\omega L}{R},\qquad(5.1)$$

it turns out that the highest practically attainable Q_L is only about 1000.

2. For low frequencies (say, below 100 Hz), the size and weight of inductors becomes very large.

3. Inductors often need ferromagnetic materials for cores. These inductors are usually nonlinear, and hence generate unwanted harmonics of the signals.

4. Inductors radiate, as well as pick up, electromagnetic waves. Hence, they tend to introduce interference noise into the circuit.

5. The quality factor Q_L is proportional to the square of the linear dimensions of the inductor.[1] Hence, inductors cannot be reduced in size without a serious decrease in Q_L. By contrast, consider a lossy capacitor. At low frequencies, its equivalent circuit is that shown in Fig. 5.1*b*. The quality factor of the capacitor is

$$Q_C = \frac{\omega C}{G}\qquad(5.2)$$

which can be shown[1] to be independent of the physical dimensions. Hence, high-Q miniature capacitors *can* be built. Q_C can be as high as 10,000 for practical capacitors.

For the listed reasons, when inexpensive monolithic op-amps first became available about 25 years ago, a major effort was made to design and build filters from op-amps, resistors, and capacitors only. These *active-RC filters* played (and still play) an important role in communication and control systems, as well as other applications. They are often realized in a hybrid construction, with monolithic op-amps and chip capacitors soldered on a board containing thick-film resistors. Thus fabricated, the active-RC filter can be made much smaller than an equivalent RLC circuit, especially at low frequencies. However, obvious physical limitations prevent the reduction of the size of such a hybrid structure to truly subminiature dimensions.

The next logical step, then, is to realize the active-RC filter as a fully integrated structure. Here, a choice must be made between bipolar and MOS technologies for fabrication. In most cases (an exception is described in Ref. 3) the choice is to use MOS technology. The main reasons[2] are the availability of high-quality capacitors in MOS integrated circuits, the ability of storing charge on a node for a substantial amount of time (up to 1 s!) without leakage, and the possibility of sensing such a charge nondestructively and continuously. Charge storage is possible since any MOS transistor connected to a node can be cut off almost completely: the off-current of a MOSFET can be less than a picoamp, and the leakage to the substrate is even smaller. Nondestructive charge sensing is possible since the low-frequency input (gate) impedance of a MOSFET is, for practical purposes, infinite.

As mentioned earlier, and described in Section 3.5, it is possible to fabricate high-quality capacitors in a variety of ways in MOS technology. However, large capacitors require considerable area (typically, 4 $mil^2 \approx 2500$ μm^2 per pF), and hence capacitors larger than about 100 pF are seldom used.

The most common application of MOS analog integrated filters has been in the voice band (0 ~ 4 kHz). Such filters require poles—and hence reciprocal RC time constants—which are of the order of 10 krad/s. Even assuming a fairly large capacitor, say 10 pF, this requires a resistance around 10 MΩ. Such a resistor would occupy an area of about 1600 $mil^2 \sim 10^6$ μm^2, which is inordinately large; the *overall* area of a typical analog MOS chip is around 20,000 mil^2. Furthermore, such an MOS resistor, whether realized as a diffusion or as a polysilicon line, tends to be nonlinear. Finally, the values of MOS resistors are somewhat inaccurate. A relative accuracy better than about 10% is hard to achieve, even though the *tracking* error between two resistors on the same chip may be kept as low as 1 ~ 2%.

Since capacitors are made in different fabrication steps than resistors, their errors do not track with those of the resistors on the same chip. The errors of the capacitors are of the same order as those of the resistors, that is, about 10% individual error and 1% tracking error; hence, the error of any RC time constant may be as large as 20%. Such errors can seldom be tolerated even in low-selectivity filters. Furthermore, the temperature and voltage coefficients of resistors and capacitors are not correlated; hence, the time constants will also vary somewhat with temperature and signal level.

We conclude that the straightforward element-by-element integration of an active-RC filter cannot satisfy the accuracy and stability requirements nor-

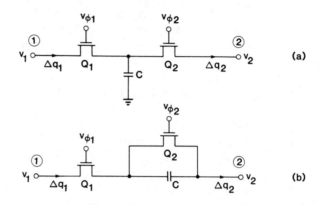

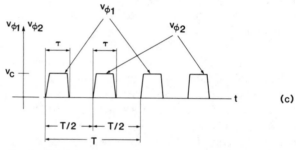

FIGURE 5.2. Simulated resistors: (*a*) shunt realization; (*b*) series realization; (*c*) clock waveforms.

mally needed. Hence, different strategies must be used. The key idea, apparently first presented by Fried[4] in the United States (see also Refs. 5–8 for publications by European authors which relate to the topic) is illustrated in Fig. 5.2. In Figs. 5.2*a* and 5.2*b* two circuits, each containing a capacitor and two MOS switches, are shown. Figure 5.2*c* illustrates the clock signals $v_{\phi 1}$ and $v_{\phi 2}$ activating the MOSFETs. Whenever $v_{\phi 1}$ equals V_C, Q_1 is in full conduction since V_C is normally large (5 ~ 10 V). Hence, during this interval Q_1 provides a low resistance between its drain and source. Whenever $v_{\phi 1}$ is zero, Q_1 acts as an open circuit. Similar conditions hold for Q_2 and $v_{\phi 2}$. Assume that v_1 and v_2 are slowly varying signals so that during a *clock interval T* (Fig. 5.2*c*) they do not change appreciably. This will be the case if the highest frequency components in v_1 and v_2 are much smaller than the *clock frequency* $f_c = 1/T$.

Consider next in Fig. 5.2*a* the charge Δq_1 entering C from the input terminal as $v_{\phi 1}$ rises to V_C. Since C was previously charged to v_2, and is now recharged through Q_1 to v_1, clearly $\Delta q_1 = C(v_1 - v_2)$. Next, $v_{\phi 1} \to 0$, Q_1 cuts off, and C holds its voltage at v_1. When now $v_{\phi 2} \to V_C$, C is recharged to v_2 again though Q_2. Clearly, $\Delta q_2 = C(v_1 - v_2) = \Delta q_1$.

Since during each clock interval T a charge $C(v_1 - v_2)$ enters at node ① and leaves at node ②, we can define the *average* current i flowing from ① to ②:

$$i \triangleq \frac{\Delta q_1}{T} = \frac{\Delta q_2}{T} = \frac{C}{T}(v_1 - v_2) \tag{5.3}$$

or $i = (1/R)(v_1 - v_2)$ where $R \triangleq T/C$. Thus the average current i and the voltage difference $v_1 - v_2$ satisfy Ohm's law, and the circuit of Fig. 5.2a behaves as a *resistor* of value T/C ohms. Physically, this is plausible for the following reasons. A resistor is different from a capacitor in that it can dissipate electrical energy but cannot store information, while the opposite statements hold for a capacitor. By recharging the capacitor C periodically, we can dissipate energy in the switches and also the memory of the capacitor is periodically destroyed. Thus, the physical behavior of a resistor is simulated.

The above argument is somewhat heuristic, and ignores the detailed variations of the actual voltages and currents. Since these cannot be neglected for exact design, they will be considered in the next section.

A similar approximate analysis can be performed for the circuit of Fig. 5.2b. Here, when $v_{\phi 2} = V_C$, C is discharged by Q_2 while Q_1 is off so that $\Delta q_1 = \Delta q_2 = 0$. Next, when Q_2 turns off and Q_1 turns on, C charges to a voltage $v_C = v_1 - v_2$. Hence, it acquires a charge $\Delta q_1 = \Delta q_2 = C(v_1 - v_2)$. Thus, the charge flows and the average current are similar to those of the circuit of Fig. 5.2a. We conclude that this circuit also performs as an approximate equivalent of a resistor of value $R = T/C$.

At this point, it is clear that active-RC filters can be modified into switched-capacitor ones, simply by replacing all resistors by equivalent switched-capacitor branches. The resulting network will thus contain only op-amps, switches, and capacitors. Assuming ideal op-amps and switches, the response depends only on the values of the capacitors. A time constant of the form $R_1 C_2$ will be replaced by $(T/C_1)C_2 = (C_2/C_1)/f_c$. Here C_1 is the value of the switched capacitor used to replace R_1, and f_c the clock frequency. Now, as mentioned earlier, while the value of an individual capacitor can only be realized with an accuracy around 10%, the *tracking* error (i.e., the error in the *ratio* of two capacitors on the same chip) can be kept as small as 1%, or even less. The clock frequency $f_c = 1/T$ can also be controlled very accurately, using crystal oscillators as clocks. Hence, the time constants of an integrated switched-capacitor filter can be realized with an accuracy of 1% or better. This is adequate for most filtering applications.

The area needed by the simulated resistor is usually much smaller than that needed for a direct realization. For the case of a resistor $R = 10$ MΩ, using $f_c = 100$ kHz, the required capacitor is $C = T/R = 10^{-12}$ F, or 1 pF. This occupies approximately 4 mil^2, rather than the 1600 mil^2 needed for the direct fabrication of R. The reduction in area is by a factor of 400!

An alternative strategy for realizing accurate and small resistors in integrated filters is to use MOS transistors in their linear (i.e., nonsaturation) regions as simulated resistors. Since the value of such a "resistor" can be controlled by changing the gate voltage, the time constants can be tuned using an on-chip control circuit with an off-chip reference, such as a crystal clock or some external RC elements. This approach is not commonly used yet, but there are promising results showing the feasibility of fabricating MOS integrated filters based on such principles.[49, 51] Hence, these circuits are briefly discussed in Section 5.9, at the end of this chapter.

5.2. SWITCHED-CAPACITOR INTEGRATORS

The basic building block of an active-RC filter is the active integrator[9-16] (Fig. 5.3). Assuming ideal components, the circuit equations are

$$v = 0, \qquad i = 0 \tag{5.4}$$

and

$$i_1 = \frac{v_{in}}{R_1} = i_2 = -C_2 \frac{dv_{out}}{dt}. \tag{5.5}$$

Hence, the output–input relation is

$$v_{out}(t) = -\frac{1}{R_1 C_2} \int_{-\infty}^{t} v_{in}(\tau) \, d\tau. \tag{5.6}$$

Using Laplace transformation and assuming zero initial conditions (i.e., $v_{C2} = v_{in} = 0$ for $t = 0-$), the relations

$$V_{out}(s_a) = \frac{-V_{in}(s_a)}{s_a R_1 C_2}$$

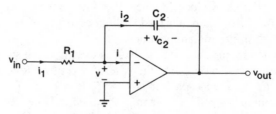

FIGURE 5.3. Active-RC integrator.

and

$$H_a(s_a) = \frac{V_{out}(s_a)}{V_{in}(s_a)} = -\frac{1/R_1C_2}{s_a} \qquad (5.7)$$

result. Here, s_a is the analog Laplace-transform variable.

A switched-capacitor equivalent of this integrator, which uses the simulated resistor of Fig. 5.2a, is shown in Fig. 5.4a. The clock waveform is illustrated in Fig. 5.4b; the input wave in Fig. 5.4c. Notice that in this figure (as well as in all subsequent ones), a shorthand notation is used for the clock signals: $v_{\phi1}$ and $v_{\phi2}$ have been replaced by ϕ_1 and ϕ_2, respectively, and $v_{\phi1} = V_C$ is indicated by $\phi_1 = 1$, and so on. This use of the logic variables (ϕ_1, ϕ_2) rather than the

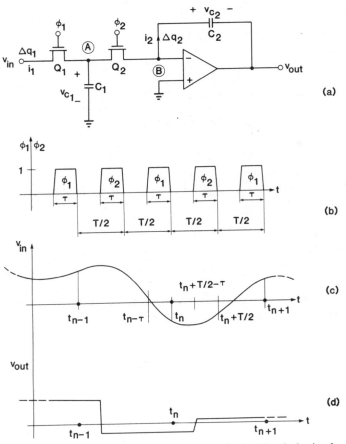

FIGURE 5.4. Shunt-switched-capacitor integrator: (a) circuit; (b) clock signals; (c) input voltage; (d) output voltage.

actual clock voltages $(v_{\phi1}, v_{\phi2})$ is common practice, since it simplifies the discussion. Also, often a symbolic switch rather than a MOSFET is drawn for the switching devices (cf., e.g., Figs. 5.7a and 5.7b).

Consider next the operating of the SC integrator. When ϕ_1 rises to 1 at $t = t_n - \tau$, Q_1 turns on and C_1 (which was earlier discharged) recharges to v_{in}. At $t = t_n$, Q_1 shuts off and C_1, having acquired a charge

$$\Delta q_1(t_n) = C_1 v_{\text{in}}(t_n) \tag{5.8}$$

is isolated. At $t = t_n + T/2 - \tau$, ϕ_2 rises to 1 and Q_2 turns on. Thus, C_1 is connected between the virtual ground at the inverting input of the op-amp and true ground, and hence it discharges. Thus, a charge

$$\Delta q_2(t_n + T/2 - \tau) = C_1 v_{\text{in}}(t_n) \tag{5.9}$$

flows through Q_2 and into C_2 (Fig. 5.4a). As a result, the charge stored in C_2 increases by Δq_2, and the voltage v_{C2} across it changes by

$$\Delta v_{C2}(t_n + T/2 - \tau) = (C_1/C_2)v_{\text{in}}(t_n). \tag{5.10}$$

If the switches are ideal, then at times $t_n - \tau$ and $t_n + T/2 - \tau$, Δq_1 and Δq_2 flow *instantaneously*. Hence, each current waveform contains a sum of impulse functions; in addition, the current i_1 through Q_1 also includes the component $C_1 \, dv_{\text{in}}/dt$ for $t_n - \tau < t < t_n$. [This component is present only if $v_{\text{in}}(t)$ changes during this period.]

The resulting waveforms are illustrated in Fig. 5.5a. For real switches, the on-resistances are not zero. Hence, the charges do not flow instantaneously; as a result, the currents do not become infinite, and v_{C2} does not change abruptly. The real responses are thus as shown in Fig. 5.5b.

Often, $v_{\text{in}}(t)$ is itself a staircase function. Then $dv_{\text{in}}/dt = 0$ (except when v_{in} jumps) and hence the $C_1 \, dv_{\text{in}}/dt$ components are not present in $i_1(t)$. Note that v_{C2} and v_{out} are staircase functions in any case, whether or not $v_{\text{in}}(t)$ is one.

From the above analysis, a difference equation can be established between the samples of $v_{\text{in}}(t)$ and $v_{\text{out}}(t)$ taken at t_{n-1}, t_n, t_{n+1}, and so on. From (5.10) and Fig. 5.4,

$$v_{\text{out}}(t_{n+1}) - v_{\text{out}}(t_n) = -v_{C2}(t_{n+1}) + v_{C2}(t_n)$$

$$= -\Delta v_{C2}(t_n + T/2 - \tau) = -(C_1/C_2)v_{\text{in}}(t_n). \tag{5.11}$$

The first and last parts of Eq. (5.11) can be equated and solved using z-transformation. This gives

$$V_{\text{out}}(z)(z - 1) = -(C_1/C_2)V_{\text{in}}(z)$$

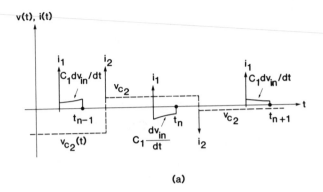

(a)

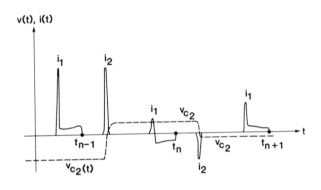

(b)

FIGURE 5.5. Voltage and current waveforms in the integrator of Fig. 5.4a: (a) ideal case; (b) practical case.

and

$$H(z) \triangleq \frac{V_{out}(z)}{V_{in}(z)} = -\frac{C_1/C_2}{z-1} = -\frac{C_1}{C_2}\frac{z^{-1}}{1-z^{-1}}. \tag{5.12}$$

It is interesting to compare the performance of the continuous-time integrator of Fig. 5.3 with that of the switched-capacitor (SC) equivalent in Fig. 5.4 for sine-wave input signals. For $s_a = j\omega_a$, Eq. (5.7) gives

$$H_a(j\omega_a) = \frac{-1/(R_1 C_2)}{j\omega_a}, \tag{5.13}$$

while, for $z = \exp(j\omega T)$, (5.12) becomes

$$H(e^{j\omega T}) = \frac{-C_1/C_2}{e^{j\omega T}-1}. \tag{5.14}$$

For $\omega T < 1$, we can use the Taylor-series expansion

$$e^{j\omega T} = 1 + j\omega T - (\omega T)^2/2 - + \cdots \tag{5.15}$$

so that

$$H(e^{j\omega T}) = \frac{-C_1/(C_2 T)}{j\omega - \omega^2 T/2 - + \cdots}. \tag{5.16}$$

We note that $H_a(j\omega_a)$ is a pure imaginary (sometimes called "lossless") function, while $H(e^{j\omega T})$ is not. Also, $H(e^{j\omega T})$ is a periodic function of ω, while $H_a(j\omega_a)$ is not a periodic function of ω_a. Hence, these two functions can be made approximately equal only in a limited frequency range. Specifically, if $\omega T \ll 1$ and $C_1 = T/R_1$, (5.16) gives

$$H(e^{j\omega T}) \approx \frac{-1/(R_1 C_2)}{j\omega} \tag{5.17}$$

which is the same as (5.13). We conclude that for signal frequencies satisfying (say) $\omega < 1/100T = f_c/100$ the SC integrator simulates its active-RC model well.

Another way of looking at the relation between the responses of the continuous-time and SC integrators is to compare (5.7) and (5.12) directly. If $C_1 = T/R_1$ is chosen, then s_a in $H_a(s_a)$ must be replaced by $(z - 1)/T$ to obtain $H(z)$. This replacement is, however, the same as the *forward-Euler transformation* (2.70) described in Chapter 2. Hence, if an active-RC filter contains only integrators of the form shown in Fig. 5.3 interconnected appropriately, and if each integrator is then replaced by an SC equivalent as shown in Fig. 5.4, the resulting SC filter response will suffer from the distortions (peaking, etc.) discussed for the forward-Euler mapping in Section 2.6.

Consider next the SC integrator (Fig. 5.6) obtained by using the equivalent resistor circuit of Fig. 5.2b. Here, C_1 discharges during $\phi_2 = 1$, and recharges during $\phi_1 = 1$. Thus, the charge flow during $t_n - \tau < t < t_n$ is

$$\Delta q_1(t) = \Delta q_2(t) = C_1 v_{in}(t). \tag{5.18}$$

Therefore, v_{C2} changes by

$$\Delta v_{C2}(t) = \Delta q_2(t)/C_2 = (C_1/C_2) v_{in}(t) \tag{5.19}$$

during this same period. Thus $v_{out} = -v_{C2}$ satisfies the difference equation

$$v_{out}(t_n) - v_{out}(t_{n-1}) = -(C_1/C_2) v_{in}(t_n). \tag{5.20}$$

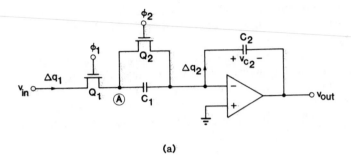

(a)

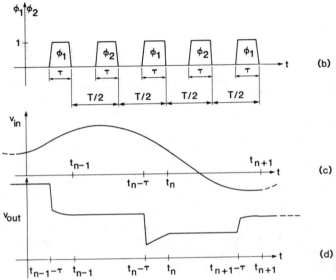

(b)

(c)

(d)

FIGURE 5.6. Series-switched-capacitor integrator: (*a*) circuit; (*b*) clock signals; (*c*) input voltage; (*d*) output voltage.

Using z-transformation,

$$V_{out}(z)(1 - z^{-1}) = -(C_1/C_2)V_{in}(z)$$

and

$$H(z) = \frac{V_{out}(z)}{V_{in}(z)} = -\frac{C_1/C_2}{1 - z^{-1}} \tag{5.21}$$

results.

For $z = \exp(j\omega T)$, therefore

$$H(e^{j\omega T}) = \frac{-C_1/C_2}{1 - e^{-j\omega T}} \cong \frac{-C_1/C_2 T}{j\omega + \omega^2 T/2} \tag{5.22}$$

where the approximation is valid for $\omega T \ll 1$. Hence, once again the integrator is not lossless. However, while the "loss term" $\omega^2 T/2$ carries a negative sign in (5.16), here it is positive.

Comparing Eqs. (5.7) and (5.21) and using $C_1 = T/R_1$ again, it becomes evident that replacing all integrators in an active-RC filter by the SC integrator of Fig. 5.6a is equivalent to replacing s_a by $(1 - z^{-1})/T$ in the transfer function. This, however, is the *backward-Euler mapping* (2.74) discussed in Section 2.6. Hence, the rounding distortion described there will affect the frequency response of the filter.

The output voltage waveform is shown in Fig. 5.6d. During the $\phi_1 = 1$ periods, the input voltage is directly coupled to the output; hence, the relation between their changes

$$\Delta v_{out}(t) = -\frac{C_1}{C_2} \Delta v_{in}(t) \tag{5.23}$$

holds in these intervals. At all other times, $v_{out}(t)$ is piecewise constant. Often, $v_{in}(t)$ is a staircase function. Then $v_{out}(t)$ will be constant for $t_n - \tau < t < t_{n+1} - \tau$, and it will jump at $t_n - \tau$, $t_{n+1} - \tau$, and so on. Thus, it will also be a staircase function.

A comparison of Eqs. (5.11) and (5.20) shows that for the circuit of Fig. 5.4, $v_{out}(t)$ depends on the *past* value of the input $v_{in}(t)$, while for that of Fig. 5.6, on the *present* value. Thus, the former circuit delays the input signal, while the latter does not. In both circuits, a *positive*-valued $v_{in}(t)$ results in a *decrease* of $v_{out}(t)$; thus, both are *inverting* integrators.

As explained in Section 2.6, for most filtering applications the bilinear *s-to-z* transformation given by Eq. (2.80) is superior to both Euler mappings. Hence, we can ask whether any SC integrator exists which corresponds to replacing s_a in (5.7) by the expression in (2.80):

$$s_a \rightarrow \frac{2}{T} \frac{z - 1}{z + 1}. \tag{5.24}$$

It turns out that in fact several such integrators exist.[11-13] Two are shown in Fig. 5.7. It can be shown that their transfer functions are given by

$$H(z) = -\frac{C_1}{C_2} \frac{z + 1}{z - 1} \tag{5.25}$$

which, for $C_1 = T/2R_1$, is the bilinear transform of $H_a(s_a)$. Note that for the circuits of Fig. 5.7, the input is sampled when $\phi_1 = 1$. For both circuits the output jumps only when $\phi_2 \rightarrow 1$. The input voltage v_{in} should be a sampled-and-held signal in both circuits.

Notice that the MOSFETs used as switches have been (symbolically) replaced by ideal switches in Fig. 5.7. This is a convenient notation which will be used in most of the diagrams in this chapter.

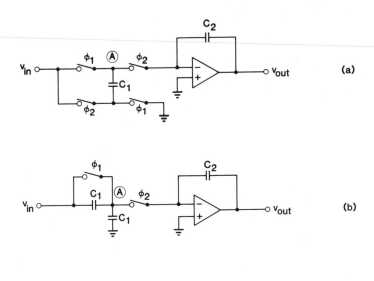

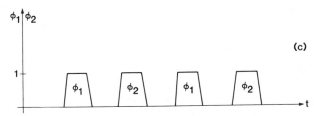

FIGURE 5.7. Bilinear switched-capacitor integrators: (a) and (b) circuits; (c) clock signals.

5.3. STRAY-INSENSITIVE INTEGRATORS

All switched-capacitor circuits described so far suffer from a common short-coming: they are sensitive to the effects of stray capacitances between the various nodes and ground. Consider, for example, node Ⓐ in Fig. 5.4a. This node is connected to the source/drain diffusions of Q_1 and Q_2 which have appreciable capacitance to the substrate; also, the leads connecting the top plate to Q_1 and Q_2 have some capacitance to the bulk. The resulting total stray capacitance C_A may be of the order of 0.05 pF, and its value is not controlled. Since C_A is in parallel with C_1, it represents an error in its value. For, say, a 1% accuracy in the value of C_1, we must hence choose $C_1 > 5$ pF. This requires, typically, that $C_2 > 50$ pF. Thus, a large area is needed on the chip for these capacitors, in order to "swamp" the stray capacitance C_A.

It should be noted that stray capacitances connected between ground and the other nodes of the circuit of Fig. 5.4a do not affect the operation significantly. This is because all these nodes are connected to low-impedance sources (v_{in}, v_{out}) or to virtual ground as is the case with node Ⓑ. Clearly, the

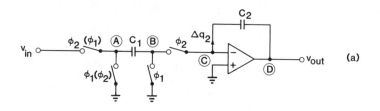

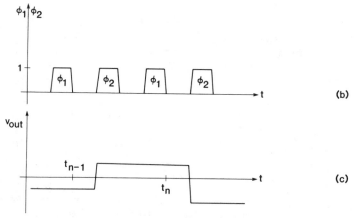

FIGURE 5.8. Stray-insensitive integrators: (*a*) circuit; (*b*) clock signals; (*c*) output voltage.

stray capacitance between node Ⓑ and ground has no voltage across it at any time. Hence, it cannot influence the operation of the circuit.

Similar arguments show that the stray capacitance between node Ⓐ and ground affects the operation of the circuits shown in Figs. 5.6 and 5.7. In fact, in some of these circuits, the stray effect introduces a capacitance where none existed before, and hence it changes the form of the transfer function.

To eliminate the harmful effects of stray capacitors, the stray-insensitive integrator[14-16] shown in Fig. 5.8a can be used. In this circuit, the stray capacitance C_A between node Ⓐ and ground is periodically charged from the input source which provides v_{in}, and then discharged to ground. C_B (which loads node Ⓑ) is grounded at both terminals,* as is C_C. Finally, C_D is driven by the low-impedance output of the op-amp, and is hence rendered harmless. In conclusion, none of the stray capacitances to ground influence the operation of the circuit.

It should be noted that the above considerations are valid only if the op-amp is ideal, that is, if it has infinite gain and zero output impedance.

---

*Except during the intervals when $\phi_1 = \phi_2 = 0$ and Ⓑ is floating. Any charge acquired from C_1 by C_B during these periods is returned to C_1 when Ⓑ returns to ground.

Otherwise, node Ⓒ will not be a virtual ground and Ⓓ will not be connected to a zero-impedance source. The effects of such nonideal op-amp characteristics will be discussed in Chapter 7.

Also, the response is *not* insensitive to strays coupling the clock signals to node Ⓒ. This phenomenon is also discussed in Chapter 7.

Consider now the operation of the circuit with the clock phases shown *without* parentheses. Then during the $\phi_1 = 1$ intervals C_1 discharges, since both of its terminals are grounded. During the $\phi_2 = 1$ intervals C_1 charges to v_{in}, and hence a charge $\Delta q_2 = C_1 v_{in}$ enters C_2. The operation is thus basically the same as that of the circuit of Fig. 5.6a, except that the two clock phases are interchanged. Hence, Eq. (5.21) remains valid for the transfer function. The circuit of Fig. 5.8 requires four, rather than two, switches; this is the price paid for the stray-insensitive property.

Let the clock phases of the first two switches now be chosen as shown *in* the parentheses. Then, during $\phi_1 = 1$, C_1 charges up to v_{in}, such that its left-side plate acquires a charge $+C_1 v_{in}$ while its right-side electrode the charge $-C_1 v_{in}$. During $\phi_2 = 1$, the $-C_1 v_{in}$ charge enteres C_2, resulting in a jump of $\Delta v_{out} = +(C_1/C_2)v_{in}$ in the output voltage. This sequence of events is similar to that described for the circuit of Fig. 5.4a where, however, $+C_1 v_{in}$ entered C_2 when $\phi_2 \rightarrow 1$. Thus the transfer function with the parenthesized clock phases is, except for a sign change, that given in (5.12):

$$H(z) = +\frac{C_1}{C_2}\frac{z^{-1}}{1 - z^{-1}}. \tag{5.26}$$

Note that for this operation of the circuit a *positive*-valued v_{in} results in an *increase* of v_{out}. Thus, the clock phases shown in parentheses make the circuit of Fig. 5.8a perform as a *noninverting* integrator.

In conclusion, this circuit can operate as an inverting delay-free integrator, with a transfer function given by (5.21), if the original clock phases are used. If the parenthesized phases are chosen, the performance is that of a noninverting delaying integrator, and $H(z)$ is given by (5.26). In both modes of operation, the response is insensitive to the effects of all stray capacitances to ground.

It is also possible to construct a circuit[17] which is compensated for by the effects of stray capacitances, and which is an inverting delaying integrator with its response given by Eq. (5.12). The circuit, however, is more complicated, and relies on the matching of some components for stray insensitivity.

Using the stray-insensitive circuit of Fig. 5.8a, the capacitance values C_1 and C_2 no longer need to be much larger than the strays. Hence, values as small as 0.1 pF can safely be used. A remaining limitation is the lateral stray capacitance *between* the lines leading to the electrodes of C_1 (and of C_2). This is of the order 2 fF; it can, however, be reduced by inserting a grounded line between the leads. In any case, the minimum permissible C_1 and C_2 values are reduced by a factor $10 \sim 50$ if the stray-insensitive configuration is used. Hence, the area required by the capacitors is reduced by the same factor.

It should be noted that there are other factors limiting the minimum capacitance usable in switched-capacitor circuits. These include the effects of random edge variations of the capacitors and clock feedthrough noise. Both become worse as the values of the capacitors are lowered. These and other nonideal effects will be discussed in Chapter 7.

5.4.   SECOND-ORDER SECTIONS; CASCADE FILTER DESIGN[19-21,37]

Using the SC integrators introduced in the previous sections, it is possible to construct simple filter sections. Of special interest are sections realizing the *biquadratic* transfer function

$$H(z) = -\frac{a_2 z^2 + a_1 z + a_0}{b_2 z^2 + b_1 z + 1}. \tag{5.27}$$

These second-order sections are often called SC *biquads*.

Recalling that $z \triangleq e^{sT}$, and assuming that the frequencies which are of interest are much lower than $f_c = 1/T$, we have

$$z \triangleq e^{sT} \cong 1 + sT. \tag{5.28}$$

Substituting into (5.27),

$$H(z) = H(e^{sT}) \cong \frac{c_2 s^2 + c_1 s + c_0}{d_2 s^2 + d_1 s + d_0} \tag{5.29}$$

results. This is a well-known transfer function; there is a wealth of information in the literature on the active-RC realization of transfer functions of this form. With appropriate modification, some of this material is thus applicable to SC filter design whenever (5.28) holds. Following Refs. 1 and 18 we rewrite the resulting $H_a(s) = H(e^{sT})$ in the form

$$H_a(s) = \frac{V_{out}(s)}{V_{in}(s)} = -\frac{K_2 s^2 + K_1 s + K_0}{s^2 + (\omega_0/Q)s + \omega_0^2}. \tag{5.30}$$

Here, ω_0 is *pole frequency* and Q is the *pole Q* associated with the poles of $H_a(s)$. Specifically, let these poles be $s_p = \sigma_p + j\omega_p$ and its conjugate. Then

$$\omega_0 \triangleq |s_p| = \sqrt{\sigma_p^2 + \omega_p^2} \tag{5.31}$$

and

$$Q \triangleq \frac{|s_p|}{2|\sigma_p|} = \frac{1}{2}\sqrt{1 + \left(\frac{\omega_p}{\sigma_p}\right)^2}. \tag{5.32}$$

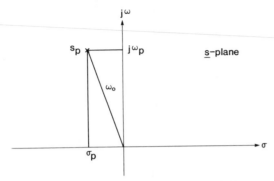

FIGURE 5.9. Complex pole in the s-plane: $\omega_0 = |s_p|$ is the pole frequency; $Q = \omega_0/2|\sigma_p|$ the pole-Q.

Figure 5.9 illustrates the meaning of ω_0 and Q. If Q is high (say 10 or more), the pole is relatively close to the $j\omega$-axis, and it causes $H_a(j\omega)$ to have a sharp peak near ω_0. Such functions tend to have widely spread element values, and higher sensitivity to element-value variations near ω_0.

To obtain the block diagram of a system which realizes $H_a(s)$, we rewrite (5.30) in the form

$$s^2 V_{\text{out}} = -\left(K_2 s^2 + K_1 s + K_0\right) V_{\text{in}} - \left(\frac{\omega_0}{Q} s + \omega_0^2\right) V_{\text{out}}. \qquad (5.33)$$

Dividing both sides by s^2, and rearranging (5.33) gives

$$V_{\text{out}} = -\frac{1}{s}\left((K_1 + K_2 s) V_{\text{in}} + \frac{\omega_0}{Q} V_{\text{out}} - \omega_0 V_1\right) \qquad (5.34)$$

where V_1 is defined as follows:

$$V_1 \triangleq -\frac{1}{s}\left(\frac{K_0}{\omega_0} V_{\text{in}} + \omega_0 V_{\text{out}}\right). \qquad (5.35)$$

Thus, V_1 is obtained by integrating the weighted sum of V_{in} and V_{out}, and V_{out} by integrating the weighted sum of V_{in}, V_{out}, and V_1. A system which can perform these operations, and thus produce V_{out} from V_{in}, is shown in Fig. 5.10a. It contains two inverting integrators to provide the $-1/s$ operations, two summers at the inputs of the integrators, and various feedback and feedforward branches to couple V_{in}, V_1, and V_{out} to the adders. V_1 is produced at the output of the first integrator.

It is clear that (5.33) could have been partitioned differently. Then two equations different from (5.34) and (5.35) would have resulted, and another system, different from that shown in Fig. 5.10a, would provide the realization.

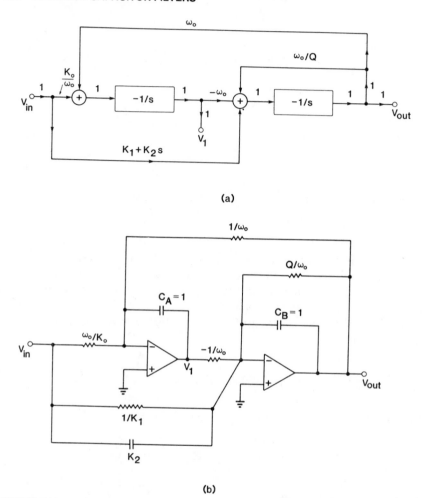

(a)

(b)

FIGURE 5.10. Switched-capacitor low-Q biquad: (a) block diagram; (b) active-RC realization; (c) circuit; (d) clock and signal waveforms.

Next, an active-RC realization will be found for the system of Fig. 5.10a. We will assume that the *input* signal of each integrator is a *current*, while its *output* signal is a *voltage*. Then the inverting integrator is realized by an op-amp with a feedback capacitor of value $C = 1$. Its transfer function is clearly $V_{out}/I_{in} = -1/sC = -1/s$, as required. For the adders, we use the inverting input nodes of the op-amps. The coupling branches have as *input* signals the op-amp *output* voltages, and their *output* signals are the *currents* entering the integrators. Thus, for example, the feedback branch with transmittance ω_0 is described by the relation $I_{out}/V_{in} = \omega_0$; it can hence be realized by a *resistor* of value $R = 1/\omega_0$. Similarly, the feedforward branch has $I_{out}/V_{in} = K_1 + K_2 s$, which corresponds to the parallel combination of a resistor of value

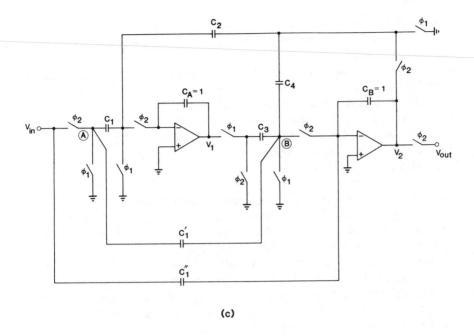

(c)

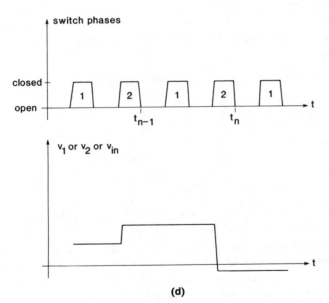

(d)

FIGURE 5.10. continued.

$1/K_1$ and a capacitor of value K_2. The resulting circuit is shown in Fig. 5.10*b*. It should be noted that this circuit is not useful as an active-RC filter, since it contains the negative resistor $-1/\omega_0$ which would require additional active elements for its realization. It serves, however, a useful purpose as a model for the final SC circuit. Thus, let each resistor be replaced by a combination of a capacitor and its charging and discharging switches. For *positive-valued* resistors, in principle the simple equivalent resistors of Fig. 5.2 could be used. We have already seen, however, that the resulting integrator stages (Figs. 5.4 and 5.6) are sensitive to stray capacitance effects. Hence, the equivalent resistor illustrated as the input branch of the integrator in Fig. 5.8 is used instead. The value of the capacitor can be found from (5.3) to be $C \cong T/R$, where R is the resistance to be simulated. For the *negative* resistor, the same circuit can be used but with the clock phases shown in parentheses in Fig. 5.8*a*. With this phasing, a positive v_{in} will result in negative charges being delivered to the op-amp, just as in a negative resistor. The capacitance is $T/|R|$.

The resulting SC biquad[19] is shown in Fig. 5.10*c*. The clock phases and the waveforms are illustrated in Fig. 5.10*d*. Note that v_{in} should change only when ϕ_2 goes high; then the other voltages v_1 and v_{out} will also change only at these same instances. In the circuit, several switches have been eliminated since they duplicated the task of other switches. Thus, say, at node Ⓑ the two switches perform as parts of all three switched resistors containing C_1', C_3, and C_4. The element values of the switched capacitors, as mentioned earlier, can be found from $C \cong T/|R|$. For the unswitched capacitors, the values remain unchanged. Thus, we have

$$C_1 \cong TK_0/\omega_0 = |H_a(0)|\omega_0 T,$$

$$C_2 \cong C_3 \cong \omega_0 T,$$

$$C_4 \cong \omega_0 T/Q, \tag{5.36}$$

$$C_1' \cong K_1 T,$$

$$C_1'' = K_2.$$

In the formula giving C_1, $H_a(0) = -K_0/\omega_0^2$ is, by (5.30), the dc gain of the stage; usually, $|H_a(0)|$ is 1 or higher for a low-pass filter.

The pole frequency ω_0 has normally a value which is nearly equal to the cut-off frequency of the filter. The latter is much smaller than the clock frequency $\omega_c = 2\pi/T$; hence, the product $\omega_0 T = 2\pi\omega_0/\omega_c$ is normally much less than 1. The values of C_1' and C_1'' usually are larger than $\omega_0 T$ but less than 1. If (as usual), $|H_a(0)| \geq 1$, and if $Q \leq 1$, then also $\omega_0 T \leq C_1, C_4 < 1$. Thus, under these circumstances, the ratio of the largest and smallest capacitors in the circuit (the so-called *capacitance spread*) satisfies

$$\frac{C_{max}}{C_{min}} \cong \frac{C_A}{C_2} \cong \frac{1}{\omega_0 T}. \tag{5.37}$$

In general, as will be shown later, it is desirable to keep C_{max}/C_{min} close to the minimum value possible. The value given in (5.37) satisfies this, as can be seen from the following argument. Clearly, to realize the pole s_p, the circuit must contain time constants of the order of $1/|s_p| = 1/\omega_0$. Thus, there must be RC products of this order. The SC realization of R results in $(T/C_1)C_2 \approx 1/\omega_0$ which requires $C_2/C_1 \approx 1/\omega_0 T$ as in (5.37). Note that (5.37) holds only if $Q \leq 1$; otherwise $C_4 < \omega_0 T$ and the spread equals $C_A/C_4 = Q/\omega_0 T > 1/\omega_0 T$. Thus, this section is more suitable for low-Q pole realization.

All the above considerations were derived based on the approximation that for all signal frequencies of interest, $|\omega T| \ll 1$ held. Under this condition, the signal voltages are nearly smooth continuous waveforms, and the "side lobes" centered at $\pm\omega_c$, $\pm 2\omega_c$, and so on are very far from the signal spectrum around $\omega = 0$. Under these conditions also the capacitance spread is very large, as (5.37) shows.

To avoid the approximation inherent in our earlier derivation, we will next derive the *exact* sampled-data transfer function $H(z)$ of the SC circuit of Fig. 5.10c. This task is simplified by the use of z-domain transfer functions for the individual integrators and coupling branches of the circuit. Let us denote (as before) the charge flow into an integrator *during* the nth clock period $(n - 1)T \leq t \leq nT$ by $\Delta q(t_n)$, and the value of a voltage at $t = nT$ by $v(t_n)$. Let the corresponding z-transforms be $\Delta Q(z)$ and $V(z)$, respectively. Then, for an integrator (containing an op-amp and its feedback capacitor C), the transfer function is obtained from

$$v_{out}(t_n) - v_{out}(t_{n-1}) = -(1/C)\,\Delta q_{in}(t_n),$$

$$V_{out}(z)[1 - z^{-1}] = -(1/C)\,\Delta Q_{in}(z), \qquad (5.38)$$

$$H(z) = \frac{V_{out}(z)}{\Delta Q_{in}(z)} = \frac{-1/C}{1 - z^{-1}}.$$

Similar transfer functions (but defined by the ratio $\Delta Q_{out}/V_{in}$) can be derived for the coupling branches. Assuming that v_{in} changes as ϕ_2 goes high, we obtain (Problem 5.5)

$$H(z) = \frac{\Delta Q}{V} = \begin{cases} C(1 - z^{-1}) \text{ for an unswitched } C \text{ (e.g., } C_1'') \\ C \text{ for a noninverting switched } C(C_1', C_2, C_1, C_4). \\ -Cz^{-1} \text{ for an inverting switched } C \text{ (e.g., } C_3) \end{cases} \quad (5.39)$$

Note that $H(z)$ for C_3 indicates a delay of a full clock period T, while in fact

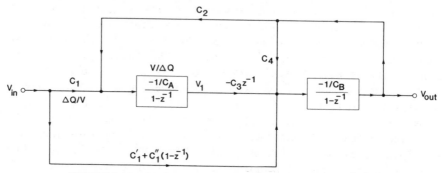

FIGURE 5.11. Block diagram of the low-Q biquad in the z-domain.

the delay between the sampling of v_1 and the subsequent charge flow is only $T/2$. However, because of the S/H nature of v_1 (Fig. 5.10d), $v_1(t_n - T/2) = v_1(t_n - T)$, so that a delay of $T/2$ has the same effect as one of T.

Using these transfer functions, a block diagram (Fig. 5.11) can be constructed, which is exactly equivalent to the circuit of Fig. 5.10c, but much simpler to analyze. The result (Problem 5.6) is

$$H(z) = \frac{V_{out}(z)}{V_{in}(z)} = -\frac{(C_1' + C_1'')z^2 + (C_1 C_3 - C_1' - 2C_1'')z + C_1''}{(1 + C_4)z^2 + (C_2 C_3 - C_4 - 2)z + 1}. \quad (5.40)$$

Thus, an exact realization of the $H(z)$ given in (5.27) is possible by matching the coefficients of the two rational functions in z. This gives

$$C_1'' = a_0,$$

$$C_1' = a_2 - C_1'' = a_2 - a_0,$$

$$C_1 = (1/C_3)(a_1 + C_1' + 2C_1'') = (a_0 + a_1 + a_2)/C_3, \quad (5.41)$$

$$C_4 = b_2 - 1,$$

$$C_2 C_3 = b_1 + 2 + C_4 = b_1 + b_2 + 1.$$

Note that $C_1' \geq 0$ requires that $a_2 \geq a_0$, and $C_4 \geq 0$ requires $b_2 \geq 1$. The latter relation indicates [as can also be seen from the denominator of $H(z)$ in (5.40)] that for any positive value of C_4, the *poles* satisfy $|z_p| < 1$. The circuit is thus always stable—an important advantage. The former relation indicates—as does the numerator of $H(z)$ in (5.40)—that the *zeros* are also inside (or, for $C_1' = 0$, on) the unit circle.

Equation (5.41) contains five relations which do not determine the six unknown capacitance values. A useful additional constraint can be determined from the requirement that the capacitance spread C_{max}/C_{min} be minimized. As explained in connection with the approximating relations (5.36) and (5.37), it is expected that C_2 and C_3 will be the smallest capacitances in the circuit. Since the last relation in (5.41) fixes C_2C_3, the smallest capacitance will have the largest possible value if $C_2 = C_3 = (b_1 + b_2 + 1)^{1/2}$ is chosen. Then all other capacitances can be found from (5.41). Note that $C_A = C_B = 1$, and hence all capacitances can be regarded as normalized values, with $C_A = C_B$ as unit capacitance.

The transfer function of (5.30) is quite general; in most specific applications, it can be simplified. If a *low-pass* response is needed, we can set $K_1 = K_2 = 0$. Then all transmission zeros are at infinite frequency. From (5.36), this leads to $C_1' = C_1'' = 0$ in the circuit of Fig. 5.10c. This, in turn, causes the numerator $H(z)$ in (5.40) to become C_1C_3z. Hence, the z-domain transmission zeros are at $z = 0$ and $z \to \infty$.

If a *high-pass* response is required, $K_0 = K_1 = 0$ can be used. Then (5.36) gives $C_1 = C_1' = 0$. The numerator of $H(z)$ in (5.40) becomes $C_1''(z^2 - 2z + 1) = C_1''(z - 1)^2$. Hence, both transmission zeros are at $z = 1$, corresponding to $s = 0$.

If a *notch (bandstop)* response is needed, $K_1 = 0$ can be set. Then, by (5.36), $C_1' = 0$. The numerator polynomial of $H(z)$ is then $C_1''[z^2 - (2 - C_1C_3/C_1'')z + 1]$. For $C_1C_3 < 2C_1''$ (which is usually the case), this polynomial has two complex conjugate zeros on the unit circle, giving the required notch.

For a *bandpass* characteristic, we can set $K_0 = K_2 = 0$ in $H_a(s)$. This leads to $C_1 = C_1'' = 0$, and hence the numerator of $H(z)$ in (5.40) becomes $C_1'z(z - 1)$. The transmission zeros are thus at $z = 0$ and $z = 1$.

A restriction on the circuit of Fig. 5.10c is that (as noted earlier) it cannot realize transmission zeros outside the unit circle. Such zeros may be needed, for example, for the realization of all-pass circuits. The restriction can be overcome if an inverting SC branch (Fig. 5.12) is added between the input terminal and node Ⓑ in the circuit of Fig. 5.10c, and C_1' is removed. Then, as can be shown, $H(z)$ is changed to

$$H(z) = -\frac{C_1''z^2 + (C_1C_3 - 2C_1'' - C_1''')z + (C_1'' + C_1''')}{(1 + C_4)z^2 + (C_2C_3 - C_4 - 2)z + 1} \quad (5.42)$$

(Problem 5.7). The magnitude of the zeros can now be greater than one.

As mentioned earlier, the circuit of Fig. 5.10c is better suited for low-Q applications. For the high-Q ($Q > 3$) case, (5.33) can be partitioned as follows

$$V_{out} = -\frac{1}{s}(K_2sV_{in} - \omega_0V_1), \quad (5.43)$$

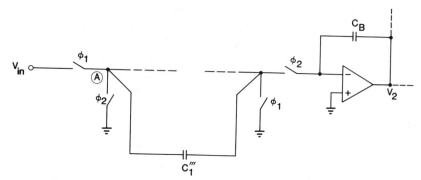

FIGURE 5.12. Added branch needed to realize zeros outside the unit circle.

where

$$V_1 = -\frac{1}{s}\left[\left(\frac{K_0}{\omega_0} + \frac{K_1}{\omega_0}s\right)V_{\text{in}} + \left(\omega_0 + \frac{s}{Q}\right)V_{\text{out}}\right]. \tag{5.44}$$

A block diagram corresponding to these relations is shown in Fig. 5.13*a*; an active-RC realization is shown in Fig. 5.13*b*. Replacing, as before, all resistors by SC branches of appropriate switching phases, the SC filter [20] of Fig. 5.13*c* results. The element values of the switched branches can be found from the approximation $C \cong T/|R|$. This gives

$$C_1 \cong \frac{K_0 T}{\omega_0} = |H_a(0)|\omega_0 T,$$

$$C_2 \cong C_3 \cong \omega_0 T,$$

$$C_4 \cong \frac{1}{Q}, \tag{5.45}$$

$$C_1' \cong \frac{K_1}{\omega_0},$$

$$C_1'' \cong K_2.$$

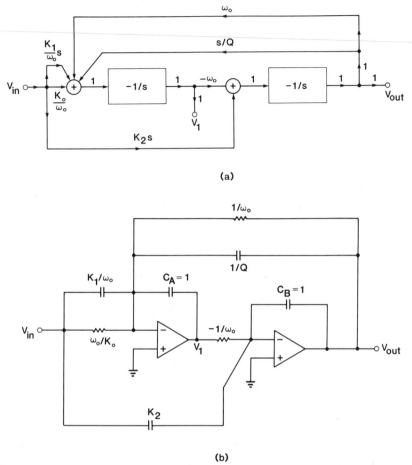

(a)

(b)

FIGURE 5.13.   Switched-capacitor high-Q biquad: (a) block diagram; (b) active-RC realization; (c) circuit; (d) clock and signal waveform.

For $Q > 1$, the capacitance spread is once again $C_A/C_2 \cong 1/\omega_0 T$.

The exact transfer function $H(z)$ can be found using the waveform diagram of Fig. 5.13d, and the block diagram of Fig. 5.14. The result is (Problem 5.8)

$$H(z) = \frac{V_{out}(z)}{V_{in}(z)} = -\frac{C_1''z^2 + (C_1C_3 + C_1'C_3 - 2C_1'')z + (C_1'' - C_1'C_3)}{z^2 + (C_2C_3 + C_3C_4 - 2)z + (1 - C_3C_4)}.$$

(5.46)

Equating the coefficients of the transfer functions in (5.27) and (5.46), five design equations can be obtained for the six unknown capacitance values.

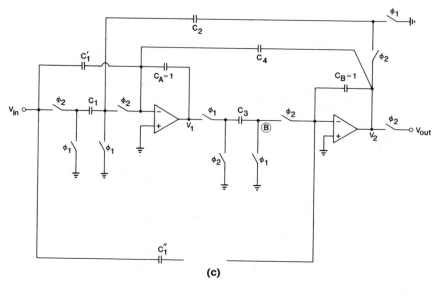

(c)

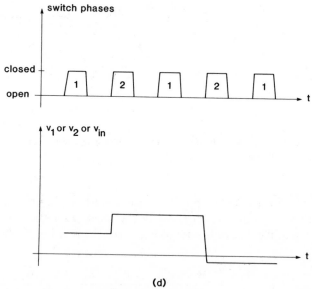

(d)

FIGURE 5.13. continued.

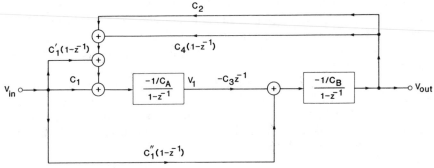

FIGURE 5.14. Block diagram of the high-Q biquad in the z-domain.

These are

$$C_1'' = \frac{a_2}{b_2},$$

$$C_1' = \frac{C_1'' - a_0/b_2}{C_3} = \frac{a_2 - a_0}{b_2 C_3},$$

$$C_1 = \frac{a_1/b_2 - C_1'C_3 + 2C_1''}{C_3} = \frac{a_0 + a_1 + a_2}{b_2 C_3}, \tag{5.47}$$

$$C_4 = \frac{1 - 1/b_2}{C_3},$$

$$C_2 C_3 = \frac{b_1}{b_2} + 2 - C_3 C_4 = \frac{1 + b_1 + b_2}{b_2}.$$

It is again possible to choose C_2 and C_3 to be equal, since by (5.45) they are likely to be the smallest capacitors in the circuit. Then the last equation in (5.47) gives $C_2 = C_3 = [(1 + b_1 + b_2)/b_2]^{1/2}$, and the other relations in (5.47) determine all other capacitance values.

As before, $H(z)$ can be simplified to obtain special responses. For a *low-pass* response, $K_1 = K_2 = 0$ can be chosen in (5.30). This leads to $C_1' = C_1'' = 0$. Hence, the numerator of $H(z)$ becomes $C_1 C_3 z$, and the transmission zeros are, as for the circuit of Fig. 5.10c, at $z = 0$ and at $z \to \infty$. For a *high-pass* response, $K_0 = K_1 = 0$ results in $C_1 = C_1' = 0$. The numerator of $H(z)$ is then again $C_1''(z - 1)^2$, and both transmission zeros are at $z = 1$ (i.e., at $s = 0$). For a *notch* response, $K_1 = 0$ can be set. This corresponds to $C_1' = 0$. Then the numerator polynomial is $C_1''[z^2 - (2 - C_1 C_3/C_1'')z + 1]$, as before. By the argument presented for the notch filter derived from the circuit of Fig. 5.10c, the zeros are (for usual values) on the unit circle. Finally, for a *bandpass* response, setting $K_0 = K_2 = 0$ leads to $C_1 = C_1' = 0$. Therefore, the numerator of $H(z)$ becomes $C_1''C_3(z - 1)$. Thus, the zeros are now at $z = 1$ and $z \to \infty$.

Since the numerator polynomial of $H(z)$ in (5.46) can be written as $C_1''[z^2 - (2 - C_1C_3/C_1'' - C_1'C_3/C_1'')z + (1 - C_1'C_3/C_1'')]$, for the usual case when $(C_1 + C_1')C_3 < 2C_1''$ its zeros are inside or on the unit circle, as are its poles. If zeros outside the unit circle are required, the inverting branch of Fig. 5.12 can again be included, and C_1' eliminated. The corresponding transfer function is (Problem 5.9)

$$H(z) = -\frac{C_1''z^2 + (C_1C_3 - 2C_1'' - C_1''')z + (C_1'' + C_1''')}{z^2 + (C_2C_3 + C_3C_4 - 2)z + (1 - C_3C_4)}. \quad (5.48)$$

Here, the constant term in the numerator $(C_1'' + C_1''')$ is greater than the leading term (C_1''). Hence, the zeros can be outside the unit circle.

It is interesting to note that some favorable properties of the transfer function of (5.46) can be deduced by inspection. If the clock frequency $\omega_c = 2\pi/T$ is much larger than the pole frequency ω_0, and the Q is high, then both poles will be close to the $z = 1$ ($s = 0$) point, and the denominator polynomial will be $z^2 - (2 - \delta_1)z + (1 - \delta_0)$, where $\delta_1, \delta_0 \ll 1$. Thus, for this important case, the deviation δ_1 of the linear coefficient from -2, and that of the constant term from $+1$ in the denominator (δ_0) are both small. Since the latter is C_3C_4, for even moderately small values of C_3 and C_4, very small δ_0 can be realized. Similarly, since $\delta_1 = (C_2 + C_4)C_3$, if $C_2 + C_4$ and C_3 are both small then δ_1 can be very small. Thus, the capacitance spread remains moderate even for high Q and ω_c/ω_0 values. A similar argument holds for the zeros of $H(z)$. By contrast, an inspection of (5.40) shows that C_4 must be very small if a high-Q, large ω_c/ω_0 response is required. This again illustrates the unsuitability of the circuit of Fig. 5.10c for such applications.

In addition, for the circuit of Fig. 5.13c the sensitivities of the constant coefficient $(1 - C_3C_4)$ to variations of C_3 or C_4 are $\partial(1 - C_3C_4)/\partial C_3 = -C_4$ and $\partial(1 - C_3C_4)/\partial C_4 = -C_3$, both of which are small. Similarly, the sensitivities of the linear coefficient $(C_2C_3 + C_3C_4 - 2)$ are C_3, $(C_2 + C_4)$, and C_3, all small. Hence the pole values are not very sensitive to the exact values of C_2, C_3, and C_4. For high-pole-Q's, this is an important advantage.

Other interesting conclusions can be obtained from Fig. 5.10b and Eq. (5.30). If the resistor Q/ω_0 is removed, then the denominator of $H_a(s)$ in (5.30) becomes $s^2 + \omega_0^2$. Thus, the natural modes are at $\pm j\omega_0$, and the circuit performs as a sine-wave oscillator, with a frequency ω_0. Replacing the resistors $\pm 1/\omega_0$ by switched capacitors of value $C_2 = C_3 = \omega_0 T$, the equation $(C_2/C_A)(C_3/C_B) = (\omega_0 T)^2$ results. Clearly, the lowest capacitance spread achievable is $C_{max}/C_{min} \approx \omega_0 T$. Also, we see that the biquad is basically a damped oscillator; we suppress the oscillation by adding the resistor Q/ω_0 (represented in the final circuit by the switched capacitor C_4) which changes the phase shift around the loop from zero to a nonzero value, and reduces the loop gain to below one, for $s = j\omega_0$.

Similar conclusions hold for the circuits of Fig. 5.13. There, the damping is achieved by adding the capacitor $1/Q$ in parallel with the resistor $1/\omega_0$. Without this capacitor, the circuit oscillates with a frequency ω_0, and again the condition $(C_2/C_A)(C_3/C_B) = (\omega_0 T)^2$ holds. Thus, the minimum capacitance spread is $\omega_0 T$, as for the circuit of Fig. 5.10.

The pole-Q is determined by the single capacitance C_4 in both circuits. In applications where the Q must be variable, this is an important advantage.

Consider now the case of a high-order transfer function $H(z)$. Factoring the numerator and denominator into second-order factors, it can be written in the form:

$$H(z) = \left(-\frac{a_2'z^2 + a_1'z + a_0'}{b_2'z^2 + b_1'z + 1} \right)\left(-\frac{a_2''z^2 + a_1''z + a_0''}{b_2''z^2 + b_1''z + 1} \right)\cdots . \quad (5.49)$$

Each factor can then be realized by one of the biquads of the form discussed above. Since each biquad provides a buffered output voltage (from the low-impedance output terminal of an op-amp with negative feedback), cascading them will not affect their transfer functions. Thus, such *cascade realization* can be used to obtain high-order transfer functions using biquads as building blocks. An attractive feature of the resulting structure is its modular configuration, which makes the design, analysis, and troubleshooting all relatively easy. Typically, the bilinear s-to-z transform, discussed in Chapter 2, Section 2.6 is used to obtain an exact transfer function $H(z)$, which can also be realized exactly by the coefficient matching methods discussed in connection with Eqs. (5.41) and (5.47).

If the order of $H(z)$ is odd, then in addition to the biquadratic factors of (5.49) we will need to include and realize also a bilinear factor of the form

$$H(z) = -\frac{a_1z + a_0}{b_1z + 1}. \quad (5.50)$$

Using the approximation $z \cong 1 + sT$, this can be approximated by

$$H_a(s) = -\frac{K_1s + K_0}{s + \omega_0}. \quad (5.51)$$

The block diagram of a continuous-time system realizing $H(s)$ is shown in Fig. 5.15a; an active-RC realization is shown in Fig. 5.15b and an SC equivalent in Fig. 5.15c with the clock phases shown without parentheses. The clock and voltage waveforms are similar to those shown in Fig. 5.13d. If the clock phases

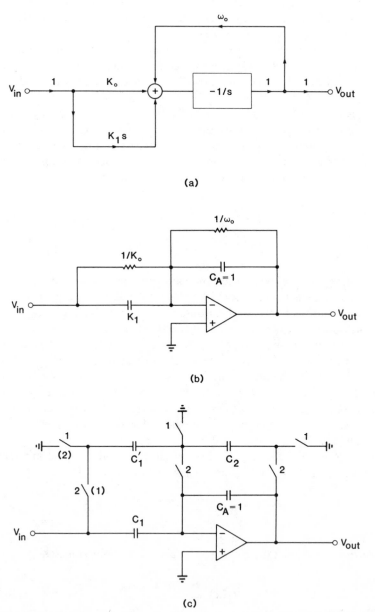

FIGURE 5.15. Switched-capacitor linear section: (*a*) block diagram; (*b*) active-RC realization; (*c*) circuit.

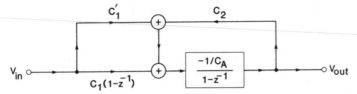

FIGURE 5.16. Block diagram of the linear section in the z-domain.

without parentheses are used, then an estimation of the element values can be obtained from $C \approx T/R$. The result is

$$C_1 = K_1,$$

$$C_1' = TK_0 = |H_a(0)|\omega_0 T, \tag{5.52}$$

$$C_2 = \omega_0 T.$$

The exact transfer function can also be obtained. Using the relations of (5.38) and (5.39) to construct the exact z-domain block diagram shown in Fig. 5.16, analysis gives

$$H(z) = \frac{V_{out}(z)}{V_{in}(z)} = -\frac{(C_1 + C_1')z - C_1}{(1 + C_2)z - 1}. \tag{5.53}$$

The element values can now be determined uniquely by coefficient matching (Problem 5.10). Clearly, both the zero and the pole are inside the unit circle, and have real positive values.

If a zero outside the unit circle is required, then the clock phases inside the parentheses should be used. The transfer function is then

$$H(z) = -\frac{C_1 z - (C_1 + C_1')}{(1 + C_2)z - 1}. \tag{5.54}$$

Negative real zero can also be obtained if the switched-capacitor branch containing C_1 is replaced by the simulated resistor of Fig. 5.2a. Similarly, replacing the SC branch containing C_2 by the circuit of Fig. 5.2a will result in a negative real pole (Problem 5.11). The value of C in Fig. 5.2a is, of course, affected by the stray capacitance in parallel with it.* This may, however, be acceptable in noncritical applications, since the resulting change for $z = e^{j\omega T}$,

*For the *input* branch, the stray-insensitive circuit of Fig. 5.27b (described later) can also be used. It inverts the phase and does not provide a continuous signal path, and hence is not suitable as a feedback branch.

$\omega T \ll 1$ is essentially a small additional constant gain (if the input branch is modified) or additional loss (if the feedback branch is affected). Problems 5.11 and 5.20 relate to the proofs of these statements.

For $H(z)$ obtained by using the bilinear s-z transformation, often a zero exists at $z = -1$ which corresponds to $s_a \to \infty$. Then, one of the input capacitors may be much smaller than the others in the circuit. The inaccuracy of this capacitor, however, also introduces only a small additional constant loss or gain, and hence is usually unimportant.*

A somewhat different timing of the switches in the biquads is also possible.[21] This may result in a reduced sensitivity of the response to the effects of finite op-amp bandwidth. It will be discussed in Chapter 7.

5.5. SWITCHED-CAPACITOR LADDER FILTERS[1, 2, 18, 21, 22]

By cascading the first- and second-order filter sections developed in the previous section, in principle any high-order transfer function $H(z)$ may be realized. In practice, however, the resulting circuit is often difficult to fabricate for very high order and selective filters. The reason is that for such filters, the transfer function $H(z)$ usually contains poles very close to the unit circle. The response of a section which realizes such high-Q poles is very sensitive to element-value variations. Hence, in the presence of normal fabrication toler-ances, the response of such a section is often unacceptable, and the yield becomes too low for economical production.

For filters which have to realize such high-Q poles, therefore, other design techniques are often used. The most successful and widely used of these alternative strategies are based on simulating the low-sensitivity response of a doubly terminated reactance two-port.[23] Consider the circuit shown in Fig. 5.17, operating at a frequency ω_0 such that the LC two-port provides optimum power matching between the generator (including R_S) and the load R_L, when all elements have their nominal values. Now let an element in the LC two-port, say C_5, change by a small amount ΔC_5. Since $|V_{\text{out}}|$ under nominal conditions has its maximum possible value, it is going to decrease regardless of whether

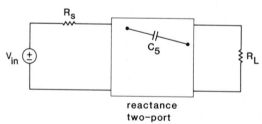

FIGURE 5.17. Doubly terminated reactance two-port with generator impedance R_S, load R_L, and internal element C_5.

*Prof. K. W. Martin, personal communication.

ΔC_5 is positive or negative. This indicates that for nominal values at ω_0

$$\frac{\partial |V_{out}|}{\partial C_5} = 0. \tag{5.55}$$

Thus, the sensitivity of the response (at a frequency where the gain is maximum) is zero. Clearly, C_5 was an arbitrary choice, and the argument is valid for *any* component of the LC two-port. It can also be shown[24] that the sensitivity at a frequency where the gain is close to being maximum is low; in fact, it is proportional to the magnitude of the reflection coefficients at the ports of the LC two-port. In addition, the sensitivity to the terminating resistances R_S and R_L is constant, independent of frequency, when the gain is maximum. Thus, a small change in (say) R_S will merely result in the multiplication of V_{out} by a constant at all frequencies where the gain is maximum (or close to it). This effect can usually be tolerated.

The above-described favorable properties have enabled the designers of LC filters over many years to build high-order filters realizing very high-Q poles. It is therefore reasonable to assume that similarly low-sensitivity SC filters can be obtained based on an LC filter model, provided the following conditions are satisfied:

1. The transfer functions of the two filters are the same.
2. The parameters of the two filters enter their respective transfer functions the same way.

It is clear that neither condition can hold exactly, since the LCR filter is a passive continuous-time circuit, while the SCF is an active sampled-data one. However, using approximations similar to those applied in the previous section, the conditions described can be at least approximately satisfied.

There have been several design techniques developed which derive an SCF from an LCR prototype. Several of these[25-27] are based on retaining the configuration (invariably a *ladder*) of the LCR model, and finding an equivalent SC realization for each L, C, and R element. Unfortunately, while the resulting SCFs need relatively few op-amps, they are not fully insensitive to stray-capacitance effects. Hence, they are seldom used in practical applications.

The most widely used strategy is based instead on the simulation of the voltage–current relations of the analog filter. It is often developed in terms of the signal flow graph (SFG) of the LCR filter. The SCF is designed as an analog computer based on the SFG. While this approach requires many operational amplifiers (the total number is equal to the order of the filter), it can lead to stray-insensitive SCFs, and is hence favored in practical applications for high-order, high-accuracy SCFs.

SFG-based SCFs (commonly called "*SC ladder filters*") can be designed on the basis of simple approximations, valid for very large clock-frequency/signal-frequency ratios. Alternatively, an exact but more complicated design strategy may be used. Both approaches will be briefly discussed in the rest of this section.

The Approximate Design of SC Ladder Filters

The design process will be illustrated by the simple example of a third-order low-pass filter. Its circuit diagram is shown in Fig. 5.18a and its loss response in Fig. 5.18b. The response of the filter can be determined from the state variables V_1, I_2, and V_3, and the input V_{in}. These are related by the equations

$$-V_1 = \frac{-1}{sC_1}\left(\frac{V_{\text{in}} - V_1}{R_S} - I_2\right),$$

$$-I_2 = \frac{-1}{sL_2}(V_1 - V_3), \tag{5.56}$$

$$V_3 = \frac{-1}{sC_3}\left(-I_2 + \frac{V_3}{R_L}\right).$$

The first and third of these relations represent the KCL applied to nodes ① and ③, while the second one represents the KVL applied to the loop formed by C_1, L_2, and C_3. Following the procedure used in the previous section, we next derive a block diagram of an analog system implementing these equations. The result is shown in Fig. 5.18c.

Next, an active-RC realization will be given for the system of Fig. 5.18c. Assuming (as in Section 5.4) that the *input* signal of each integrator is a *current*, while the *output* signal is a *voltage*, the circuit of Fig. 5.18d results. Note that this circuit satisfies Condition 1 given above; its transfer function is exactly the same as that of the LCR filter of Fig. 5.18a. Furthermore, the capacitors correspond to the reactances of the original circuit, and enter the transfer function the same way. The same holds for the resistor R_L. By contrast, the role of R_S is played by *two* resistors in the circuit of Fig. 5.18d, and also four new elements of values ± 1 have been added. Thus, the sensitivity relations are not exactly the same as for the LCR circuit. However, calculations show that all sensitivities are still very low in the passband of the filter, as long as the nominal loss is small, say 0.1 dB or less.

Next, as before, we replace each resistor R by a switched capacitor of value $T/|R|$. This gives the circuit shown in Fig. 5.18e. The element values are given by the approximating relations

$$C_S \cong T/R_S,$$

$$C \cong T, \tag{5.57}$$

$$C_L \cong T/R_L.$$

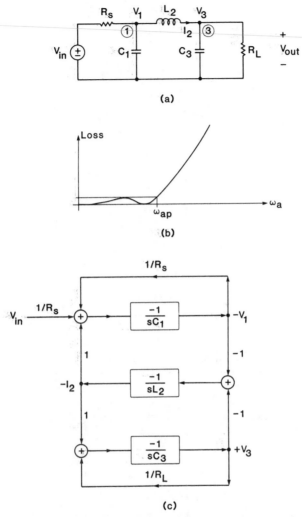

FIGURE 5.18. Switched-capacitor ladder filter: (a) LCR prototype filter; (b) loss response; (c) block diagram; (d) active-RC realization; (e) final circuit.

The approximation made in the above process results in an error in the response. This error can be estimated from Eq. (5.16) which (except for a sign change) describes the response of the noninverting delaying integrator forming the middle stage of the circuit of Fig. 5.18e, and from Eq. (5.22) which describes the first and third stages. Rewriting Eq. (5.22) in the form

$$H\left(e^{j\omega T}\right) = \frac{-C_1/T}{j\omega C_2 + \omega^2 TC_2/2} \tag{5.58}$$

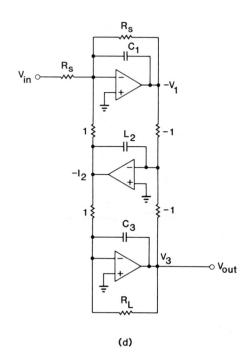

(d)

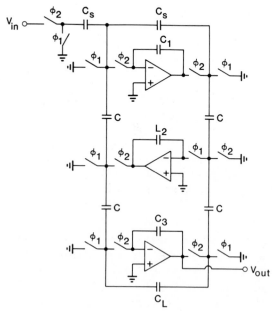

(e)

FIGURE 5.18. continued.

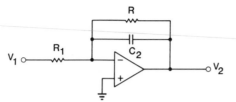

FIGURE 5.19. Lossy active-RC integrator.

it can be regarded as the response of a lossy analog integrator (Fig. 5.19) with its resistances given by

$$R_1 = \frac{T}{C_1}$$

and   (5.59)

$$R = \frac{2}{\omega^2 T C_2}.$$

Thus, the feedback capacitor, combined with R, has a quality factor

$$Q \triangleq \omega C_2 R = \frac{2}{\omega T}.$$ (5.60)

Since normally $\omega T = 2\pi f/f_c \ll 1$, $Q \gg 1$ results.

Since the feedback capacitances in the first and last stages represent the first (C_1) and last (C_3) capacitances of the LCR circuit of Fig. 5.18a, it is clear that the approximation error in these two integrators can be represented by attributing the Q given in (5.60) to C_1 and C_3 in the LCR filter.

Similarly, an analysis based on (5.16) results in the conclusion that the approximation error in the integrator of the center stage can be represented by attributing a quality factor $-Q = -2/(\omega T)$ to the inductor L_2 in the LCR filter (Problem 5.12). Note that for a capacitor the Q is represented by a *parallel* resistor, while for an inductor by a series one.

With the approximation errors represented by Q factors in the LCR circuit, their effects can be estimated for $\omega T \ll 1$ by using existing formulas[24, 28, 29] for Q effects in passive filters. Alternatively, computer programs capable of analyzing ladder filters can be used.

To a first-order approximation, it can be expected that the effects of the equal and opposing Q factors in the L's and C's will cancel, and hence the distortion will be minor. This conclusion can also be reached in a completely different way. Consider two cascaded integrators, one inverting and the other noninverting. Their combined transfer function is, by Eqs. (5.21) and (5.26),

$$H(z) = -\frac{(C_1/C_2)_{\text{inv}}}{1 - z^{-1}} \frac{(C_1/C_2)_{\text{noninv}} z^{-1}}{1 - z^{-1}}$$

$$= \frac{-Kz^{-1}}{(1 - z^{-1})^2} = \frac{-K}{(z^{1/2} - z^{-1/2})^2}. \tag{5.61}$$

Here, K is the product of the capacitance ratios. For $z = \exp(j\omega T)$,

$$H(e^{j\omega T}) = \frac{K}{4 \sin^2(\omega T/2)} \tag{5.62}$$

is pure real, as would be the transfer function of two cascaded ideal integrators. In fact, as (5.61) reveals, the $H(z)$ of the cascade can be regarded as the same as that of two integrators (one inverting, the other noninverting) derived from the ideal analog integrator via the LDI transformation (2.89). (Here, z is replaced by $z^{1/2}$, which corresponds to $T \to T/2$, and is inconsequential.) Since $\omega T \ll 1$, such LDI integrators have lossless (i.e., pure imaginary) transfer functions for $z = e^{j\omega T}$.

It should be noted that the above discussion is valid only for a cascade or a loop of two integrators. At the two terminations, the argument no longer holds, since the corresponding stages are *lossy* integrators, that is, they contain switched capacitors (C_S, C_L in Fig. 5.18e) in their feedback paths. Thus, the LDI transformation does *not* relate the $H(z)$ of the SCF to the $H_a(s_a)$ of the LCR circuit of Fig. 5.18a. This is just as well, since (as pointed out in connection with Eq. (2.92) of Section 2.6) such an $H(z)$ would be unstable. Hence, an approximation error, obtainable as described above, remains.

The alternation of the integrator types (and hence of the signs of the Q's in the LCR model) is a very useful feature,[22] to be preserved whenever possible to reduce the error. Further improvements can be obtained by modifying slightly the terminating sections,[30, 31] or by iterative computer-aided optimization of the element values.

The procedure described above is readily generalized for higher-order low-pass filters. Figure 5.20 illustrates the process for a fifth-order circuit. The detailed derivations are left for the reader, as an exercise.

Very often, the SCF is required to produce transmission zeros on the unit circle near the passband edge. Then the LCR ladder model must provide finite $j\omega_a$-axis transmission zeros, unlike the circuit of Fig. 5.18a whose three zeros (loss poles) were all at infinite frequency.

A third-order low-pass filter with transmission zeros at $\omega_{a1} = -\omega_{a2} = [L_2C_2]^{-1/2}$ and $\omega_{a3} \to \infty$ is shown in Fig. 5.21a; its loss response is shown in Fig. 5.21b. Compared to the circuit of Fig. 5.18a, the filter has an additional element C_2. It conducts a current $I_{C_2} = sC_2(V_1 - V_3)$ from node ① to node

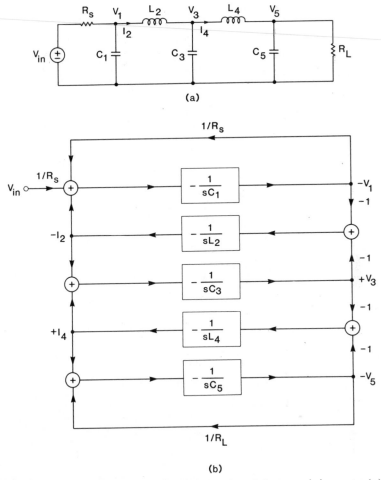

(a)

(b)

FIGURE 5.20. Fifth-order switched-capacitor filter *without* finite transmission zeros: (*a*) LCR prototype circuit; (*b*) block diagram; (*c*) active-RC realization; (*d*) switched-capacitor circuit.

③. Hence, Eq. (5.56) becomes

$$-V_1 = \frac{-1}{s(C_1 + C_2)}\left[\frac{V_{in} - V_1}{R_S} + sC_2V_3 - I_2\right],$$

$$-I_2 = \frac{-1}{sL_2}[V_1 - V_3], \qquad\qquad (5.63)$$

$$V_3 = \frac{-1}{s(C_2 + C_3)}\left[-sC_2V_1 - I_2 + \frac{V_3}{R_L}\right].$$

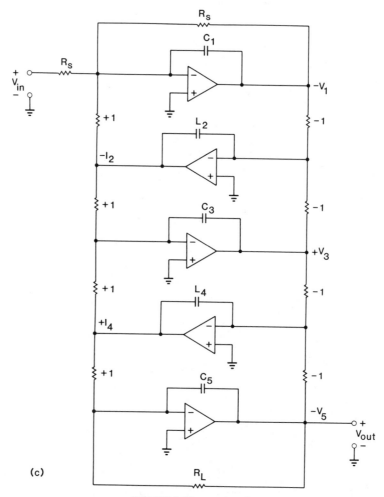

FIGURE 5.20. continued.

The corresponding block diagram is shown in Fig. 5.21c, while the active-RC realization is shown in Fig. 5.21d. Finally, Fig. 5.21e shows the SCF obtained when all resistors are replaced by switched capacitors. The element values are given by

$$
\begin{aligned}
C_S &\cong T/R_S, \\
C_A &= C_1 + C_2, \\
C &\cong T, \\
C_B &= L_2, \\
C_C &= C_2 + C_3, \\
C_L &\cong T/R_L.
\end{aligned}
\tag{5.64}
$$

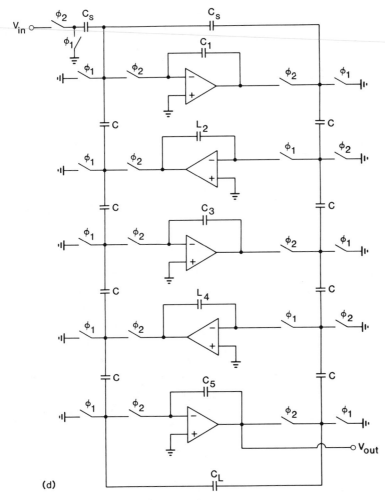

FIGURE 5.20. continued.

Note that this circuit is less closely related to its LCR model than that of Fig. 5.18a, since the role of the single capacitor C_2 of the model filter is performed here by *four* capacitors. Even so, for low passband ripple α_p (Fig. 5.21b), all passband sensitivities remain quite low—typically, about 0.1 dB or less for a 1% element-value change. The alternation of the signs of the Q-factors of the integrators remains valid, and hence the approximation error for $\omega T \ll 1$ is small. In the simulation of the approximation error, the $\pm Q$ should be associated with C_1, L_2, and C_3 only, as before; C_2 is lossless.

As before, the design process can readily be extended to higher-order filters. Figure 5.22 illustrates the design steps for a fifth-order circuit. The reader should derive the equations describing each step, as an exercise.

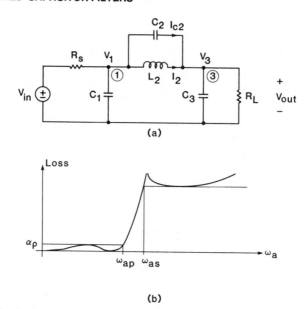

(a)

(b)

FIGURE 5.21. Low-pass switched-capacitor filter *with* finite transmission zeros: (*a*) LCR prototype circuit; (*b*) loss response; (*c*) block diagram; (*d*) active-RC realization; (*e*) switched-capacitor circuit.

The situation is slightly more complicated for bandpass filters.[32,33] As an illustration, consider the fourth-order bandpass prototype filter shown in Fig. 5.23*a*. Using the same procedure as for the circuit of Fig. 5.21*a*, we obtain the state equations

$$-V_1 = \frac{-1}{s(C_1 + C_2)}\left[\frac{V_{in} - V_1}{R_S} + sC_2V_3 - I_1 - I_2\right],$$

$$-I_1 = \frac{-V_1}{sL_1},$$

$$-I_2 = \frac{-V_1 + V_3}{sL_2},$$

$$I_3 = \frac{V_3}{sL_3},$$

$$V_3 = \frac{-1}{s(C_2 + C_3)}\left[-sC_2V_1 + I_3 - I_2 + \frac{V_3}{R_L}\right].$$

The corresponding block diagram is shown in Fig. 5.23*b*. This system, however, has a stability problem at dc, as will be demonstrated next. Assuming $V_{in} = 0$, consider the gain of the path from node Ⓐ to node Ⓑ, with the rest

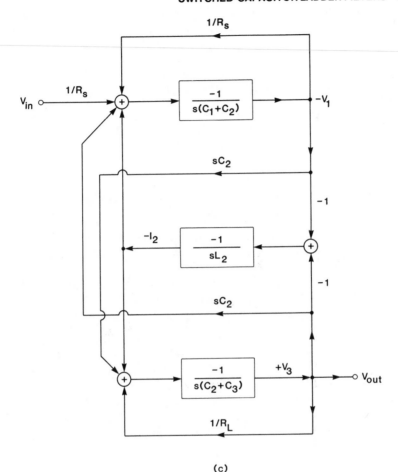

(c)

FIGURE 5.21. continued.

of the circuit below Ⓐ and Ⓑ disconnected. A simple calculation gives

$$H_{AB} \triangleq \frac{V_B}{V_A} = \frac{-sL_1}{s^2L_1(C_1 + C_2) + sL_1/R_S + 1}.$$

By symmetry, the path gain from node Ⓒ to node Ⓓ is

$$H_{CD} \triangleq \frac{V_C}{V_D} = \frac{-sL_3}{s^2L_3(C_2 + C_3) + sL_3/R_L + 1}.$$

When $s \to 0$, $H_{AB} \to H_{CD} \to 0$, and the two capacitive paths sC_2 also become open circuits. Hence, there will be no dc feedback paths around the center integrator which provides $-I_2$. Thus, its offset voltage V_{off} (due to internal asymmetries, external clock feedthrough, leakage currents, etc.) will be amplified by the full open-circuit dc gain A of the op-amp. Since, typically, $V_{\text{off}} \sim 10 \text{ mV}$

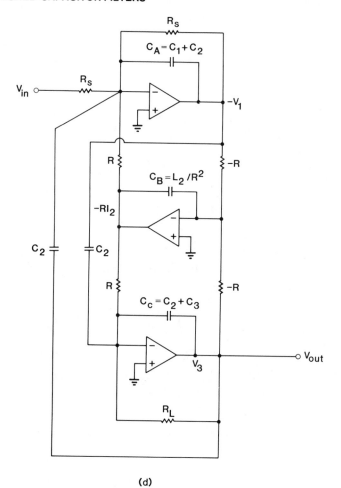

(d)

FIGURE 5.21. continued.

and $A \geq 1000$, this will cause the op-amp to saturate, and the circuit will not function. Furthermore, similar calculations reveal that the integrators providing $-I_1$ and I_3 also function without negative dc feedback, and hence are likely to saturate.

To explore the situation in more detail, let the integrators in Fig. 5.23b be modeled by the circuit of Fig. 5.24, where the finite gain A and offset voltage V_{off} are included. Assuming dc (quiescent) conditions in the system, the capacitor current i must be zero, and we obtain for the voltages

$$v = \frac{\sum_{i=1}^{3} G_i v_i}{\sum_{i=1}^{3} G_i},$$

$$v_0 = A(V_{off} - v),$$

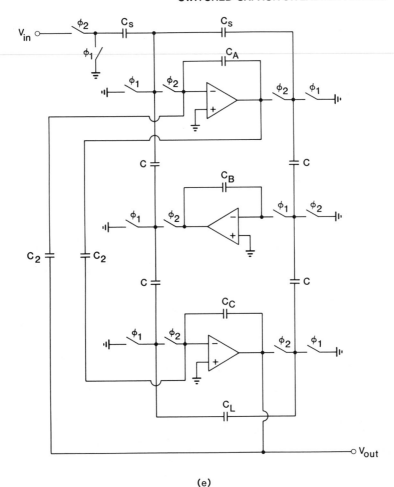

(e)

FIGURE 5.21. continued.

where $G_i \triangleq 1/R_i$. Using these relations for the integrators representing L_1, L_2, and L_3, the dc equations

$$-i_1 = A_1 V_{\text{off1}} + A_1 v_1,$$

$$-i_2 = A_2 V_{\text{off2}} + \frac{A_2(v_1 - v_3)}{2},$$

$$i_3 = A_3 V_{\text{off3}} - A_3 v_3$$

result. Each of these relations describe a branch containing a dc voltage source (of value $V_{\text{off1,3}}$ for the L_1 and L_3 branches and of value $2V_{\text{off2}}$ for the L_2 branch) in series with a small negative resistor (of value $-1/A_{1,3}$ for L_1 and L_3, of value $-2/A_2$ for L_2). Ideally, the sum of the dc sources in the L_1, L_2,

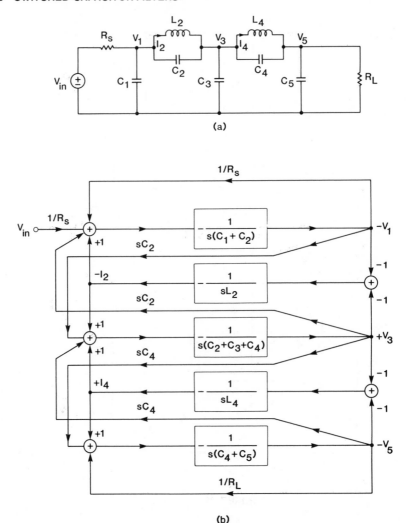

FIGURE 5.22. Fifth-order switched-capacitor low-pass ladder filter *with* finite transmission zeros: (*a*) LCR prototype; (*b*) block diagram; (*c*) active-RC realization; (*d*) switched-capacitor circuit.

and L_3 loop is zero; in fact, the cancellation is imperfect and a loop current of order AV_{off} will result. This will cause saturation of the op-amps.

The problem is clearly caused by our insistence of modeling the inductor currents in the $L_1 - L_2 - L_3$ loop separately. Assume, by contrast, that we choose to simulate only the two inductive currents $I_①$ and $I_③$ entering nodes ① and ③, respectively.[32, 33] For $v_{\text{in}} = 0$ and $s \to 0$, these are the *only* currents to these nodes, and hence by the Kirchhoff current law (KCL) both must become zero. This property precludes any instability of the active circuit model for dc conditions.

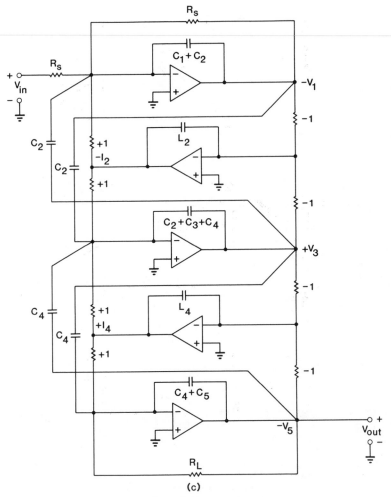

FIGURE 5.22.   continued.

From the circuit diagram of Fig. 5.23a, we can then write

$$-V_1 = \frac{-1}{s(C_1 + C_2)}\left[\frac{V_{\text{in}} - V_1}{R_S} + sC_2V_3 - I_①\right],$$

$$-I_① \triangleq -(I_1 + I_2) = \frac{-1}{sL_{12}}\left[V_1 - \frac{L_1V_3}{L_1 + L_2}\right],$$

$$V_3 = \frac{-1}{s(C_2 + C_3)}\left[-sC_2V_1 - I_③ + \frac{V_3}{R_L}\right],$$

$$-I_③ \triangleq I_3 - I_2 = \frac{-1}{sL_{12}}\left[-V_3 + \frac{L_1V_1}{L_1 + L_2}\right],$$

where $L_{12} \triangleq L_1 \| L_2 = L_1L_2/(L_1 + L_2)$.

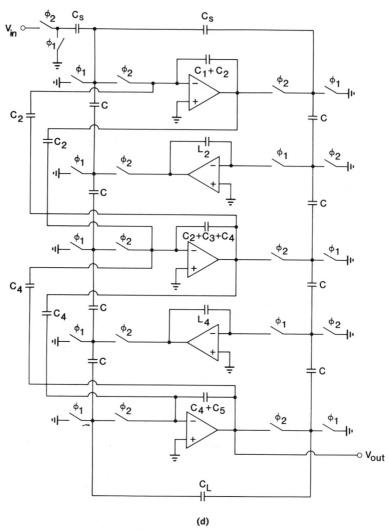

(d)

FIGURE 5.22. continued.

These equations can be modeled by the system of Fig. 5.23c. The corresponding active-RC circuit is shown in Fig. 5.23d; the final switched-capacitor filter is shown in Fig. 5.23e.

As a check, one can calculate the gain functions of the feedback paths for all integrators in the block diagram of Fig. 5.23c. They all have positive values for $s = 0$, indicating stability at dc.*

Note that in the circuit of Fig. 5.23e, the integrators representing capacitors (i.e., the stages generating $-V_1$ and V_3) are inverting ones, while the integra-

---

*A price paid for the dc stability is a small increase in the sensitivity to element-value variations.

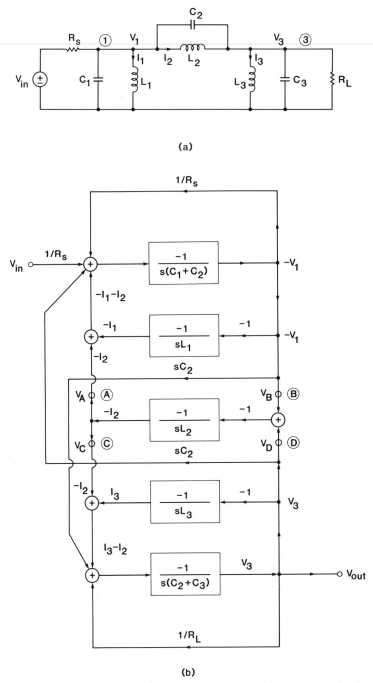

FIGURE 5.23. Switched-capacitor bandpass ladder filter: (*a*) LCR prototype circuit; (*b*) dc-unstable block diagram; (*c*) dc-stable block diagram; (*d*) active-RC realization; (*e*) switched-capacitor circuit.

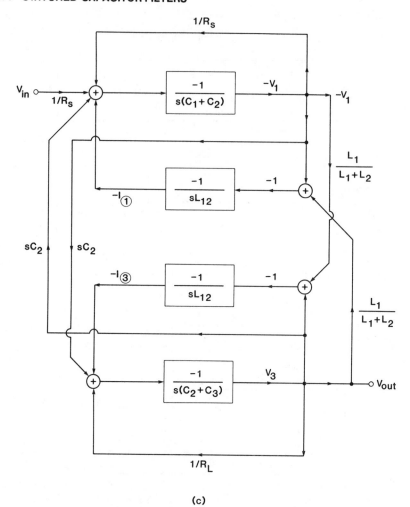

(c)

FIGURE 5.23. continued.

tors representing inductors (i.e., the ones generating $-I①$ and $-I③$) are noninverting stages. As discussed earlier, this feature results in a first-order cancellation of the approximation errors, and hence gives an improved frequency response.

The circuit of Fig. 5.23a, and hence that of Fig. 5.23e, is readily modified to represent *high-pass filters* with nonzero transmission zeros; C_1 and C_2 should then both be set to zero. The extension to higher-order filters is also straightforward, and is left as an exercise to the reader. As an illustration, Fig. 5.25 shows the resulting circuit for a sixth-order bandpass filter[33] derived from a third-order low-pass filter by reactance transformation.[1] Note that $R = 1$ and $C = 1$ were used in deriving the circuit of Fig. 5.25b.

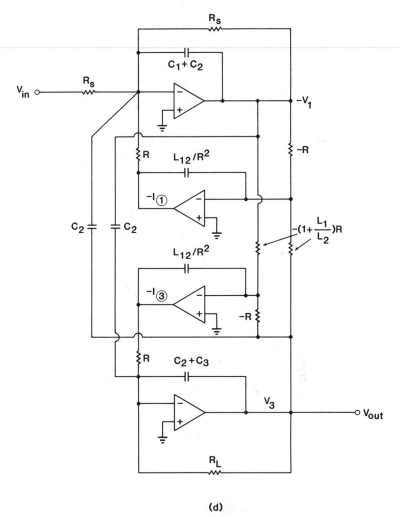

(d)

FIGURE 5.23.   continued.

In conclusion, the approximate design of SC ladder filters can be performed in the following steps:

1. A doubly terminated LC two-port is designed from the SCF specifications. The specifications can be prewarped using the relation

$$\omega_a = \frac{2}{T}\sin\left(\frac{\omega T}{2}\right) \tag{5.65}$$

which represents the frequency transformation due to the LDI transfor-

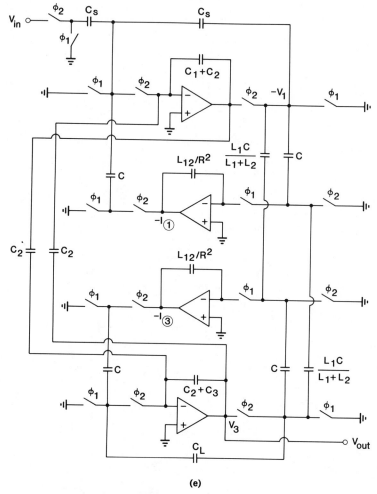

(e)

FIGURE 5.23. continued.

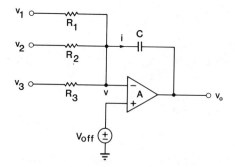

FIGURE 5.24. Multi-input integrator with finite-gain nonzero-offset op-amp.

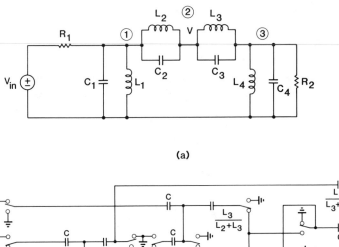

(a)

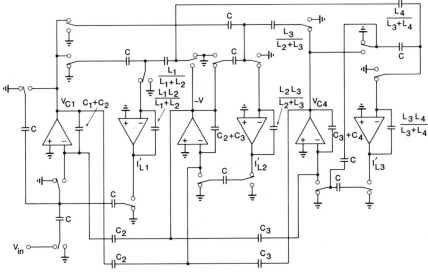

(b)

FIGURE 5.25. Sixth-order bandpass filter: (a) LCR prototype; (b) switched-capacitor realization.

mation implicit in the design procedure. Equation (5.65) follows from (2.91) of Section 2.6, with T replaced by $T/2$.

2. The state equations of the resulting LCR circuit are found. The signs of the voltage and current variables must be chosen such that inverting and noninverting integrators alternate in the implementation. If inductive loops exist, then the inductive node currents can be used as variables instead of the individual inductor currents.*

---

*If the LCR circuit contains capacitive cutsets, then its *dual* should be modeled, and the resulting inductive loops treated as described above.

3. The block diagram or signal flow graph is constructed from the state equations. It is then transformed (directly or via the active-RC circuit) into the SCF.

4. If necessary, additional circuit transformations can be performed to improve the response of the SCF. These include a modification of the terminating sections[19, 30, 31] which improves the passband response. Also, for filters with high clock rates, it is advantageous to perform transformations which eliminate unnecessary capacitive couplings between op-amps.[34] This results in faster settling time.

The Exact Design of SC Ladder Filters[35, 36]

The errors inherent in the LDI design techniques can be eliminated by using a ladder synthesis based on the bilinear s_a-to-z transformation of Eq. (2.80) given in Section 2.6:

$$s_a = \frac{2}{T} \frac{z-1}{z+1}. \tag{5.66}$$

As described in Section 2.6, this transformation changes a stable LCR filter into a stable SCF, and (since it maps the $j\omega_a$-axis on the unit circle) preserves the flatness of passbands and stopbands. Hence, it is more useful in most applications than any other mapping.

As before, the design technique will be illustrated by the synthesis of a third-order SCF. We start again with the LCR model of Fig. 5.21a, described by the state equations in (5.63). These equations can be rewritten if capacitor C_2 is split into the parallel combination of a positive capacitance

$$C_2' \triangleq C_2 + C_{L_2} \tag{5.67}$$

and a small negative capacitance

$$-C_{L2} = \frac{-T^2}{4L_2}. \tag{5.68}$$

The resulting circuit is shown in Fig. 5.26a. [The purpose of this simple manipulation is to facilitate the application of Eq. (5.66), as will become evident soon.] With the current I_2 redefined as that flowing through the parallel combination of L_2 and $-C_{L_2}$, the state equations become

$$-V_1 = \frac{-1}{s_a(C_1 + C_2')}\left[\frac{-V_1 + V_{in}}{R_S} + s_a C_2' V_3 - I_2\right],$$

$$-I_2 = \left[s_a C_{L2} - \frac{1}{s_a L_2}\right][V_1 - V_3], \tag{5.69}$$

$$V_3 = \frac{-1}{s_a(C_2' + C_3)}\left[-s_a C_2' V_1 - I_2 + \frac{V_3}{R_L}\right].$$

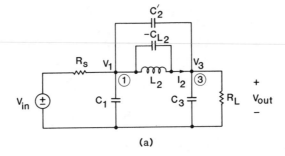

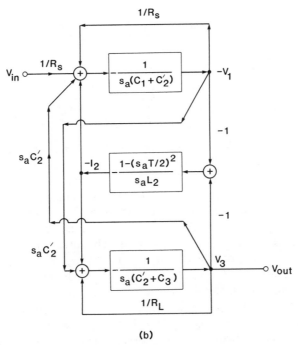

(b)

FIGURE 5.26. Bilinear switched-capacitor ladder filter: (*a*) LCR prototype circuit; (*b*) block diagram.

The corresponding block diagram is shown in Fig. 5.26*b*. Note that the transfer function of the center block is, from (5.68),

$$H_2(s_a) = s_a C_{L_2} - \frac{1}{s_a L_2} = -\frac{1 - s_a^2 L_2 C_{L_2}}{s_a L_2} = -\frac{1 - (s_a T/2)^2}{s_a L_2}. \quad (5.70)$$

To transform the block diagram into an SC circuit, the charge versus voltage relations of all blocks and branches will first be found in the s_a-domain, and then transformed via (5.66) into the *z*-domain. The *z*-domain relations will then be realized by SC components. To realize the input branch $1/R_S$, we note that in the analog system the charge flowing through this branch is given by the relation $Q_{in}(s_a) = V_{in}/s_a R_S$. Using (5.66), the corresponding charge in the SCF input branch is given by

$$Q_{in}(z) = \frac{T}{2} \frac{z+1}{z-1} \frac{V_{in}(z)}{R_S}. \tag{5.71}$$

This relation can be rewritten in the form

$$(1 - z^{-1})Q_{in}(z) = \frac{T}{2R_S}(1 + z^{-1})V_{in}(z) \tag{5.72}$$

or, in the time domain,

$$q_{in}(t_n) - q_{in}(t_{n-1}) = \frac{C_S}{2}[v_{in}(t_n) + v_{in}(t_{n-1})]. \tag{5.73}$$

The LHS of (5.73) is $\Delta q_{in}(t_n)$, the incremental charge passing through the branch for $t_{n-1} < t < t_n$. The branch relation (5.73) can be satisfied by either of the branches shown in Fig. 5.27. Both circuits require a sampled-and-held input voltage which changes at the time when ϕ_1 goes high (Fig. 5.27c). For the circuit of Fig. 5.27a, the bridging capacitor $C_S/2$ delivers a charge $(C_S/2)[v_{in}(t_n) - v_{in}(t_{n-1})]$, while the shunt switched capacitor delivers a charge $C_S v_{in}(t_{n-1})$, during the $t_{n-1} < t < t_n$ interval. The total charge increment Δq_{in} thus satisfies (5.73). For the circuit of Fig. 5.27b, the top branch again contributes a charge $(C_S/2)[v_{in}(t_n) - v_{in}(t_{n-1})]$, while the switched capacitor provides $-C_S v_{in}(t_n)$. The total charge is thus the *negative* of that required by (5.73). The resulting 180° phase change of V_{out} is usually unimportant.

Note that the circuit of Fig. 5.27a is *not* fully stray insensitive; the stray capacitance at the top plate of C_S is in parallel with it, and increases C_S by a small amount. It can, however, be shown (Problem 5.13) that this merely contributes a small nearly constant gain to the response, which is usually tolerable.

The circuit of Fig. 5.27b *is* fully stray insensitive; it is, however, more complicated than that of Fig. 5.27a. The switch in series with $C_S/2$ does not change the charge which this branch contributes; however, it assures that the charge flows only when $\phi_2 \to 1$ rather than at the time $\phi_1 \to 1$ when $v_{in}(t)$ changes. This assures that the output voltage of the first op-amp changes only once in each clock interval.

The charge versus voltage relation of the coupling branches marked $s_a C_2'$ is simply $Q/V = C_2'$ which is frequency independent. They can thus be realized simply by coupling capacitors of value C_2'. Similarly (as in earlier circuits), the first and last blocks have voltage/charge ratios $-1/(C_1 + C_2')$ and

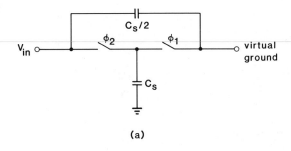

(a)

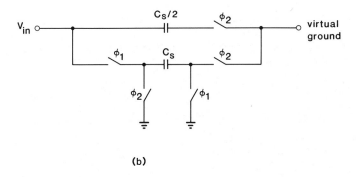

(b)

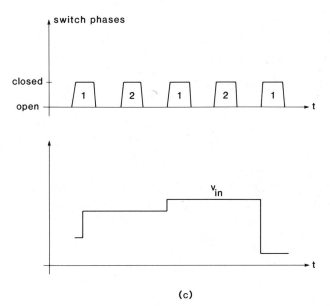

(c)

FIGURE 5.27. Input branch circuits: (*a*) stray-sensitive circuit; (*b*) stray-insensitive circuit; (*c*) waveforms.

$-1/(C_2' + C_3)$, respectively, which correspond to op-amps with feedback capacitors $C_1 + C_2'$ and $C_2' + C_3$, respectively, in both s- and z-domains.

A remaining block yet unrealized is the center one. Its output voltage is

$$-I_2 = -\frac{1 - (s_a T/2)^2}{s_a L_2}(V_1 - V_3). \tag{5.74}$$

Hence, it contributes a charge

$$Q(s_a) = \frac{-I_2}{s_a} = -\frac{1 - (s_a T/2)^2}{s_a^2 L_2}(V_1 - V_3) \tag{5.75}$$

to the adders at the input terminals of the first and third integrators. Replacing s_a by the RHS of (5.66), we get

$$\frac{Q(z)}{V_1(z) - V_3(z)} = -\frac{1 - [(z-1)/(z+1)]^2}{(4L_2/T^2)[(z-1)/(z+1)]^2} = -\frac{4C_{L_2}z}{(z-1)^2}. \tag{5.76}$$

The z-transform of the *incremental* charge $\Delta q(t_n)$, $\Delta Q(z)$, is thus

$$\Delta Q(z) = (1 - z^{-1})Q(z) = \frac{4C_{L_2}}{z-1}(V_3 - V_1). \tag{5.77}$$

Referring back to Eqs. (5.38) and (5.39), it is clear that the transfer function of (5.77) which can be written in the form

$$\frac{\Delta Q}{V_3 - V_1} = \frac{4C_{L_2}z^{-1}}{1 - z^{-1}} = (-Cz^{-1})\left(\frac{-4C_{L_2}/C^2}{1 - z^{-1}}\right)(C) \tag{5.78}$$

can be produced by cascading an inverting switched capacitor C, an op-amp with a feedback capacitor $C^2/4C_{L_2}$, and a noninverting switched capacitor C. The resulting circuit is shown in Fig. 5.28.

Finally, the feedback branches $1/R_S$ and $1/R_L$ could, in principle, be realized the same way as the input branch $1/R_S$. A simpler realization is also possible, however, for these branches. Consider the branch of Fig. 5.29. By (5.39), its transfer function is

$$\frac{\Delta Q}{V} = -\frac{C_S}{2}(1 - z^{-1}) + C_S = \frac{C_S}{2}(1 + z^{-1}). \tag{5.79}$$

This is the required transfer function of the branch marked $1/R_S$, as can be seen, for example, from (5.72). The negative capacitance $-C_S/2$ can be absorbed in the feedback capacitor $C_1 + C_2'$ of the first stage which is usually much larger.

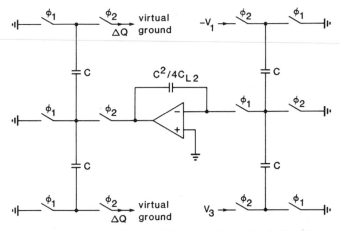

FIGURE 5.28.   Internal stage of the switched-capacitor ladder filter.

Putting together the z-domain branches and building blocks, the complete SC ladder filter shown in Fig. 5.30a results. From the construction of the circuit it follows that any op-amp output voltage $V_i(z)$ in the SCF can be obtained from the corresponding block output voltage $V_i(s_a)$ in the analog system of Fig. 5.26b by replacing s_a using (5.66) by its bilinear transform. Similarly, any charge $Q_k(z)$ in a capacitive branch can be found from the charge $Q_k(s_a)$ in the corresponding path in the system of Fig. 5.26b via the bilinear s_a-to-z transformation. Thus, the same statement is valid for the relation between

$$H_a(s_a) = \frac{V_{out}(s_a)}{V_{in}(s_a)} \quad \text{and} \quad H(z) = \frac{V_{out}(z)}{V_{in}(z)},$$

without any approximation.

It can be expected that the sensitivities of the SCF to the variations of the elements corresponding to LCR filter components will be low. This is indeed the case.[36, 39] Also, the sensitivities to the coupling capacitances C are found to

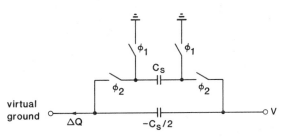

FIGURE 5.29.   Realization of the damping branch $1/R_S$ in the first integrator of Fig. 5.26b.

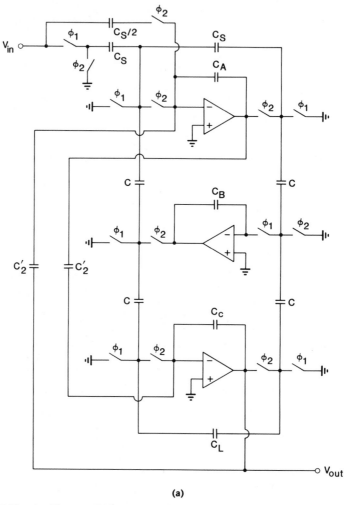

(a)

FIGURE 5.30. (a) The complete bilinear ladder circuit equivalent to the LCR circuit of Fig. 5.26a. (b) A fifth-order bilinear ladder filter.

be low as well.[36,39] Thus, the circuit is free from the approximation errors inherent in the LDI ladder, but shares its low-sensitivity properties.

As before, the extension of the process to higher-order filters is straightforward (if tedious). As an illustration, Fig. 5.30b illustrates a fifth-order low-pass circuit,[36] obtainable from the LCR prototype of Fig. 5.22a. The circuit uses the input branch shown in Fig. 5.27a; as explained above, this circuit is a practical one in spite of not being fully stray insensitive. As before, the reader is urged to derive this circuit independently, and to find the element values in terms of those of its LCR prototype (Fig. 5.22a).

To generalize the design process to bandpass circuits, consider the filter section shown in Fig. 5.31a. Clearly, the voltage $-V_2$ can be produced as

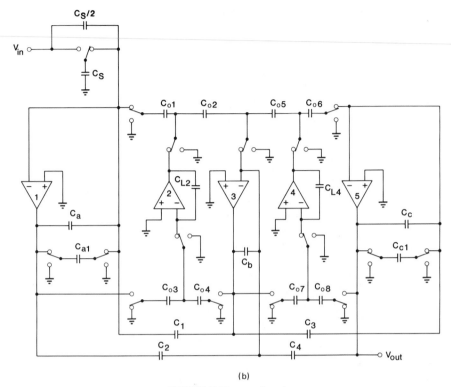

(b)

FIGURE 5.30.   continued.

shown in Fig. 5.31b, provided the current I_2 remains the same as that flowing through C_2 in Fig. 5.31a. Thus, we must have

$$I_2 = \left(sC_1 + \frac{1}{sL_1} \right)(V_1 - V_2) + \left(sC_3 + \frac{1}{sL_3} \right)(V_3 - V_2) - \frac{V_2}{sL_2}. \quad (5.80)$$

Since (as illustrated by the realization of the central block in Fig. 5.26b) transfer functions of the forms sC and $(1/s - T^2s/4)/L$ can readily be realized after the bilinear $s \to z$ transformation, we partition I_2 as follows

$$I_2(s) = s\left[(C_1 + C_{L_1})V_1 + (C_1 + C_3 + C_{L_1} + C_{L_3})(-V_2) + (C_3 + C_{L_3})V_3\right]$$

$$+ \left(\frac{1}{s} - \frac{T^2s}{4} \right)\left[\Gamma_1V_1 + (\Gamma_1 + \Gamma_2 + \Gamma_3)(-V_2) + \Gamma_3V_3\right]. \quad (5.81)$$

Here (as before) we denoted $C_{L_i} \triangleq T^2/4L_i$ and also introduced, for convenience, $\Gamma_i \triangleq 1/L_i$. The total charge $Q_2(s)$ into C_2 is $I_2(s)/s$. Thus, $Q_2(s)$

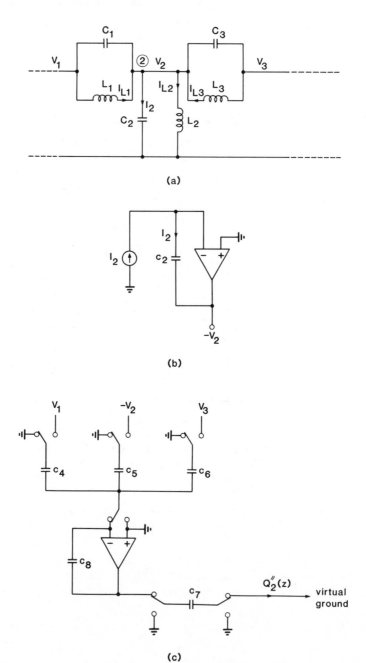

FIGURE 5.31. Bilinear ladder branches for bandpass filters: (a) LC prototype section; (b) stage providing $-V_2$; (c) stage providing $Q_2''(z)$.

contains two terms. The first one is of the form

$$Q_2'(s) = (C_1 + C_{L_1})V_1 + (C_1 + C_3 + C_{L_1} + C_{L_3})(-V_2)$$

$$+ (C_3 + C_{L_3})V_3. \qquad (5.82)$$

This can be provided in either the s- or the z-domain simply by connecting fixed (i.e., unswitched) capacitors of appropriate value between the voltages V_1, $-V_2$, and V_3, and the virtual ground of the op-amp of Fig. 5.31b.

After the $s \rightarrow z$ transformation, the second term in $Q_2(s)$ becomes

$$Q_2''(z) = \left[\left(\frac{T}{2} \frac{z+1}{z-1} \right)^2 - \frac{T^2}{4} \right] [\Gamma_1 V_1 + (\Gamma_1 + \Gamma_2 + \Gamma_3)(-V_2) + \Gamma_3 V_3]$$

$$= \frac{T^2 z^{-1}}{(1 - z^{-1})^2} [\Gamma_1 V_1 + (\Gamma_1 + \Gamma_2 + \Gamma_3)(-V_2) + \Gamma_3 V_3]. \qquad (5.83)$$

The corresponding transfer functions $\Delta Q_2''/V_1$, $\Delta Q_2''/(-V_2)$ and $\Delta Q_2''/V_3$ are all in the form of the expression in Eq. (5.78). Hence, a possible realization is that shown in Fig. 5.31c, where the conditions

$$\frac{c_4 c_7}{c_8} = \frac{T^2}{L_1} = 4C_{L_1},$$

$$\frac{c_5 c_7}{c_8} = T^2 \left(\frac{1}{L_1} + \frac{1}{L_2} + \frac{1}{L_3} \right) = 4(C_{L_1} + C_{L_2} + C_{L_3}), \qquad (5.84)$$

$$\frac{c_6 c_7}{c_8} = \frac{T^2}{L_3} = 4C_{L_3}$$

must hold.

Replacing the current source I_2 in Fig. 5.31b by the fixed capacitors providing $Q_2'(z)$ and the circuit of Fig. 5.31c, we obtain the SC circuit which corresponds to the LC section shown in Fig. 5.31a. The final circuit is shown in Fig. 5.32, which also contains a diagram of the LC circuit being simulated, and the design equations. The reader should be careful to distinguish in the latter between upper-case symbols C_i (which refer to elements in the LC section) and lower-case c_i referring to element values in the SCF.

Comparing the parameters in the LC and SC sections, we note that the output voltage of the top op-amp is the negative of node voltage V_2, after the bilinear $s \rightarrow z$ transformation. The output voltage of the bottom op-amp is $Q_2''(z)/C_7$ which is the bilinear z-transform of $Q_2''(s)/C_7$. The latter is proportional to the sum of the currents flowing into node ② through the

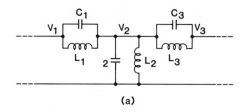

(a)

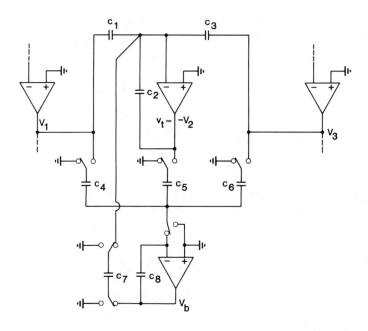

(b)

$$C_{Li} = T^2/(4L_i)$$
$$c_1 = C_1 + C_{L1}$$
$$c_2 = C_1 + C_2 + C_3 + C_{L1} + C_{L2} + C_{L3}$$
$$c_3 = C_3 + C_{L3}$$
$$c_4 = 4\frac{c_8}{c_7}C_{L1}$$
$$c_5 = 4\frac{c_8}{c_7}(C_{L1} + C_{L2} + C_{L3})$$
$$c_6 = 4\frac{c_8}{c_7}C_{L3}$$

c_7, c_8 arbitrary

(c)

FIGURE 5.32. Bilinear ladder, internal section: (*a*) LC prototype; (*b*) SC realization; (*c*) design equations.

inductors:

$$I_2''(s) = I_{L_1}(s) - I_{L_2}(s) + I_{L_3}(s)$$

$$= \frac{1}{s}\left[\Gamma_1 V_1 + (\Gamma_1 + \Gamma_2 + \Gamma_3)(-V_2) + \Gamma_3 V_3\right]. \qquad (5.85)$$

When $s \to 0$, the currents entering node ② through capacitors C_1, C_2, and C_3 become zero. Hence, $I_2''(s)$ is the total node current, and (by the KCL) therefore $I_2''(0) = 0$. This remains valid regardless of the values of the individual currents $I_{L_1}(0)$, $I_{L_2}(0)$, and $I_{L_3}(0)$. Thus, the output of the bottom op-amp will not saturate due to offset voltages, even if inductive loops exist in the prototype circuit.

The realization of the input and output sections is based on similar considerations, and is left to the reader as an exercise (Problems 5.14 and 5.15). The resulting circuits and their design equations are shown in Figs. 5.33 and 5.34.

Several variations of the circuits of Figs. 5.32–5.34 are also possible. For example, in Fig. 5.32b the capacitors c_4 and c_6 may be omitted, and instead two new capacitors $\hat{c}_4$ and $\hat{c}_6$ switched from the outputs of the lower op-amps in the preceding and the following sections (along with c_7) to the input of the top op-amp. The output voltage of the bottom op-amp changes, but that of the top op-amp remains $-V_2$ for appropriate choice of element values.

An important property of the section shown in Fig. 5.32b can be deduced from its transfer function. Using the transfer functions of Eqs. (5.38) and (5.39), it is easy to write the KCL equations for the inverting input terminals of the top and bottom op-amps. They are

$$c_1(1 - z^{-1})V_1 + c_3(1 - z^{-1})V_3 + c_2(1 - z^{-1})V_t + c_7 V_b = 0,$$
$$\qquad (5.86)$$
$$-c_4 z^{-1} V_1 - c_6 z^{-1} V_3 - c_5 z^{-1} V_t + c_8(1 - z^{-1})V_b = 0.$$

Here, V_t is the output voltage of the top op-amp, and V_b that of the bottom one. Solving (5.86) gives

$$V_t = -\frac{N_1 V_1 + N_3 V_3}{D}$$

and $\qquad\qquad\qquad\qquad\qquad\qquad\qquad\qquad\qquad\qquad\qquad\qquad (5.87)$

$$V_b = \frac{z^{-1}(1 - z^{-1})\left[(c_2 c_4 - c_1 c_5)V_1 + (c_2 c_6 - c_3 c_5)V_3\right]}{c_8 D}.$$

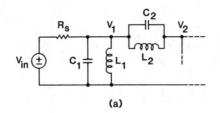

(a)

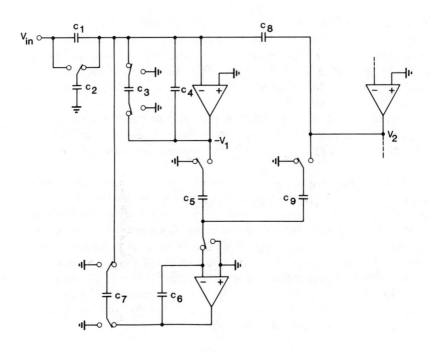

(b)

$$C_{Li} \triangleq \frac{T^2}{4L_i}, \quad C_S \triangleq \frac{T}{R_S}$$

$$c_1 = \frac{C_S}{2}, c_2 = c_3 = C_9$$

$$c_4 = C_1 + C_2 + C_{L1}$$
$$+ C_{L2} - C_S/2$$

$$c_5 = 4\frac{c_6}{c_7}(C_{L1} + C_{L2})$$

$$c_8 = C_2 + C_{L2}$$

$$c_9 = 4\frac{c_6}{c_7}C_{L2}$$

$$c_6, c_7 \text{ arbitrary}$$

(c)

FIGURE 5.33. Bilinear ladder, input section: (*a*) LCR prototype; (*b*) SC realization; (*c*) design equations.

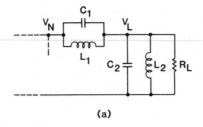

(a)

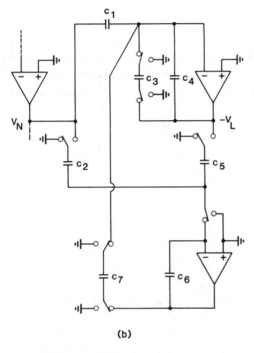

(b)

$$C_{L1} \triangleq \frac{T^2}{4L_i}, \quad C_L \triangleq \frac{T}{R_L}$$

$$c_1 = C_1 + C_{L1}$$

$$c_2 = 4\frac{c_6}{c_7}C_{L1}$$

$$c_3 = C_L$$
$$c_4 = C_1 + C_2 + C_{L1} + C_{L2} - C_L/2$$

$$c_5 = 4\frac{c_6}{c_7}(C_{L1} + C_{L2})$$

c_6, c_7 are arbitrary

(c)

FIGURE 5.34. Bilinear ladder, output section: (a) LCR prototype; (b) SC realization; (c) design equations.

Here, the expressions N_1, N_3, and D are defined as follows:

$$N_1(z) = c_1 c_8 \left[(1 - z^{-1})^2 + \frac{c_4 c_7}{c_1 c_8} z^{-1} \right],$$

$$N_3(z) = c_3 c_8 \left[(1 - z^{-1})^2 + \frac{c_6 c_7}{c_3 c_8} z^{-1} \right], \tag{5.88}$$

$$D(z) = c_2 c_8 \left[(1 - z^{-1})^2 + \frac{c_5 c_7}{c_2 c_8} z^{-1} \right].$$

All zeros and poles of the transfer functions V_t/V_1, V_t/V_3, V_b/V_1, and V_b/V_3 are located on the unit circle. This follows from the fact that the highest- and lowest-order coefficients of all polynomials are equal to one and the coefficient of the linear term is less than two. Using the bilinear s-to-z transformation, Eq. (5.87) becomes

$$V_t = -\frac{\left[(c_1 c_8 - c_4 c_7/4) s^2 + c_4 c_7/T^2 \right] V_1 + \left[(c_3 c_8 - c_6 c_7/4) s^2 + c_6 c_7/T^2 \right] V_3}{(c_2 c_8 - c_5 c_7/4) s^2 + c_5 c_7/T^2},$$

and $\tag{5.89}$

$$V_b = \frac{s(1/T - s/2)\left[(c_2 c_4 - c_1 c_5) V_1 + (c_2 c_6 - c_3 c_5) V_3 \right]}{(c_2 c_8 - c_5 c_7/4) s^2 + c_5 c_7/T^2}.$$

The first equation shows that the phase shifts between V_t and V_1, as well as between V_t and V_3, are either 0° or 180° for $s = j\omega$. This is the same as for the LC prototype section of Fig. 5.32a, and remains valid regardless of the element values c_i. Thus, the stage simulates a lossless LC section no matter whether the elements are accurate or not. This indicates (by the arguments[23] presented earlier in this section) that the sensitivities of the loss to variations of the c_i will be low in the passband of the SCF.

The second expression in (5.89) shows that V_b is stable (in fact, zero) both at $s = 0$ and $s \to \infty$, corresponding to $\omega = 0$ and $\omega_c/2$ in terms of the radian frequency of the SCF.

As can be seen from Eq. (5.89), the circuit of Fig. 5.32b can simulate the behavior of any LC ladder section which has a T configuration (Fig. 5.35), and in which the internal node voltage V_2 is related to the external node voltages V_1 and V_3 by

$$V_2 = \frac{(as^2 + b) V_1 + (cs^2 + d) V_3}{es^2 + f}. \tag{5.90}$$

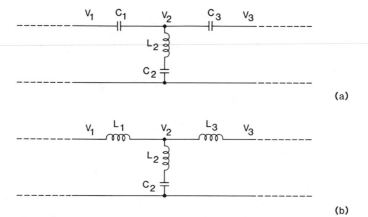

FIGURE 5.35. Filter sections which can be simulated by the SC section of Fig. 5.32b: (a) high-pass section; (b) low-pass section.

In addition to the circuit of Fig. 5.32a, this relation is also satisfied by the circuits shown in Fig. 5.35. Thus, the stage of Fig. 5.32b can be used to simulate these sections as well. For the section of Fig. 5.35a

$$V_2 = \frac{sC_1V_1 + sC_3V_3}{sC_1 + sC_3 + 1/(sL_2 + 1/sC_2)} = \frac{(s^2L_2C_2 + 1)(C_1V_1 + C_3V_3)}{s^2L_2C_2(C_1 + C_3) + (C_1 + C_2 + C_3)}. \tag{5.91}$$

Equating coefficients in both the numerators and denominators of $-V_2$ and V_t, in Eqs. (5.89) and (5.91), design equations can be obtained for the c_i, $i = 1, 2, \ldots, 8$. Since there are only five equations, three c_i values can be assigned arbitrarily. A possible set of solutions is

$$c_1 = c_4 = C_1,$$

$$c_2 = C_1 + C_3 + \frac{C_{L_2}C_2}{C_{L_2} + C_2},$$

$$c_3 = c_6 = C_3,$$

$$c_5 = C_1 + C_2 + C_3, \tag{5.92}$$

$$c_7 = \frac{TC_2}{R(C_{L_2} + C_2)},$$

$$c_8 = \frac{C_2L_2}{RT}.$$

Here, $C_{L_2} \triangleq T^2/4L_2$ as before, and R is an arbitrary resistance.

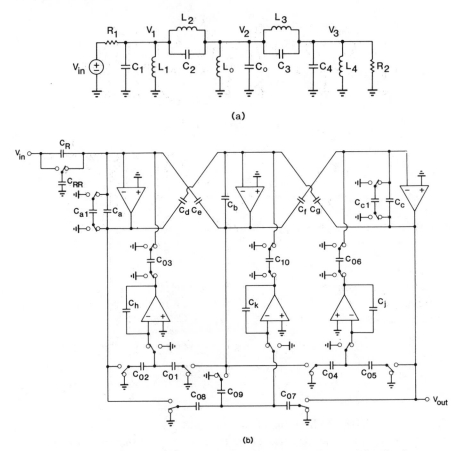

(a)

(b)

FIGURE 5.36. Sixth-order bandpass filter: (a) LCR prototype; (b) SC realization.

The verification of Eq. (5.92), and the derivation of the design equations for the circuit of Fig. 5.35b are left as an exercise for the reader (Problems 5.16 and 5.17).

Note that in deriving the relations (5.92) we only aimed at simulating the value of $-V_2$ by V_t. The output voltage V_b of the bottom op-amp was not matched to any voltage or current in the RLC ladder. Thus, this design process (in contrast to those described earlier) simulates only the node voltages, not all the state variables of the RLC filter. It can thus be derived also from the node equations of the RLC ladder.[20]*

As an illustration of the process, Fig. 5.36a shows the LCR prototype of a sixth-order bandpass filter; using Figs. 5.32–5.34, the circuit of Fig. 5.36b results.[39] The reader should derive the formulas giving the element values of

---

*An alternative derivation of the exact design of SC ladder filters was recently given by Datar and Sedra.[58]

the SC filter in terms of the prototype. The correct results are

$$C_a = C_1 + C_2 + C_{L1} + C_{L2} - C_{R1},$$

$$C_{a1} = 2C_{R1},$$

$$C_b = C_0 + C_2 + C_3 + C_{L0} + C_{L2} + C_{L3},$$

$$C_c = C_3 + C_4 + C_{L3} + C_{L4} - C_{R2},$$

$$C_{c1} = 2C_{R2},$$

$$C_d = C_e = C_2 + C_{L2},$$

$$C_f = C_g = C_3 + C_{L3},$$

$$C_{RR} = 2C_R = 2C_{R1},$$

$$C_{01} = \frac{4C_h}{C_{03}}C_{L2},$$

$$C_{02} = \frac{4C_h}{C_{03}}(C_{L1} + C_{L2}),$$

$$C_{07} = \frac{4C_k}{C_{10}}C_{L3},$$

$$C_{08} = \frac{4C_k}{C_{10}}C_{L2},$$

$$C_{09} = \frac{4C_k}{C_{10}}(C_{L0} + C_{L2} + C_{L3}),$$

$$C_{04} = \frac{4C_j}{C_{06}}C_{L3},$$

$$C_{05} = \frac{4C_j}{C_{06}}(C_{L3} + C_{L4}).$$

Here, $C_{Ri} = T/2R_i$ and $C_{Li} = T^2/4L_i$. The values of C_{03}, C_{06}, C_{10}, C_h, C_j, and C_k are optional parameters. Many of the switches perform the same task and can be shared; as an example, the toggle switches that are connected to the right-hand side of C_{01}, the left-hand side of C_{04}, and the upper plate of C_{09} have identical functions and can be merged together, saving two toggle switches. It is easy to see that a total of six toggle, or 12 single switches, can be saved in the SC filter.

As shown above, for circuits containing inductive loops the ladder circuits designed by the above process (often called *bilinear ladder* SCF) will *not* exhibit the dc instability phenomenon described for the LDI circuit. However, for filters for which the $z = -1$, that is, $f = f_c/2$ point is in the passband, such as a high-pass or bandstop filter, the above design procedure leads to an unstable circuit. This can be readily understood from Eqs. (5.71) and (5.72) which show that for $f = f_c/2$ the input branch does not permit any charge Q_{in} into the circuit. Hence, the rest of the circuit must exhibit infinite gain (i.e., become oscillatory) if the filter is designed for zero loss at this frequency. This phenomenon can be avoided by using a simple scaling process[36,39] described next. It eliminates the instability at $f_c/2$, but may make the circuit sensitive to stray-capacitance effects.

To illustrate the process, consider the fourth-order LCR high-pass filter shown in Fig. 5.37a. Dividing all impedances in the circuit by $sT/2$, all

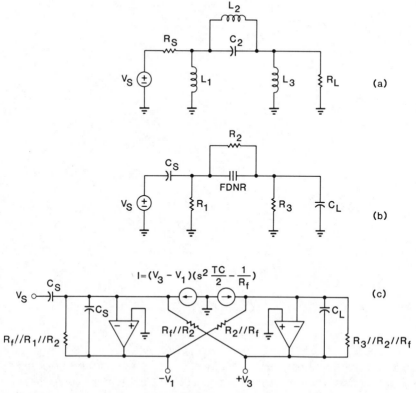

FIGURE 5.37. Bilinear high-pass ladder filter: (*a*) LCR prototype; (*b*) RC–FDNR equivalent; (*c*) active-RC realization; (*d*) SC realization.

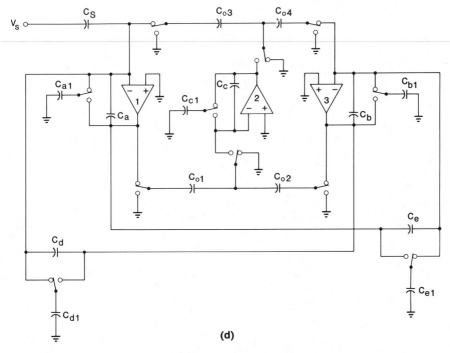

(d)

FIGURE 5.37. continued.

resistors are transformed into capacitors and all inductors into resistors, while any capacitor C becomes a frequency-dependent negative resistor (FDNR), as shown in Fig. 5.37b. The transformation leaves $H_a(s_a)$ unchanged. The FDNR has an admittance $s^2TC/2$; this can be written in the form $2C/T + (TC/2)(s^2 - 4/T^2)$. Thus, the FDNR can be realized as the parallel combination of a resistor $R_f = T/2C$ and a branch with an admittance $Y_t = (2C/T)(s^2T^2/4 - 1)$. Using Eq. (5.66), the latter can be readily transformed into the transfer function $(\Delta Q/V)_f = -4C/(z + 1)$. As already shown in Eqs. (5.71)–(5.73), the resistor R_f can be realized by one of the circuits of Fig. 5.27 in the z-domain. It is also easy to verify (Problem 5.22) that the circuit of Fig. 5.38 has the $(\Delta Q/V)_f$ function required for the admittance Y_f, if $C_2 = 2C_f$ and $C_1C_3/C_2 = 2C$ are chosen.

Using these results, the circuit of Fig. 5.37 can be transformed first to that of Fig. 5.37c, and then to that of Fig. 5.37d. Note that the final SC circuit is not fully stray insensitive: the grounded capacitors $C_{a1}, C_{b1}, \ldots,$ are affected by the stray capacitances in parallel with them. However, if (say) $C_a > C_{a1}$, then the feedback branch of the first op-amp may be replaced by that of Fig. 5.39, where $C_x = C_a - C_{a1}$ and $C_{x1} = C_{a1}$. Other useful transformations can also be used[39,40] to reduce the number of stray-sensitive branches and to improve the performance of the circuit. The detailed derivation of the element values is left

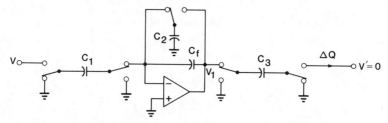

FIGURE 5.38. SC branch for the realization of the transfer function $\Delta Q / V = -4C/(z + 1)$.

for the reader. The resulting equations are

$$C_S = \frac{T}{2R_S},$$

$$C_L = \frac{T}{2R_L},$$

$$C_a = \frac{T}{2R_S} + \frac{T^2}{4}\left(\frac{1}{L_1} + \frac{1}{L_2}\right) + C_2,$$

$$C_{a1} = 2(C_a - C_S),$$

$$C_b = \frac{T}{2R_L} + \frac{T^2}{4}\left(\frac{1}{L_2} + \frac{1}{L_3}\right) + C_2,$$

$$C_{b1} = 2(C_b - C_L),$$

$$C_d = C_e = \frac{C_{d1}}{2} = \frac{C_{e1}}{2} = C_2 + \frac{T^2}{4L_2},$$

$$C_{01} = C_{02},$$

$$C_{03} = C_{04},$$

$$C_c = \frac{C_{01} \cdot C_{03}}{4C_2},$$

$$C_{c1} = 2C_c.$$

The values of C_{01} and C_{03} are optional parameters.

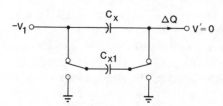

FIGURE 5.39. Equivalent realization for the feedback branch of op-amp 1 in Fig. 5.37d.

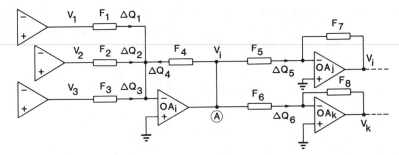

FIGURE 5.40. SC filter section.

For very narrow-band bandpass filters (say, with relative bandwidths of 1% or less) none of the design techniques described in this chapter are very effective, since they result in circuits which are too sensitive to finite op-amp gain effects, strays, and element-value variations. Then, design based on the *N-path filter concept* may be used. This concept, and the corresponding design techniques, are discussed in Section 5.8.

5.6. THE SCALING OF SWITCHED-CAPACITOR FILTERS[19, 36]

The actual performance of an SCF can be improved, and the area which it occupies on its silicon chip reduced, by scaling operations. To understand how such operations can be carried out, consider a segment of an SCF containing the ith op-amp OA_i and all branches connected to it. This is schematically illustrated in Fig. 5.40, in which the coupling branches may contain capacitors and switches in any arrangement. Each coupling branch in a stray-insensitive SCF is connected between an op-amp output and a virtual ground; each can be described by a transfer function $F_k(z) = \Delta Q_k(z)/V_l(z)$. Here, $\Delta Q_k(z)$ is the z-domain representation of the charge-per-clock-period sequence $q_k(nT) - q_k(nT - T)$, and $V_l(z)$ is the z-transform of the voltage sequence $v_l(nT)$ at the input of the branch.

Now let all branches connected to the *output* terminal of OA_i be modified such that their $\Delta Q/V$ transfer functions (F_4, F_5, F_6) are multiplied by a positive real constant factor k_i. This can be achieved simply by multiplying all capacitors in these branches by k_i, since (for given voltages connected across these branches) their charges are proportional to these capacitances.

Since the input branches and their voltages were left unchanged, the charge flowing in the feedback branch

$$\Delta Q_4(z) = -\Delta Q_1(z) - \Delta Q_2(z) - \Delta Q_3(z) \qquad (5.93)$$

remains at its original value. The output voltage of OA_i (assuming ideal op-amps) thus becomes $V_i'(z) = \Delta Q_4(z)/[k_i F_4(z)] = V_i(z)/k_i$, where the prime indicates values obtained *after* the multiplier k_i has been introduced. Thus, $V_i \rightarrow V_i/k_i$ due to the factor k_i. By contrast, the values of ΔQ_5 and ΔQ_6 remain unchanged since, for example,

$$\Delta Q'_5(z) = F'_5(z)V'_i(z) = k_i F_5(z)\frac{V_i(z)}{k_i}$$

$$= F_5(z)V_i(z) = \Delta Q_5(z). \tag{5.94}$$

In conclusion, multiplying all capacitors which are connected (or switched) to the output terminal of the ith op-amp OA_i changes its output voltage V_i to V_i/k_i, while leaving all charges flowing from (and to) OA_i to (and from) the rest of the circuit unchanged. Thus, by this operation we can increase or decrease V_i by a constant scale factor without affecting any other voltages or charges in the SCF.

The described scaling operation is useful in improving the dynamic range of SC circuits. As defined in Section 4.1 of Chapter 4, the dynamic range is $20 \log_{10}(v_{in,max}/v_{in,min})$. Here, $v_{in,max}$ is the maximum input signal which the SCF can handle without excessive nonlinear distortion; in an SCF, $v_{in,max}$ is usually limited by the saturation of the op-amps. The minimum useful input signal $v_{in,min}$ is the smallest value for which the SCF output voltage v_{out} can be distinguished from the output noise $v_{n,out}$.

In the following discussions it will be assumed that the dynamic range of the SCF is to be optimized using scaling. Consider the SC circuit schematically illustrated in Fig. 5.41. Let the amplitude of the largest output signal which the op-amps in the filter (assumed, for simplicity, to be identical) can handle be V_{max}. Also, let the passband gain of the filter be A_p. Since v_{out} (produced by the output op-amp OA_n) is at most V_{max}, we have

$$v_{in,max} \leq V_{max}/A_p. \tag{5.95}$$

Let the frequency responses of all op-amp outputs $V_1, V_2, \ldots, V_i, \ldots, V_n$ be as shown in Fig. 5.42, where a fifth-order low-pass filter is assumed. Then, as the amplitude of v_{in} is increased from a small value to V_{max}/A_p at a frequency near ω_2 (the peak frequency of V_2), OA_2 will saturate before OA_5 does since

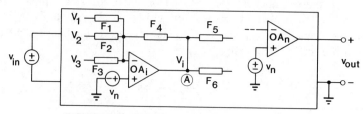

FIGURE 5.41. SC filter with noisy internal op-amp.

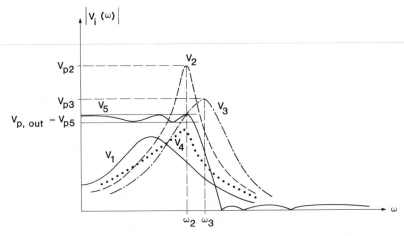

FIGURE 5.42.   Op-amp output voltage responses for a low-pass filter.

$|V_2| > |V_5|$ for $\omega \sim \omega_2$. Hence, now $v_{in,max} = V_{max}/A_2$, where A_2 is the gain from the filter input to the output of OA_2 at $\omega = \omega_2$. Clearly, from Fig. 5.42, $A_2 = |V_{p2}/v_{in}|$ while $A_p = |V_{p5}/v_{in}|$, so $A_2 = A_p|V_{p2}/V_{p5}|$. Hence, now the maximum permissible input voltage is

$$v_{in,max} = \frac{V_{max}}{A_2} = \frac{V_{max}}{A_p}\frac{V_{p5}}{V_{p2}} < \frac{V_{max}}{A_p} \qquad (5.96)$$

since $V_{p5}/V_{p2} < 1$.

Using scaling as described above, $V_2(\omega)$ can be reduced in amplitude. In particular, if $k_2 = V_{p2}/V_{p5}$ is used, then the scaled response $V_2'(\omega)$ will have its peak value V_{p2}' equal to V_{p5}. Hence, as far as OA_2 is concerned, the maximum value of v_{in} for which saturation can be avoided is V_{max}/A_p, as given in (5.95).

A similar argument shows that for OA_3, $k_3 = V_{p3}/V_{p5}$ can be chosen. Then OA_3 will saturate at the same v_{in} amplitude as OA_5 and the scaled OA_2. Clearly, as (5.96) shows, if both OA_2 and OA_3 are scaled as described above, the value of $v_{in,max}$ is increased by a factor k_2. Choosing k_2 and k_3 larger than V_{p2}/V_{p5} and V_{p3}/V_{p5}, respectively, no longer increases $v_{in,max}$, and in fact (as will be shown next) actually reduces the dynamic range.

Scaling not only affects $v_{in,max}$ but also the output noise level and hence $v_{in,min}$. To find the contribution of the noise voltage source v_n in OA_i to v_{out} before and after scaling, we set, in Fig. 5.41, $V_1 = V_2 = V_3 = 0$. Then the stage of OA_i has the equivalent circuit of Fig. 5.43, where Y_4 is the admittance of the feedback branch and Y_{123} denotes the sum of the admittances of the input branches earlier described by F_1, F_2, and F_3. If the op-amp gain is high then the voltage v at the inverting input terminal is nearly equal to v_n, and a simple

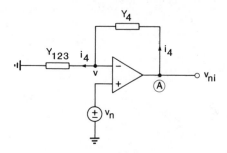

FIGURE 5.43. Noisy amplifier stage with zero input voltages.

calculation gives

$$V_{ni}(\omega) = \left(\frac{Y_{123}(\omega)}{Y_4(\omega)} + 1 \right) V_n(\omega)$$

for the output noise of the stage. At the filter output terminal, this noise appears as $H_i(\omega)V_{ni}(\omega)$, where $H_i(\omega)$ is the transfer function from node Ⓐ to the output. Here $H_i(\omega)$ includes, of course, all paths from Ⓐ to the output node via all op-amps in the circuit.

Next, let the scaling by k_i be performed on all branches connected to node Ⓐ. Then, as explained earlier, the signal voltage V_i changes to $V_i' = V_i/k_i$, but all other voltages, including V_{out}, remain unchanged. This implies that the transfer function $H_i(\omega)$ changes to $H_i'(\omega) = k_i H_i(\omega)$; this is due to the scaling of F_5 and F_6 (Fig. 5.41). The noise voltage at Ⓐ changes to

$$V_{ni}'(\omega) = \left(\frac{Y_{123}(\omega)}{k_i Y_4(\omega)} + 1 \right) V_n(\omega).$$

This is because scaling results in the multiplication of all capacitances in the feedback branch by k_i; this changes Y_4 to $k_i Y_4(\omega)$. The new noise contribution to V_{out} is hence

$$H_i'(\omega)V_{ni}'(\omega) = H_i(\omega)\left(\frac{Y_{123}(\omega)}{Y_4(\omega)} + k_i \right) V_n(\omega).$$

Notice that this increases *at most linearly* with increasing k_i, if $|Y_{123}/Y_4| \ll k_i$. A similar derivation reveals that the scaling also affects the noise contributions from op-amps j and k (Fig. 5.40). Prior to scaling, op-amp j contributed $H_j(\omega)[Y_5(\omega)/Y_7(\omega) + 1]V_n(\omega)$ to the output noise. After scaling, $Y_5(\omega)$ becomes $k_i Y_5(\omega)$, while $H_j(\omega)$ and $Y_7(\omega)$ remain unchanged. Hence, the new output noise contribution is $H_j(\omega)[k_i Y_5(\omega)/Y_7(\omega) + 1]V_n(\omega)$. A similar expression describes the noise contribution from op-amp k. Again, for $|Y_5| \gg |Y_7|/k_i$, the output noise contribution from OA_j increases linearly with increasing k_i; otherwise, the increase is slower. In conclusion, the operation of scaling using $k_i > 1$ *increases* the output noise from OA_i and from the op-amps receiving inputs from OA_i, but only by a factor which is at most k_i.

Since the noise sources in the op-amps are uncorrelated, the total output noise prior to scaling is given by

$$\bar{v}_{n,\text{out}} = \left[\bar{v}_{n,1}^2 + \bar{v}_{n,2}^2 + \cdots + \bar{v}_{n,i}^2 + \cdots + \bar{v}_{n,n}^2 \right]^{1/2}$$

where $\bar{v}_{n,i}$ denotes the RMS noise at the output due to the noise of OA_i. After scaling, we have

$$\bar{v}_{n,\text{out}}' \leq \left[\bar{v}_{n,1}^2 + \cdots + k_i^2 \bar{v}_{n,i}^2 + \cdots + k_i^2 \bar{v}_{n,j}^2 + \cdots \right.$$
$$\left. + k_i^2 \bar{v}_{n,k}^2 + \cdots + \bar{v}_{n,n}^2 \right]^{1/2} < k_i \bar{v}_{n,\text{out}}.$$

Hence, the output noise grows with increasing k_i, but the rate of increase is slower than linear.

Now let $V_i(\omega)$ be the only op-amp output with a peak which is larger than the value $V_{p,\text{out}}$ of $V_{\text{out}}(\omega)$ in the passband of the SCF (such as for V_2 or V_3 in Fig. 5.42). Then, as discussed earlier, scaling by $k_i = V_{pi}/V_{p,\text{out}}$ increases $v_{\text{in,max}}$ by a factor k_i. As shown just above, the output noise $\bar{v}_{n,\text{out}}$ (and hence $v_{\text{in,min}}$) also increases, but by less than a factor k_i. Hence, the dynamic range $v_{\text{in,max}}/v_{\text{in,min}}$ is increased.

If more than one V_i has a peak higher than $V_{p,\text{out}}$, then $v_{\text{in,max}}$ is multiplied by the largest k_i, while a number of the noise contributions $\bar{v}_{n,i}$ in the output noise get increased by various k_i factors. However, this increase is by a factor less than that of $v_{\text{in,max}}$ and hence the dynamic range is increased.

If the peak value V_{pi} of V_i is *less* than $V_{p,\text{out}}$, then scaling by $k_i = V_{pi}/V_{p,\text{out}}$ < 1 *increases* $|V_i(\omega)|$ such that after scaling $V_{pi}' = V_{p,\text{out}}$. The noise from OA_i and all op-amps receiving their inputs from OA_i is now *decreased*, while $v_{\text{in,max}}$ is unaffected by the scaling since OA_i does not saturate before the output op-amp does. Hence, the dynamic range is always increased by this operation. Making $k_i < V_{pi}/V_{p,\text{out}}$ would reduce the noise gain even more, but now $V_{pi}' > V_{p,\text{out}}$ and hence $v_{\text{in,max}}$ would be reduced to $k_i V_{\text{max}}/A_p$. As shown above, the latter effect would outweigh the former. Hence, the dynamic range would be reduced for $k_i < V_{pi}/V_{p,\text{out}}$.

The conclusion is that the optimum value of the scaling factor for each stage is $k_i = V_{pi}/V_{p,\text{out}}$, where (as shown in Fig. 5.42) V_{pi} is the peak output voltage $V_i(\omega)$ of op-amp i and $V_{p,\text{out}}$ is the peak (passband) value of $V_{\text{out}}(\omega)$. In other words, *for maximum dynamic range, all op-amp outputs should be scaled such that each (at its own maximum frequency) saturates for the same input voltage level.*

As an illustration, Fig. 5.44 shows schematically the frequency responses of the op-amp outputs in a sixth-order bandpass SCF before and after scaling.[38]

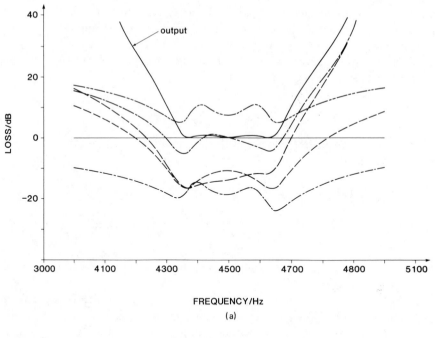

(a)

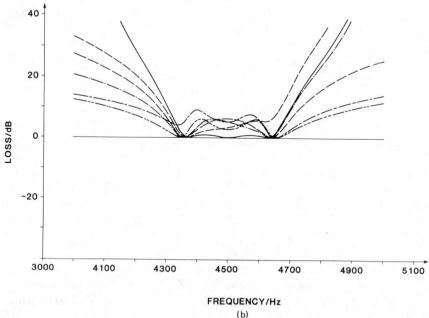

(b)

FIGURE 5.44. Op-amp output responses for a sixth-order bandpass filter: (*a*) responses before scaling; (*b*) responses after scaling.

Identical saturation voltages were assumed for all op-amps, and hence the maxima were all equal after scaling for maximum dynamic range.

Another scaling operation is also useful. Referring again to Fig. 5.40, let the transfer functions $F_j(z) \triangleq \Delta Q_j / V_j$ of all branches connected to the *input* terminal of OA_i be multiplied by a positive real constant m_i. In Fig. 5.40, the affected functions are F_1, F_2, F_3, and F_4. As a result of the scaling, the charge flows ΔQ_n, $n = 1, 2, 3, 4$ all become $\Delta Q'_n = m_i \Delta Q_n$. However, the output voltage V_i remains unchanged, since

$$V'_i = \frac{\Delta Q'_4}{F'_4} = \frac{m_i \Delta Q_4}{m_i F_4} = \frac{\Delta Q_4}{F_4} = V_i. \tag{5.97}$$

Thus, the output charges ΔQ_5 and ΔQ_6 also remain the same. Hence, the described scaling by m_i leaves all op-amp output voltages (including V_i) in the SCF unchanged. Only the charges in the scaled branches get multiplied by the scale factor m_i.

This scaling operation is highly effective in reducing the capacitance spread and total capacitance of an SCF, in the following manner. Even for a stray-insensitive SCF (which is assumed throughout this section), there is a minimum capacitance value C_{min} which can still be reliably and accurately fabricated. For example, the stray capacitance between the lines connecting the plates of the capacitance may be of order 1 fF $= 10^{-3}$ pF. Thus, a 1% accuracy requires $C_{min} \geq 0.1$ pF.

Minimizing the total capacitance in the branches realizing F_1, F_2, F_3, and F_4 can therefore be accomplished in the following manner. The smallest capacitor $C_{i, min}$ among all capacitors contained in these four branches is located; then all capacitances in the four branches are multiplied by $m_i = C_{min}/C_{i, min}$. As a result, the smallest capacitance becomes C_{min}, and (as discussed above) all op-amp voltages remain unaffected.*

If (as is usually the case) both scalings are performed, the scaling for optimum dynamic range should be performed first, and scaling for minimum capacitance afterwards. This is because the latter does not affect any of the V_i, and hence the optimality achieved by the first scaling is not destroyed by the second one. This statement is not true if the order of the two scalings is reversed: the scaling for optimum dynamic range upsets the minimum-capacitance property.

According to the preceding discussions, the two scaling processes are performed in the following steps:

1. *Scaling for Maximum Dynamic Range.*
   (a) Set $V_{in}(\omega)$ to the largest value for which the output op-amp does not saturate.

---

*Alternatively, the value of the feedback capacitor can be set by this scaling step to the minimum value permitted by noise and clock feedthrough effects.

(b) Calculate the maximum values V_{pi} for all internal op-amp output voltages. These values usually occur near the passband edge (or edges) of the SCF.

(c) Multiply all capacitors connected or switched to the *output* terminal of op-amp i by $k_i = V_{pi}/V_{i,\max}$, where $V_{i,\max}$ is the saturation voltage of op-amp i.

(d) Repeat for all internal op-amps.

2. *Scaling for Minimum Capacitance.*

(a) Divide all capacitors in the SCF into nonoverlapping sets. Capacitors in the ith set S_i are connected or switched to the *input* terminal of op-amp i.

(b) Multiply all capacitors in S_i by $m_i = C_{\min}/C_{i,\min}$, where $C_{\min}$ is the smallest capacitor which the fabrication technology permits, and $C_{i,\min}$ is the smallest capacitor in S_i.

(c) Repeat for all sets S_i, including that associated with the output op-amp.

An important aspect of scaling concerns its effect on the sensitivity of the response to element-value variations. Consider, for example, a capacitor C_r in branch 2 of the section shown in Fig. 5.40. A small change δC_r in C_r results in ΔQ_2 changing by $\delta \Delta Q_2$; the change is approximately given by

$$\delta \Delta Q_2 = \delta(V_2 F_2) \cong V_2 \frac{\partial F_2}{\partial C_r} \delta C_r. \tag{5.98}$$

The corresponding change in ΔQ_4 is, of course, $-\delta \Delta Q_2$, and hence the change in V_i is

$$\delta V_i = \frac{-\delta \Delta Q_2}{F_4} = -\frac{V_2}{F_4} \frac{\partial F_2}{\partial C_r} \delta C_r. \tag{5.99}$$

Next let the branches 1, 2, 3, and 4 all be scaled by a factor m_i, as described in Step 2(b) above. Then F_1, F_2, F_3, and F_4 all get multiplied by m_i, as do all element values in the corresponding branches. Thus, C_r becomes $C_r' = m_i C_r$. If next this capacitance changes by the same *relative* amount as before so that $\delta C_r' = m_i \delta C_r$, then the change in ΔQ_2 is

$$\delta \Delta Q_2' = \delta(V_2 F_2') = V_2 \frac{\partial(m_i F_2)}{\partial(m_i C_r)}(m_i \delta C_r) = m_i \delta \Delta Q_2. \tag{5.100}$$

The corresponding change in V_i is now

$$\delta V_i' = \frac{-\delta \Delta Q_2'}{F_4'} = \frac{-m_i \delta \Delta Q_2}{m_i F_4} = \delta V_i. \tag{5.101}$$

Thus, the output voltage of op-amp i changes by the same amount as before scaling. Since the output branches (numbered 5 and 6) are unaffected by scaling, all other voltages, including V_{out}, change by the same amount. We conclude that the sensitivity to relative changes of the element values in the input branches (numbered 1, 2, and 3 in Fig. 5.31) is the same before and after scaling.

A trivial modification of the above proof shows that the argument holds also for the sensitivity to elements in branch 4, the feedback branch. Now again both $\delta\Delta Q_4$ and F_4 get multiplied by m_i and hence V_i remains unchanged.

Considering the effect of scaling for maximum dynamic range on the sensitivity, let an element C_r in branch 5 change. Then

$$\delta\Delta Q_5 = \delta(V_i F_5) = V_i \frac{\partial F_5}{\partial C_r}\delta C_r. \tag{5.102}$$

After the scaling of branches 4, 5, and 6 by k_i as described in Step 1(c) above, if C_r changes by the same relative amount so that $\delta C_r' = k_i \delta C_r$, ΔQ_5 changes by

$$\delta\Delta Q_5' = (V_i/k_i) \frac{\partial(k_i F_5)}{\partial(k_i C_r)}(k_i\delta C_r) = \delta\Delta Q_5. \tag{5.103}$$

Thus the change in all outside voltages (including V_{out}) due to the variation of C_r remains the same as in the unscaled circuit.

Finally, for a change of an element in branch 4, the corresponding change in V_i (with ΔQ_4 fixed) is

$$\delta V_i = \Delta Q_4 \frac{\partial(1/F_4)}{\partial C_r}\delta C_r \tag{5.104}$$

and the resulting changes in the output charge flows are

$$\delta\Delta Q_n = F_n\delta V_i = \Delta Q_4 F_n \frac{\partial(1/F_4)}{\partial C_r}\delta C_r \tag{5.105}$$

where, in the circuit of Fig. 5.40, $n = 5$ and 6.

After scaling branches 4, 5, and 6 by k_i, the change in the output charges due to a change $\delta C_r' = k_i\delta C_r$ in the scaled value $k_i C_r$ is given by

$$\delta\Delta Q_n' = \Delta Q_4 k_i F_n \frac{\partial(1/k_i F_4)}{\partial(k_i C_r)}k_i\delta C_r$$

$$= \delta\Delta Q_n, \qquad n = 5, 6. \tag{5.106}$$

Hence, once more the change to the rest of the circuit is the same as before scaling.

In conclusion, a fixed small relative change (say, by 0.1%) in any element value has the same effect on the output voltage V_{out} after scaling as before.

Therefore, the semilogarithmic and logarithmic sensitivities of an SCF,

$$\frac{\partial V_{out}}{\partial (\ln C_r)} = C_r \frac{\partial V_{out}}{\partial C_r}$$

and

(5.107)

$$\frac{\partial (\ln V_{out})}{\partial (\ln C_r)} = \frac{C_r}{V_{out}} \frac{\partial V_{out}}{\partial C_r},$$

remain unchanged by scaling.

The situation is different with respect to the effects of the finite gain of the op-amps used. There, scaling for optimum dynamic range has an important effect. Consider the stage of Fig. 5.45a, where the op-amp is assumed to have a finite gain A. It can easily be shown that the output voltage V_{out} remains unchanged if the op-amp is replaced by an infinite-gain one (Fig. 5.45b) with an additional feedback branch (Problem 5.19). Scaling for minimum capacitance affects all branches shown the same way, and hence V_{out} (which depends on the *ratios* of transfer functions) remains unchanged. By contrast, scaling for

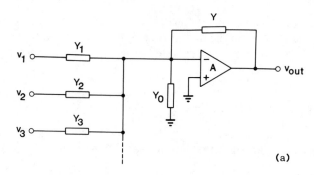

(a)

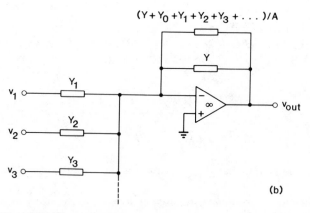

(b)

FIGURE 5.45. The influence of finite op-amp gain: (*a*) actual circuit; (*b*) equivalent circuit.

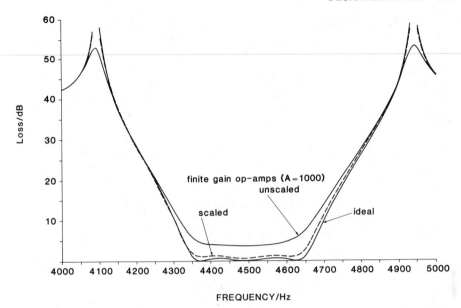

FIGURE 5.46. The influence of scaling on the finite op-amp gain effect in a sixth-order bandpass SCF. Bottom curve: loss response with infinite-gain op-amps; top curve: response with 60-dB op-amp gains before scaling; middle curve: response with 60-dB op-amp gains, after scaling.

optimum dynamic range results in the feedback transfer function F being multiplied by k_i, while F_0, F_1, F_2, and F_3 remain unchanged. Thus, to minimize the effect *for this stage* k_i should be large. However, the output branches of the stage then have large $k_i F_5$ and $k_i F_6$ values which adversely affect the performance of the *next* stages. Hence, an optimum choice of k_i exists also from the viewpoint of sensitivity to finite op-amp gain. Theory as well as experience indicates that the scaling for optimum dynamic range also improves the sensitivity to finite op-amp gain effects. As an illustration, Fig. 5.46 shows the responses of a sixth-order bandpass SCF with op-amp gains of 1000, before and after scaling for optimum dynamic range.[38] Clearly, the sensitivity is much reduced by scaling.

5.7. DESIGN EXAMPLES

To illustrate the application of the design techniques described in the earlier sections of this chapter, three design examples will be worked out in detail. All three involve the realization of the transfer functions derived earlier in Section 2.6 for a low-pass filter satisfying the following specifications:

Passband	0–1 kHz, loss ≤ 0.05 dB.
Stopband	1.5–25 kHz, loss ≥ 38 dB.
Sampling rate	$f_c = 50$ kHz.

It was shown in Section 2.6 that if the bilinear s-to-z transformation is used, the fifth-order continuous-time elliptic filter CO5 42° 10% meets these specifications. In what follows, the circuit design of three different SC filters will be performed from this prototype, using three of the design strategies discussed in Chapter 5.

Cascade Design

In this realization, the z-domain transfer function derived in Section 2.5

$$H(z) = C \frac{z + 1}{z + d_0} \frac{z^2 + c_1 z + 1}{z^2 + e_1 z + f_1} \frac{z^2 + c_2 z + 1}{z^2 + e_2 z + f_2}$$

is used, where the values of the constants are

$$C = 3.8719271 \times 10^{-3},$$

$$c_1 = -1.962247471,$$

$$c_2 = -1.916465445,$$

$$d_0 = -0.906284158,$$

$$e_1 = -1.871739343,$$

$$f_1 = 0.88543246,$$

$$e_2 = -1.949416807,$$

$$f_2 = 0.968447477.$$

The design may begin with the linear section from the transfer function

$$H_0(z) = \frac{z + 1}{z - 0.9063}.$$

Utilizing the derivations given in connection with Eqs. (5.50)–(5.54) and Figs. 5.15–5.16, we obtain the circuit of Fig. 5.47a, where the input branch (C_S and $C_S/2$) was realized by the circuit of Fig. 5.27b. The waveforms are as shown in Fig. 5.27c. As was shown earlier in Eqs. (5.71)–(5.73), the $\Delta Q/V$ transfer function of this branch is $-(C_S/2)(1 + z^{-1})$. Thus, it can be used to realize the numerator (zero) of $H_0(z)$. The overall transfer function of the stage can be found from the block diagram of Fig. 5.47b (Problem 5.23). It is

$$H_0(z) = \frac{C_S/2}{C_D + C_E} \frac{z + 1}{z - C_E/(C_D + C_E)}.$$

Choosing arbitrarily $C_S = 2$, we obtain $C_E = 0.9063$ and $C_D = 1 - C_E =$

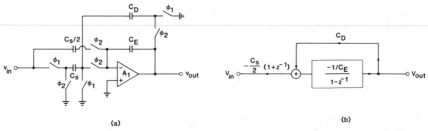

FIGURE 5.47. Section for the realization of $H_0(z) = (z + 1)/(z - 0.9063)$: ($a$) circuit diagram; ($b$) block diagram for analysis.

0.0937. These values will, of course, be modified by the scaling process. First, however, we shall obtain the complete unscaled circuit.

In realizing the biquad sections, the first design decision involves the pairing of zeros and poles. Computations (not included here) reveal that slightly lower element-value sensitivities and better dynamic range is attainable with the pairing already indicated above, than with the other possible pairing.* Hence, we are going to keep it unchanged. As a look at the s-domain values given in Section 2.6 reveals, the s-domain pole-Q of the first biquadratic transfer function is

$$Q_1 = \frac{\left(a_1^2 + b_1^2\right)^{1/2}}{2|a_1|} \cong 0.99,$$

while that of the second is around $Q_2 = 4.33$. Hence, the first biquad will be treated as a "low-Q" section, while the second as a "high-Q" one. Accordingly, the first biquad will use the circuit of Fig. 5.10c, and the design will proceed from Eqs. (5.40) and (5.41). To facilitate this, the transfer function will be written in the form

$$H_1(z) = \frac{z^2 + c_1 z + 1}{(1/f_1)z^2 + (e_1/f_1)z + 1}.$$

Then, Eqs. (5.27) and (5.41) give the values

$$C_1'' = a_0 = 1,$$
$$C_1' = a_2 - a_0 = 0,$$
$$C_2 = C_3 = \sqrt{b_1 + b_2 + 1} = \sqrt{(e_1 + 1)/f_1 + 1} \approx 0.12436,$$
$$C_1 = (a_0 + a_1 + a_2)/C_3 \cong 0.30358,$$
$$C_4 = b_2 - 1 = 1/f_1 - 1 \cong 0.12939,$$
$$C_A = C_B = 1.$$

*The capacitance spread is, however, slightly larger for the pairing selected.

By contrast, the circuit of the high-Q section is that shown in Fig. 5.13c, and the design equations are (5.46) and (5.47).

Writing the transfer function again in the form

$$H_2(z) = \frac{z^2 + c_2 z + 1}{(1/f_2)z^2 + (e_2/f_2)z + 1},$$

comparison with (5.27) and (5.47) gives

$$C_1'' = a_2/b_2 = f_2 \cong 0.96845,$$

$$C_1' = (a_1 - a_0)/b_2 C_3 = 0,$$

$$C_2 = C_3 = \sqrt{(1 + b_1 + b_2)/b_2} = \sqrt{f_2 + e_2 + 1}$$

$$\cong 0.13795,$$

$$C_1 = (a_0 + a_1 + a_2)/b_2 C_3 = (2 + c_2)f_2/C_3$$

$$\cong 0.58645,$$

$$C_4 = (1 - 1/b_2)/C_3 = (1 - f_2)/C_3 \cong 0.22873.$$

Note that in writing the transfer functions $H_i(z)$ we disregarded the constant factors multiplying them. These factors will be adjusted appropriately during scaling.

Next, the important issue of ordering the three filter sections must be decided. In order to attenuate higher-frequency out-of-band signals and input noise, and (conceivably) also to ease the requirements of the antialiasing filter preceding the SCF, the low-pass linear section is placed first in the cascade. Since experience indicates that placing the high-Q section to the center of the filter tends to result in good signal-to-noise ratio, this section [described by $H_2(z)$] will follow, with the low-Q section at the end of the cascade. The circuit diagram is thus as that shown in Fig. 5.48.*

Next, the scaling of the element values was performed. For the input (linear) section, the maximum value V_{p1} of the output voltage occurs at dc, where the gain is $H_0(1) = C_S/C_D \cong 21.345$. Assuming an overall passband gain of 1 for the complete filter, this peak is to be reduced to 1. As discussed in Section 5.6, this can be accomplished by multiplying all capacitors connected or switched to the output terminal of op-amp A_1 by 21.345. This gives $C_D = C_S = 2$, $C_E \cong 19.345$, $C_1'' = 20.672$, and $C_1 \cong 12.518$. Next, all capacitors at the input terminal of A_1 should be scaled so that the smallest (i.e., $C_S/2$) equals 1. This

---

*The necessary sample-and-hold stage at the input is omitted. It should sample a continuous-time input voltage when ϕ_1 is high.

is already the case; hence the final values are

$$C_S = 2(C_S/2) = C_D = 2$$

and

$$C_E = 19.345.$$

Progressing to the next stage (the high-Q biquad), computer simulation shows that the peak output voltage V_{p2} of op-amp 2 occurs around $f_{p2} = 1.10$ kHz. The voltage there is $V_{p2} \cong 177.05$ for $V_{\text{in}} = 1$. To reduce this value to 1, C_A and C_3 must be multiplied by V_{p2}, giving $C_A \cong 177.05$ and $C_3 \cong 24.424$. Next, the capacitors at the output of A_3 are scaled so as to make its output voltage peak $V_{p3} = 1$. Computer analysis shows that the peak occurs near 1.07 kHz, and its value is $V_{p3} \cong 180.80$. To reduce this to 1, C_B, C_2, and C_4 must be multiplied by V_{p3}, giving $C_B \cong 180.80$, $C_2 \cong 24.941$, and $C_4 \cong 41.354$.

To obtain minimum total capacitance, the capacitors at the input terminal of A_2, that is, C_1, C_2, C_4, and C_A are scaled so as to make the smallest one, C_1, equal to 1. This is achieved by dividing their values by C_1 giving

$$C_1 = 1,$$

$$C_2 \cong 1.9926,$$

$$C_4 \cong 3.3036,$$

$$C_A \cong 14.144.$$

Similarly, the capacitors at the input of A_3, that is, C_1'', C_3, and C_B should be scaled to make the smallest capacitance in this group (i.e., C_1'') equal to 1. The result is

$$C_1'' = 1,$$

$$C_3 = 1.1815,$$

$$C_B = 8.7466.$$

Repeating the scaling process for the low-Q section gives the values

$$\overline{C}_A \cong 17.666,$$

$$\overline{C}_B \cong 7.7286,$$

$$\overline{C}_1 \cong 1.9926,$$

$$\overline{C}_2 = 1,$$

$$\overline{C}_3 \cong 2.1116,$$

$$\overline{C}_4 = 1,$$

$$\overline{C}_1'' \cong 6.3085.$$

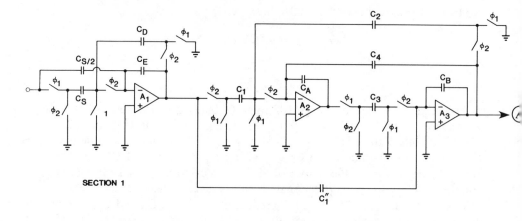

SECTION 2 (HIGH Q)

FIGURE 5.48. Cascade realization of a fifth-order elliptic switched-capacitor filter.

The reader is urged to rederive these values (Problem 5.24).

An inspection of the element values reveals that the largest capacitance C_E is contained in the linear section. Hence, a considerable saving in the total capacitance can be obtained by simplifying the input stage. It can easily be shown that simply omitting the capacitor $C_S/2$ from the input branch replaces $H_0(z)$ by

$$\hat{H}_0(z) = \frac{C_S}{C_D + C_E} \frac{1}{z - C_E/(C_D + C_E)}.$$

Comparison with the previous transfer function $H_0(z)$ reveals that this change is equivalent to including an extra factor $F(z) = 2/(z + 1)$ in the transfer function. In the frequency domain, the magnitude of $F(z)$ is

$$|F(e^{j\omega T})| = \left| \frac{2}{e^{j\omega T} + 1} \right| = \frac{1}{|\cos(\omega T/2)|}.$$

At the passband limit ($f = 1$ kHz), $\omega T/2 = 0.02\pi = 3.6°$ and $|F(e^{j\omega T})| \cong 1.002$. At the second minimum of the loss response, near 4 kHz (cf. Fig. 2.19), $|F(e^{j\omega T})| \cong 1.032$, representing a decrease of only about 0.3 dB in attenuation. The stopband loss then decreases. At $f = f_c/2$ (i.e., $z = -1$), $|\hat{H}_0(z)|$ is $C_S/(2C_E + C_D)$ which corresponds to only 26.2 dB. This effect is, however, counteracted by the $\sin(\omega T/2)/(\omega T/2)$ factor due to the sample-and-hold operation, and may be acceptable.

In the absence of the capacitance $C_S/2$, C_S is now the smallest capacitance in the stage, and hence we can rescale all values. The result is

$$C_S = C_D = 1$$

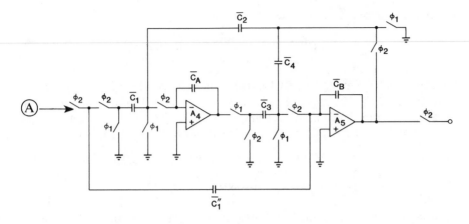

SECTION 3 (LOW Q)

FIGURE 5.48. continued.

and

$$C_E = 9.6725.$$

The total capacitance of the stage is now less than half of the earlier value.

To complete the design, the minimum practical capacitance value $C_{\min}$ (say, 0.5 pF) is assigned to the smallest capacitors presently of value 1, and all other capacitance values are multiplied by $C_{\min}$. Also, the sensitivity of the circuit to nonideal effects (finite op-amp gain and bandwidth, element-value variations, clock feedthrough, noise, etc.) should be checked. For our circuit, computer simulations revealed that a 70-dB dc gain for the op-amps results in a passband loss distortion of only about 0.01 dB, which is negligible. Similarly, for a 3-MHz op-amp bandwidth, the finite bandwidth did not affect the response perceptibly. The highest passband sensitivities to capacitance variations were around 0.2 dB/1% element-value change. This may not be low enough in some applications. Then, a ladder realization may be needed. This, however, requires a more difficult procedure, both in terms of paper design and of layout, testing, and troubleshooting.

Bilinear Ladder Design

This process uses the *circuit* of the elliptic filter C05, $\theta = 42°$, $p = 10\%$ as its starting point. The circuit diagram is that shown in Fig. 5.22a; the element values are $R_S = R_L = 1$, $C_1 = 0.85535$, $C_2 = 0.15367$, $L_2 = 1.20763$, $C_3 = 1.48438$, $C_4 = 0.46265$, $L_4 = 0.89794$, and $C_5 = 0.63702$. This circuit has its passband limit at $\hat{\omega}_{ap} = 1$ rad/s; hence, as discussed in Section 2.6, it should be frequency denormalized using the frequency unit

$$\omega_{ap} \cong \frac{2}{T}\tan\frac{\omega_p T}{2} = 2f_c\tan\frac{\pi f_p}{f_c} \cong 6291.4667 \text{ rad/s.}$$

As explained in Ref. 1, Section 1-4, this is accomplished by multiplying each inductor by $L_0 = Z_0/\omega_{ap}$ and each capacitor by $C_0 = 1/Z_0\omega_{ap}$, where L_0 is an arbitrary impedance unit. choosing for simplicity $C_0 = 1$, we obtain $Z_0 = 1/\omega_{ap}$ and $L_0 = 1/\omega_{ap}^2$. This results in the denormalized element values

$C_1 = 0.85535$, $\quad C_2 = 0.15367$, $\quad L_2 = 1.20763 \times L_0 = 3.05090 \times 10^{-8}$,
$C_3 = 1.48438$, $\quad C_4 = 0.46265$, $\quad L_4 = 0.89794 \times L_0 = 2.26851 \times 10^{-8}$,
$C_5 = 0.63702$, $\quad R_S = R_L = Z_0 = 1.58945 \times 10^{-4}$.

Splitting C_2 and C_4 each into a positive and negative capacitance, as explained in Section 5.5 ("The Exact Design of SC Ladder Filters"), and writing the state equations for the resulting network, we obtain

$$-V_1 = -\frac{1}{sC_1'}\left(\frac{1}{R_s}(V_{in} - V_1) - I_2 + sC_2'V_3\right),$$

$$-I_2 = -\left(\frac{1}{sL_2} - sC_{L2}\right)(V_1 - V_3),$$

$$V_3 = \frac{1}{sC_3'}(-I_2 - sC_2'V_1 - sC_4'V_5 + I_4),$$

$$I_4 = \left(\frac{1}{sL_4} - sC_{L4}\right)(V_3 - V_5),$$

$$-V_5 = \frac{-1}{sC_5'}\left(I_4 + sC_4'V_3 - \frac{V_5}{R_L}\right),$$

where

$$C_{L2} = \frac{T^2}{4L_2} = 0.003278,$$

$$C_2' = C_2 + C_{L2} = 0.15695,$$

$$C_1' = C_1 + C_2' = 1.01230,$$

$$C_{L4} = \frac{T^2}{4L_4} = 0.0044082,$$

$$C_4' = C_4 + C_{L4} = 0.46706,$$

$$C_3' = C_3 + C_2' + C_4' = 2.10839,$$

$$C_5' = C_5 + C_4' = 1.10408.$$

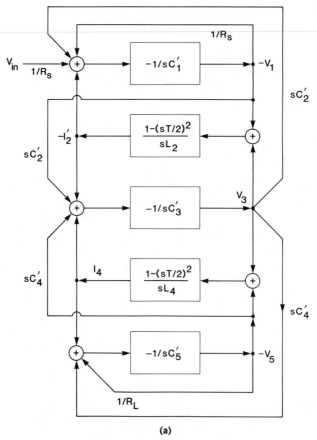

(a)

FIGURE 5.49. Bilinear ladder realization of a fifth-order elliptic switched-capacitor filter: (*a*) block diagram in the *s*-domain; (*b*) circuit diagram.

The block diagram of a system which simulates these equations is shown in Fig. 5.49*a*. Using the SC realization of the various branches and blocks as described in Section 5.5 (the second subsection), the circuit of Fig. 5.49*b* results. In this figure,

$$C_{11} = \frac{C_S}{2}, \quad C_{12} = C_S,$$

$$C_{01} = C_{02} = C_{03} = C_{04} = C,$$

$$C_{05} = C_{06} = C_{07} = C_{08} = C',$$

$$C_{21} = C_{22} = C_2',$$

$$C_{41} = C_{42} = C_4',$$

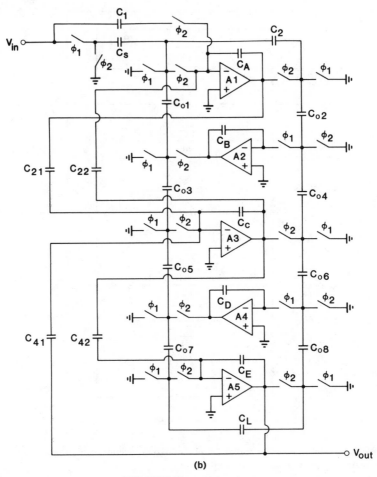

(b)

FIGURE 5.49.   continued.

where

$$C_S = \frac{T}{R_S} = 0.1258293,$$

$$C_A = C_1 + C_2 + C_{L2} - \frac{C_S}{2} = 0.94938,$$

$$C_B = \frac{C^2}{4C_{L2}} = \frac{C_{L2}}{4} = 0.0008195,$$

$$C_C = C_3' = 2.10839,$$

$$C_D = \frac{C'^2}{4C_{L4}} = \frac{C_{L4}}{4} = 0.001102,$$

$$C_E = C_5' - \frac{C_L}{2} = 1.041165,$$

$$C_L = \frac{T}{R_L} = 0.12583.$$

We chose here (arbitrarily) $C = C_{L2} = 0.003278$ and $C' = C_{L4} = 0.004408$. These values will be readjusted in any case by the scaling process, to be carried out next.

At this point the maximum output voltages in the unscaled circuit can be calculated for all op-amps. The resulting values are

$$V_{p1} \cong 0.92 \text{ V},$$

$$V_{p2} \cong 34 \text{ V},$$

$$V_{p3} \cong 0.764 \text{ V},$$

$$V_{p4} \cong 28.86 \text{ V},$$

$$V_{p5} \cong 0.5 \text{ V},$$

where an input voltage $V_{\text{in}} = 1$ V was assumed. To achieve a common peak value of 1 V, dynamic range scaling must be performed. For the stage containing A_1, this requires that C_4, C_2, C_{21}, and C_{02} all be multiplied by V_{p1}. Similarly, in the next stage, C_3, C_{01}, and C_{03} must be multiplied by V_{p2}, and so on. Having performed this operation on the output capacitors for all five stages, the minimum-capacitance scaling is used to readjust (to a minimum value of 1) all input capacitances. Thus, for example, C_A is first multiplied by $V_{p1} = 0.92$ and then divided by $C_S/2 \cong 0.06295$, since $C_S/2$ is the smallest capacitor at the input of op-amp A_1 after dynamic range scaling. The resulting new value of C_A is 13.872. Performing both scaling steps, the element values of the SCF in Fig. 5.49b become

C_1	=	1.00000	C_{05}	=	1.14172
C_2	=	1.83854	C_{06}	=	1.52861
C_3	=	2.00000	C_{07}	=	2.02212
C_A	=	13.87171	C_{08}	=	1.00000
C_{01}	=	1.77112	C_E	=	8.27441
C_{02}	=	1.20275	C_D	=	14.43078
C_{03}	=	1.00000	C_L	=	1.00000
C_{04}	=	1.00000	C_{41}	=	2.09575
C_B	=	11.11901	C_{42}	=	5.67396
C_C	=	14.46156	C_{21}	=	1.29480
			C_{22}	=	1.90667

As before, all capacitance values should be multiplied next by the minimum practical capacitance C_{min} to obtain the actual values.

Computations show that for ideal components the nominal passband ripple (0.044 dB) and minimum stopband loss (39.5 dB) are achieved by this SC filter. For a realistic op-amp dc gain value (70 dB), the passband ripple increases to 0.06 dB—still very low. As before, a finite-op-amp bandwidth of 3 MHz does not have noticeable effects.

A calculation of the element-value sensitivities of the response reveals that all are below 0.09 dB/%; in fact, if the effects consisting of constant gain shifts are neglected, the maximum sensitivity is less than 0.05 dB/%. This is much lower than the values achieved for the cascade realization. On the other hand, as already pointed out, the SC ladder circuit (due to its multifeedback configuration) is more difficult to lay out, test, and troubleshoot than a cascade of low-order buffered sections.

LDI Ladder Design

The design is again performed from the circuit diagram of the fifth-order elliptic filter used for the bilinear ladder circuit. To denormalize, we can now use simply $\omega_{ap} = \omega_p$ as the frequency unit and $Z_0 = 1 \ \Omega$ as the impedance unit. This gives $C_0 = 1/(2\pi 10^3) \ F$ and $L_0 = 1/(2\pi 10^3)$ H as the capacitance and inductance units, respectively. The denormalized element values are then

$$R_S = 1 \ \Omega,$$

$$C_1 = 136.13318 \ \mu F,$$

$$C_2 = 24.45734 \ \mu F,$$

$$L_2 = 192.20028 \ \mu H,$$

$$C_3 = 236.24641 \ \mu F,$$

$$C_4 = 73.633034 \ \mu F,$$

$$L_4 = 142.91159 \ \mu H,$$

$$C_5 = 101.38488 \ \mu F,$$

$$R_L = 1 \ \Omega.$$

The state equations of the circuit can be written in the form

$$-V_1 = -\frac{1}{s(C_1 + C_2)}\left(\frac{-V_1 + V_{\text{in}}}{R_S} + sC_2V_3 - I_2\right),$$

$$-I_2 = \frac{V_3 - V_1}{sL_2},$$

$$V_3 = -\frac{1}{s(C_2 + C_3 + C_4)}(-I_2 - sC_2V_1 - sC_4V_5 + I_4),$$

$$I_4 = \frac{V_3 - V_5}{sL_4},$$

$$-V_5 = -\frac{1}{s(C_4 + C_5)}\left(I_4 + sC_4V_3 - \frac{V_5}{R_L}\right).$$

These equations may be modeled by the system of Fig. 5.22b. The latter can be implemented by the active-RC circuit of Fig. 5.22c. Finally, replacing each resistor R by a switched capacitor $C = T/R$ (where $T = 20 \ \mu s$ is the clock period) the circuit of Fig. 5.22d results. The element values are

$$C_1 + C_2 = 160.59 \ \mu F,$$

$$C_S = \frac{T}{R_S} = 20 \ \mu F,$$

$$C = \frac{T}{1} = 20 \ \mu F,$$

$$C_2 + C_3 + C_4 = 334.34 \ \mu F,$$

$$C_4 + C_5 = 175.018 \ \mu F,$$

$$C_L = \frac{T}{R_L} = 20 \ \mu F.$$

Calculating the frequency responses of all five op-amps output voltages assuming a 1 V input signal, the maxima V_{pi} are found next. The results are given below:

for A_1: $V_{p1} = 0.927$ V at 1.182 kHz.
for A_2: $V_{p2} = 1.198$ V at 1.121 kHz.
for A_3: $V_{p3} = 0.857$ V at 1.061 kHz.
for A_4: $V_{p4} = 1.105$ V at 1.061 kHz.
for A_5: $V_{p5} = 0.501$ V at 967 Hz.

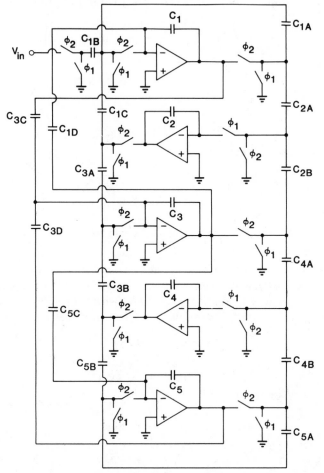

FIGURE 5.50. Circuit for the LDI ladder realization of a fifth-order elliptic switched-capacitor filter.

Using the values of the V_{pi}, we can perform the dynamic range scaling, followed by minimum-capacitance scaling. As a result, the final circuit of Fig. 5.50 is obtained. The element values are

$$C_1 = 8.03214,$$

$$C_{1A} = 1,$$

$$C_{1B} = 1.07930,$$

$$C_{1C} = 1.29263,$$

$$C_{1D} = 1.13212,$$

$$C_2 = 13.42236,$$

$$C_{2A} = 1.08053,$$

$$C_{2B} = 1,$$

$$C_3 = 12.97271,$$

$$C_{3A} = 1.08390,$$

$$C_{3B} = 1,$$

$$C_{3C} = 1.02540,$$

$$C_{3D} = 1.66885,$$

$$C_4 = 15.76379,$$

$$C_{4A} = 1.71203,$$

$$C_{4B} = 1,$$

$$C_5 = 8.75121,$$

$$C_{5A} = 2.20614,$$

$$C_{5B} = 1,$$

$$C_{5C} = 6.29664.$$

Figure 5.51 shows the passband response of the circuit. The ripple is around 0.095 dB, much higher than the nominal 0.044 dB but acceptable in many applications. The minimum stopband loss is about 40.5 dB, nearly 1 dB higher than the theoretical value. These loss values remain nearly unchanged when finite dc gain (70 dB) is assumed for the op-amps, or when the unity-gain frequency is lowered to 3 MHz.

The maximum passband sensitivity to capacitance value variations is about 0.08 dB/%; even less, if flat-loss effects are disregarded.

5.8.   SWITCHED-CAPACITOR *N*-PATH FILTERS

As pointed out at the conclusion of Section 5.5, narrow-band bandpass filters cannot be realized using the design techniques discussed earlier in this chapter. To understand the difficulties encountered,[41] compare the frequency responses

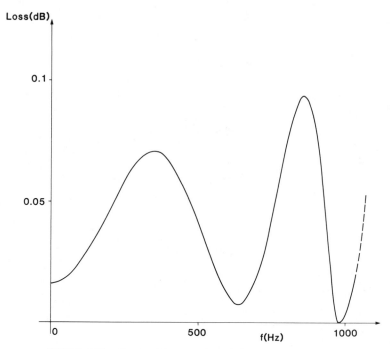

FIGURE 5.51. Passband loss response of the LDI filter circuit.

of a low-pass and a bandpass filter with identical bandwidths as well as passband and stopband specifications (Fig. 5.52a). The corresponding locations of the poles of $H(s)$ are shown in Fig. 5.52b. The pole-Q of the dominant pole is, by Eq. (5.32), for the low-pass filter approximately $Q_{LP} \cong (B/2)/2|\sigma_1|$, where B is the bandwidth and σ_1 the real part of the dominant pole. The zero/pole pattern of the narrow-band bandpass filter can, to a good approximation, be obtained simply by shifting the pole/zero pattern of the low-pass circuit by $\pm j\omega_0$, where ω_0 is the center frequency of the bandpass response (see, e.g., Ref. 1, pp. 584–585). As Fig. 5.52b illustrates, the dominant poles of the bandpass filters have pole-Q's given by $Q_{BP} \simeq (\omega_0 \pm B/2)/2|\sigma_1| \simeq \omega_0/2|\sigma_1| \simeq (2\omega_0/B)Q_{LP}$. For a narrow-band filter, we have $\omega_0 \gg B/2$ and hence $Q_{BP} \gg Q_{LP}$. If, say, $\omega_0 = 50B$, then $Q_{BP} \simeq 100Q_{LP}$. Thus, the pole-Q of 4.33 used in the "high-Q" section of the cascade filter in the preceding section would become 433 in a bandpass filter with a similar response centered at 100 kHz. As a rough estimate, the required op-amp dc gain A is about $200Q_{max}$; this gives $A = 866$ for the low-pass filter, which is easy to achieve, but $A = 86{,}600$ (i.e., nearly 100 dB) for the bandpass circuit, which is impractical. Similarly, the sensitivities to capacitance variations can be shown to be proportional to the pole-Q, and hence will increase 100 times if the relative bandwidth is $B/\omega_0 = 0.02$. Thus, a typical low-pass sensitivity of 0.05

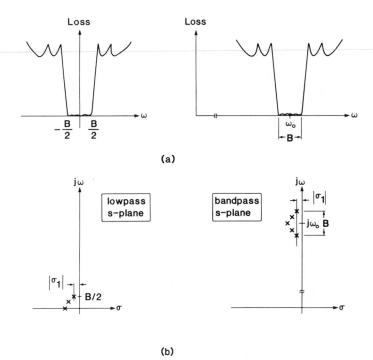

FIGURE 5.52. Comparison of a low-pass filter and a bandpass filter with identical bandwidths: (*a*) loss responses; (*b*) pole locations.

dB/% will become 5 dB/%, which would require impractically tight tolerances for the capacitances.

An effective way to solve these problems is to use *N-path circuits*. The basic concept[42] is illustrated (for $N = 3$) in Fig. 5.53. In general, the low-pass filters (LPF) used in the N signal paths can be continuous-time circuits; however, for fully integrated implementation, it is convenient to choose switched-capacitor low-pass circuits for the path filters (Fig. 5.54*a*). For sampled-and-held input signals, the system can then be analyzed and designed entirely in the *z*-domain.[43, 44] Assume that the sampling period of the input v_{in} and the output v_{out} is T, while the internal clock signals of the path filters have a clock period NT and are shifted by T with respect to each other (Fig. 5.54*b*). Denote the sequence of signal values (assumed to occur every T seconds) at the input of SCF1 by v_{in}^1, that at the input of SCF2 by v_{in}^2, and so on. Then, for $N = 3$, we have

$$v_{\text{in}}^1(nT) = v_{\text{in}}(0), 0, 0, v_{\text{in}}(NT), 0, 0, \ldots,$$

$$v_{\text{in}}^2(nT) = 0, v_{\text{in}}(T), 0, 0, v_{\text{in}}(NT + T), \ldots, \qquad (5.108)$$

$$v_{\text{in}}^3(nT) = 0, 0, v_{\text{in}}(2T), 0, 0, v_{\text{in}}(NT + 2T), \ldots .$$

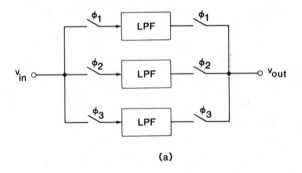

(a)

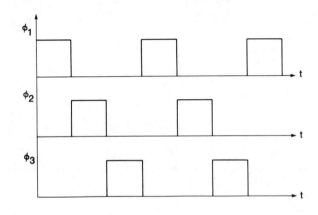

(b)

FIGURE 5.53. N-path filter, for $N = 3$: (a) block diagram; (b) clock signals.

Note that the input values are regarded as being equal to zero at those sampling instances when the input switch of the SCF is open. Clearly, for any N,

$$v_{in}(nT) = \sum_{k=1}^{N} v_{in}^{k}(nT).$$ (5.109)

Also, in the z-domain, from (5.108),

$$V_{in}^{1}(z) = \sum_{n=0}^{\infty} v_{in}^{1}(nT)z^{-n} = \sum_{m=0}^{\infty} v_{in}(mNT)z^{-mN},$$

$$V_{in}^{2}(z) = \sum_{n=0}^{\infty} v_{in}^{2}(nT)z^{-n} = \sum_{m=0}^{\infty} V_{in}(mNT + T)z^{-mN}z^{-1},$$ (5.110)

$$V_{in}^{3}(z) = \sum_{n=0}^{\infty} v_{in}^{3}(nT)z^{-n} = \sum_{m=0}^{\infty} V_{in}(mNT + 2T)z^{-mN}z^{-2},$$

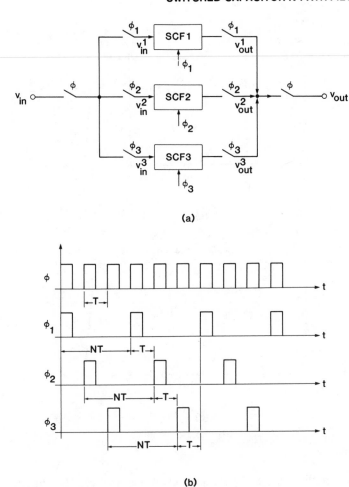

(a)

(b)

FIGURE 5.54. Switched-capacitor 3-path filter: (*a*) block diagram; (*b*) clock signals; (*c*) frequency response; (*d*) prefilter response and combined prefilter + 3-path filter response.

and, from (5.109),

$$V_{in}(z) = \sum_{n=0}^{\infty} v_{in}(nT)z^{-n} = \sum_{k=1}^{N} V_{in}^{k}(z). \qquad (5.111)$$

Exactly parallel definitions can be used for the path output voltage sequences $v_{out}^{k}(nT)$ and their z-transforms $V_{out}^{k}(z)$. These are related the same way to the overall output $v_{out}(nT)$ and its transform $V_{out}(z)$ as the v_{in}^{k} and V_{in}^{k} to v_{in} and V_{in}.

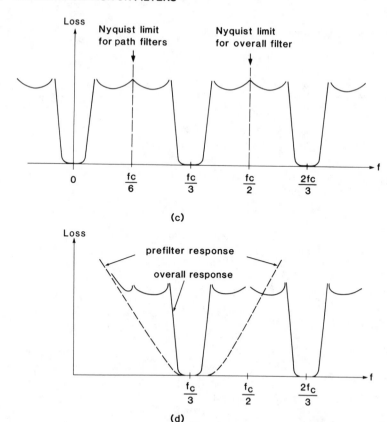

(c)

(d)

FIGURE 5.54. continued.

Since the path filters (SCF) are assumed to be identical, they have the same transfer functions

$$H_P(z) = \frac{V_{out}^1(z)}{V_{in}^1(z)} = \frac{V_{out}^2(z)}{V_{in}^2(z)} = \cdots . \qquad (5.112)$$

The overall filter transfer function is therefore

$$H(z) = \frac{V_{out}(z)}{V_{in}(z)} = \frac{V_{out}^1(z) + V_{out}^2(z) + \cdots + V_{out}^N(z)}{V_{in}^1(z) + V_{in}^2(z) + \cdots + V_{in}^N(z)}$$

$$= \frac{H_P(z)V_{in}^1(z) + H_P(z)V_{in}^2(z) + \cdots + H_P(z)V_{in}^N(z)}{V_{in}^1(z) + V_{in}^2(z) + \cdots + V_{in}^N(z)}$$

$$= H_P(z). \qquad (5.113)$$

   Thus, the transfer function of the overall filter is the same as that of each path. However, the sampling rate of the path filters is only $1/NT = f_c/N$, where f_c is the sampling rate of the overall filter (Fig. 5.54*b*). Consider the case when the SCFs are sampled-data low-pass filters, each with the loss response schematically indicated (for $N = 3$) in Fig. 5.54*c*. Since the path sampling rate is $f_c/N = f_c/3$, the passband centered at $f = 0$ will be replicated at $f_c/3, 2f_c/3, f_c, \ldots$. However, if only a single path filter would be used, then the passband centered at $f_c/3$ could *not* be used for bandpass filtering. This is because any input signal of a frequency higher than $f_c/6$ (which is half the clock rate of the path filter) would be aliased (folded back) to a frequency symmetrically located below $f_c/6$. By contrast, for the overall filter the Nyquist limit of the signal spectrum is $f_c/2$; hence, the narrow passband centered at $f_c/3$ is safely within the Nyquist range. Hence, by using a bandpass prefilter which eliminates the part of the input spectrum which would fall into the passbands centered at $f = 0$ and $2f_c/3$ (as well as above $2f_c/3$), a narrow-band bandpass filter response results. Fig. 5.54*d* illustrates a possible prefilter response and the resulting overall filter characteristics.

   It is instructive to examine also the location of the poles and zeros in the z-plane. Consider one of the path filters SCF. Its sampling period is NT (Fig. 5.54*b*) and hence its performance can be analyzed in terms of a z-variable defined by

$$z_P \triangleq e^{s(NT)} = \left(e^{sT}\right)^N = z^N. \qquad (5.114)$$

Clearly, z_P can be regarded as the z-variable of a single path filter. Since the path filters have narrow-band low-pass responses, their pole/zero patterns are in the vicinity of the $z_P = 1$ point (Fig. 5.55*a*). In terms of the z-variable, any value of z_P given by $r_P e^{j\phi_P}$ corresponds to N values given by

$$z_k = r_P^{1/N} e^{j(\phi_P + k2\pi)/N}, \qquad k = 0, 1, 2, \ldots, N - 1. \qquad (5.115)$$

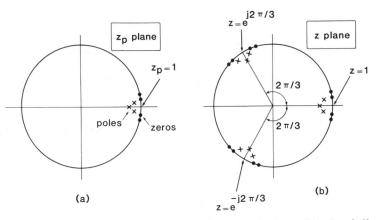

FIGURE 5.55.   Pole and zero locations for (*a*) single path filter and (*b*) 3-path filter.

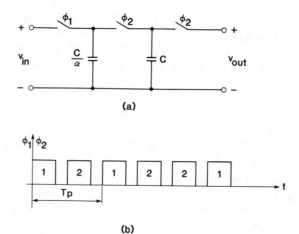

(a)

(b)

FIGURE 5.56. A simple passive path filter: (a) circuit diagram; (c) clock signals.

Thus, the z_P-plane pole/zero pattern of Fig. 5.55a is transformed into the z-plane pattern shown (for $N = 3$) in Fig. 5.55b. The pattern centered around $z = \exp(j2\pi/3)$ corresponds to the passband centered at $f_c/3$, shown in Fig. 5.54c.

As a straightforward application of the above principles, consider the simple path filter[43] shown in Fig. 5.56a. Assuming that the sampling instants t_s occur at the end of each $\phi_1 = 1$ interval, the transfer function can readily be shown (Problem 5.25) to be

$$H(z_P) = \frac{z_P^{1/2}}{1 + (\alpha + 1)(z_P - 1)}. \tag{5.116}$$

It is also easy to show that for $z_P = \exp(j\omega T_P)$ and $\omega T_P \ll 1$, the frequency response of the circuit is given by

$$H(e^{j\omega T_P}) \simeq \frac{1}{1 + j\omega(\alpha + 1)T_P}. \tag{5.117}$$

Hence, the circuit has a response similar to that of a first-order RC low-pass filter (Fig. 5.57), with $R = (\alpha + 1)T_P/C$. The filter response can be made steeper by cascading several such sections with alternating clock phasing and

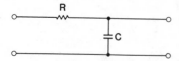

FIGURE 5.57. First-order RC prototype circuit for the path filter.

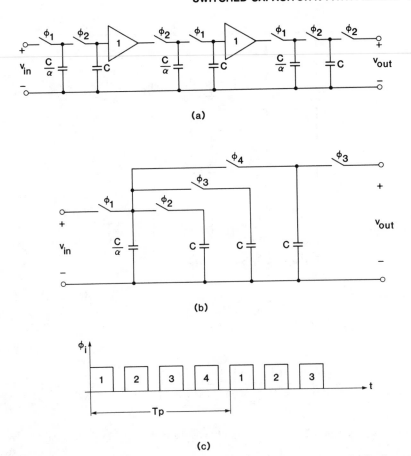

FIGURE 5.58. Third-order path filter: (*a*) cascade realization; (*b*) charge-redistributing realization; (*c*) clock signals.

separating them by unity-gain buffers (Fig. 5.58*a*). However the same transfer function, $[H(z_P)]^3$, can be also be obtained by the much simpler passive circuit shown in Fig. 5.58*b*. To show that the responses of the two circuits are the same, note that in the circuit of Fig. 5.58*a* the capacitances C/α and C in the input stage share charge, and then the resulting voltage v_1 is transferred by the buffer to the capacitor C/α in the second stage. However, the capacitor C/α in the input stage is already at the voltage v_1, and hence can be reused as the first capacitor element in the *second* stage, without any buffering. Thus, only the second capacitor C must be provided N times. Note that the new circuit needs four (rather than two) clock phases, and is only half as fast as the circuit of Fig. 5.58*a*. These disadvantages disappear, however, when the circuit is used in an *N*-path filter, as illustrated in Fig. 5.59 which shows a 4-path circuit utilizing the path filter of Fig. 5.58*b*. As can be seen from the earlier

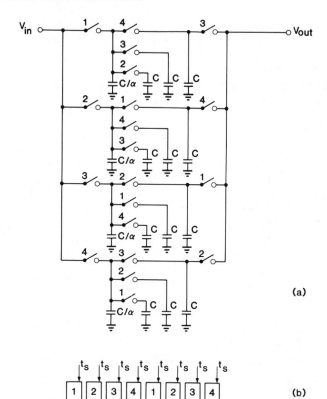

FIGURE 5.59.   4-path filter using the path filter of Fig. 5.58*b*: (*a*) circuit diagram; (*b*) clock signals.

discussions, this circuit has the same transfer function as that of Fig. 5.58*a*, that is, $[H(z_P)]^3$. Thus, from (5.116),

$$H(z_P) = \frac{z_P^{3/2}}{\left[(\alpha + 1)z_P - \alpha\right]^3} = \frac{z^6}{\left[(\alpha + 1)z^4 - \alpha\right]^3} \qquad (5.118)$$

where (5.114) was used, with $N = 4$.

By allowing a unity-gain buffer, the form of $H(z)$ can be made much more general. Figure 5.60 shows an active N-path filter based on the same principle as that of Fig. 5.59. The path transfer function, in terms of z_P, is now (Problem 5.26)

$$H(z) = \frac{-z_P^{3/2}/\left[\alpha^2(\alpha + 1)\right]}{-(1 + 1/\alpha)^2 z_P^3 + (3 + 4/\alpha)z_P^2 - \left[3 + 1/\alpha + 1/(\alpha + 1)\right]z_P + 1} \qquad (5.119)$$

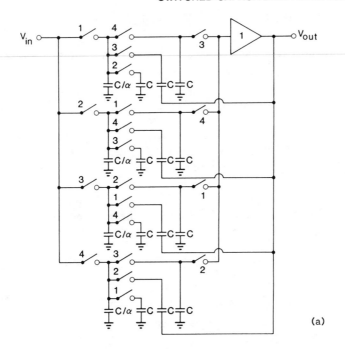

(a)

(b)

FIGURE 5.60. 4-path filter with unity-gain buffer: (a) circuit diagram; (b) clock signals.

which (for $\omega T_P \ll 1$) can approximate a Butterworth low-pass filter response. To show this, we use the approximation $z_P = e^{sT_P} \cong 1 + sT_P$. Then the denominator of $H(z_P)$ can be written in the form

$$\frac{-1}{\alpha^2(\alpha+1)}\left[(\alpha+1)^3(sT_P)^3 + 2\left((\alpha+1)^2 + \frac{\alpha+1}{2}\right)(sT_P)^2\right.$$

$$\left. + 2\left((\alpha+1) + \frac{1}{2}\right)(sT_P) + 1\right]. \quad (5.120)$$

The polynomial in the square brackets can be compared with the third-order normalized Butterworth polynomial (see Ref. 1, Section 12-2)

$$\left(\frac{s}{\omega_{3dB}}\right)^3 + 2\left(\frac{s}{\omega_{3dB}}\right)^2 + 2\left(\frac{s}{\omega_{3dB}}\right) + 1. \quad (5.121)$$

Setting $\alpha + 1 = 1/(\omega_{3\mathrm{dB}}T_P)$, and noting that since $\omega_{3\mathrm{dB}}T_P = \omega_{3\mathrm{dB}}/f_P \ll 1$ also $\alpha \gg 1$, it is easy to show that the two polynomials will have (approximately) the same coefficients. Since the bandwidth B (in Hz) of the final bandpass response is $2f_{3\mathrm{dB}}$, the formula for α is

$$\alpha = \frac{1}{\omega_{3\mathrm{dB}}T_P} - 1 = \frac{1}{\pi B T_P} - 1. \tag{5.122}$$

The overall transfer function of the 4-path filter can of course be obtained from (5.119) simply by replacing z_P by z^4.

To illustrate the performance achievable with the circuit of Fig. 5.60, a bandpass SCF was designed with a passband center frequency $f_{\mathrm{center}} = 65.925$ kHz and a 3-dB bandwidth of $B = 120$ Hz. The corresponding values of the fast clock frequency f_c and α are[41]

$$f_c = Nf_P = Nf_{\mathrm{center}} = 4f_{\mathrm{center}} = 263.7 \text{ kHz}$$

and

$$\alpha = \frac{f_{\mathrm{center}}}{\pi B} - 1 \simeq 174.$$

The corresponding computed loss characteristics of the 4-path filter is shown in Fig. 5.61 and the details of the passband are shown in Fig. 5.62. These figures also illustrate the sensitivity of the response to $\pm 10\%$ variations of the small capacitance C/α. (The large capacitors C can be matched to each other accurately.) Figure 5.63 demonstrates the effects of $\pm 10\%$ variations in the gain A of the buffer. Neither imperfection affects the center frequency, which is fixed by f_P; the tolerances of αC merely change the bandwidth somewhat, while those of A introduce a small, nearly constant, gain or loss into the response. Both effects are usually acceptable in practical applications.

An important shortcoming of all N-path filters is that any asymmetry among the path filters introduces spurious output signals. Specifically, sampling the path input voltages v_{in}^i at the slow clock rate $f_P = 1/T_P$ introduces the "mirror-frequency" signals $nf_P \pm f_{\mathrm{in}}$ ($n = 1, 2, \ldots, [N/2])^*$ into the path output voltages v_{out}^i. These signals are, ideally, equally large and (due to the delay T among the path clock signals) their phases differ by $2\pi/N$. Hence, their phasors cancel when they are added to obtain v_{out}. (Fig. 5.64 illustrates the ideal situation for $N = 3$.) However, any change in the amplitude or phase response of a path filter will upset this balance, and as a result the mirror-frequency signals will appear in v_{out}. In addition, each path filter is subject to some clock feedthrough noise, which occurs at the low (path) clock rate f_P and its integer multiples. Again, for identical paths these noise signals cancel at all

---

*Here, $[N/2]$ denotes the integer part of $N/2$.

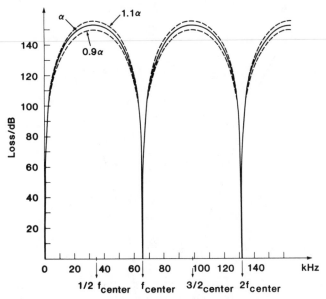

FIGURE 5.61. Computed responses of the 4-path filter of Fig. 5.60 with $f_{center} = 65.925$ kHz, $B = 120$ Hz, and $\alpha = 274 \pm 10\%$ (from Ref. 41, © 1982 IEEE).

frequencies, except at 0, $Nf_P = f_c$, $2f_c$, $3f_c$, and so on. Any path asymmetry will, however, cause the nominally cancelling components to reappear, including (most inconveniently) a noise occurring at f_P which is normally the center frequency of the passband of the overall filter. This phenomenon is illustrated in Fig. 5.65, taken from Ref. 44. This noise can be reduced by using a fully differential configuration for each path filter, or for the overall *N*-path filter.[45]

It is, of course, also possible to use any one of the SCFs discussed earlier in this chapter as the path filter. When this is done, it is expedient to time share all memoryless elements, that is, op-amps and the switched capacitors which are fully discharged in each cycle. This is possible, since (as Fig. 5.54*b* shows) only one path filter is active at any given time. Such timesharing is advantageous not only because it saves components, but also because it eliminates some potential sources of path asymmetry, since now the same op-amps and switched capacitors can be used in all paths.[46] The process is illustrated, for a 4-path circuit with third-order Chebyshev filters used as path filters, in Fig. 5.66.[46] As the figure illustrates, the switches activated by ϕ_1, ϕ_2, ϕ_3, and ϕ_4 which are normally needed at the input as well as at the output, can now be omitted since the input and output terminals are shared by the four path filters as are all coupling capacitors C_u and all op-amps. Unfortunately, the remaining asymmetry in the feedback capacitors and the multiplexing switches associated with them is still sufficient to introduce a significant amount of noise at the center of the passband.[46]

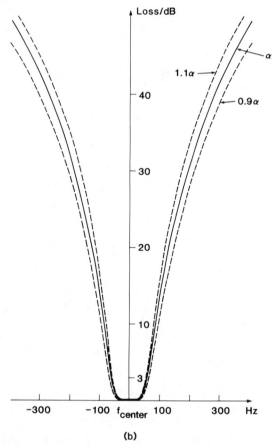

(b)

FIGURE 5.62. Loss response[41] of the 4-path filter around f_{center}.

The uniformity of the paths can be improved, at the cost of an added capacitor, by using a single feedback capacitor in each integrator for all signal paths, and by employing extra capacitors (one for each path) to store the signal charges when they are not processed in the circuit. Figure 5.67a illustrates (for $N = 3$) a possible circuit[41,47] of a single multiplexed integrator. Here, C is the common feedback capacitor shared by all paths, and the C_i ($i = 1, 2, 3$) are the storage capacitors used for the signal charges of path i. The circuit operates as follows. When ϕ_1 and ϕ_3 are high, the circuit is part of path 1. During this time, C_1 discharges, and its charge is transferred via the virtual ground terminal of the op-amp into C. At the same time, through C_0 a charge increment $C_0 v_{in}$ (part of the path 1 signal flow) also enters C, augmenting its signal charge. Next, when ϕ_2 and ϕ_4 become high, C transfers the augmented signal charge back into C_1. This charge will then be stored there until it is again the turn of path 1 to process the input signal and to produce an output

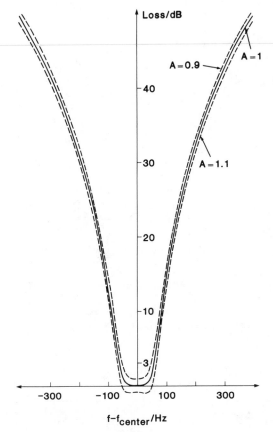

FIGURE 5.63. The effects of $\pm 10\%$ buffer gain variations of the loss response.[41]

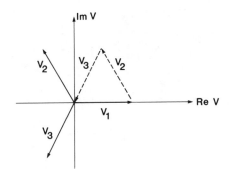

FIGURE 5.64. Phasor diagram for the slow-clock-rate output voltages of 3-path filter: ideally, $V_1 + V_2 + V_3 = 0$.

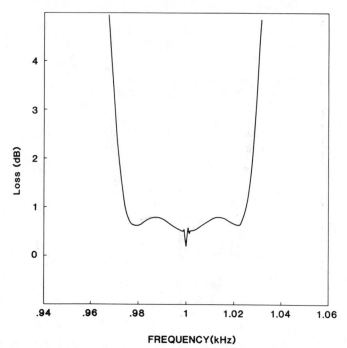

FREQUENCY(kHz)

FIGURE 5.65. Measured passband response of an *N*-path filter showing band-center noise peak (from Ref. 44 © 1981 IEEE).

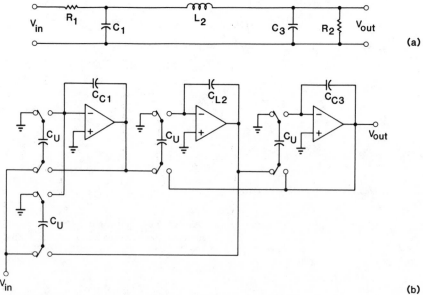

FIGURE 5.66. 4-path Chebyshev filter (from Ref. 46, © 1980 IEEE): (*a*) LCR prototype for the path filter; (*b*) SC low-pass path filter; (*c*) 4-path circuit; (*d*) clock signals.

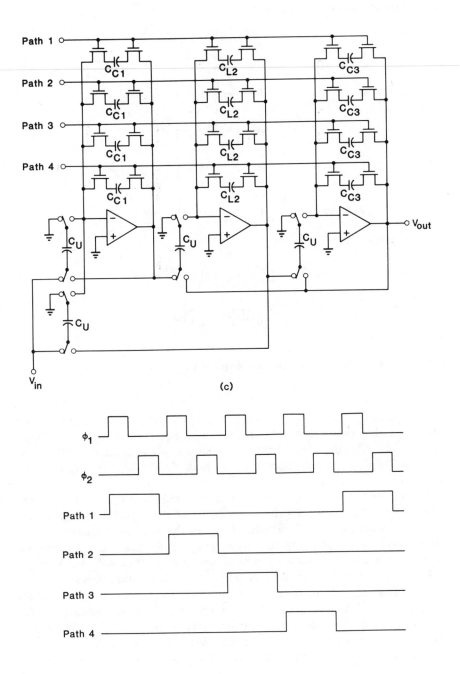

FIGURE 5.66. continued.

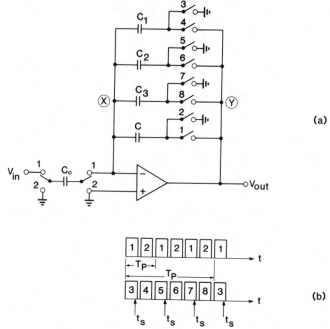

FIGURE 5.67. Multiplexed (RAM-type) integrator: (*a*) circuit diagram; (*b*) clock signals.

sample. Next, ϕ_1 and ϕ_5 rise, and the charge in C_2 (the path 2 storage element) is transferred to C where it is incremented by $C_0 v_{in}$. Then, ϕ_2 and ϕ_6 go high, causing the updated path 2 charge to be stored in C_2. Similarly, during the intervals when ϕ_7 and subsequently ϕ_8 are high, the circuit performs as a part of path 3.

Clearly, for this circuit, the *same* integrator is used by all paths. The storage capacitors C_i need not be matched accurately to C or to each other, since they are merely used as memory devices. However, the clock feedthrough charges due to the switches charging and discharging the C_i must be small and/or closely matched. Otherwise, again a clock feedthrough noise will appear at f_P.

A similar circuit,[38] which needs only five clock phases and is less sensitive to clock feedthrough effects, is shown in Fig. 5.68. Here, the storage capacitors are disconnected from the rest of the circuit when the switches operated by ϕ_3, ϕ_4, and ϕ_5 open; the rest of the circuitry is common to all three paths and hence cannot introduce clock feedthrough noise into the passband.

To illustrate the use of the multiplexed integrators of Figs. 5.67 or 5.68 in a 3-path filter, consider again the third-order elliptic low-pass filter prototype of Fig. 5.21a, reproduced (with $R_S = R_L = R$) in Fig. 5.69a. Using the exact design techniques described in Section 5.5 (the second subsection), the circuit of Fig. 5.69b is obtained, where the switches are illustrated during the $\phi_1 = 1$

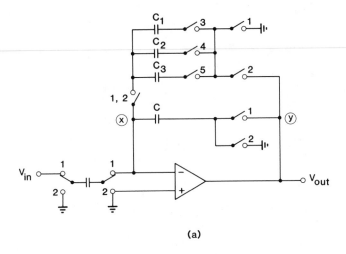

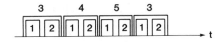

(a)

(b)

FIGURE 5.68. Multiplexed RAM-type integrator with reduced clock feedthrough: (a) circuit diagram; (b) clock signals.

interval.* In this circuit, in addition to the op-amp feedback capacitors C_a, C_{L2}, and C_b, there are also other nonswitched (and hence memoried) capacitors, namely C_1 and C_2. These coupling capacitors, which are needed for the realization of the finite transmission zeros, cannot be multiplexed among the paths and hence would require $N = 3$ capacitors and associated switches for realization in the N-path filter. Fortunately, the network equivalence[47] illustrated in Fig. 5.70 can be used to replace all unswitched coupling capacitors by switched (and hence memoryless) ones. It can readily be shown that, with the sampling instances chosen as shown in Fig. 5.70b, the input–output relations of both circuits are given by

$$V_0(z) = -\frac{C_1(1 - z^{-1})V_1(z) + \Delta Q(z)}{C_a + C_{a1} - C_a z^{-1}} \qquad (5.123)$$

(Problem 5.27).

*The switching scheme in this figure is complementary to that of Fig. 5.30a. This is, of course, immaterial as far as the transfer function is concerned.

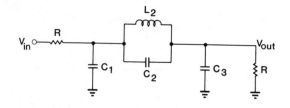

(a)

(b)

FIGURE 5.69. Elliptic 3-path bandpass filter: (a) LCR low-pass prototype circuit for the path filter; (b) low-pass SC filter derived from the LCR prototype; (c) 3-path filter circuit.

    With the unswitched coupling capacitors gone, the only memoried elements remaining are the op-amp feedback capacitors C_a, C_{L2}, and C_b. These should then be replaced by the feedback branch connecting nodes ⓧ and ⓨ in the circuit of Fig. 5.67a (or that of Fig. 5.68a). The resulting filter, in which the feedback branches A, B, and C are shown only symbolically, is illustrated in Fig. 5.69c.

    As mentioned earlier, the multiplexed integrators of Figs. 5.67 and 5.68 are less sensitive to clock feedthrough asymmetries than the N-path circuits shown in the preceding figures. Nevertheless, they are not totally immune to such

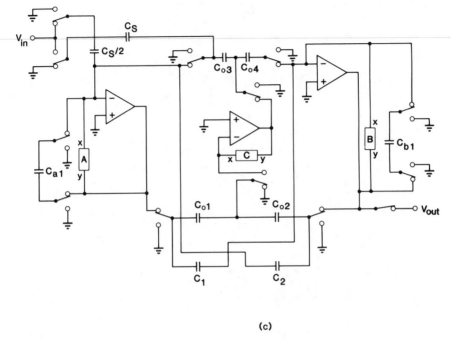

(c)

FIGURE 5.69. continued.

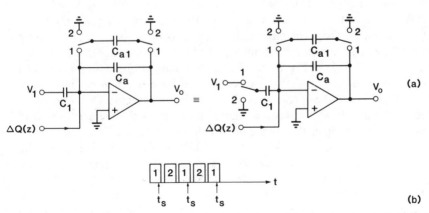

FIGURE 5.70. A network equivalence—both circuits shown have the input–output relation given in Eq. (5.123): (*a*) the equivalent circuits; (*b*) the clock signals and sampling instances t_s.

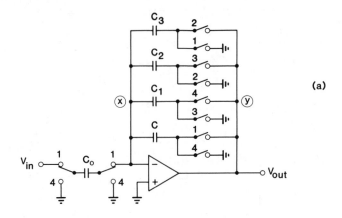

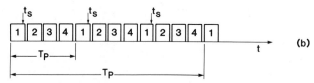

FIGURE 5.71. Multiplexed integrator with circulating memory: (*a*) circuit diagram; (*b*) clock signals.

effects; the remaining band-center noise is typically 60–80 dB below the maximum signal level. If this cannot be tolerated, then a different circuit, based on the "pseudo-N-path" principle[48] may be used. In such a circuit, each memoried element (i.e., unswitched capacitor) of the path filter is replaced by a *circulating* memory. This forces the signal charges belonging to all paths to travel through the *same* delay stages of the memory. All path asymmetry is thus eliminated from the circuit, and hence so is the band-center noise peak caused by such asymmetry. A multiplexed integrator using this principle[43] is shown (for $N = 3$) in Fig. 5.71a. While the circuit configuration is identical to that shown in Fig. 5.67a, the clock phases have a different sequence (Fig. 5.71b) and hence the operation is also completely different. During $\phi_1 = 1$, the feedback capacitor C receives a signal charge from the storage capacitor C_3; assume, for example, that this charge belongs to path 2. The charge in C is incremented by an input charge $C_0 v_{in}$, which is also in path 2. Next, during $\phi_2 = 1$, the charge in C_2 (which belongs to path 3) is transferred to C_3; during $\phi_3 = 1$, the charge (of path 1) is shifted from C_1 to C_2; and during $\phi_4 = 1$, the updated path-2 charge is deposited in C_1 while C_0 is discharged. Then, when ϕ_1 goes high again, the charge now in C_3 (which belongs to path 3) enters C, and is updated by $C_0 v_{in}$, and so on. Clearly, as mentioned earlier, the path charges all follow exactly the same route through the circuit. Hence, there cannot exist any path asymmetry, and, as a result, clock feedthrough noise only occurs at integer multiples of the fast clock frequency $f_c = 1/T$, which are not in the passband.

Clearly, it requires $N(N + 1) = 12$ clock pulses to perform a full cycle of operation. Thus, each path signal is updated only once in every 12 clock pulses. Hence, the path clock interval $T_P = NT$ contains $N(N + 1)$ pulses, while the clock interval T of the overall bandpass filter contains $N + 1 = 4$ clock pulses (Fig. 5.71*b*). Comparison with Figs. 5.67 and 5.68 reveals that the pseudo-*N*-path circuit is only half as fast as those shown there. This is a price paid for the complete elimination of the band-center clock feedthrough noise.*

To illustrate the use of the circulating-memory *N*-path integrator in filter design,[41] consider the fifth-order Chebyshev low-pass prototype filter shown in Fig. 5.72*a*. Using the LDI design process, this circuit can be transformed into the "active ladder" SC low-pass filter shown in Fig. 5.72*b*, where each feedback branch contains a single unswitched capacitor, of value C_A for branch *A*, C_B for branch *B*, and so on.

To transform this circuit (which may be regarded as a low-pass path filter) into the *N*-path bandpass one, each feedback branch must be replaced by the feedback circuit (shown between nodes ⓧ and ⓨ) in Fig. 5.71*a*. The value of C is C_A in branch *A*, C_B in branch *B*, and so on. The values of C_1, C_2, and C_3 are not critical; choosing them equal to C is usually satisfactory.[41] Figure 5.72*c* shows how the two capacitors switched to the output of each op-amp can be combined into one, to save a few components in the circuit.

To understand why the *N*-path and pseudo-*N*-path circuits have a reduced sensitivity as compared to a direct narrow-band bandpass realization, we rewrite eq. (5.113) in the form

$$V_{\text{out}}(z) = H_P^1(z)V_{\text{in}}^1(z) + H_P^2(z)V_{\text{in}}^2(z) + \cdots + H_P^N(z)V_{\text{in}}^N(z). \quad (5.124)$$

Here, $H_P^k(z)$ is the transfer function of the kth path filter. Nominally, $H_P^1(z) = H_P^2(z) = \cdots = H_P^N(z)$. If a parameter P (element value, op-amp gain, etc.) in the kth path filter changes from its nominal value by a small amount ΔP, then the output voltage changes by

$$\Delta V_{\text{out}}(z) = \Delta H_P^k(z)V_{\text{in}}^k(z) \simeq \frac{\partial H_P^k(z)}{\partial P}\Delta P V_{\text{in}}^k(z). \quad (5.125)$$

Hence, for $z = \exp(j\omega T)$, the change in the frequency response of the overall filter is the same as for the kth path filter acting alone. Since the latter is a low-pass filter, its pole-Q's Q_{LP} are low, and hence, its sensitivities can also be made low. Thus, the sensitivities are lowered by about a factor $Q_{BP}/Q_{LP} \sim 2\omega_0/B$, as discussed at the beginning of this section.

As an illustration of the advantages achievable through the use of *N*-path filters, consider the example of a bandpass filter with its passband in the

---

*Another drawback is a somewhat higher sensitivity to finite op-amp gain, due to the larger number of transitions for each charge packet.

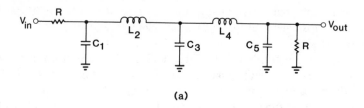

(a)

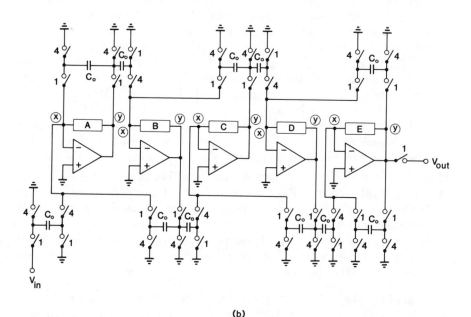

(b)

FIGURE 5.72. *N*-path Chebyshev filter with circulating-memory type integrators: (*a*) LCR prototype low-pass filter; (*b*) *N*-path filter circuit; (*c*) *N*-path circuit after merging capacitors at each op-amp output.

1900–2100 Hz range.[41] A bandpass ladder filter (using the circuit of Fig. 5.36*b*) requires 6 op-amps; the maximum passband element-value sensitivity was then found to be 1.2 dB/1% capacitance change. The passband loss distortion introduced by the finite op-amp gain (A_{dc} = 60 dB) was 3.8 dB. The realization of the same filter as an *N*-path filter needed only 3 op-amps (Fig. 5.69*b*). The maximum passband sensitivity was now only 0.065 dB/%, while the 60 dB op-amp gain introduced a passband loss distortion of only 0.27 dB. As these numbers indicate, the direct-realization circuit is not practical for such a narrow-band response, while the *N*-path filter is quite feasible.

These major advantages of the *N*-path filter should be balanced against some specific shortcomings. Some of these, such as the band-center noise peak and mirror-frequency noise caused by path asymmetry, and the multiphase

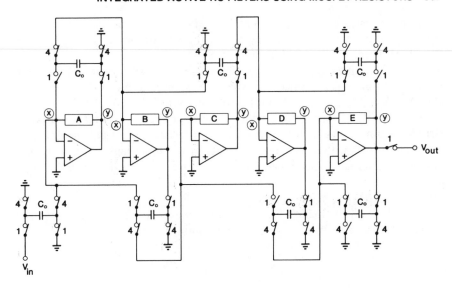

(c)

FIGURE 5.72.   continued.

clock signals needed by the circuit, have already been discussed. In addition, the wide-band noise generated by the switches and the op-amps of the circuit gets aliased into the narrow passband, severely limiting the dynamic range achievable by the N-path filter. This noise sensitivity may rule out the use of this filter in some applications (such as carrier-frequency telephone filtering) where very strict specifications exist on the dynamic range.

5.9.   INTEGRATED ACTIVE-RC FILTERS USING MOSFET RESISTORS

As the discussions of the preceding sections of Chapter 5 demonstrated, switched-capacitor filters are capable of high performance, adequate for many filtering applications. Nevertheless, they have several drawbacks compared to an equivalent continuous-time realization. A few are listed below.[49]

1. The dynamic range of SC filters is always reduced by clock feedthrough noise and by the aliasing of the wide-band op-amp and switch noise into the passbands.

2. SC filters usually need antialiasing filtering at their inputs, and smoothing filtering at their outputs.

3. The switches and the associated clock lines can make the layout of the circuit complicated.

These problems, as well as the desirability of on-chip antialiasing and smoothing filters, indicates a need for integrated continuous-time active-RC filters. On the other hand, as discussed in Section 5.1, the direct realization of resistors needed in an integrated active-RC filter using polysilicon or diffused lines requires a large area, since the sheet resistance is low (of order 20–50 Ω/square) and the accuracy poor ($\pm 10\%$ or worse).

By contrast, the channel of an MOS transistor operating in its nonsaturated (triode) region shows an approximately linear resistance between the drain and source terminals with a sheet resistance about 100 times higher than the above values. Furthermore, the value of this resistance is a function of the gate-to-source voltage v_{GS}. The latter can therefore be used to control the value of the resistor realized by the device.

As given in Table 3.2 and Eq. (3.7) of Chapter 3, the drain current of an NMOS device in its triode region is approximately

$$i_D \cong \mu_n C_{ox} \frac{W}{L}\left(v_{GS} - V_T - \frac{v_{DS}}{2}\right)v_{DS}. \tag{5.126}$$

A more detailed and accurate expression, which explicitly incorporates the body effect and the variation of the charge density along the channel, can also be derived.[50] The resulting equation, Eq. (8.1.12) in Ref. 50, contains terms of the form $(v_D - v_B + 2|\phi_p|)^{3/2}$ and $(v_S - v_B + 2|\phi_p|)^{3/2}$. Here, ϕ_p is the potential associated with p-type silicon, briefly discussed at the end of Section 3.3, and v_B the substrate (bulk) voltage. Assuming $v_D \approx v_S \approx 0$, Taylor-series expansion used in this expression gives[51]

$$i_D = \mu_n C_{ox} \frac{W}{L}$$

$$\times \left[(v_G - V_T)(v_D - v_S) - \frac{1}{2}\left(1 + \frac{\gamma}{2\sqrt{2|\phi_p| - v_B}}\right)(v_D^2 - v_S^2)\right] \tag{5.127}$$

where γ is the body-effect constant defined in Section 3.3. Since for $v_D \approx v_S \approx 0$ the bulk potential v_B must be negative, we can define

$$m = 1 + \frac{\gamma}{2\sqrt{2|\phi_p| + |v_B|}} \tag{5.128}$$

and rewrite (5.127) in the form

$$i_D = k\left[2(v_G - V_T)(v_D - v_S) - m(v_D^2 - v_S^2)\right]. \tag{5.129}$$

Here, as in Section 3.3, $k \triangleq (\mu_n C_{ox}/2)W/L$ was used. Thus, the conductance between the drain and source is

$$\frac{i_D}{v_{DS}} = \frac{i_D}{v_D - v_S} = 2k(v_G - V_T) - km(v_D + v_S). \qquad (5.130)$$

The first term is independent of v_D and v_S; however, the second term is not, and hence nonlinear distortion results when the device is used as a resistor in an otherwise linear circuit. For the typical values[49] $v_G - V_T = 2$ V, $k = 5$ μA/V^2, $\gamma \cong 0.4$ V$^{1/2}$, $|\phi_p| = 0.3$ V, $v_D = 1$ V, $v_S = 0$ V, and $v_B = -5$ V, the first (linear) term on the RHS of (5.130) is 20 μA/V, while the nonlinear second term is around -5.4 μA/V, far from being negligible. It is of course possible to calculate also the higher-order terms in the Taylor-series expansion, but these turn out to be very small,[49] about two orders of magnitude below those included in (5.129). Hence, we shall use this relation as the basis for our discussions from now on.

As a simple example of the application of the MOSFET resistor,[49] consider an RC integrator (Fig. 5.73a). If the resistor is replaced by a MOSFET (Fig. 5.73b), then using (5.129)

$$v_{\text{out}} = -\frac{1}{C} \int_{-\infty}^{t} i_D(\tau) \, d\tau$$

$$= -\frac{2k(v_G - V_T)}{C} \int_{-\infty}^{t} v_{\text{in}}(\tau) \, d\tau + \frac{mk}{C} \int_{-\infty}^{t} v_{\text{in}}^2(\tau) \, d\tau \quad (5.131)$$

results. Clearly, the first term on the RHS represents the ideal response, and the second the nonlinear distortion. For usual values and an input signal of 2 V peak-to-peak, the total harmonic distortion (THD) exceeds 7%.[49] This is unacceptable in most applications.

A conceptually simple way[49] to remedy this situation is to use a fully differential balanced circuit for the integrator (Fig. 5.73c). The input terminals of the op-amp will (due to the negative dc feedback necessarily provided somewhere else in the circuit) be at the same common-mode potential v_x. Hence the output terminals will be at the voltages

$$v_+ = v_x - \frac{1}{C} \int_{-\infty}^{t} i_+ \, d\tau = v_x - \frac{1}{2RC} \int_{-\infty}^{t} v_{\text{in}} \, d\tau + \frac{mk}{4C} \int_{-\infty}^{t} v_{\text{in}}^2 \, d\tau$$

and $\qquad (5.132)$

$$v_- = v_x - \frac{1}{C} \int_{-\infty}^{t} i_- \, d\tau = v_x + \frac{1}{2RC} \int_{-\infty}^{t} v_{\text{in}} \, d\tau + \frac{mk}{4C} \int_{-\infty}^{t} v_{\text{in}}^2 \, d\tau.$$

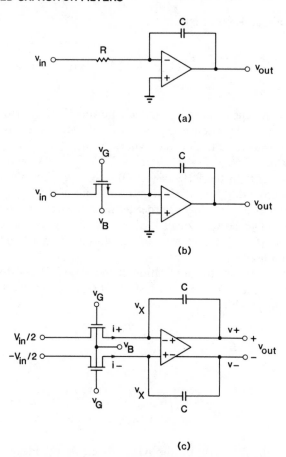

(a)

(b)

(c)

FIGURE 5.73. Active-RC integrators: (a) integrator with conventional resistor; (b) integrator using MOSFET; (c) differential integrator using MOSFETs.

Here, the usual notation

$$R \triangleq \frac{1}{2k(v_G - V_T)} \tag{5.133}$$

has been introduced for the linear resistance being realized.

Since $v_{\text{out}} = v_+ - v_-$, the input–output relation

$$v_{\text{out}} = \frac{1}{RC} \int_{-\infty}^{t} v_{\text{in}}(\tau) \, d\tau \tag{5.134}$$

results. In this expression, the terms representing nonideal effects have canceled, and the resulting input–output relation is the same as for the ideal integrator

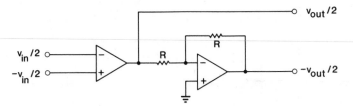

FIGURE 5.74. Composite op-amp with balanced output voltages, constructed from two single-ended-output op-amps.

of Fig. 5.73*a*. (As a discussion including higher-order terms readily shows,[49] all terms containing *even* powers of v_{in} cancel due to the differential configuration.)

When the integrator is to function as a part of an active-RC filter, its two output voltages are used as the balanced input voltages of some other integrators. Hence, $v_+ = -v_-$ must be enforced. This can be achieved, for example,[49] by using the composite op-amp circuit of Fig. 5.74.

In Ref. 49, a fifth-order Chebyshev low-pass filter, designed using such fully balanced integrators with composite op-amps, is described. It had a 3-kHz passband limit and a 0.1-dB ripple. It was fabricated in CMOS technology. For better matching of the time constants, all resistors (i.e., MOSFETs) were designed to be identical, and the time constant ratios were realized only by capacitance ratios. For reasonably small-signal voltages (say, up to 3 V peak-to-peak), the resulting THD was below -50 dB, adequate for many applications. The dynamic range (~ 95 dB) was excellent, due to the noise cancellation in the differential circuit.

Even though the feasibility of MOS integrated active-RC filters was demonstrated by this circuit, it also revealed some difficulties. In order to obtain the large time constants required by the low-frequency (3 kHz) cutoff, and still be able to use practical capacitance values, the common value of the MOSFET resistances had to be chosen quite high, several MΩ large. This necessitated using a very low aspect ratio, $W/L = 0.01$, for these devices; $W = 4$ μm and $L = 400$ μm were the dimensions chosen. For such sizes, the parasitic capacitances to the substrate and to the gate affect the performance significantly. Computer-aided optimization of the op-amp feedback capacitances could reduce this effect, but the achievable passband ripple grew to 0.2 dB.

A drawback of the differential circuit technique is that it requires many extra components; the circuit is about twice as complicated (and hence requires nearly twice as much chip area) as a single-ended one. An alternative approach is hence to carry out the nonlinearity correction individually for each simulated resistor. Let the gate voltage of the MOSFET be made equal to

$$v_G = v_c + m\frac{v_D + v_S}{2} \qquad (5.135)$$

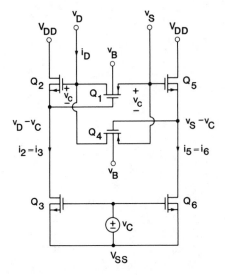

FIGURE 5.75. CMOS circuit for realizing a floating linear resistor.

by using some appropriate auxiliary circuitry. Here, v_c is a control voltage, needed to adjust the value of the resistance realized. Substituting this value of v_G into (5.129),

$$i_D = k\left[2\left(v_c + m\frac{v_D + v_S}{2} - V_T\right)(v_D - v_S) - m\left(v_D^2 - v_S^2\right)\right]$$

$$= 2k(v_c - V_T)(v_D - v_S) \tag{5.136}$$

results. Thus, the device now behaves as an ideal linear resistor of value

$$R = \frac{1}{2k(v_c - V_T)}. \tag{5.137}$$

To achieve the value v_G given by (5.135), several strategies can be used. Since for appropriate substrate bias v_B the value of m is usually in the range of $1.05 \sim 1.1$, the approximation $m \simeq 1$ can often be used. Then, $v_G \cong v_c + (v_D + v_S)/2$ is to be generated. This can be achieved by using the circuit[52] of Fig. 5.75. The circuit functions as follows: Devices Q_1, Q_2, and Q_3 are matched to Q_4, Q_5, and Q_6, respectively. The channels of Q_1 and Q_4 are connected in parallel, and form the simulated resistor R. Assuming for simplicity also that Q_2 is matched to Q_3, and that both devices operate in saturation, the common current $i_2 = i_3$ forces the source voltage of Q_2 (via the $Q_1 - Q_2$ feedback loop) to have the value $v_D - v_c$. Similarly, if Q_5 and Q_6 are matched devices, then the source voltage of Q_5 will be $v_S - v_c$. Since Q_1 and Q_4 are matched devices connected in parallel, they can be considered as a single device with an

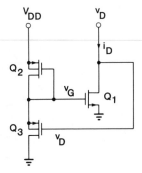

FIGURE 5.76. CMOS circuit for realizing a grounded linear resistor.

aspect ratio $W/L = 2(W/L)_1 = 2(W/L)_4$ and a gate voltage

$$v_G = \frac{v_{G_1} + v_{G_4}}{2} = \frac{v_D + v_S}{2} - v_c. \tag{5.138}$$

This (apart from an unimportant change in the sign of v_c) is the required gate voltage. Thus, this circuit realizes a floating linear resistor whose value can be changed with the control voltage v_c.

Because of the approximations made in the device model, and also due to the inevitable mismatches in the realization, there will remain some nonlinearity in the resistance actually realized. Its value can be expected to be around $\pm 2\%$ for $|v_D - v_S| \leq 4$ V.[52]

Often, all resistors needed in an active filter have one grounded terminal (usually at virtual ground); this is the case, for example, for the active-RC filter discussed in Ref. 49. Then, the circuit shown in Fig. 5.76 (which can be regarded as a modified version* of the $Q_1 - Q_2 - Q_3$ portion of the circuit of Fig. 5.75) can be used. This circuit functions as follows. Both Q_2 and Q_3 operate in saturation, and carry equal currents. Hence, by the appropriate formula in Table 3.2 (neglecting channel-length modulation effects) we can write

$$-i_3 = k_3(v_D - v_G - V_{TP})^2 = -i_2 = k_2(v_G - V_{DD} - V_{TP})^2. \tag{5.139}$$

This equation can be solved for v_G. The result is

$$v_G = \frac{v_D}{1 + \alpha} + \frac{\alpha V_{DD}}{1 + \alpha} + \frac{1 - \alpha}{1 + \alpha}|V_{TP}|. \tag{5.140}$$

Here, the notation

$$\alpha = \sqrt{\frac{k_2}{k_3}} = \sqrt{\frac{(W/L)_2}{(W/L)_3}} \tag{5.141}$$

*The circuit of Fig. 5.75 can only be realized using p-well CMOS, while that of Fig. 5.76 requires n-well CMOS technology.

was used. Defining the control voltage of the resistor as

$$v_c = \frac{\alpha V_{DD} + (1 - \alpha)|V_{TP}|}{1 + \alpha} \tag{5.142}$$

and comparing (5.140) with (5.135) for $v_S = 0$ gives

$$\alpha = \frac{2 - m}{m}. \tag{5.143}$$

So, by (5.141) and (5.128), the device sizes should satisfy

$$\frac{(W/L)_2}{(W/L)_3} = \alpha^2 = \left(\frac{2 - m}{m}\right)^2 = \left(\frac{2\sqrt{2|\phi_p| + |v_B|} - \gamma}{2\sqrt{2|\phi_p| + |v_B|} + \gamma}\right)^2. \tag{5.144}$$

For large $|v_B|$, $m \approx 1$, and we have $(W/L)_2 \approx (W/L)_3$. However, if $v_B \approx 0$, then m is typically around 1.3, and $(W/L)_2 \approx 0.35(W/L)_3$ is optimal.

The value of the linear resistor realized by the circuit is, by (5.137), (5.142), and (5.143), given by

$$R = \frac{1}{2k(v_c - V_{Tn})} = \frac{1}{2k\{[\alpha V_{DD} + (1 - \alpha)|V_{TP}|]/(1 + \alpha) - V_{Tn}\}}$$

$$= \frac{1}{2k[(1 - m/2)V_{DD} + (m - 1)|V_{TP}| - V_{Tn}]}. \tag{5.145}$$

The value of R may thus be controlled by changing V_{DD}.*

Since we used in our derivations the current–voltage relation (5.127) which is valid only in the triode region of Q_1, the condition $v_G - V_{Tn} > v_D$ must hold; hence, by (5.140) and (5.143), the circuit of Fig. 5.76 requires for linearity the condition

$$0 \le v_D < V_{DD} + \left(\frac{1}{\alpha} - 1\right)|V_{TP}| - \left(\frac{1}{\alpha} + 1\right)V_{Tn}$$

$$= V_{DD} + \frac{m - 1}{1 - m/2}|V_{TP}| - \frac{V_{Tn}}{1 - m/2}. \tag{5.146}$$

For typical values ($V_{DD} = 5$ V, $|V_{TP}| = V_{Tn} = 1$ V, $m = 1.3$) $v_D < 3$ V results. This allows a fairly large dynamic range for the circuit.

To obtain a *floating* MOS resistor, the circuit of Fig. 5.77 can be used. In this circuit, equating the currents of Q_2 and Q_3 (and assuming that they are

---

*Alternatively, V_{DD} may be fixed and a floating control voltage source v_z connected between the gate and the source of Q_2. Note that v_z will not draw any current, and hence is easily realized.

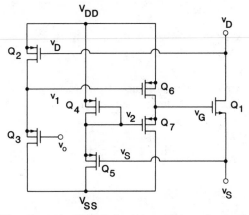

FIGURE 5.77. Alternative CMOS circuit for realizing a floating linear resistor.

both in saturation) gives

$$v_1 = \alpha V_{DD} - \alpha v_D + (1 - \alpha)|V_{TP}| + v_0 \qquad (5.147)$$

where α is defined, as before, by (5.141). Similarly, equating the saturation-region currents of Q_4 and Q_5 gives

$$v_2 = \frac{v_S}{1 + \beta} + \frac{\beta V_{DD}}{1 + \beta} + \frac{1 - \beta}{1 + \beta}|V_{TP}| \qquad (5.148)$$

where $\beta \triangleq \sqrt{(W/L)_4/(W/L)_5}$. Finally, equating the saturation-region currents of Q_6 and Q_7 gives, for $(W/L)_6 = (W/L)_7$,

$$v_1 - V_{DD} = v_2 - v_G. \qquad (5.149)$$

Hence, from (5.147)–(5.149), we obtain

$$v_G = V_{DD} + v_2 - v_1 = \left(1 + \frac{\beta}{1 + \beta} - \alpha\right)V_{DD} - v_0 + \frac{v_S}{1 + \beta}$$

$$+ \alpha v_D + \left(\frac{1 - \beta}{1 + \beta} - (1 - \alpha)\right)|V_{TP}|. \qquad (5.150)$$

Choosing $\alpha = 1/(1 + \beta) = m/2$, (5.150) becomes

$$v_G = (2 - m)V_{DD} + \frac{m}{2}(v_S + v_D) + \left(\frac{3m}{2} - 2\right)|V_{TP}| - v_0. \qquad (5.151)$$

Comparison with (5.135) shows that the linearity condition is now satisfied.

The effective control voltage is

$$v_c = (2 - m)V_{DD} + \left(\frac{3m}{2} - 2\right)|V_{TP}| - v_0.$$ (5.152)

The value of R can be then obtained from (5.137); it can be tuned, for example, by changing v_0. The device sizes must satisfy

$$\frac{(W/L)_2}{(W/L)_3} = \alpha^2 = \left(\frac{m}{2}\right)^2 = \left(\frac{1}{2} + \frac{\gamma}{4\sqrt{2|\phi_p| + |v_B|}}\right)^2$$ (5.153)

and

$$\frac{(W/L)_4}{(W/L)_5} = \beta^2 = \left(\frac{2}{m} - 1\right)^2 = \left(\frac{2\sqrt{2|\phi_p| + |v_B|} - \gamma}{2\sqrt{2|\phi_p| + |v_B|} + \gamma}\right)^2.$$ (5.154)

For $v_B \approx 0$ and $m \approx 1.3$, the ratios $(W/L)_2/(W/L)_3$ and $(W/L)_4/(W/L)_5$ will be in the 0.3–0.5 range.

Linearization of the floating MOSFET channel resistance can also be achieved by adding the geometric mean of v_D and v_S to both v_G and v_B.[53, 54] To see why, we note that for a MOSFET with terminal voltages v_{G_0}, v_{B_0}, and

$$v_{D_0} = v_{S_0} = \frac{v_D - v_S}{2},$$ (5.155)

the distortion term in i_D, given by $km(v_{D_0}^2 - v_{S_0}^2)$ in (5.129), is zero. Adding $(v_D + v_S)/2$ to all terminal voltages does not, of course, alter i_D and thus the linearity. It does, however, change v_{D_0} to v_D, v_{S_0} to v_S, and v_{G_0} and v_{B_0} to $v_G + (v_D + v_S)/2$ and $v_B + (v_D + v_S)/2$, respectively (Fig. 5.78a). Thus, this circuit implements a linear resistor. To realize this scheme, the MOSFET must be embedded in a well whose potential may be varied at the signal rate. A possible implementation,[54] which requires four level shifters supplementing two parallel-connected MOSFETs, is shown in Fig. 5.78b.

For NMOS technology, the circuit of Fig. 5.79a can provide a floating resistor.[51, 55] Of the two parallel-connected depletion-mode devices, for $v_D > v_S$, Q_1 will always be in the triode region. This is because $v_{GS1} - V_T = v_D - v_S + |V_T| > v_{DS1} = v_D - v_S$, since V_T is negative for a depletion-mode NMOS transistor. Q_2 will, however, be in the triode region only if $v_D - v_S < |V_T|$. In that case, $v_{GS2} - V_T = |V_T| > v_{DS2} = v_D - v_S$, as required. Thus, for $0 < v_D - v_S < |V_T|$, both devices will be in their triode regions. Because of the symmetry of the circuit, this also holds if $0 < v_S - v_D < |V_T|$. In conclusion, for $|v_D - v_S| < |V_T|$, the drain currents of both Q_1 and Q_2 will satisfy (5.129).

Assume next that $v_D > v_S$ always holds; this will be the case, for example, if the resistor feeds a virtual ground and all signal voltages are positive. Then the

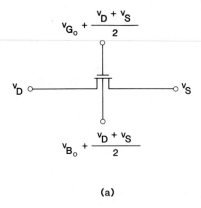

(a)

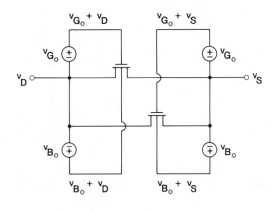

(b)

FIGURE 5.78.   Floating MOSFET resistor using level shifters.

resistor current is

$$i_D = i_{D_1} + i_{D_2} = k_1\left[2(v_D - v_S + |V_T|)(v_D - v_S) - m(v_D^2 - v_S^2)\right]$$

$$+ k_2\left[2|V_T|(v_D - v_S) - m(v_D^2 - v_S^2)\right] \qquad (5.156)$$

$$= 2(k_1 + k_2)|V_T|(v_D - v_S) + \left[k_1(2 - m) - k_2 m\right](v_D^2 - v_S^2).$$

The first term on the RHS represents the current of a linear resistor of value

$$R = \frac{1}{2(k_1 + k_2)|V_T|}, \qquad (5.157)$$

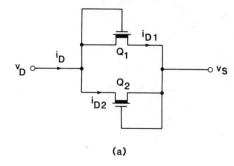

(a)

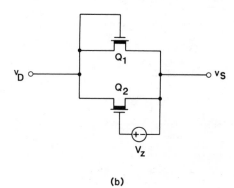

(b)

FIGURE 5.79. NMOS floating resistor realizations: (*a*) basic circuit; (*b*) circuit with extended linear range.

while the second term represents the nonlinear distortion. For

$$\frac{k_2}{k_1} = \frac{(W/L)_2}{(W/L)_1} = \frac{2 - m}{m} = \frac{2\sqrt{2|\phi_p| + |v_B|} - \gamma}{2\sqrt{2|\phi_p| + |v_B|} + \gamma} \qquad (5.158)$$

the nonlinear term vanishes. For typical values ($V_T = -3$ V, $v_B = 0$, $\gamma = 0.4$ $V^{1/2}$, $|\phi_p| = 0.3$ V), (5.158) gives $k_2/k_1 \approx 0.6$. Experiments show[51] that using this design technique, resistors in the $1 \sim 50$ kΩ range can readily be realized with a nonlinearity below 0.5%. The resulting value of the resistance $R = m/(4k_1|V_T|)$ can be controlled by choosing $(W/L)_1$ appropriately.

If the resistor must process bipolar voltages, that is, if $v_D < v_S$ can also occur, then $k_2 = k_1$ minimizes the worst case distortion. Defining the relative nonlinearity NL as the ratio of the nonlinear and linear terms in i_D, from (5.156),

$$NL = \left| \frac{k_1(2 - 2m)(v_D^2 - v_S^2)}{4k_1|V_T|(v_D - v_S)} \right| = \left| (1 - m)\frac{v_D + v_S}{2V_T} \right|. \qquad (5.159)$$

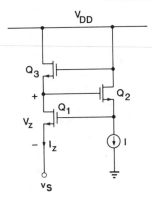

FIGURE 5.80. The realization of the floating voltage source V_z for the floating NMOS resistor.

Thus, if the magnitude of the common-mode voltage $(v_D + v_S)/2$ equals $|V_T|$, the nonlinearity is $NL = |1 - m| = \gamma/(2\sqrt{2|\phi_p|} + |v_B|)$. Depending on the process parameters and v_B, $NL = 0.05 \sim 0.3$ may result.

For the case of unipolar operation $v_D > v_S$, the linear range is $0 < v_D - v_S < |V_T|$. This may not be adequate for the large-signal operation of the circuit. Then, a voltage source V_z can be inserted into the gate lead of Q_2, as illustrated in Fig. 5.79b. The effect is to replace $|V_T|$ for Q_2 by $|V_T| + V_z$, and hence to extend the range of linear operation. The remaining large-signal distortion is (for small V_z) a rapidly decreasing function of V_z.[51] Linear operation was achieved for $v_S = 0$ and $0 \le v_D = 10$ V, with a nonlinearity less than 0.3% of the full-scale value.[51] In Ref. 51, an NMOS integrated noninverting amplifier with an accurately controlled gain of 16 is also described. The circuit used two of the resistors shown in Fig. 5.79b, as input and feedback resistors. The floating voltage sources were realized by the circuit shown in Fig. 5.80. It can readily be shown that the voltage generated by this circuit is

$$V_z = V_{T_1} + V_{T_2} + \sqrt{\frac{I_z}{k_1}} + \sqrt{\frac{I}{k_2}} \qquad (5.160)$$

(Problem 5.30). Here, it is assumed that all devices operate in saturation. By choosing $(W/L)_1 \gg (W/L)_2$ and $I \gg I_z$, the third term can be made negligible compared to the fourth, and the internal impedance $\partial V_z/\partial I_z$ of the floating battery will be small.*

The remaining problem to be discussed is the generation of the control voltage v_c, which [as shown, e.g., in Eq. (5.137)] determines the exact value of the linearized resistance R. There are many techniques for deriving v_c, differing not only in details, but also in principle. Among the available alternatives are

---

*Note that if the resistor of Fig. 5.79b is connected to a virtual ground, so that $v_S = 0$, then V_z can simply be grounded, and the floating supply of Fig. 5.80 is not necessary.

the following. One can:

1. Provide a separate tuning circuit for each resistor; that is, derive individual v_c values for each resistor.
2. Match all resistors to each other and to a reference resistor using a single v_c value.
3. Match all resistors to each other and to those in a reference filter section, using a single v_c value.

Other alternatives also exist. In addition, one has the option to:

1. Keep the resistor(s) connected in the filter circuit while being tuned.
2. Use two resistors for every one needed in the active-RC filter: while one is tuned, the other is connected to the filter.
3. Use two filters; while one is being tuned, the other is processing signals.

Some typical tuning circuits based on resistance matching are shown in Fig. 5.81. Consider the circuit of Fig. 5.81a. Assume that the voltage-controlled resistor $R(v_c)$ satisfies Eq. (5.137), reproduced below:

$$R(v_c) = \frac{1}{2k(v_c - V_T)}. \tag{5.161}$$

The KCL for node (A) gives, for ideal op-amp,

$$V\left(\frac{1}{R(v_c)} - \frac{1}{R_r}\right) + C\frac{dv_c}{dt} = 0. \tag{5.162}$$

Hence, using (5.161),

$$\frac{dv_c}{dt} = -\frac{V}{C}\left(\frac{1}{R(v_c)} - \frac{1}{R_r}\right) = -\frac{2kV}{C}\left[v_c - \left(V_T + \frac{1}{2kR_r}\right)\right]. \tag{5.163}$$

Thus, if $1/R(v_c) > 1/R_r$, then v_c will decrease. This, by (5.161), will reduce $1/R(v_c)$. Once $R(v_c) = R_r$ is reached, dv_c/dt becomes zero, and v_c is held constant. Clearly, the circuit adjusts the value of $R(v_c)$ to that of the reference resistor R_r (which can be an external device). The time constant of v_c during tuning is $C/2kV$, which can be reduced by making C small, V large, and the MOSFETs realizing $R(v_c)$ wide.

The circuit just described requires two accurately matched voltage supplies $\pm V$. This can be avoided by using the concept of Fig. 5.81b, which has the SC realization[56] shown in Fig. 5.81c. As explained earlier in the chapter, the inverting switched capacitor C_r is approximately equivalent to a negative resistor $R_r = -T/C_r = -1/f_cC_r$, where $f_c = 1/T$ is the clock frequency of ϕ_1

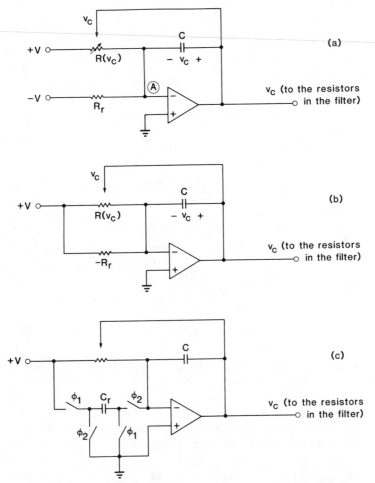

FIGURE 5.81.   Tuning circuits for MOSFET resistors: (a) circuit using matched voltage supplies of values $\pm V$; (b) a hypothetical circuit with negative resistor; (c) circuit with a switched-capacitor realization of the negative resistor.

and ϕ_2. Note that in this circuit v_c contains f_c and its harmonics in addition to the desired average (dc) value needed to tune the controlled resistors. If these resistors are in the filter during tuning, then these high-frequency components have to be filtered out of v_c. Note also that for accurate $R(v_c)$ the circuit of Fig. 5.81c requires a precision clock and an accurate (possibly external) capacitor C_r, instead of the precision voltages and resistor needed by the circuit of Fig. 5.81a. If C_r is an *on-chip* capacitor, then the on-chip *time constant* $R_i C_i$ can be matched to $T = 1/f_c$.

A different tuning principle can be based on detecting the phase, rather than the amplitude, of some reference clock signal v_R. A possible system[56] is shown

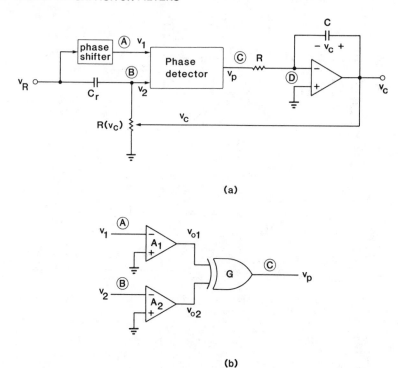

(a)

(b)

FIGURE 5.82. Tuning circuit based on phase detection: (*a*) block diagram; (*b*) a phase detector circuit; (*c*) phase detector input and output waveforms.

in Fig. 5.82*a*. Here, the phase detector generates an output voltage v_p whose average (dc) value v_{p0} is zero when the phases of its two input voltages v_1 and v_2 differ by 90°; v_{p0} is positive if $\Delta\theta = \angle v_1 - \angle v_2 > 90°$ and negative if $\Delta\theta < 90°$.* Any departure of $\Delta\theta$ from 90° thus generates a dc current charging the active-RC integrator, and hence changing v_c so as to restore the 90° difference between the phases of v_1 and v_2.

A possible circuit for the phase detector[56] is shown in Fig. 5.82*b*. It contains two open-loop op-amps A_1 and A_2 acting as comparators, and an exclusive-OR gate G with symmetrical output voltage levels. A_1 and A_2 provide square-wave output voltages v_{01} and v_{02} from their sine-wave inputs. The output voltage v_p of G is positive if v_{01} and v_{02} differ in sign and negative if their signs agree (Fig. 5.82*c*). The average value v_{p0} of v_p thus depends on the phase difference $\Delta\theta$ between v_{01} and v_{02} (or, equivalently, between v_1 and v_2). For a 90° difference, v_{p0} is zero. Otherwise, a current v_{p0}/R enters the integrator, and v_c changes according to the formula

$$\frac{dv_c}{dt} = -\frac{v_{p0}}{RC} \tag{5.164}$$

*This monotone response is valid only in the vicinity of $\Delta\theta = 90°$ phase difference.

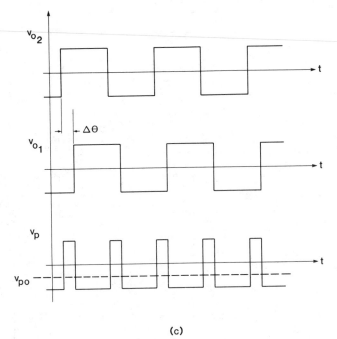

(c)

FIGURE 5.82.   continued.

which can be obtained by writing the KCL for node (D). The change is such as to make the phase difference $\Delta\theta$ equal to 90°. The speed of the tuning is given by (5.164).

In equilibrium, the phase difference between v_1 and v_2 is thus 90°. Then, the locked value of $R(v_c)$ is determined by the condition[56]

$$R(v_c) = \frac{-\tan\theta_R}{\omega_R C_R} \tag{5.165}$$

(Problem 5.31). Here, ω_R is the radian frequency of v_R, and θ_R is the phase angle produced by the phase shifter. Note that if C_R is an on-chip capacitor, then the on-chip time constants $R_i C_i$ can be tuned to the clock frequency ω_R.

It is also possible to use a whole duplicated segment of the filter in the tuning circuit. For example, the active-RC ladder filter discussed earlier in this chapter contains as its basic building blocks two-integrator loops. Hence, by forming a voltage-controlled oscillator (VCO) from a two-integrator loop and tuning it to an external clock frequency, the correct v_c can be derived. As an example, Fig. 5.83 illustrates the principle for a fifth-order ladder filter.[57] Here, each block I_k contains an active-RC integrator with a transfer function ω_k/s; that is, the output voltage of each block is

$$V_{\text{out}}(s) = \frac{\omega_k}{s}\left[V_{\text{in}}^+(s) - V_{\text{in}}^-(s)\right]. \tag{5.166}$$

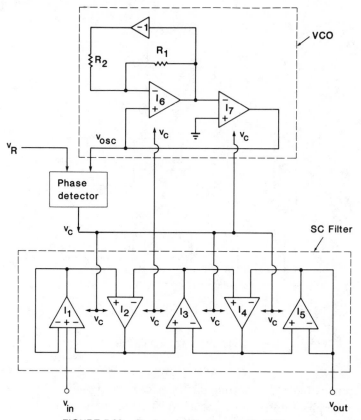

FIGURE 5.83. Tuning circuit using on-chip VCO.

In equilibrium, the VCO oscillates at a frequency $\omega_{osc} = \sqrt{\omega_6 \omega_7} = \omega_R$ (Problem 5.32). The corresponding value of v_c also tunes the integrators $I_1 \sim I_5$ (matched to I_6 and I_7 by properly choosing the corresponding capacitances) to have their appropriate gain constants $\omega_1 \sim \omega_5$. The exact values of R_1 and R_2 do not affect ω_{osc}, and hence need not be precisely controlled.

As mentioned earlier, there exist many different strategies and circuits for the tuning process. The few circuits described in Figs. 5.81–5.83 represent only some basic approaches. The interested reader is referred to Refs. 49 to 57, as well as Refs. 1 to 9 quoted in the recent paper[49] of Banu and Tsividis, for additional information.

PROBLEMS

5.1. Plot schematically all voltages and currents in the circuit of Fig. 5.4. What is the mathematical expression for $i_2(t)$?

5.2. Prove that the transfer function of the circuits of Fig. 5.7 is given by Eq. (5.25). Plot $v_{out}(t)$ for a sine-wave input voltage.

5.3. Derive the transfer function of the circuit of Fig. 5.6a if a stray capacitance C_A exists between node Ⓐ and ground.

5.4. Derive the transfer function of the circuit of Fig. 5.7a if a stray capacitance C_A exists between node Ⓐ and ground.

5.5. Derive the transfer functions $H(z) = \Delta Q/V$ given in (5.39). [*Hint:* Express $\Delta q(t_n)$ in terms of $v(t_n)$ and $v(t_{n-1})$; use z-transformation.]

5.6. Derive the transfer function of (5.40) from the block diagram of Fig. 5.11. (*Hint:* Express V_1 in terms of V_{in} and V_{out}, and V_{out} in terms of V_{in} and V_1!)

5.7. Show that the transfer function of the SC biquad becomes that given in (5.42) when C_1''' is added (Fig. 5.12) and C_1' is set to zero.

5.8. Derive the transfer function of (5.46) from the block diagram of Fig. 5.14.

5.9. Show that by including the branch of Fig. 5.12 in the circuit of Fig. 5.13c, and eliminating C_1', the transfer function becomes that given in (5.48).

5.10. Determine the element values of Fig. 5.15c, for both parenthesized and original clock phases, in terms of the coefficients of $H(z)$ given in (5.50).

5.11. Find the transfer function of the circuit of Fig. 5.15c if (a) the SC branch containing C_1' is replaced by the circuit of Fig. 5.2a and (b) the SC branch containing C_2 is replaced by the branch of Fig. 5.2a (left and right terminals interchanged).

5.12. Show that the approximation error due to using $z \cong 1 + j\omega T$ in the design of the central integrator in Fig. 5.18e can be represented by attributing a negative quality factor $-Q = -2/\omega T$ to L_2 in Fig. 5.18a

5.13. Calculate the effect of a stray capacitance δC_s in parallel with the shunt capacitor C_s in Fig. 5.27a. Show that the change in the loss is, to a very good approximation, 0.08686 dB/% change in C_s and is nearly independent of the frequency. (*Hint:* Calculate $|\Delta Q_{in}|^2$ with and without δC_s. The change in loss is then $10 \log_{10}[|\Delta Q_{in}|^2_{\delta C_s \neq 0}/|\Delta Q_{in}|^2_{\delta C_s = 0}]$.)

5.14. Derive the design relations for the *input* section of a bilinear SC bandpass ladder filter, as given in Fig. 5.33.

5.15. Derive the design relations for the *output* section of a bilinear SC bandpass ladder filter, as given in Fig. 5.34.

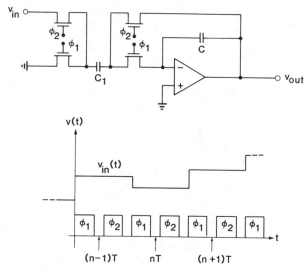

FIGURE 5.84. A first-order SC circuit (used in Problem 5.18).

5.16. Verify the design equations of (5.89) and (5.92) for the simulation of the circuit of Fig. 5.35a.

5.17. Derive the design equations for the simulation of the section of Fig. 5.35b.

5.18. For the circuit shown in Fig. 5.84 find: (a) the difference equation between v_{in} and v_{out}; (b) the transfer function $V_{out}(z)/V_{in}(z)$; (c) the frequency response $|H(e^{j\omega T})|$; (d) for what values of C_1/C the circuit is stable; (e) what function the circuit will perform if $C_1 = C$; (f) what the steady-state output voltage is if $v_{in} = 0$ and the op-amp has an input offset voltage v_{os}.

5.19. Show that the circuit equivalences[37] of Fig. 5.45 hold.

5.20. Replace the SC input branch containing C_1' in the circuit of Fig. 5.15c by the simulated resistor of Fig. 5.2a. (a) Where is the zero of the new circuit? (b) Where does the zero move if a stray capacitance C_s is in parallel with C? (c) Let $C_s \ll C$. Show that the effect of C_s is (to a first approximation) an additional flat gain C_s/C (in nepers). (*Hint:* Calculate the extra charge entering the virtual ground due to C_s!) (d) Repeat the above calculations if the feedback SC branch containing C_2 is replaced by the branch of Fig. 5.2a, with its left and right terminals interchanged. Where is the pole of $H(z)$ without and with C_s? How is the gain affected by C_s? (e) Using second-order approximations, show that the gain or loss variation due to C_s is $C_s/C + (C_s/C)^2 [1 + (\pi f/f_c)^2]/2$ in nepers. Plot this function for $C_s = 0.1C$ and $|f| \le f_c/10$. What conclusions can you draw?

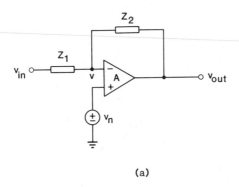

(a)

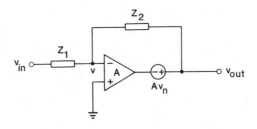

(b)

FIGURE 5.85.   Two equivalent noisy circuits (for Problem 5.21).

5.21. Show that the identity illustrated in Fig. 5.85 holds, and hence the noise source v_n in Fig. 5.85b is a valid representation of the noise generated by the op-amp. [*Hint:* Show that the relations between v_{in}, v_{out}, and the voltage v at the inverting op-amp input terminal are the same for the two circuits. Hence (for a given v_{in}) v and v_{out} are the same in the two circuits.]

5.22. Find the $\Delta Q/V$ transfer function of the circuit of Fig. 5.38.

5.23. Find the transfer function $H(z)$ of the stage of Fig. 5.47a. Assume that the waveforms are as shown in Fig. 5.27c.

5.24. Perform the scaling of the capacitance values in the last stage of the filter shown in Fig. 5.48. The initial values are $\bar{C}_1'' = 1$, $\bar{C}_1 = 0.30357$, $\bar{C}_2 = \bar{C}_3 = 0.12436$, $\bar{C}_4 = 0.12939$, $\bar{C}_A = \bar{C}_B = 1$. The peak values of the op-amp output voltages are $V_{p4} \simeq 503.57$ and $V_{p5} \simeq 230.14$; the desired values are $V_{p4} = V_{p5} = 1$.

5.25. Prove that the transfer function of the passive SCF shown in Fig. 5.56 is given by Eq. (5.116).

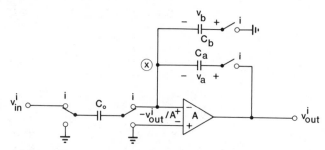

FIGURE 5.86. Charge transfer in N-path filters (for Problem 5.28).

5.26. Show that the z_p-domain transfer function of a path filter of the 4-path circuit shown in Fig. 5.60 is given by Eq. (5.119).

5.27. Show that the input–output relations of both circuits shown in Fig. 5.70 are given by Eq. (5.123), and that these circuits (with the indicated timing of the sampling instances t_s) are hence equivalent.

5.28. The circuit of Fig. 5.86 illustrates the charge transfer in an N-path filter stage containing a finite-gain op-amp. Assume that C_a is charged to a voltage v_a^- before the switches close, and C_b to a voltage v_b^-. Calculate the following: (a) the new output voltage v_{out}^i, after the switches marked by i have closed; (b) the new values of the voltages across all capacitors; (c) the percentage error in v_{out}^i due to the finite op-amp gain A, if $C_a = C_b = 5C_0$ and $A = 10^3$.

5.29. Prove formulas (5.147)–(5.152) for the circuit of Fig. 5.77. What is the range of v_D and v_S for which the linearity conditions are satisfied?

5.30. The circuit of Fig. 5.80 can be used to generate a floating voltage V_z. Assuming that all devices are in saturation, calculate: (a) the exact expression for V_z in terms of k_1, k_2, k_3, v_S, V_{DD}, I, and the processing parameters; (b) the internal resistance $\partial V_z/\partial I_z$ of the source voltage; (c) the sensitivity of V_z to v_S and V_{DD}; (d) in view of the above formulas, which parameters should be chosen large and which should be chosen small, if we wish V_z to be determined by k_1, k_2, k_3, and I, but insensitive to all other parameters?

5.31. Prove Eq. (5.165) for the circuit of Fig. 5.82.

5.32. Show that the poles (i.e., natural modes) of the VCO shown in Fig. 5.83 satisfy the equation

$$s_{1,2}^2 + \frac{R_1 - R_2}{R_1 + R_2}\omega_6 s_{1,2} + \omega_6\omega_7 = 0.$$

What are the criteria for oscillation?

REFERENCES

1. G. C. Temes and J. W. LaPatra, *Introduction to Circuit Synthesis and Design*, McGraw-Hill, New York, 1977, Chap. 7.

2. R. W. Brodersen, P. R. Gray, and D. A. Hodges, *Proc. IEEE*, **67**, 61–75 (1979).

3. M. Sharif-Bakhtiar, A New Technique for High Frequency Monolithic Filter Design, Ph.D. dissertation, UCLA (1982).

4. D. L. Fried, *IEEE J. Solid-State Circuits*, **SC-7**, 302–304 (1972).

5. W. Poschenrieder, *Proc. NTG Symp.*, Stuttgart, 220–237 (1966).

6. A. Fettweis, in G. Biorci (Ed.), *Network and Switching Theory*, Academic, New York, 1968, pp. 382–446.

7. R. Boite and J. V. P. Thiran, *IEEE Trans. Circuit Theory*, **CT-15**, 447–454 (1968).

8. H. Weinrichter, *Arch. Elektr. Übertrag.*, **26**, 293–305 (1972).

9. J. T. Caves, M. A. Copeland, C. F. Rahim, and S. D. Rosenbaum, *IEEE J. Solid-State Circuits*, **SC-12**, 592–599 (1977).

10. B. J. Hosticka, R. W. Broderson, and P. R. Gray, *IEEE J. Solid-State Circuits*, **SC-12**, 600–608 (1977).

11. C. Rahim, M. A. Copeland, and C. H. Chan, *IEEE J. Solid-State Circuits*, **SC-13**, 906–909 (1978).

12. G. C. Temes and I. A. Young, *Electron. Lett.*, **14**, 287–288 (1978).

13. G. C. Temes, *Electron. Lett.*, **14**, 361–362 (1978).

14. K. Martin and A. S. Sedra, *Electron. Lett.*, **15**, 365–366 (1979).

15. R. Gregorian and W. Nicholson, *IEEE J. Solid-State Circuits*, **SC-14**, 970–980 (1979).

16. B. White, G. Jacobs, and G. Landsburg, *IEEE J. Solid-State Circuits*, **SC-14**, 991–997 (1979).

17. P. E. Fleischer, A. Ganesan, and K. R. Laker, *Electron. Lett.*, **17**, 929–931 (1981).

18. A. S. Sedra and P. O. Brackett, *Filter Theory and Design: Active and Passive*, Matrix, Champaign, IL, 1978, Chap. 9.

19. K. Martin, *IEEE Trans. Circuits and Systems*, **CAS-27**, 237–244 (1980).

20. K. Martin and A. S. Sedra, *IEEE Trans. Circuits and Systems*, **CAS-27**, 469–475 (1980).

21. R. Gregorian, K. W. Martin, and G. C. Temes, *Proc. IEEE*, **71**, 941–966 (1983).

22. G. M. Jacobs, D. J. Allstot, R. W. Brodersen, and P. R. Gray, *IEEE Trans. Circuits & Systems*, **CAS-25**, 1014–1021 (1978).

23. H. J. Orchard, *Electron. Lett.*, **2**, 224–225 (1966).

24. G. C. Temes and H. J. Orchard, *IEEE Trans. Circuits & Systems*, **CAS-24**, 567–571 (1977).

25. A. Fettweis, D. Herbst, B. Hoefflinger, J. Pandel, and R. Schweer, *IEEE Trans. Circuits & Systems*, **CAS-27**, 527–528 (1980).

26. B. Hosticka and G. Moschytz, *IEEE Trans. Circuits & Systems* **CAS-27**, 569–573 (1980).

27. J. A. Nossek and G. C. Temes, *IEEE Trans. Circuits & Systems* **CAS-27**, 481–491 (1980).

28. G. C. Temes, *IRE Trans. Circuit Theory*, **CT-9**, 385–400 (1962).

29. M. L. Blostein, *IRE Trans. Circuits Theory*, **CT-14**, 21–25 (1967).

30. K. Haug, *Arch. Elektr. Ubertragung*, **35**, 279–287 (1981).

31. T. C. Choi and R. W. Brodersen, *IEEE Trans. Circuits & Systems*, **CAS-27**, 545–552 (1980).

32. P. R. Gray, unpublished lecture, IEEE International Conference on Circuits and Systems, Chicago, Illinois, April 27–29, 1981.

33. R. Gregorian and S.-C. Fan, Proc. 15th Asilomar Conf. on Circuits, Systems & Computers, 281–284 (1981).

34. T. H. Hsu, Improved Design Techniques for Switched-Capacitor Ladder Filters, Ph.D. dissertation, UCLA (1982).

35. M. S. Lee and C. Chang, *IEEE Trans. Circuits & Systems*, **CAS-28**, 265–270 (1981).

36. M. S. Lee, G. C. Temes, C. Chang, and M. B. Ghaderi, *IEEE Trans. Circuits & Systems*, **CAS-28**, 811–822 (1981).

37. P. E. Fleischer and K. R. Laker, *Bell Syst. Tech. J.*, **58**, 2235–2269 (1979).

38. G. Müller, ANT (West Germany), personal communication (1983).

39. M. B. Ghaderi, New Design Techniques for Switched-Capacitor Bandpass Filters, Ph.D. dissertation, UCLA (1981).

40. A. Hedayati, New Techniques for the Design of Bilinear Switched-Capacitor Ladder Filters, Ph.D. dissertation, UCLA (1984).

41. M. B. Ghaderi, J. A. Nossek, and G. C. Temes, *IEEE Trans. Circuits & Systems*, **CAS-29**, 557–572 (1982).

42. L. W. Franks and I. W. Sandberg, *Bell Syst. Tech. J.*, **39** 1321–1350 (1960).

43. J. A. Nossek, *NTZ Archiv*, **3**, 351–358 (1981).

44. M. S. Lee and Ch. Chang, Proceedings of IEEE International Conference on Circuits and Systems, Chicago, Illinois, April 27–29, pp. 166–169 (1981).

45. E. Hebenstreit and R. Schreiber, *Digest of Tech. Papers*, Fourth European Solid State Circuits Conference, Amsterdam, Netherlands, 199–201 (1978).

46. D. J. Allstot and K. S. Tan, Proceedings of IEEE International Conference on Circuits and Systems, Houston, Texas, April 28–30, pp. 313–316 (1980).

47. M. B. Ghaderi, G. C. Temes, and J. A. Nossek, Ref. 44, pp. 519–522.

48. A. Fettweis and H. Wupper, *IEEE Trans. Circuit Theory*, **CT-18**, 403–405 (1971).

49. M. Banu and Y. Tsividis, *IEEE J. Solid-State Circuits*, **SC-18**, 644–651 (1983).

50. R. S. Muller and T. I. Kamins, *Device Electronics for Integrated Circuits*, Wiley, New York, 1977, Sec. 8.1.

51. J. N. Babanezhad and G. C. Temes, *IEEE J. Solid-State Circuits*, **SC-19**, (1984).

52. M. Banu and Y. Tsividis, *Electron. Lett.*, **18**, 678–679 (1982).

53. J. N. Babanezhad, Novel Techniques for Analog MOS Realization of Precision Computational Circuits, Ph.D. Dissertation, UCLA (1985).

54. M. Banu and Y. Tsividis, Proceedings of IEEE International Conference on Circuits and Systems, Newport Beach, California, pp. 602–605 (1983).

55. I. S. Han and S. B. Park, *Proc. IEEE*, **72**, 1655–1657 (1984).

56. R. L. Geiger, P. E. Allen, and D. T. Ngo, *IEEE Trans. Circuits & Systems*, **CAS-29**, 306–315 (1982).

57. K.-S. Tan and P. R. Gray, *IEEE J. Solid-State Circuits*, **SC-13**, 814–821 (1978).

58. R. B. Datar and A. S. Sedra, *IEEE Trans. Circuits & Systems*, **CAS-30**, 888–898 (1983).

Chapter Six ─────────────────────────────────

NONFILTERING APPLICATIONS OF SWITCHED-CAPACITOR CIRCUITS

Although analog filtering applications received the most attention recently among switched-capacitor circuits, there are many other analog signal processing and generation tasks which can be performed using the same fabrication technology and circuit elements as those used in the filters. In this chapter, some simple circuits are discussed which can be applied to carry out such nonfiltering analog signal processing functions. These include voltage amplifiers, digital-to-analog (D/A) as well as analog-to-digital (A/D) converters, voltage comparators, modulators, rectifiers and peak detectors, as well as relaxation and sine-wave oscillators. No attempt was made to include all available circuits, or even all possible areas of applications. Nor did we try to give an in-depth description of such important topics as A/D and D/A signal conversion. Rather, we attempted merely to illustrate the versatility of switched-capacitor circuits, and show how the MOS technology and devices originally developed for digital logic circuits can be utilized to perform such entirely different tasks.

Naturally, in most applications it is advantageous to fabricate the largest possible number of system building blocks on the same chip. Hence, it is to be expected that many more signal processing functions, currently performed by discrete components, will be carried out in the future by novel on-chip MOS circuitry. Thus, the basic concepts described in this chapter are expected to have many future applications.

6.1. SWITCHED-CAPACITOR GAIN STAGES

One of the most common functions in analog signal processing is voltage amplification. The circuit most often used to perform this function is shown schematically in Fig. 6.1. The input–output relation is clearly

$$\frac{V_{out}}{V_{in}} = -\frac{Z_2}{Z_1}. \tag{6.1}$$

If $Z_2 = kZ_1$, where k is a constant, then a fixed gain is achieved. In bipolar technology, Z_1 and Z_2 are usually chosen as precision resistors; in MOS technology, it is preferable (as discussed in Section 5.1) to use a combination of capacitors and switches. The simplest arrangement is to choose Z_1 and Z_2 as fixed (i.e., unswitched) capacitors. However, then the inverting input terminal of the op-amp is floating, and even a minute leakage current will charge the small stray capacitance between this terminal and ground to a dc voltage which saturates the op-amp. Another simple choice for Z_1 and Z_2 is a switched-capacitor branch, such as was shown between the input node and node Ⓒ in Fig. 5.8 of Chapter 5. However, if such a branch realizes the feedback impedance Z_2, then during one-half of every clock cycle the feedback branch is open circuited. During this time, the op-amp may drift into saturation, and latch up.

Since the fixed capacitor provides continuous feedback but allows the leakage current to accumulate in the stray capacitance at the op-amp input terminal, while the switched-capacitor branch discharges the stray capacitance but does not provide continuous feedback, it follows that a parallel combination of both kinds of branches should be used. A circuit[1] in which both Z_1 and Z_2 are realized this way, and the switches at the op-amp input are appropriately combined, is shown in Fig. 6.2. Calculation of the z-domain transfer function (Problem 6.1) reveals that

$$H(z) = \frac{V_{out}(z)}{V_{in}(z)} = -\frac{C_1}{C_2}. \tag{6.2}$$

Thus, $H(z)$ is a frequency-independent constant as desired.

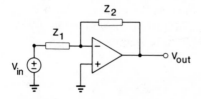

FIGURE 6.1. Voltage amplifier circuit.

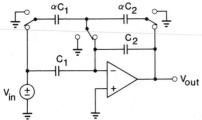

FIGURE 6.2.   Switched-capacitor voltage amplifier.

Note that (6.2) holds regardless of the form (continuous or sampled-and-held) of the input signal $v_{in}(t)$, *if* all components are ideal. Thus, in the ideal case there are no aliasing considerations restricting the bandwidth of the signal and the clock frequency. In fact, of course, parasitic capacitances associated with the switches (as discussed in Section 3.6) allow a clock-frequency feedthrough noise to appear in the signal path. To allow the easy separation of the signal and this clock feedthrough noise, the clock frequency $f_c = 1/T$ should be much higher than the highest signal frequency.

The circuit of Fig. 6.2 can readily be generalized to the case when there are several input signals $v_1, v_2, \ldots, v_n$ (Fig. 6.3). The input–output relation is then

$$V_{out}(z) = -\sum_{i=1}^{n} \frac{C_i}{C_f} V_i(z). \tag{6.3}$$

If $C_1 = C_2 = \cdots = C_n$, $V_{out}(z)$ is proportional to the sum of the input voltages.

An important disadvantage of the circuits of Figs. 6.1–6.3 is that the offset voltage of the op-amp affects the output voltage v_{out}. Denoting the input-

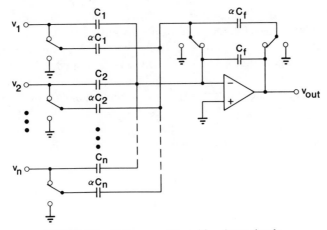

FIGURE 6.3.   Voltage amplifier with n input signals.

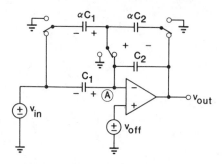

FIGURE 6.4. Switched-capacitor voltage amplifier showing the input-referred op-amp offset voltage v_{off}.

referred offset voltage by v_{off}, the equivalent circuit of Fig. 6.4 can be used to represent this effect. To analyze the steady-state dc output voltage V_{os} due to v_{off}, we assume that the op-amp is otherwise ideal, with infinite gain and bandwidth, and that $v_{\text{in}} = 0$. The voltage at node Ⓐ is then exactly v_{off}, and (in steady state) the voltages across C_1 and C_2 are v_{off} and $v_{\text{off}} - V_{\text{os}}$, respectively; both are constant values. When αC_1 switches into the circuit as indicated in Fig. 6.4, it absorbs a charge $\delta q_1 = \alpha C_1 v_{\text{off}}$ from node Ⓐ, while αC_2 at the same time absorbs a charge $\delta q_2 = \alpha C_2 (v_{\text{off}} - V_{\text{os}})$. Since the charges of C_1 and C_2 do not change, $\delta q_1 + \delta q_2 = 0$ must hold, and

$$\alpha C_1 v_{\text{off}} = \alpha C_2 (V_{\text{os}} - v_{\text{off}})$$

and   (6.4)

$$V_{\text{os}} = \left(1 + \frac{C_1}{C_2}\right) v_{\text{off}} = (1 + |A_v|) v_{\text{off}}.$$

Here, $|A_v| = C_1/C_2$ is the gain of the amplifier; typical values may be $|A_v| = 5 \sim 20$. As discussed in Chapter 7, v_{off} is usually around 10 mV. Thus, $V_{\text{os}} \approx 0.1$ V can occur. This value will also vary with temperature and time. In many applications, such large output offset is unacceptable.

To reduce the effects of the op-amp offset voltage, the compensated circuit[2,4] of Fig. 6.5 (in which the switches are represented by single MOSFETs) can be used. When $\phi_1 = $ "1," the op-amp has its inverting input terminal shorted to its output node, and hence performs as a unity-gain voltage follower, with output voltage v_{off}. Hence, capacitor αC charges to $v_{\text{off}} - v_{\text{in}}$, while C changes to v_{off}. When next ϕ_2 goes high, αC recharges to v_{off} and C to $v_{\text{off}} - v_{\text{out}}$. If the time when this happens is $t = nT$, by charge conservation at node Ⓐ

$$\alpha C\{v_{\text{off}} - [v_{\text{off}} - v_{\text{in}}(nT - T/2)]\}$$

$$+ C\{[v_{\text{off}} - v_{\text{out}}(nT)] - v_{\text{off}}\} = 0.   (6.5)$$

In this equation, v_{off} can be canceled and $v_{\text{out}}(nT) = \alpha v_{\text{in}}(nT - T/2)$ results.

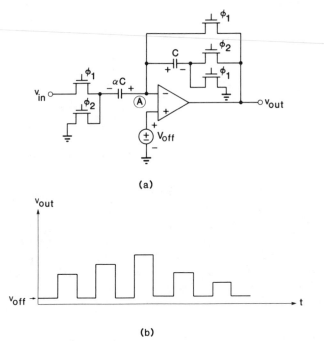

(a)

(b)

FIGURE 6.5.   Offset-compensated noninverting voltage amplifier: (a) circuit; (b) output waveform.

Hence, the transfer function is $H(z) = \alpha z^{-1/2}$; that is, a positive gain of α and a delay $T/2$ are provided by the stage, and the output offset voltage is $V_{os} = 0$. Note also that the circuit of Fig. 6.5 (as all other circuits discussed in the chapter) is fully stray insensitive.

By interchanging the clock phases at the input terminals, an inverting voltage amplifier can also be obtained (Fig. 6.6). By an analysis similar to that performed for the circuit of Fig. 6.5, it can be shown (Problem 6.3) that the transfer function is now $H(z) = -\alpha$. Thus, this circuit is a delay-free inverting amplifier with a gain α. As before, v_{off} is canceled by the switching arrange-

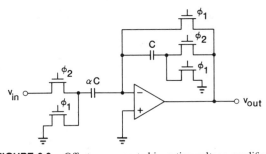

FIGURE 6.6.   Offset-compensated inverting voltage amplifier.

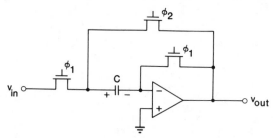

FIGURE 6.7. Offset-compensated half-period delay circuit.

ment, and does not enter v_{out} if the op-amp gain is infinite. (See Problems 6.1–6.3 for the finite-gain case.)

As mentioned earlier, in the circuits of Figs. 6.5 and 6.6 the output voltage is v_{off} whenever ϕ_1 is high, and the output is hence valid only when ϕ_2 is high. Thus, for example, for the circuit of Fig. 6.5a the output waveform is as schematically illustrated in Fig. 6.5b. Clearly, the op-amp must have a high slew rate and fast settling time, especially if the clock rate is high. At the cost of a few additional components,[3] this disadvantage of offset compensation can be eliminated (Problem 6.4).

A simple offset-free half-period delay circuit[24] is shown in Fig. 6.7. Here, the capacitor C acquires a voltage $v_{in} - v_{off}$ when ϕ_1 is high. When ϕ_2 becomes high, the output voltage is $v_{in} - v_{off} + v_{off} = v_{in}$.

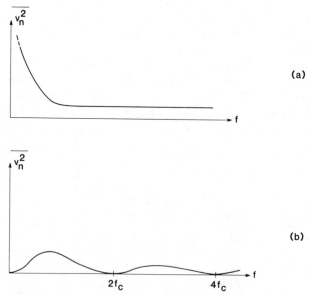

FIGURE 6.8. Noise power for a switched-capacitor voltage amplifier: (a) without offset compensation; (b) with offset compensation.

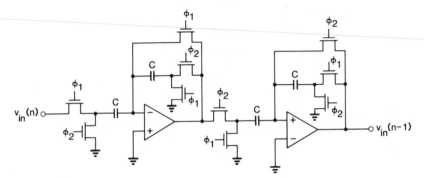

FIGURE 6.9. Switched-capacitor delay stage with offset compensation.

Low-frequency noise signals, which do not change substantially during a clock period T, are similarly canceled by offset compensation. Thus, the troublesome $1/f$ noise discussed in Section 3.7 is greatly reduced. Figures 6.8a and 6.8b illustrate the output noise spectra without and with offset compensation, respectively. Note that cancellation occurs also at $2f_c, 4f_c, \ldots$, which are equivalent to dc for the sampled noise.

By cascading two noninverting voltage amplifiers with complementary clock phases (Fig. 6.9), a clock-controlled full-period delay stage is obtained. Such stages, in conjunction with switches and digital control logic, can perform a variety of sampled-data signal processing functions.[5] A simpler delay stage results if the circuit of Fig. 6.7 is used in the cascade.

In addition to the dc offset voltage due to the op-amp, some output offset is also caused by clock feedthrough. This effect, and some methods which can be used to reduce it, will be discussed in detail in Chapter 7.

6.2. PROGRAMMABLE CAPACITOR ARRAYS: DIGITAL-TO-ANALOG AND ANALOG-TO-DIGITAL CONVERTERS

An important advantage of switched-capacitor circuits is that they can be made digitally variable and thus also programmable. This is accomplished by replacing some capacitors in the circuit by *programmable capacitor arrays* (PCAs). Such a binary-programmed array[6] is shown in Fig. 6.10. In the figure, the triangular symbols denote inverters, and $d_0, d_1, \ldots, d_7$ are binary-coded (high or low, 1 or 0) digital signals. Thus, if (say), d_7 is *high*, the left-side switching transistor associated with capacitor C is "on," while the right-side one is "off." Hence, C is connected between node $\textcircled{x}$ and $\textcircled{x'}$. If d_7 is *low*, the right-side transistor is "on," and it connects the right-side terminal of C to ground rather than to $\textcircled{x'}$. Thus, C never floats, and the total capacitance loading node $\textcircled{x}$ is constant. The value of the capacitance between $\textcircled{x}$ and $\textcircled{x'}$

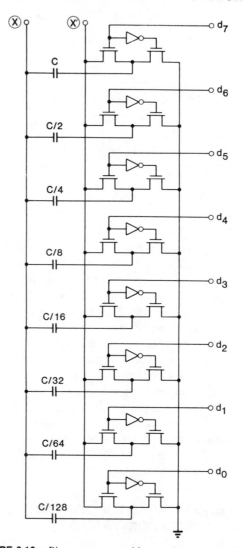

FIGURE 6.10. Binary-programmable capacitor array (PCA).

in the eight-bit PCA of Fig. 6.10 is thus clearly

$$C_t = \sum_{i=0}^{7} \frac{C}{2^{7-i}} d_i = 2^{-7} C \sum_{i=0}^{7} 2^i d_i, \tag{6.6}$$

while the total capacitance loading node $\textcircled{x}$ is $2C(1 - 2^{-8})$ (Problem 6.5).

Care must be taken in the design of the PCA to minimize noise injection from the substrate into the circuit. The bottom plate of the capacitor (which is in the substrate or right above it) should hence never be connected to the inverting input terminal of an op-amp; otherwise, the noise from the power supply which biases the substrate will be coupled to the op-amp's input, and amplified by the op-amp.

If the PCA is getting programmed (i.e., the binary values change) *while* the circuit carries signals, then the terminal (x') should be a virtual ground. This will reduce switching noise, since no dc potentials are altered by the switching. If, on the other hand, the array is programmed at times when no signals are being processed, then it is more advantageous to have the terminal (x) connected to the input of the op-amp. This is because the capacitance loading (x) (as discussed above) is constant. Having a constant capacitive load at the input terminal makes the compensation of the op-amp easier.

Depending on the available area and the desired accuracy, the largest number of bits (i.e., capacitors) in a PCA is $6 \sim 10$. For a larger number (n) of bits, it is usually expedient to replace a single large PCA by two ($n/2$)-bit arrays, even if this requires an extra op-amp.[7]

An obvious application of PCAs is the realization of programmable filters[8] and amplifiers. An example of the latter is illustrated in Fig. 6.11. It was obtained by replacing the input capacitor αC in the offset-free voltage amplifier circuit of Fig. 6.5 by a PCA. Using

$$\alpha C \to \sum_{i=1}^{n} b_i 2^{n-i} C = 2^n C \sum_{i=1}^{n} 2^{-i} b_i, \tag{6.7}$$

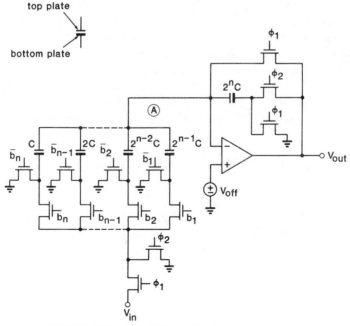

FIGURE 6.11. A multiplying digital-to-analog converter.

the transfer function is found to be

$$H(z) = z^{-1/2} \sum_{i=1}^{n} 2^{-i} b_i. \tag{6.8}$$

Thus, ignoring the half-period delay indicated by the $z^{-1/2}$ factor, the input–output relation is

$$v_{out} = v_{in} \sum_{i=1}^{n} b_i 2^{-i}. \tag{6.9}$$

Thus, the output is the product of the analog voltage v_{in} and the binary-coded digital signal $(b_1, b_2, \ldots, b_n)$. If v_{in} is a temperature-stabilized constant reference voltage V_{ref} then the circuit functions as a *digital-to-analog converter* (DAC), producing an analog output voltage v_{out} which is V_{ref} times the value of the digital signal. If v_{in} is an analog signal, then v_{out} gives the product of v_{in} and the digital signal. Thus, it performs the role of a *multiplying DAC* (MDAC).

Note that (as discussed earlier) the orientation of all capacitors is such that their *top* plates (indicated by light lines) are connected to the op-amp input terminal. This reduces substrate noise voltage injection. Also, due to the presence of the switching devices driven by $\bar{b}_1$, $\bar{b}_2$, and so on, the total capacitance connected to the op-amp input is constant, which makes its compensation an easier task.

If the digital input signal is *bipolar*, that is, if it has either a positive or a negative sign as indicated by a sign bit b_0, then the DAC shown in Fig. 6.12 can be used. If $b_0 = 0$ indicating that the digital signal is positive, then the circuit functions exactly the same way as the MDAC of Fig. 6.11. If, however, $b_0 = 1$ so that the digital signal is negative, then (as can easily be deduced) in the input branch ϕ_1 and ϕ_2 exchange roles. Now the circuit functions as the *inverting* voltage amplifier of Fig. 6.6. Thus, the input–output relation is

$$v_{out} = -v_{in} \sum_{i=1}^{n} b_i 2^{-i} \tag{6.10}$$

as required by the negative digital signal.

For $n \geq 8$, the capacitance spread C_{max}/C_{min} of the circuit is at least $2^8 = 256$, and the total capacitance is at least $511C$. This can lead to inaccuracy (if C is chosen small) or excessive chip area (if C is large). To avoid either of these undesirable conditions, a DAC containing two cascaded programmable gain stages[7,9] may be used.

Analog-to-digital converters (ADCs) can also be realized using analog MOS circuitry. A circuit[6] which uses a PCA and a *voltage comparator* VC is shown (for five-bit conversion) in Fig. 6.13a. The comparator is basically an inverting

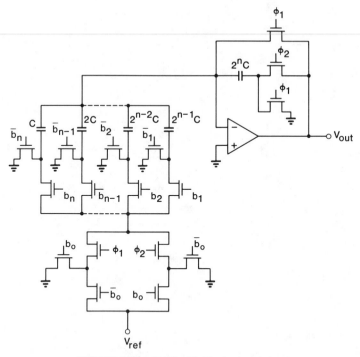

FIGURE 6.12. DAC with bipolar input signal.

op-amp without feedback. Its output is thus normally latched to the positive or negative supply, depending whether its input voltage is negative or positive. The operation of the ADC is performed in three stages. In the first one, S_0 is closed and the bottom plates of all capacitors are connected to v_{in}. This results in a charge proportional to v_{in} stored in all capacitors. In the second stage S_0 is first opened and all bottom plates are then grounded (Fig. 6.13b). This causes the top plate potential to become $-v_{in}$. In the final stage, the bits of the digital output are found, one by one. To find the most significant bit (MSB) b_1, the bottom plate of the largest capacitor (C) is switched by S_7 and S_1 to V_{ref} (Fig. 6.13c). The top-plate potential v_x is now raised to $-v_{in} + V_{ref}/2$. If $v_{in} > V_{ref}/2$, then this value will be less than zero. Therefore, the comparator output will be positive, corresponding to a logic "1," and this will be the value assigned to b_1. Otherwise, $b_1 = 0$. Next, S_1 will return to ground if $b_1 = 0$ (but stay at V_{ref} if $b_1 = 1$), and S_2 will be switched to V_{ref}. The value of v_x then becomes

$$v_x = -v_{in} + \left(\frac{b_1}{2} + \frac{1}{4} \right) V_{ref} \tag{6.11}$$

as can easily be shown (Problem 6.6). If $v_x > 0$, b_2 will be assigned the value of

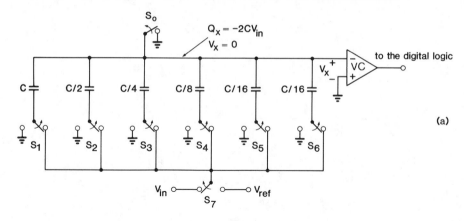

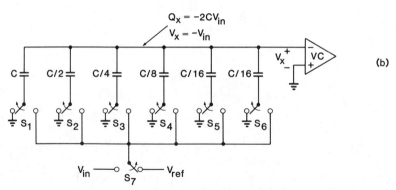

FIGURE 6.13. A successive-approximation analog-to-digital converter (from Ref. 6, © 1975 IEEE): (a) conceptual circuit diagram, shown in the first (sample) stage of operation; (b) circuit in the second (hold) stage; (c) the approximation stage; (d) the final configuration for the output 01001.

0, and S_2 will return to ground; otherwise, $b_2 = 1$ and S_2 will stay at V_{ref}. Next, b_3, b_4, and b_5 are found in a similar manner. Figure 6.13d shows the final positions of $S_1 - S_5$ after obtaining the digital output 01001.

The described circuit represents only the basic concept of the ADC. One of the factors ignored is the offset voltage of the comparator. This can be greatly reduced by using offset compensation ("autozero") circuits, similar to those shown in Figs. 6.5 and 6.6. Figure 6.14 illustrates the basic concept.[6] When S_1 is closed and S_2 grounded, the capacitor C is charged up to the offset voltage V_{off} of the comparator. Next, S_1 opens and S_2 samples v_{in}. Now v_{in} and V_{off} (stored in C) appear in series at the input of the comparator. Thus, $v_x = v_{in} + V_{off}$. The output v_{out} of the comparator will be high if $v_x < V_{off}$ and low if $v_x > V_{off}$; these conditions correspond to v_{out} being high if $v_{in} < 0$ and low if

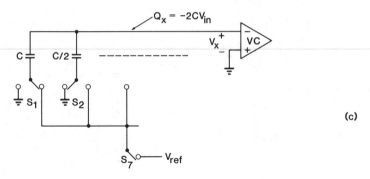

(c)

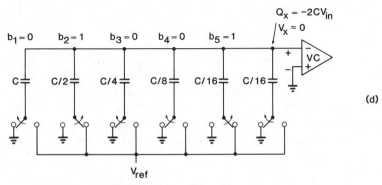

(d)

FIGURE 6.13. continued.

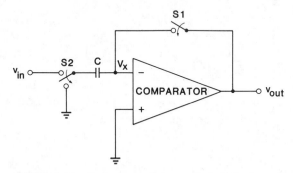

FIGURE 6.14. Comparator offset cancellation by autozeroing.

$v_{in} > 0$, as in an offset-free comparator. Another compensation method is described in Problem 6.9.

Comparators are often built by cascading simple inverter stages. If the gain of each stage is low, then several stages may be required, and the feedback path provided by S_1 may lead to instability. Then it is more expedient to connect S_1 to an intermediate stage (Fig. 6.15). The offset of A_1 is eliminated

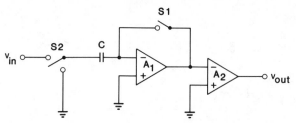

FIGURE 6.15. Reduction of input offset voltage by capacitive storage for a multistage comparator.

by the autozero circuit; the input offset of A_2 is not, but when it is referred back to the input of A_1, it gets divided by the gain of A_1 and, hence, greatly reduced (Problem 6.7). It is also possible to use separate autozeroing circuits for the two circuits. Because of their importance in several applications, comparators are discussed in detail in the next section.

For higher accuracy or smaller-chip area, the ADC may utilize a combination of a multitap resistive divider and a capacitor array.[10,11] Such a circuit[11] is shown in Fig. 6.16. The operation of the circuit is again performed in three stages. In the first stage, S_F is closed and the bottom plates of all capacitors are connected to v_{in}. Thus, all capacitors are charged to $v_{in} - V_{off}$, where V_{off} is the offset (threshold) voltage of the comparator. Next, S_F is opened, and a

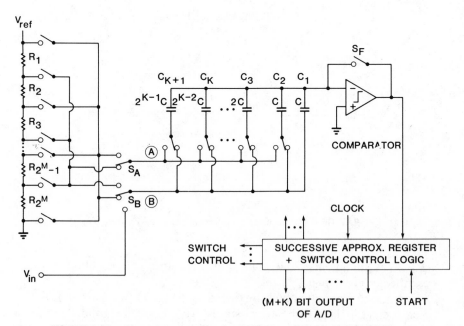

FIGURE 6.16. Successive-approximation ADC (from Ref. 11, © 1979 IEEE).

search is performed among the resistor string taps to find the segment within which this stored voltage sample lies. Nodes Ⓐ and Ⓑ are then switched to the terminals of the resistor R_i which defines this segment. In the final stage, the bottom plates of $C_{k+1}, C_k, \ldots, C_1$ are switched successively back and forth between Ⓐ and Ⓑ until the input voltage of the comparator converges to V_{off}. The sequence of comparator outputs during the successive approximations gives the binary code for v_{in}. Due to the first step in which V_{off} was subtracted from v_{in}, the offset voltage of the comparator does not affect its output.

The operations of the described ADCs are based on successive approximation. For faster operation, a parallel or "flash" configuration may be used. In such circuit, for an n-bit conversion, $2^n - 1$ comparators are used. Each compares v_{in} with a tap voltage from a multitap voltage divider (usually, a resistor chain). The output bits $b_1, b_2, \ldots, b_n$ are then computed simultaneously from the comparator outputs by a digital logic circuit. In Ref. 12, for example, a six-bit converter operating at 20 MHz is described; it uses simple offset-compensated comparators.

Other approaches, based on cascaded one-bit ADCs have also been suggested.[13,14] They are aimed at a compromise performance which combines the speed of flash converters with the economy of the successive-approximation ones.

6.3. MOS COMPARATORS

As already briefly described in the preceding section, a comparator is a circuit which can detect whether its input signal is positive or negative,* and produce a very large output voltage with an appropriate sign. The most important application for a comparator occurs in analog-to-digital conversion, although (as will be shown later in this chapter) other systems may also require voltage comparison. For, say, a 10-bit ADC with a signal amplitude around 1 V, the accuracy of the comparator must be 1 mV or better. Since the input-referred dc offset error of MOS op-amps is around 10 mV, and since the clock feedthrough noise associated with the switches can cause even higher offset errors, clearly some special circuit techniques are required to achieve such an accuracy. These techniques will be discussed in this section.

There are two widely used comparator configurations. The first one[26] contains a cascade of inverter stages (Fig.6.17a) and the second[27,28] a differential-input operational amplifier (Fig. 6.17b). Usually, either kind of amplifier stage is followed by a *latch*, which is essentially a bistable multivibrator. The latch provides a large and fast output signal, whose amplitude and waveform are independent of those of the input signal, and is hence well suited for the logic circuits usually following the latch. If no latch is used, the output v_{out}

---

*More generally, a comparator compares *two* input voltages, and the sign of its output voltage indicates which of the two is larger.

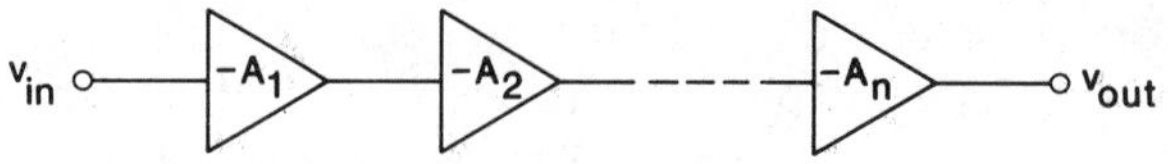

(a)

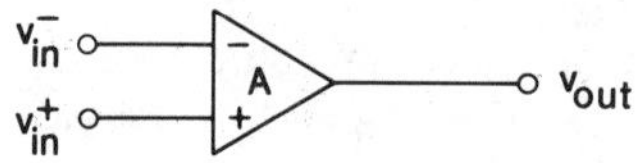

(b)

FIGURE 6.17.   Comparator configurations: (a) cascade of inverter stages; (b) differential-input amplifier.

should have a large swing, say, from -5 to $+5$ V as the input changes from -1 to $+1$ mV. Thus the required gain is around 10 V/2 mV $=$ 5000, or 74 dB. If a latch is used, then v_{out} only needs to be higher than the combined offset and threshold voltages of the latch; this value is around 0.2 V or less. Hence, now a gain of 200 is adequate.

Considering the system of Fig. 6.17a first, let the inverters be realized in CMOS technology. Then a possible configuration for the input inverter is shown in Fig. 6.18a; the clock waveforms and the node voltage v_A are illustrated in Fig. 6.18b. The operation is the following: At $t = 0$, switches S_2 and S_3 are closed. S_2 connects the left-side terminal of the autozeroing capacitor C to ground, while S_3 shorts nodes Ⓐ and Ⓑ. As a result, these nodes assume a voltage which can be found from the intersection of the input–output dc characteristics of the inverter and the 45° line representing the $v_A = v_B$ condition. Figure 6.19 illustrates the situation: Fig. 6.19a shows the inverter, and Fig. 6.19b (center curve) illustrates an example of input–output characteristics (with S_3 open) for $V_{DD} = -V_{SS} = 5$ V and for the threshold voltages $V_{Tn} = -V_{Tp} = 1$ V. The intersection of this curve with the $v_A = v_B$ line occurs at the origin; this is in the middle of the linear range where Q_1 and Q_2 are both in saturation and the gain of the inverter is a maximum. This favorable bias condition is quite insensitive to the variations of the threshold voltages, as is illustrated by curve ① (drawn for $V_{Tn} = 0.7$ V, $V_{Tp} = -1.2$ V) and curve ② (for $V_{Tn} = 1.2$ V, $V_{Tp} = -0.7$ V); the intersection point is in both cases near the middle of the linear range. (See Problem 6.16 for a graphical identification of the linear range for all three curves.)

Returning to the circuit of Fig. 6.18a, clearly during the time interval between $t = 0$ and $t = t_1$, the capacitor C charges to a voltage $v_{AB} - 0 = v_{AB}$, where v_{AB} is the intersection (self-bias) voltage illustrated in Fig. 6.19b. (There, for nominal threshold voltages, $v_{AB} \simeq 0$.) Next, at t_1, ϕ_3 goes low. This results

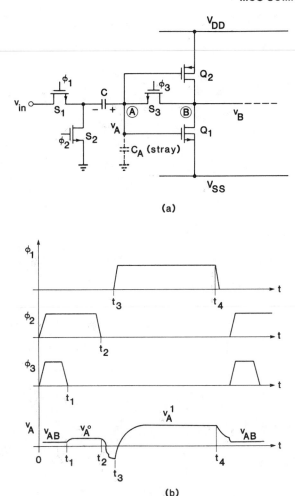

FIGURE 6.18. Input inverter for a CMOS cascade comparator: (*a*) circuit diagram; (*b*) waveforms.

in nodes Ⓐ and Ⓑ being disconnected. Also, part of the charge in the channel of S_3 enters C; in addition, through the gate-to-source overlap capacitance of S_3, additional clock feedthrough charges enter C. The dimensions of C and S_3 must be determined such that even with the change in v_A due to these charges, the $Q_1 - Q_2$ inverter should still remain safely in its linear range. Let the resulting node voltages at Ⓐ and Ⓑ be denoted by v_A^0 and v_B^0, respectively.

Next (at $t = t_2$), S_2 opens. Now, apart from the small stray capacitance C_A, node Ⓐ is floating and hence v_A and v_B can drift to any value. However, since C is nearly open circuited at Ⓐ, there will be only minimal clock

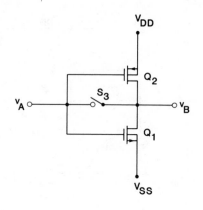

(a)

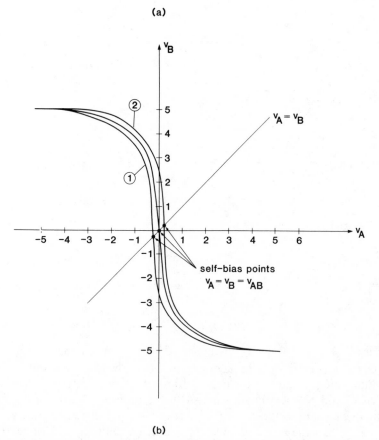

(b)

FIGURE 6.19. Bias conditions for the input inverter: (*a*) circuit diagram; (*b*) input–output dc characteristics for various threshold voltages (see text).

feedthrough into C due to the cutoff of S_2; also, this clock feedthrough charge is stored in C_A and will be returned to C when S_1 closes. Hence, this clock feedthrough has no significant effect.

At $t = t_3$, ϕ_1 goes high and C (charged earlier to v_A^0 during the $t_1 < t \le t_2$ interval) is connected between the input terminal and node Ⓐ. Hence, the node voltage v_A now becomes $v_A^1 = v_{in} + v_A^0$, and the voltage difference $v_A^1 - v_A^0$ has the same sign (and magnitude) as v_{in}. The last statement holds regardless of the values of C_A and v_{AB}, the magnitude of the clock feedthrough, and any other parasitic effects. Thus, by monitoring the change $v_B^1 - v_B^0$ (which is the amplified version of $v_A^1 - v_A^0$) we can decide whether $v_{in} > 0$ or $v_{in} < 0$ holds.

A more complete diagram of a possible circuit is shown schematically in Fig. 6.20a, with the clock signals illustrated in Fig. 6.20b. The operation is as follows. As explained above, by the end of the ϕ_3 pulse, C_1 is charged to a voltage v_A^0 which is suitable for biasing the $Q_1 - Q_2$ inverter in its linear range. If Q_3 is matched to Q_1, and Q_4 is matched to Q_2, then the second inverter has the same bias point, and hence is also biased in its linear range. (Note, however, that the clock feedthrough voltage due to the opening of S_3 is amplified by Q_1 and Q_2 and then connected to the gates of Q_3 and Q_4; hence, the second stage is more vulnerable to this effect.) During this time, the $Q_5 - Q_6$ inverter is also biased in its linear range by S_4, and C_2 is precharged to the corresponding bias voltage. S_4 opens only *after* S_3 does, so that the output voltage v_C of the $Q_5 - Q_6$ inverter is not affected by the amplified clock feedthrough transient due to S_3. It is affected, however, by the opening of S_4. This transient can, in turn, be rendered ineffective by C_3 and S_5. When S_5 is closed, the latch is locked in a self-biased balanced state. Thus, when S_5 opens, the voltage acquired by C_3 would keep a perfectly symmetrical latch in an unstable balanced position. For the latch to function, however, the enabling ("strobe") signal ϕ_6 must also go high.

By $t = t_4$, all precharging operations are complete. At this point, S_2 is opened and the input node Ⓐ now floats. When next S_1 closes, v_D at the input of the latch will rise (if $v_{in} < 0$) or fall (if $v_{in} > 0$) from its earlier balanced value. Hence, when ϕ_6 rises, the latch will switch to the appropriate one of its two stable states.

Obviously, the system shown in Fig. 6.20 is only one example of the many different possibilities. (It is, in fact, a CMOS equivalent of the circuit described in Ref. 26.) Other circuits may contain only two inverter stages, or may use source followers as buffers between the various stages, or may use capacitive coupling also between the first and second stages, and so on. In most cases, however, the biasing and autozeroing is accomplished using the precharging operations illustrated in Fig. 6.20.

The speed of the cascaded inverter stages is limited by the RC time constants represented by the output resistance R_0 of the ith inverter and the input capacitance C_{in} of the next (i.e., $i + 1$st) one. The former is the parallel combination of the drain resistances of the PMOS and NMOS devices in the

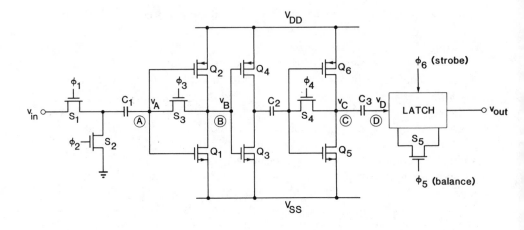

(a)

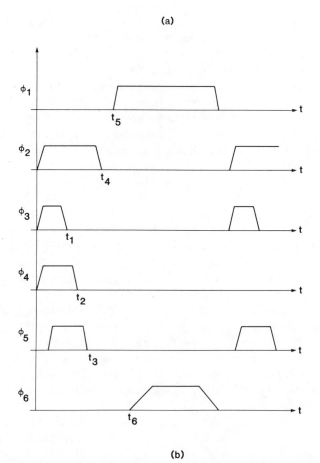

(b)

FIGURE 6.20. CMOS cascade comparator: (a) circuit diagram; (b) clock signals.

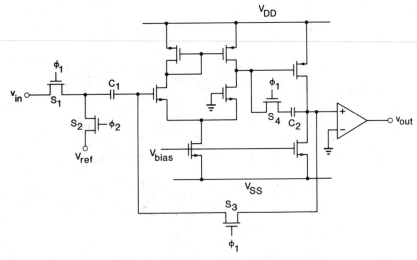

FIGURE 6.21. Op-amp type comparator.

ith stage, the latter can be approximated by the sum of the C_{gd} values in the next stage, multiplied by $(1 + |A_{i+1}|)$ due to the Miller effect. The dc gain A of the inverter is, of course, the sum of the transconductances divided by the sum of the drain conductances of the two devices in the stage. Typical values are $R_0 \sim 100$ kΩ, $C_{in} \sim 0.5$ pF, and $A \sim 10$.

Next, the alternative system shown in Fig. 6.17b will be briefly considered. A typical circuit[28] is shown in Fig. 6.21. The input section illustrated is simply a two-stage amplifier with an RC compensating branch $S_4 - C_2$. This branch is effective, however, only during the $\phi_1 = 1$ half-period, when C_1 is precharged through S_1 and S_3, and hence the op-amp functions in a feedback configuration. During the $\phi_2 = 1$ interval, the amplifier is in an open-loop configuration and hence compensation (which slows down its operation) is not needed.

For high-accuracy applications, it is preferable to use fully differential amplifiers, since in such circuits clock feedthrough effects, power-supply noise, $1/f$ noise, and so on tend to cancel. Such comparators are described in Refs. 27, 29, and 30. As an illustration, Fig. 6.22 shows a fully differential stage.[29] In this circuit, the two autozeroing capacitors C_1 and C_2 are precharged during the $\phi_1 = 1$ interval to $v_{in}^- - v_s$ and to $v_{in}^+ - v_s$, respectively, where v_s is the self-biased input voltage of the amplifier. When next $\phi_2 \to 1$, a voltage step $v_{in}^+ - v_{in}^-$ is generated at node Ⓐ, and a step $v_{in}^- - v_{in}^+$ at node Ⓑ. As a result, an amplified differential output voltage $v_{out}^+ - v_{out}^-$ appears. There is also a clock feedthrough signal at each input node, but this appears as a common-mode signal, and is hence suppressed. To improve the CMRR, V_{bias_2} can be obtained from $v_{out}^+ + v_{out}^-$ using a negative feedback circuit.

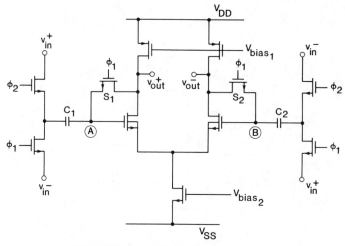

FIGURE 6.22. Fully differential comparator.

A high-gain (> 80 dB) comparator, using a folded-cascode first stage followed by a cascode second stage and a switched compensation capacitor, is described in Ref. 31.

As mentioned earlier, the amplifier of the comparator is usually followed by a latch. The latter can be simply a cross-coupled bistable multivibrator. Some possible latch circuits are shown in Figs. 6.23–6.25. The latch[33] of Fig. 6.23 is shown with a differential input; for single-ended input v_{in}^- or v_{in}^+ can be replaced by a threshold voltage or can be generated by self-biasing. During

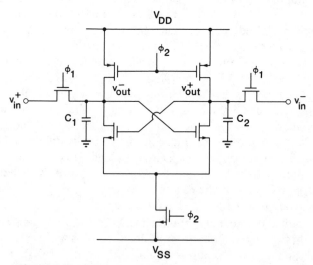

FIGURE 6.23. Direct-coupled latch with differential input signals.

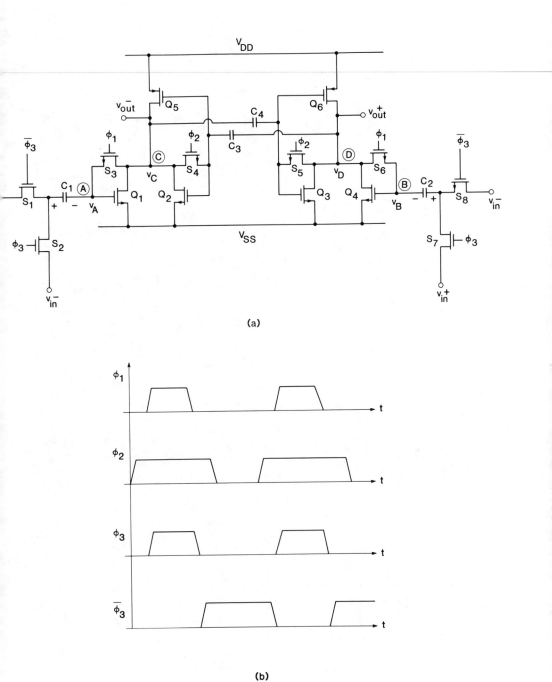

FIGURE 6.24. Capacitively coupled latch with autozeroing input circuitry: (*a*) circuit diagram; (*b*) clock signals.

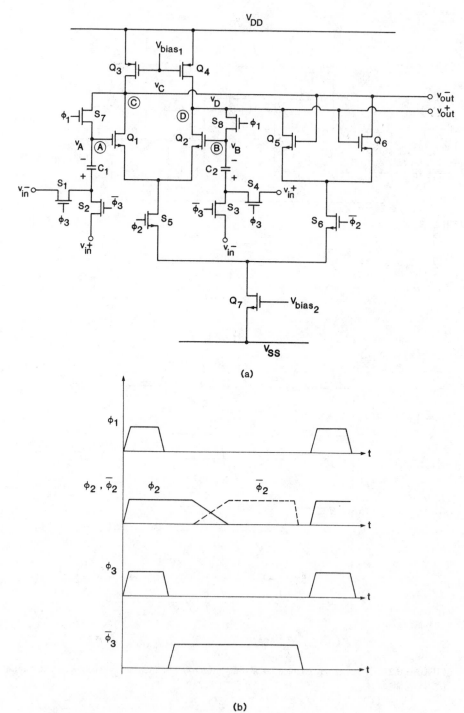

(a)

(b)

FIGURE 6.25. Preamplifier-latch combination: (*a*) circuit diagram; (*b*) clock signals.

$\phi_1 = 1$, C_1 and C_2 are precharged to the input voltages. Subsequently, the ϕ_2 (strobe) signal enables the load and current source devices, and the latch assumes one of its stable states depending on the sign of $v_{in}^+ - v_{in}^-$.

The latch of Fig. 6.24 includes also the self-biasing and autozeroing circuitry.[29] The operation is as follows. When ϕ_2 goes high, S_4 and S_5 short circuit the gates and drains of Q_2 and Q_3, respectively. This action biases the two inverters $Q_2 - Q_5$ and $Q_3 - Q_6$ (which form the multivibrator) in their linear regions. It also precharges the capacitors C_3 and C_4 such that any asymmetry between the two inverters is compensated for by the slightly different bias voltages provided by C_3 and C_4. During this time, the multivibrator has a loop gain less than 1, and hence it does not switch into either one of its stable stages. When next ϕ_1 and ϕ_3 go high, S_2 and S_3 precharge C_1 to $v_{in}^- - v_A$, while S_6 and S_7 precharge C_2 to $v_{in}^+ - v_B$. Next, ϕ_1 goes low, and the input devices Q_1 and Q_4 are released from their self-biased states, leaving C_1 and C_2 floating. Then, ϕ_3 goes low and $\bar{\phi}_3$ high: v_A now changes by $v_{in}^+ - v_{in}^-$, while v_B changes by $v_{in}^- - v_{in}^+$ from their self-biased values. This also causes corresponding changes in v_C and v_D inside the multivibrator. When finally ϕ_2 goes low unleashing the multivibrator, the voltage differences between v_A and v_B as well as v_C and v_D cause it to go to the appropriate stable state: v_C will be high and v_D low if $v_{in}^+ < v_{in}^-$ holds, or v_C low and v_D high if $v_{in}^+ > v_{in}^-$ is valid.

The sequence in which the various clock phases rise and fall is important for the proper operation of this latch circuit (as it is for almost any other!). The reader is urged to analyze the operation if (say) ϕ_2 goes low before ϕ_3 goes high, and so on, to convince himself (or herself) of the validity of this statement.

Yet another latch[32] is shown in Fig. 6.25. In this circuit, transistors Q_1, Q_2, Q_3, Q_4, and Q_7 act as a differential preamplifier when S_5 is closed. On the other hand, when S_6 is closed, Q_3, Q_4, Q_5, Q_6, and Q_7 form a bistable multivibrator. When ϕ_1, ϕ_2, and ϕ_3 are all high, the preamplifier is self-biased and C_1 and C_2 are precharged to $v_{in}^- - v_A$ and $v_{in}^+ - v_B$, respectively. Then, ϕ_1 goes low, allowing the amplifier to function and leaving C_1 and C_2 floating at nodes Ⓐ and Ⓑ. Next, the bottom terminals of C_1 and C_2 are switched to v_{in}^+ and v_{in}^-, respectively; this causes an amplified voltage difference between nodes Ⓒ and Ⓓ. At this point, S_5 slowly opens and S_6 slowly closes. This causes the multivibrator to come to life and to assume one of its stable states. The state chosen is determined by the sign of $v_C - v_D$.

Even if the amplifier and the latch are built from the same types of inverters, the rise and fall times of the amplifier will be much longer than those of the latch. To understand this phenomenon, consider the simple multivibrator of Fig. 6.26a. Its small-signal equivalent circuit in the saturation range of all devices is shown in Fig. 6.26b, where $g_m = g_{m1} + g_{m3} = g_{m2} + g_{m4}$ and $g_d = g_{d1} + g_{d3} = g_{d2} + g_{d4}$; also, C is the capacitance loading nodes Ⓐ and Ⓑ. It can easily be shown (Problem 6.17) that the natural modes (poles) of the circuit of Fig. 6.26 are $s_{1,2} = \pm g_m/C$. Hence, its transients are exponential

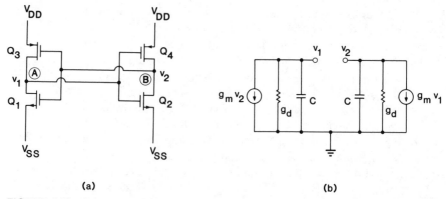

(a)   (b)

FIGURE 6.26. Direct-coupled multivibrator: (a) circuit diagram; (b) small-signal equivalent circuit.

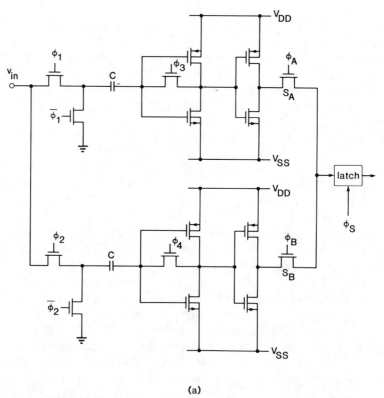

(a)

FIGURE 6.27. Fast comparator system with two amplifiers and a single latch: (a) circuit diagram; (b) clock signals.

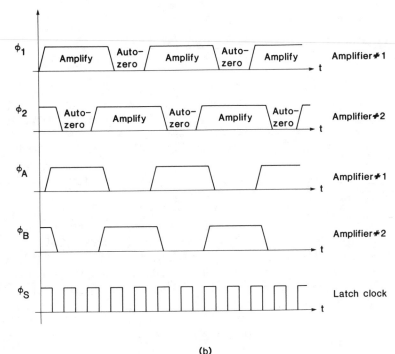

FIGURE 6.27. continued.

functions with a time constant $\tau = C/g_m$. In the absence of positive feedback, if the inverters $Q_1 - Q_3$ and $Q_2 - Q_4$ are simply cascaded as in the amplifier, the time constant is $\tau' = C/g_d$. The ratio of the time constants is $\tau/\tau' = g_d/g_m = 1/A$, where A is the gain of the inverter.[33] Since typically $A = 10$, the latch can be about an order of magnitude faster than the amplifier driving it. It is possible to take advantage of the speed of the latch by using *two* amplifiers to feed a single latch (Fig. 6.27a). In this system,[34] the two amplifiers have the same configuration as the two input stages in the circuit of Fig. 6.20a. They alternate in autozeroing and amplifying, but the amplifying periods have a duty cycle *longer* than 50% (Fig. 6.27b), so that the input of the latch can receive a *continuous-time* input signal. To assure this, the intervals during which the switches S_A and S_B connect the amplifiers to the latch overlap. Thus, the latch clock frequency (which is the effective overall clock frequency of the comparator) can be different from the amplifier clock rates. This system can thus operate about 10 times faster than the usual single-amplifier version, since the limiting factor is now the speed of the latch, not that of the amplifier.

6.4. MODULATORS, RECTIFIERS, AND DETECTORS

An important class of nonlinear circuits produces replicas of the spectrum of the input signal, shifted along the ω-axis. To shift the spectrum $M(\omega)$ of a

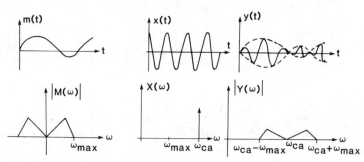

FIGURE 6.28. Amplitude modulation: waveforms and spectra of the modulating signal $m(t)$, the carrier $x(t)$, and the modulated signal $y(t)$, for sine-wave carrier.

signal $m(t)$ by an amount ω_{ca}, we can form the product $y(t) = m(t)\cos \omega_{ca}t$. Now (as given in Table 2.4, Chapter 2) the Fourier transform of $\cos \omega_{ca}t$ is $\pi\delta(\omega - \omega_{ca}) + \pi\delta(\omega + \omega_{ca})$; multiplication in the time domain corresponds to convolution in the frequency domain; and convolution by $\delta(\omega \pm \omega_{ca})$ merely shifts the spectrum by $\pm\omega_{ca}$. Hence, the spectrum of $y(t) = m(t)\cos \omega_{ca}t$ is

$$Y(\omega) = \tfrac{1}{2}M(\omega + \omega_{ca}) + \tfrac{1}{2}M(\omega - \omega_{ca}). \qquad (6.12)$$

Thus, $M(\omega)$ is indeed shifted by $\pm\omega_{ca}$. Figure 6.28 illustrates the waveforms and spectra of $m(t)$, $x(t) = \cos \omega_{ca}t$ and $y(t) = m(t)x(t)$. It was assumed in Fig. 6.28, and is the usual case, that ω_{ca} is larger than ω_{max}, the maximum frequency of $M(\omega)$ so that no aliasing occurs due to the operation. As the waveform of $y(t)$ illustrates, the amplitude of $x(t)$ (usually called the *carrier*) is varied (modulated) by $m(t)$, the *modulating signal*. Hence, the operation is called amplitude modulation, and the circuit which accomplishes it is called an *amplitude modulator*.

More generally, the carrier $x(t)$ can be any periodic signal. If its period is $T_{ca} = 2\pi/\omega_{ca}$, then its Fourier-series expansion is

$$x(t) = \sum_{n=-\infty}^{\infty} a_n e^{jn\omega_{ca}t}. \qquad (6.13)$$

Since the Fourier transform of $e^{jn\omega_{ca}t}$ is $2\pi\delta(\omega - n\omega_{ca})$ as given in Table 2.4, using the same argument which led earlier to Eq. (6.12) we now obtain for the spectrum of the modulated signal $y(t)$ the expansion

$$Y(\omega) = \sum_{n=-\infty}^{\infty} a_n M(\omega - n\omega_{ca}). \qquad (6.14)$$

Note that the carrier signal $x(t)$ does *not* enter the output signal directly.

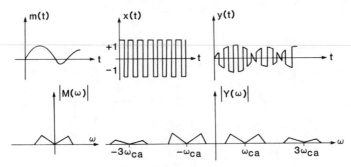

FIGURE 6.29. Waveforms and spectra for square-wave carrier.

Hence, this process is a *suppressed-carrier* modulation. Usually, the first sidebands at $\omega \pm \omega_{ca}$ represent the desired output signal.

A periodic carrier signal which is readily generated using SC circuitry is a square wave alternating between two equal values $\pm V$. For such an $x(t)$, the coefficients a_n are zero for even values of n, and their magnitudes decrease as $1/n$ for odd n. Hence, using for simplicity $V = 1$, the waveforms and spectra of Fig. 6.29 are obtained.

An easy way to perform modulation with a square-wave carrier is to switch the polarity of the input signal $m(t)$ periodically. Consider, for example, the SC circuit of Fig. 6.30, with all switches clocked at a rate $\omega_c = 2\pi/T$. If the clock phases *without* parentheses are used at the input terminal, then the circuit performs as a damped integrator with a negative dc gain: $-C_1/C_2$. If, by contrast, the phases *in* parentheses are used, then the circuit (while still a damped integrator) has a positive dc gain $+ C_1/C_2$. (There is, of course, then also an additional half-period signal delay $T/2$.) Thus, by interchanging the clock phases, the polarity of the input signal is effectively reversed. Switching clock phases every $T_{ca}/2$ seconds is hence equivalent to multiplying the input

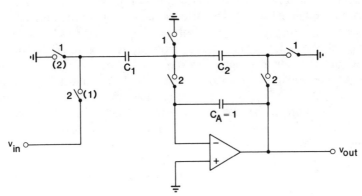

FIGURE 6.30. A damped integrator with a dc gain of $-C_1/C_2$ *or* $+C_1/C_2$.

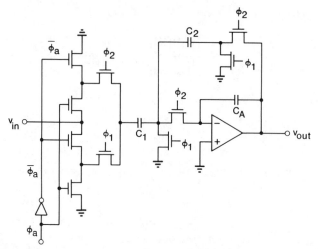

FIGURE 6.31.   A switched-capacitor modulator with two clock signals.

by a square-wave signal with peak values ± 1 and frequency $\omega_{ca} = 2\pi/T_{ca}$.[15] A stray-insensitive SC modulator circuit which performs according to this principle is shown in Fig. 6.31. The clock phases ϕ_1 and ϕ_2 are operated at a fast clock rate ω_c, while the phase ϕ_a changes at the slow carrier-frequency rate ω_{ca}. Normally, ω_c is much larger (by a factor of 30 or more) than ω_{ca}.

The extra signal delay $T/2$, caused by the circuit when ϕ_a is low, results in an asymmetry of the effective carrier signal. It introduces a dc offset into $x(t)$ and, hence, as Eqs. (6.13) and (6.14) show, causes a feedthrough of the modulating signal $m(t)$ into $Y(t)$. This can be prevented if a sample-and-hold stage precedes the modulator, so that v_{in} changes only when phase ϕ_1 goes high. In this case, v_{in} does not change during the $T/2$ delay interval, and hence the delay has no effect. It is also possible to use a more complicated switching scheme to eliminate the effect of the delay.*

Special care must be taken in the design of the modulator of Fig. 6.31 (and of all similar circuits) to avoid intermodulation products which overlap with the desired modulated signal. To illustrate the potential problems, assume that the clock rates are $f_{ca} = \omega_{ca}/2\pi = 3$ kHz and $f_c = \omega_c/2\pi = 115$ kHz, and the spectrum of the input signal extends from dc to 1 kHz. Then, the desired spectrum of the modulated signal lies at $f_{ca} \pm 1$ kHz, or in the 2–4 kHz range.

Due to the square-wave carrier clock signal ϕ_a, odd-indexed sidebands centered at $\omega_{ca}, 3\omega_{ca}, 5\omega_{ca}, \ldots$ will be generated. Their amplitudes decrease as $1/n$; thus, the 39th sideband centered at 39×3 kHz = 117 kHz, will be at a level which is only $20 \log_{10} 39 \cong 31.8$ dB lower than that of the desired first sideband. When next the 39th sideband is sampled at the fast clock rate $f_c = 115$ kHz, its spectrum will be replicated and shifted to $117 + nf_c$, where $n = 0, \pm 1, \pm 2, \ldots$. For $n = -1$, the replica will occupy the frequency range

---

*Also, the circuit acts as a low-pass filter on v_{in}.

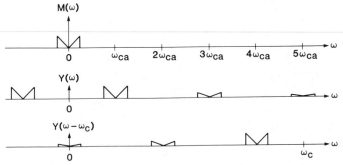

FIGURE 6.32. Spectra of the modulating signal $m(t)$, the modulated signal $y(t)$, and the first replica $Y(\omega - \omega_c)$, for $\omega_c = 5\omega_{ca}$.

1–3 kHz, where it will overlap with the desired modulated signal spectrum located in the 2–4 kHz band.

To avoid this situation, the fast clock rate ω_c should be an *integer multiple* of the slower carrier frequency ω_{ca}, so that $\omega_c = N\omega_{ca}$, N an integer. Then, if the bandwidth of $m(t)$ is restricted to the $-\omega_{ca}/2 < \omega < \omega_{ca}/2$ range and N is *odd*, the replicated sidebands will not overlap. The situation is illustrated for $N = 5$ in Fig. 6.32, which shows the spectrum of $m(t)$, the spectrum $Y(\omega)$ after modulation, and the first replica $Y(\omega - \omega_c)$ generated by the fast clock. Clearly, the sidebands of the replica interlace with those of $Y(\omega)$, and no intermodulation distortion occurs in the band of interest centered at ω_{ca}. The same holds for the other replicas $Y(\omega + \omega_c)$, $Y(\omega \pm 3\omega_c)$, and so on. Thus, a bandpass filter can be used to eliminate all unwanted sidebands, and retain the band centered at ω_{ca}.

A similar argument shows that if N is *even* the sidebands overlap in such a way that no distortion occurs, and only the amplitude of the modulated signal is altered due to added sidebands. Here (as can readily be shown) a *low-pass* filter can eliminate the unwanted modulation products. This is usually a simpler circuit than the bandpass filter needed for odd N values.

An alternative method for avoiding intermodulation distortion caused by the simultaneous presence of the two clock signals is to use a single clock. As an illustration, Fig. 6.33 shows a suppressed-carrier modulator which uses the carrier clock signal ϕ_a also as the clock signal of the integrator. In this circuit, when $\phi_a = $ "0" the input signal is applied directly to the output buffer amplifier B and thence to the output. During this period, the inverter stage (containing op-amp A and the two capacitors C) is reset and its offset voltage stored for cancellation as in the circuit of Fig. 6.6. When $\phi_a = $ "1," v_{in} is inverted by op-amp A and then applied to the buffer B. An analog continuous-time filter (illustrated schematically by the RC ladder) should be used to suppress the higher-order sidebands so that an SC post filter can next be used to extract the spectrum centered at $\pm \omega_{ca}$ without aliasing effects. The circuit can also be

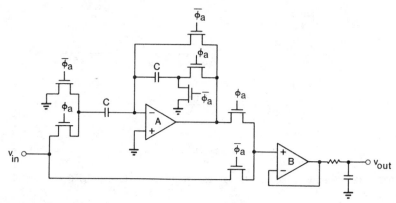

FIGURE 6.33.   An alternative switched-capacitor modulator which requires a single clock signal.

used as a single-sideband (SSB) filter, if a bandpass SC filter is used as a postfilter to pass one sideband and reject the other.

Switched-capacitor circuits can also be used as *full-wave rectifiers* and as *peak detectors*. A full-wave rectifier converts an input signal $v_{in}(t)$ to its absolute value $|v_{in}(t)|$. A simple way of implementing an SC full-wave rectifier is to add a comparator to one of the modulators discussed earlier. Consider, for example, the modulator of Fig. 6.31. If the carrier clock signal ϕ_a is derived from $v_{in}(t)$ such that $\phi_a =$ "0" for $v_{in} > 0$ and $\phi_a =$ "1" for $v_{in} < 0$, then the circuit will invert negative signals, but not positive ones. Thus, $v_{out} \propto |v_{in}|$. In actual fact, the modulator stage formed by C_1, C_2, C_A, and the op-amp is a low-pass filter and hence it modifies the waveform of v_{in} even for $v_{in} > 0$. However, if the 3-dB bandwidth of the filter is much wider than the spectrum of v_{in} then this band-limiting effect will be negligible.

The circuit[15] of a switched-capacitor full-wave rectifier based on the modulator of Fig. 6.31 is shown in Fig. 6.34a. Here, A is set to "1" if $v_{in} > 0$ and to "0" if $v_{in} < 0$, while B is set to $\bar{A}$, by the comparator and the latch which follows it, each time ϕ_1 goes high. The signals A and B then set the polarity of the transfer function so that it inverts negative input signals, but not positive ones.

As in the D/A converters, the comparator can be made offset free here too by using autozeroing. Figure 6.34b shows a possible arrangement. When ϕ_2 goes high, C stores the offset voltage V_{off}. Next, when ϕ_1 goes high, $v_{in} + V_{off}$ is applied to the input of the amplifier. The output v_{out} is thus independent of V_{off} as long as the amplifier gain is high enough (Problem 6.10).

As discussed earlier in connection with the modulator of Fig. 6.31, the signal path from v_{in} to v_{out} is delay free if $A =$ "0" (i.e., if $v_{in} < 0$), but has a delay $T/2$ if $A =$ "1" (i.e., if $v_{in} > 0$). This causes a polarity-dependent jitter. The jitter will be eliminated if v_{in} is a sampled-and-held signal which changes only at the leading edge of the ϕ_1 clock phase so that the delay has no effect.

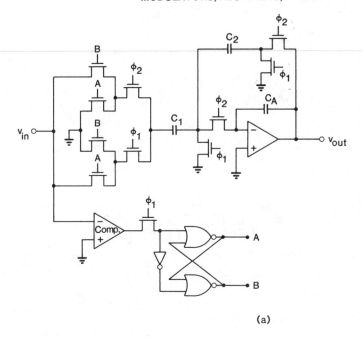

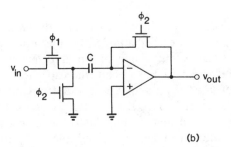

(a)

(b)

FIGURE 6.34. Switched-capacitor full-wave rectifier: (*a*) complete circuit; (*b*) an offset-compensated comparator.

The output voltage v_{out} of the circuit of Fig. 6.34*a* also includes an error term due to the offset voltage of the op-amp in the upper (signal) path. This term is equal to $(1 + C_1/C_2)V_{\text{off}}$, where V_{off} is the input-referred op-amp offset voltage. If this error is unacceptable, the offset-compensation scheme of Figs. 6.5 and 6.6 can be used. The resulting circuit (which now incorporates offset cancellation for both amplifiers) is shown in Fig. 6.35. As explained earlier, the op-amp output must now slew back and forth between $|v|$ and V_{off}. Thus v_{out} is valid only when ϕ_2 is high; also, the op-amp must have high slew rate and fast settling time. The circuits described in Ref. 3 can be used to reduce the resulting requirements on the op-amp, at the cost of a few additional elements.

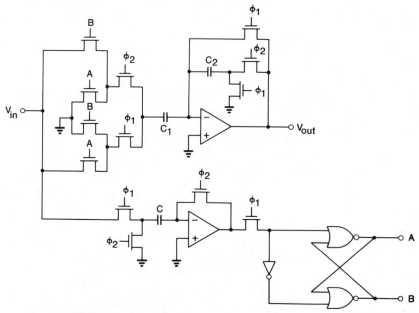

FIGURE 6.35.   Offset-canceling switched-capacitor full-wave rectifier.

A *peak detector* is a circuit whose output holds the largest positive (or, if so specified, negative) voltage earlier attained by the input signal. Figure 6.36 illustrates the input and output signals of a peak detector. A possible realization is shown in Fig. 6.37. In this circuit, the capacitor C stores the largest earlier value V_{max} of v_{in}. The op-amp A acts as a comparator, with the present values of v_{in} and V_{max} as its inputs. If $v_{in} > V_{max}$, then the op-amp output goes negative and (when ϕ_1 and then ϕ_2 turn high) the logic signal A is set high. This turns on the switch S at the input, and allows v_{in} to recharge C. (The same result could have been achieved, of course, simply by using a floating diode instead of S. Unfortunately, however, it is difficult to realize a floating diode in MOS technology.)

If $v_{in} < V_{max}$, then $A = "0"$ and C continues to hold V_{max}, which remains the value of v_{out}.

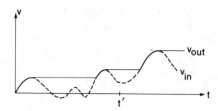

FIGURE 6.36.   Input and output signals of a peak detector.

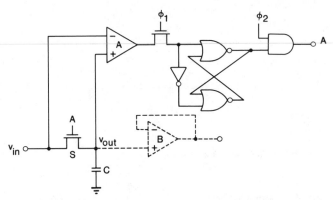

FIGURE 6.37.   Switched-capacitor peak detector circuit.

Since the comparison of v_{in} and V_{max} takes place when $\phi_1 =$ "1," but C is recharged when $\phi_2 =$ "1," the operation is exact only if v_{in} is a sampled-and-held signal which changes at the leading edges of $\phi_1 =$ "1" half-periods. However, if the clock frequency is much higher than the highest frequency in v_{in}, then the error due to the delay between comparison and recharging will be small.

The output voltage v_{out} is the voltage across C. If necessary, a buffer (shown as a unity-gain amplifier B in Fig. 6.37) should be used, since the circuit only functions with open-circuited output terminals; otherwise, C will be discharged. In some applications, it *is* desirable at times to discharge C and thus reset the detector. For instantaneous reset, a switch can be connected in parallel with C. For a slow reset, a small capacitor can be periodically connected across C (say, during $\phi_1 =$ "1") and then discharged during $\phi_2 =$ "1."

By cascading a full-wave rectifier with a peak detector, it is possible to detect the maximum value of $|v_{in}(t)|$.

A continuous-time MOS peak detector[16] is shown in Fig. 6.38. The op-amp again acts as a comparator, with $v_{out} = V_{max}$ and v_{in} as its inputs. If $v_{in} > V_{max}$, the op-amp output goes high and M1 conducts, charging C until $v_{out} \approx v_{in}$ is reached. If $v_{in} < V_{max}$, then the op-amp output is low, M1 is cut off, and $v_{out} = V_{max}$ is held by C. As before, a buffer is needed if the output must drive a load. To reset v_{out}, a switch or a current source or an SC resistor can be connected in parallel with C.

6.5.   SWITCHED-CAPACITOR OSCILLATORS

There are several strategies available for fabricating integrated MOS oscillators. The simplest method is to use digital scalers to divide the frequency of a

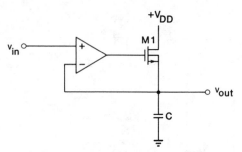

FIGURE 6.38. Continuous-time peak detector.

high-frequency master-clock signal. This results in good frequency stability, since the master clock is usually crystal controlled. However, the frequency is restricted to be a subharmonic of the master-clock frequency, and it cannot be readily changed by a control voltage as is required in a phase-locked loop (PLL). Also, if an SC filter is used to suppress the higher harmonics of the divided signal, then its output voltage will contain high-frequency components which can cause jitter in the system.[17]

Another possibility for designing an MOS oscillator is to use an astable multivibrator.[17-19] The frequency may then be controlled by a voltage, for example, by incorporating a voltage-controlled current source (voltage-to-current converter) in the circuit. However, since it is difficult to fabricate stable and accurate current sources without using off-chip components, the frequency of oscillation cannot be accurately controlled, and tends to be temperature sensitive.

It is also possible to design an active-RC oscillator, and then replace all resistors by SC branches, such as the ones discussed in Sections 5.1 and 5.2. The frequency can now be independent of the clock frequency; the output spectrum will, however, contain the clock-frequency and higher-frequency products. Since the oscillation frequency depends only on capacitance ratios, it can be made highly accurate, predictable, and stable. A parasitic-sensitive oscillator based on an active-RC phase-shift oscillator[20] is illustrated schematically in Fig. 6.39. (It is possible also to realize stray-insensitive versions of this circuit.) Assuming that the clock frequency f_c is much larger than the oscillation frequency f_0, that the dc open-circuit voltage gain $A = g_m r_d$ of the MOSFET is much greater than one, and that $\alpha \ll 1$, the critical value of g_m required for oscillation turns out to be

$$g_{m \text{ crit}} \cong 12 f_c \alpha C (1 + 14/A) \qquad (6.15)$$

and the oscillation frequency is

$$f_0 \cong \frac{\sqrt{3}}{2\pi} \alpha f_c (1 + 6/A) \qquad (6.16)$$

(Problem 6.11).

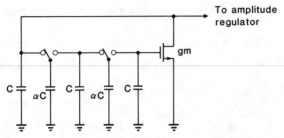

FIGURE 6.39. A SC phase-shift oscillator.

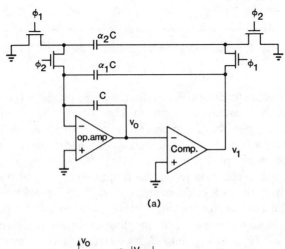

(a)

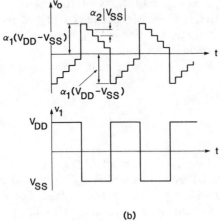

(b)

FIGURE 6.40. A SC relaxation oscillator: (*a*) circuit diagram; (*b*) waveforms.

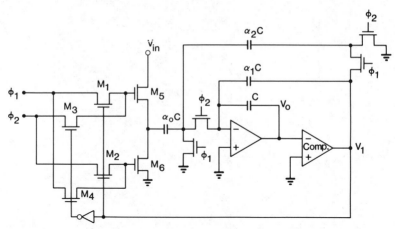

FIGURE 6.41. A voltage-controlled SC oscillator based on the relaxation oscillator of Fig. 6.40.

It is also possible to design SC oscillators based on the relaxation oscillator principle.[21] Such a circuit is shown in Fig. 6.40a. The circuit operates as follows. Assume that the comparator output voltage v_1 is locked to the negative supply voltage V_{SS}. The inverting switched capacitor $\alpha_2 C$ will then feed a positive charge packet $\delta q_2 = \alpha_2 C |V_{SS}|$ into the integrating capacitor C during each clock cycle. As a result, the output voltage v_0 of the first op-amp will decrease in each clock cycle, in steps of $\alpha_2 |V_{SS}|$ (Fig. 6.40b). When $v_0 < 0$ is reached, the comparator output voltage switches to the positive supply voltage V_{DD}. Now the unswitched coupling capacitor $\alpha_1 C$ feeds a positive charge $\delta q_1 = \alpha_1 C (V_{DD} - V_{SS})$ into C. As a result, v_0 drops by $\alpha_1 (V_{DD} - V_{SS})$ (Fig. 6.40b). Thereafter, $\alpha_2 C$ feeds negative charge packets $-\alpha_2 C V_{DD}$ into C, and v_0 will take positive steps of size $\alpha_2 V_{DD}$, until $v_0 > 0$ is reached. Then v_1 drops to V_{SS}, and the cycle is repeated.

For $\alpha_2 \ll \alpha_1$ and $V_{SS} = -V_{DD}$, the oscillation frequency is given by

$$f_0 \cong \frac{\alpha_2}{4\alpha_1} f_c \qquad (6.17)$$

where f_c is the clock frequency (problem 6.12).

By adding a switched feed-in capacitor $\alpha_0 C$ (Fig. 6.41), the circuit can be converted into VCO with v_{in} as its control voltage. Let v_{in} be positive, and $v_1 = V_{SS} < 0$. Then M_1 and M_2 are off, while M_3 and M_4 are on; as a result, M_5 is on when $\phi_2 = $"1" and M_6 is on when $\phi_1 = $"1." Thus the switched capacitor $\alpha_0 C$ feeds a *positive* charge $\alpha_0 C v_{in}$ into C whenever ϕ_2 goes high. This charge is added to the positive charge $\delta q_2 = \alpha_2 C |V_{SS}|$ fed to C by $\alpha_2 C$. Hence, the steps of v_0 become larger than they would be without $\alpha_0 C$, and v_0

reaches the zero level sooner. When v_1 switches to V_{DD}, M_3 and M_4 turn off and M_1 and M_2 turn on. Now M_5 is on when $\phi_1 = $ "1," and M_6 is on when $\phi_2 = $ "1." As a consequence, $\alpha_0 C$ now feeds *negative* charge packets into C as does $\alpha_2 C$. Again, $v_0 = 0$ is reached sooner. Hence, f_0 is *increased*.

A similar argument shows that for $v_{\text{in}} < 0$ the oscillation frequency *decreases*. For $\alpha_0, \alpha_2 \ll \alpha_1$ and $V_{SS} = -V_{DD}$,

$$f_0 \cong \frac{\alpha_2}{4\alpha_1} f_c + \frac{\alpha_0 v_{\text{in}}}{4\alpha_1 V_{DD}} f_c \qquad (6.18)$$

gives the oscillation frequency (Problem 6.13).

An oscillator which does not require a master clock is shown in Fig. 6.42a. The op-amp (used as a comparator) and its feedback resistors R_1 and R_2 form

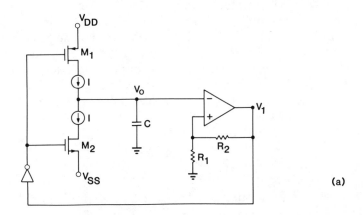

(a)

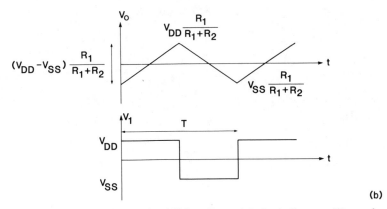

(b)

FIGURE 6.42.   A continuous-time MOS oscillator: (a) circuit diagram; (b) waveforms.

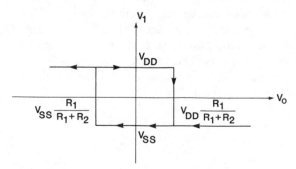

FIGURE 6.43. The dc transfer characteristics of the Schmitt trigger in Fig. 6.42a.

a Schmitt trigger with the dc transfer characteristics shown in Fig. 6.43; its output voltage v_1 controls the switches M_1 and M_2. Let $v_1 = V_{DD}$, the positive supply voltage; this means that the op-amp has a positive input voltage, and $v_0 < V_{DD}R_1/(R_1 + R_2)$. Since v_1 is high, M_2 is cut off and M_1 is turned on, charging C and thus increasing v_0 (Fig. 6.42b). When v_0 reaches $V_{DD}R_1/(R_1 + R_2)$, the op-amp output v_1 changes to V_{SS}. This causes M_1 to turn off and M_2 to turn on, discharging C and decreasing v_0 until it reaches $V_{SS}R_1/(R_1 + R_2)$. It can readily be shown (Problem 6.14) that for $V_{SS} = -V_{DD}$ the oscillation frequency is

$$f_0 = \frac{1 + R_2/R_1}{4C} \frac{I}{V_{DD}}. \tag{6.19}$$

To make the circuit function as a voltage-controlled oscillator (VCO), the two current sources I can be replaced by voltage-controlled current sources. This can be accomplished by using the voltage-to-current converter shown in Fig. 6.44. The op-amp is in a negative feedback loop, and thus forces the voltage across R_{in} to equal v_{in}. Hence, the current through the MOS device must be v_{in}/R_{in}.

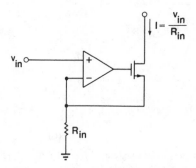

FIGURE 6.44. A voltage-to-current converter.

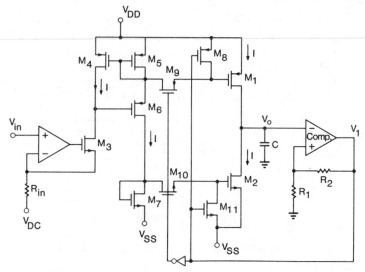

FIGURE 6.45. A VCO based on the circuits of Figs. 6.42 and 6.44.

The use of the converter in the VCO is illustrated in Fig. 6.45. Here, the current I in M_3 and M_4 is mirrored into the $M_5 - M_7$ bias chain and then into M_1 or M_2, depending on whether M_9 or M_{10} is conducting at the moment, that is, whether v_1 is high or low.* M_8 pulls up the gate of M_1 to cut it off when M_9 is off; M_{11} pulls down the gate of M_2 to cut it off when M_{10} is off. If $V_{SS} = -V_{DD}$, the oscillation frequency is

$$ f_0 = \frac{1 + R_2/R_1}{4R_{in}C} \frac{v_{in} - V_{DC}}{V_{DD}}. \tag{6.20} $$

Equation (6.20) can easily be obtained by replacing I in Eq. (6.19) by $(v_{in} - V_{DC})/R_{in}$. Here, V_{DC} is a bias voltage which (for $V_{DC} < 0$) enables the VCO to oscillate when $v_{in} = 0$. Alternatively, instead of applying V_{DC}, a constant-current source can be connected to the drain of M_4 to force it to conduct even if M_3 is cut off.

An SC *sine-wave* oscillator can also be obtained by combining an SC biquad (such as the ones described in Section 5.4) with a nonlinear amplitude-controlling circuit. The latter should be connected to the biquad in such a way that the pole-Q becomes negative if the peak value V_p of the output voltage v_{out} is less than a reference value V_r, but positive if $V_p > V_r$. For the circuit of Fig. 5.13c, a possible realization[22] of this scheme is shown in Fig. 6.46. To understand its operation, it should be recalled—see Eq. (5.45)—that the

*Note that M_9 and M_{10} are complementary-type devices, as are M_8 and M_{11}.

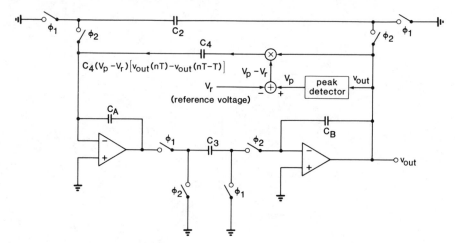

FIGURE 6.46.   A sine-wave SC oscillator.

pole-Q is determined by C_4; in fact, for the normalization $C_A = C_B = 1$ used earlier, $Q \cong 1/C_4$. In the circuit of Fig. 6.46, the charge fed back by C_4 in each clock cycle is $(V_p - V_r)C_4 \Delta v_{out}$, rather than $C_4 \Delta v_{out}$ as in the circuit of Fig. 6.13c. Thus, C_4 has effectively been replaced by $(V_p - V_r)C_4$, and thus now $Q \cong 1/[(V_p - V_r)C_4]$. If $V_p > V_r$, then $Q > 0$ and the peak amplitude V_p will decrease from cycle to cycle until $V_p = V_r$ is reached; if, however, $V_p < V_r$, the amplitude will increase to V_r. Thus, the added circuitry stabilizes the peak amplitude such that $V_p \cong V_r$.

Detailed analysis[22,23] of the circuit reveals that the value of C_4 which leads to the fastest settling of the amplitude-control process is $C_4 \cong 1/\pi V_r$, where the normalization $C_A = C_B = 1$ and the values $C_2 = C_3 = 2 \sin(\omega_0 T/2)$ are assumed (Problem 6.15).

The subtractor and multiplier blocks in Fig. 6.46 can both be realized by a single gain-controlled amplifier.[22]

An alternative sine-wave oscillator,[25] also based on the biquad of Fig. 5.13c, is shown in Fig. 6.47. Here, the two op-amps OA1 and OA2, along with the capacitors C_A, C_B, C_1, C_2, C_3, and C_4 form the stable high-Q biquad described earlier in Chapter 5. With MOSFET M1 initially off, C_5 in series with C_6 forms a positive feedback path which guarantees that the circuit will start oscillating when it is first turned on. When the oscillation is in progress, the output voltage v_1 of OA1 is limited into a square wave by the comparator op-amp OA3, and changed into the logic signals X and $\bar{X}$ by the two inverters. C_7 and C_8, along with the diode-connected NMOS enhancement devices M2, M3, M4, and M5, now perform as a full-wave rectifier which pumps positive charge into the stray capacitance C_s loading the gate of M1. This causes the gate voltage v of M1 to rise rapidly, turning M1 on. Now M1 grounds C_5 and C_6, breaking

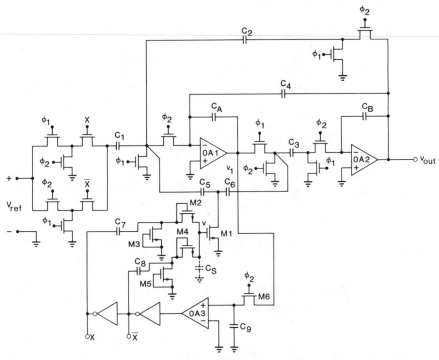

FIGURE 6.47. Sine-wave oscillator with hard-limited feedback path and startup circuit.

the positive feedback loop formed by them. However, another feedback loop is now in operation, activated by the logic signals X and $\bar{X}$. It contains the dc reference voltage V_{ref}, C_1, and the eight switches in the input circuit which are associated with them. With the oscillation in progress, X is in phase with v_1, and $\bar{X}$ in opposite phase. When $v_1 > 0$, X is high and hence C_1 is connected in an inverting mode. Therefore, for $V_{ref} > 0$ it feeds a negative charge $-C_1 V_{ref}$ in each clock cycle into C_4. This causes v_1 to rise further. Similarly, when $v_1 < 0$, $\bar{X}$ is high, and C_1 is connected as a noninverting switched capacitor. Hence, it will pump positive charges $C_1 V_{ref}$ into C_4, lowering v_1. This constitutes positive feedback. The negative feedback provided by C_4, however, keeps the amplitude of oscillation from rising. The sample-and-hold circuit of M6 and C_9 provides a clock cycle delay for the positive feedback loop, and thus prevents parasitic oscillations.

In steady state, the equivalent circuit of the oscillator of Fig. 6.47 is that shown in Fig. 6.48a, with $v_{in}(t)$ and $v_{out}(t)$ as shown in Fig. 6.48b. Since the biquad itself is stable, it simply acts as a filter-amplifier processing $v_{in}(t)$. Hence, the amplitude of the output voltage is simply determined by V_{ref} and

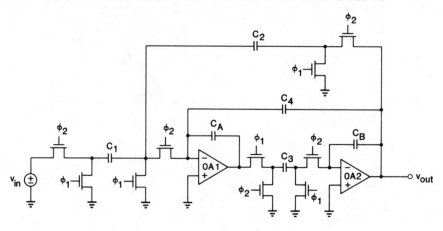

(a)

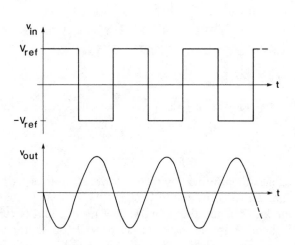

(b)

FIGURE 6.48. Equivalent circuit representing the operation of the oscillator of Fig. 6.47: (*a*) circuit diagram; (*b*) input and output waveforms.

the gain of the biquad at the oscillation frequency f_0. The latter is, for high pole-Q, approximately equal to the pole frequency of the biquad.

The fundamental component of the output voltage is a sine wave with frequency f_0, whose amplitude is given by

$$V_0 = \frac{4V_{\text{ref}}}{\pi} \frac{C_1}{C_2} Q. \tag{6.21}$$

Here, Q is the z-domain pole-Q, defined analogously to Eq. (5.32), with $|\sigma_p|$ replaced by the distance between the pole and the frequency axis (i.e., the unit circle), and $|s_p|$ by the angle $\omega_0 T$ of the pole. Thus, with $r \triangleq |z_p|$,

$$Q \triangleq \frac{\omega_0 T}{2(1 - r)} . \qquad (6.22)$$

For the biquad of Figs. 6.47 and 6.48, Q is given by[25]

$$Q \simeq \frac{1}{C_4} \left(\frac{C_A C_B C_2}{C_3} \right)^{1/2} . \qquad (6.23)$$

The proof of Eqs. (6.23) and (6.21) is implied in Problems 6.20 and 6.21.

In addition to the fundamental sine-wave with frequency f_0, the output will contain also the odd harmonics. This is because the input square wave $v_{in}(t)$ of Fig. 6.48 contains all odd-harmonic components, and these are only partially suppressed by the biquad. It can be shown (Problem 6.22) that the amplitude V_n of the nth harmonic is given by[25]

$$V_n = \frac{V_0}{n^3 Q} , \qquad n \text{ odd.} \qquad (6.24)$$

Ideally, there are no even harmonics in v_{out}, but any imperfections causing an asymmetry between the positive and negative signal swings will introduce some even-harmonic distortion.

The oscillatior of Fig. 6.47 can also be used to provide an amplitude-modulated sine wave, if V_{ref} contains the modulating signal. In addition, in the circuits of Figs. 6.46 and 6.47, by changing the values of C_A or C_B or C_2 or C_3, the frequency of oscillation can be tuned to implement frequency-shift keying, programmable-frequency oscillators, capacitance-to-frequency conversion, and so on.

PROBLEMS

6.1. Calculate the transfer function $H(z) = V_{out}/V_{in}(z)$ of the voltage amplifier of Fig. 6.2. Repeat your calculations assuming a constant *finite* gain for the op-amp.

6.2. Calculate $V_{out}(z)$ for the circuit of Fig. 6.3. Assume first infinite op-amp gain; then repeat the calculation with a constant finite gain.

6.3. Analyze the circuit of Fig. 6.6 to find its transfer function $H(z) = V_{out}/V_{in}$. Assume first infinite, then finite op-amp gain.

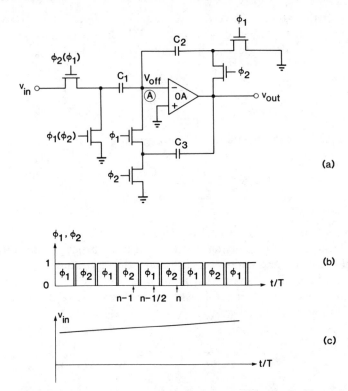

(a)

(b)

(c)

FIGURE 6.49.   An offset-compensated voltage amplifier (for Problem 6.4).

6.4. Figure 6.49 shows an offset-compensated voltage amplifier which does not require a high-slew-rate op-amp.[3] Analyze the circuit for both choices (shown in parentheses and without parentheses) of the input branch clock phases. How much does v_{out} vary between the two intervals $\phi_1 =$ "1" and $\phi_2 =$ "1"? Plot the output voltage v_{out} for both choices.

6.5. Calculate the total capacitance loading node $\widehat{x}$ in an n-bit PCA as shown in Fig. 6.10.

6.6. By analyzing the circuit of Fig. 6.13a five times, calculate the values of v_x in the five conversion cycles if $v_{in} = 0.7V_{ref}$.

6.7. Calculate the effective input offset voltage of the autozeroed comparator of Fig. 6.15 from the individual gains and offset voltages of its two stages A_1 and A_2.

6.8. Analyze the effects of the stray capacitances C_{st} and C_{sb} between the top and bottom plates of the capacitors and ground in Fig. 6.13a on the digital output.

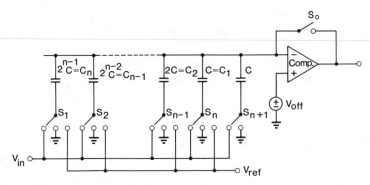

FIGURE 6.50. Comparator offset elimination method (for Problem 6.9).

6.9. The circuit of Fig. 6.50 can be used to eliminate the effect of the comparator offset voltage in the D/A converter of Fig. 6.13a. Analyze the operation of the circuit.

6.10. Analyze the operation of the offset-compensated comparator of Fig. 6.24 for a finite amplifier gain A.

6.11. Prove Eqs. (6.15) and (6.16) for the circuit of Fig. 6.39. (*Hints*: Assume $\alpha \ll 1$, $A \gg 1$, and $f_0 \ll f_c$; replace the switched capacitors by equivalent resistors.)

6.12. Prove Eq. (6.17) for the circuit of Fig. 6.40.

6.13. Prove Eq. (6.18) for the circuit of Fig. 6.41.

6.14. Show that the oscillation frequency of the circuit of Fig. 6.42a is given by Eq. (6.19).

6.15. Show that (for $V_p = V_r$) the circuit of Fig. 6.46 has natural nodes (poles) at $z_{1,2} = e^{\pm j\omega_0 T}$ if $C_A = C_B = 1$ and $C_2 = C_3 = 2\sin(\omega_0 T/2)$. Hence, the circuit is then a sine-wave oscillator with oscillation frequency ω_0.

6.16. For the inverter of Fig. 6.19a with the input–output characteristics shown in Fig. 6.19b, prove that the limits of the linear range (where Q_1 and Q_2 are both in saturation) are the intersections of the characteristics with the two 45° lines $v_B = v_A + |V_{Tp}|$ and $v_B = v_A - V_{Tn}$. Draw these lines for the curves of Fig. 6.19b, and identify the linear ranges.

6.17. For the multivibrator of Fig. 6.26a: (a) derive the small-signal equivalent circuit of Fig. 6.26b; (b) using Laplace-transform analysis, find the natural modes of the circuit; (c) find the natural modes for the case when the two inverters (Q_1/Q_3 and Q_2/Q_4) are cascaded without closed-loop feedback; (d) what conclusions can be drawn from the relative magnitudes of the natural modes of the two circuits?

6.18. The circuit of Fig. 6.20 is to be fabricated using a CMOS process with the following parameters: $V_{Tn} = -V_{Tp} = 1$ V; $\mu_n = 3\mu_p = 670$ cm^2/Vs; $t_{ox} = 800$ Å; $\lambda_n = 0.012$ V^{-1}, $\lambda_p = 0.02$ V^{-1}. Design the input inverter such that for $v_A = 0$ V also $v_B = 0$ V and that $i_{D1} = i_{D2} = 50$ μA. How much is the gain of the stage? (*Hint:* Use the formulas of Tables 3.2 and 3.3!)

6.19. In the circuit of Fig. 6.20a, the clock feedthrough capacitance between the gate of S_3 and node Ⓐ is 15 fF. The clock voltage is 10 V peak-to-peak. How large must C_1 be if the first two inverters (Q_1/Q_2 and Q_3/Q_4) are to operate with all devices in saturation, in spite of the clock feedthrough voltage at node Ⓐ? Assume the W and L values obtained in Problem 6.18 for the two input inverters.

6.20. Using the definition of the z-domain pole-Q given in Eq. (6.22), show that the pole-Q of the biquad of Fig. 6.48 is given by Eq. (6.23). Assume $\omega_0 T \ll 1$ and $Q \gg 1$. [*Hints:* Show using the methods of Section 5.4 that the denominator polynomial of $H(z)$ is

$$D(z) = z^2 + \left(\frac{C_3}{C_A C_B}(C_2 + C_4) - 2 \right) z + \left(1 - \frac{C_3 C_4}{C_C C_B} \right).$$

Equate this to $(z - re^{j\omega_0 T})(z - re^{-j\omega_0 T})$, and use the approximations indicated by $\omega_0 T \ll 1$ and $Q \gg 1$.]

6.21. Prove that the fundamental component of v_{out} of the circuit of Fig. 6.48 is given by Eq. (6.21). Assume, as in Problem 6.20, $\omega_0 T \ll 1$ and $Q \gg 1$. [*Hint:* Calculate $H(z)$; substitute $z_0 = \exp(j\omega_0 T)$.]

6.22. Prove that the amplitude of the nth odd harmonic of the output voltage v_{out} of the circuit of Fig. 6.48 is given by Eq. (6.24). What should the value of Q be if the total harmonic distortion

$$\left[\sum_{n=3}^{\infty} V_n^2 \right]^{1/2}$$

is to be less than 0.01% of the fundamental? Assume $\omega_0 T \ll 1$ and $Q \gg 1$.

6.23. The transfer function of a linear interpolator with a sampled-and-held input signal is given by the expression $H(z) = (1 - z^{-r})/(1 - z^{-1})$. Here r, an integer, is the interpolation ratio.[35-37] Figure 6.51 illustrates the input and output waveforms for $r = 4$, and Fig. 6.52 shows four circuits capable of providing this transfer function. (a) Analyze each

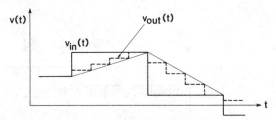

FIGURE 6.51. Input and output waveforms for a linear interpolator with a sampled-and-held input signal. The interpolation ratio is $r = 4$. A delay of rT, necessary for realizability, has been omitted for clarity.

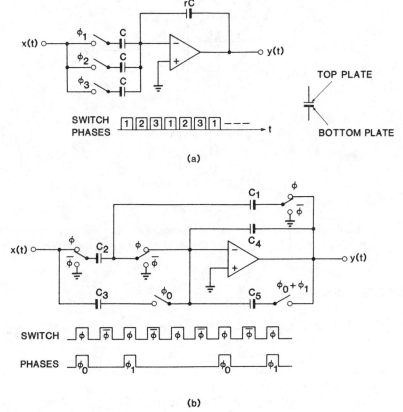

FIGURE 6.52. Linear interpolator circuits (for Problem 6.23): (a) an impractical circuit; (b)–(d) some practical circuits.

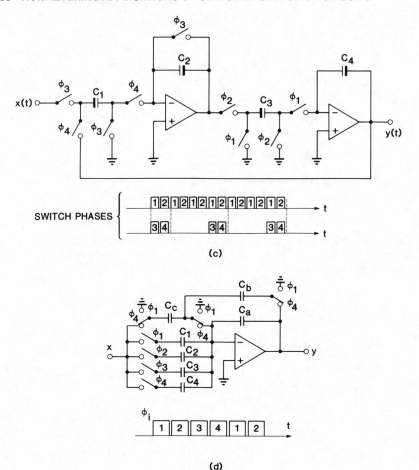

FIGURE 6.52. continued.

circuit. For what values of the elements do these circuits realize $H(z)$? What is r? (b) The circuit of Fig. 6.52a is not practical. Why?

REFERENCES

1. T. Foxall, R. Whitbread, L. Sellars, A. Aitken, and J. Morris, *1980 ISSCC Digest Tech. Papers*, 90–91 (1980).
2. R. Gregorian, *Microelectronics J.*, **12**, 10–13 (1981).
3. K. Haug, G. C. Temes, and K. Martin, Proceedings of the International Symposium on Circuits and Systems, pp. 1054–1057 (1984).
4. R. Gregorian, K. Martin, and G. C. Temes, *Proc. IEEE*, **71**, 941–966 (1983).
5. R. H. McCharles and D. A. Hodges, *IEEE Trans. Circuits Syst.*, **CAS-25**, 490–497 (1978).

6. J. McCreary and P. R. Gray, *IEEE J. Solid-State Circuits*, **SC-10**, 371–379 (1975).

7. R. Gregorian and G. Amir, *1981 ISCAS Proc.*, 733–736 (1981).

8. D. J. Allstot, R. W. Brodersen, and P. R. Gray, *1979 ISSCC Digest Tech. Papers*, 76–88 (1979).

9. R. Gregorian, *Microelectronics J.*, **12**, 10–13 (1981).

10. T. Redfern, J. Connolly, S. Chin, and T. Frederiksen, *IEEE J. Solid-State Circuits*, **SC-14**, 912–920 (1979).

11. B. Fotouhi and D. A. Hodges, *IEEE J. Solid-State Circuits*, **SC-14**, 920–926 (1979).

12. A. Dingwall, *IEEE J. Solid-State Circuits*, **SC-14**, 926–932 (1979).

13. R. McCharles and D. A. Hodges, *IEEE Trans. Circuits Syst.*, **CAS-25**, 490–497, (1978).

14. K. Martin, Proceedings of the 15th Asilomar Conference on Circuits Systems and Computers, pp. 489–492 (1981).

15. K. Martin and A. S. Sedra, *IEEE Trans. Circuits Syst.*, **CAS-28**, 576–584 (1981).

16. C. Hewes, D. Mayer, R. Hester, W. Eversole, R. Hiri, and R. Pettengill, International Conference on the Application of CCD's, 3A (1978).

17. Y. Haque and V. Saletore, Proceedings of the 15th Asilomar Conference on Circuits, Systems and Computers, pp. 303–307 (1981).

18. H. Khorramabadi, *NMOS Phase Lock Loop*, University of California, Berkeley, Int. Memo UCB/ERL, M77/67 (1977).

19. W. Steinhagen and W. Engl, *IEEE J. Solid-State Circuits*, **SC-13**, 799–805 (1978).

20. E. Vittoz, *IEEE J. Solid-State Circuits*, **SC-14**, 662–624 (1979).

21. K. Martin, *IEEE J. Solid-State Circuits*, **SC-16**, 412–414 (1981).

22. Fr. Krummenacher, personal communication.

23. W. B. Mikhael and S. Tu, Proceedings of the International Symposium on Circuits and Systems, pp. 1149–1152 (1983).

24. Y. A. Haque, R. Gregorian, R. W. Blasco, R. A. Mao, and W. E. Nicholson, *IEEE J. Solid-State Circuits*, **SC-14**, 961–969 (1979).

25. P. E. Fleischer, A. Ganesan, and K. R. Laker, *IEEE J. Solid-State Circuits*, **SC-20**, 641–647 (1985).

26. Y. S. Yee, L. M. Terman, and L. G. Heller, *IEEE J. Solid-State Circuits*, **SC-13**, 63–66 (1978).

27. D. J. Allstot, *IEEE J. Solid-State Circuits*, **SC-17**, 1080–1087 (1982).

28. R. Gregorian and J. G. Gord, *IEEE J. Solid-State Circuits*, **SC-17**, 692–700 (1983).

29. K. W. Martin, Project Reports. Microelectronics Innovation and Computer Research Opportunities (MICRO) Program, University of California, pp. 99–102 (1983).

30. J. L. McCreary and J. B. Hunt, *IEEE J. Solid-State Circuits*, **SC-16**, 689–694 (1981).

31. H. S. Lee, D. A. Hodges, and P. R. Gray, IEEE International Solid-State Circuits Conference, pp. 64–65 (1984).

32. K. Martin, personal communication.

33. W. C. Black, Jr., personal communication.

34. Y. Fujita, E. Masuda, S. Sakamoto, T. Sakaue, and Y. Sato, IEEE International Solid-State Circuits Conference, pp. 56–57 (1984).

35. M. B. Ghaderi, G. C. Temes, and S. Law, *IEE Proc.*, **128**, Pt. G, 213–215 (1981).

36. M. B. Ghaderi, New Design Techniques for Switched-Capacitor Bandpass Filters, Ph.D. dissertation, Chapter 7, UCLA (1981).

37. T.-H. Hsu, Improved Design Techniques for Switched-Capacitor Ladder Filters, Ph.D. dissertation, Chapter 8, UCLA (1982).

Chapter Seven

NONIDEAL EFFECTS IN SWITCHED-CAPACITOR CIRCUITS

As discussed earlier, switched-capacitor circuits are usually intricate systems containing many components and multiple feedback loops, and are realized normally in a fully integrated form. As a result of the integrated realization, many unavoidable parasitic effects occur in the circuit. If these nonideal effects are not considered carefully, and are not minimized or eliminated by appropriate design techniques, then they can make the fabricated circuit inoperable, or at least deficient in performance. One such parasitic effect, involving stray capacitances, has already been discussed in Section 5.3, where we have also described an efficient method for avoiding its detrimental effects on the response of switched-capacitor integrators.

In this chapter, additional nonideal effects will be considered. We shall describe the origin of each effect, analyze its influence on the performance of the circuit, and (whenever available) discuss methods for their elimination.

Because of their great impact, the effects described in the following should *not* be considered minor practicalities or mundane second-order phenomena. Any design effort in which these potential problems are not considered is likely to turn into a futile exercise which yields an inoperative product. Hence, the importance of the topics in this chapter is comparable to those in any of the preceding ones.

7.1. NONIDEAL EFFECTS IN THE SWITCHES

In Section 3.6, we have already briefly discussed the realization of MOSFET switches, including the effects of their stray capacitances and leakage. In the following, these and other nonidealities will be analyzed in more detail.

Nonzero "On"-Resistance

As mentioned in Section 3.6, the instantaneous value of the clock signal on the gates of the MOSFETs used as switches is usually the supply voltage (V_{DD} or V_{SS}) of the circuit. Thus, normally $|v_{GS} - V_T| > |v_{DS}|$ when the device is "on"; it is thus in its nonsaturated region and therefore behaves as a linear resistor of value

$$R_{on} = \frac{1}{2k(v_{GS} - V_T)} \tag{7.1}$$

as given earlier in Eq. (3.34). In a typical application, the switch may be used to charge a capacitor C to an input voltage v_{in} (Fig. 7.1). For the NMOS switch shown, the transistor will operate in its linear (triode) region if the clock signal $\phi = V_{DD} > \max(v_{in} + V_{Tn}, v_{out} + V_{Tn})$. Then, by (3.33),

$$i_{out} = C\frac{dv_{out}}{dt} = k\left[2(v_{GS} - V_{Tn})v_{DS} - v_{DS}^2\right], \tag{7.2}$$

where V_{Tn} is the NMOS threshold voltage, while $v_{DS} = v_{in} - v_{out}$ and $v_{GS} = V_{DD} - v_{out}$. Assuming the initial condition $v_{out}(0) = 0$ (i.e., an initially discharged capacitor) and a constant v_{in}, Eq. (7.2) may be solved to find $v_{out}(t)$. Ignoring the body effect, that is, assuming that V_{Tn} is independent of v_{out}, Eq. (7.2) can easily be solved (Problem 7.1). The result is

$$t_{ch} = \frac{C}{2k(V_{DD} - v_{in} - V_{Tn})} \ln\left(\frac{v_{in}(v_{out} + v_{in} + 2V_{Tn} - 2V_{DD})}{(v_{out} - v_{in})(2V_{DD} - v_{in} - 2V_{Tn})}\right), \tag{7.3}$$

where t_{ch} is the time required to charge C to the value v_{out}. The charging time needed to reach, say, $v_{out} = 0.999v_{in}$ clearly depends on the value of v_{in}. As v_{in} approaches $V_{DD} - V_{Tn}$, t_{ch} approaches infinity, since the channel resistance R_{on}, by (7.1), becomes infinite. Thus signals reaching this level will be clipped. To prevent this in NMOS technology a "boot-strapped" clock driver (Fig. 7.2a) may be used. In this circuit, the input clock signal ϕ_1 rises first, and pulls the output clock signal V_ϕ up to $V_{DD} - V_{Tn}$. Next, ϕ_2 rises with a delay $t_2 - t_1$ and adds a step $V_{DD}C/(C + C_L)$ to V_ϕ, where C_L is the load capacitance. The maximum value V_{max} of V_ϕ can, for $V_{Tn} \ll V_{DD}$ and $C_L \ll C$, be close to $2V_{DD}$. To avoid an undershoot ($V_\phi < 0$), ϕ_2 should return to zero before ϕ_1 does (Fig. 7.2b).

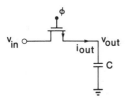

FIGURE 7.1.   NMOS transistor used as a bidirectional switch.

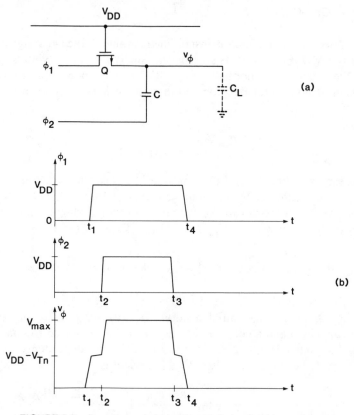

FIGURE 7.2.   Bootstrapped clock driver: (*a*) circuit; (*b*) waveforms.

In CMOS technology, as discussed in Section 3.6, clock feedthrough considerations often require the use of a transmission gate (Fig. 7.3) containing two
complementary devices. This also avoids the clipping problem, since now at
least one of the two devices will always conduct whenever ϕ is high, regardless
of the signal level v_{in}. The on-resistance R_{on} performance of NMOS, PMOS,
and CMOS switches is compared in Fig. 7.4 (from Ref. 6). As the curves show,
for an *n*-channel device with the processing parameters given in the figure and

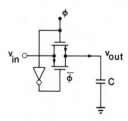

FIGURE 7.3.   CMOS transmission gate.

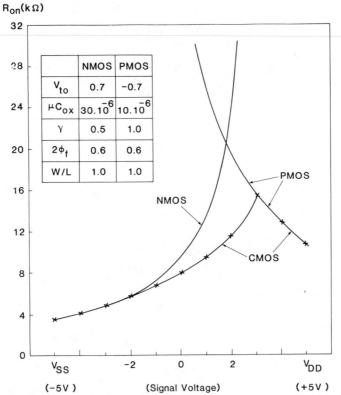

	NMOS	PMOS
V_{to}	0.7	-0.7
μC_{ox}	30.10^{-6}	10.10^{-6}
γ	0.5	1.0
$2\phi_f$	0.6	0.6
W/L	1.0	1.0

FIGURE 7.4.   Variations in the small-signal ON resistance of several commonly used switch types vs changes in the voltage level v_{in} (from Ref. 6, © 1983 IEEE).

a gate voltage of $V_{DD} = 5$ V, the switch will cut off when the signal voltage exceeds 2 V. Similarly, the PMOS switch cuts off when $V_{in} < 0$. The composite (CMOS) switch, by contrast, has a maximum on-resistance of 16 kΩ, and hence conducts for all values of v_{in}.

It should be noted that, unlike the simplified analysis leading to Eq. (7.3), the curves of Fig. 7.4 *do* include the body effect, computed using the parameters indicated in the table contained within the figure.

The small-signal settling time of the circuit of Fig. 7.1 (or of Fig. 7.3) to within 0.1% of the final value of v_{out} is approximately $7R_{on}C$. For a given clock frequency, the maximum value of R_{on} can thus be found, and from R_{on} the minimum permissible value of the aspect ratios W/L of the MOSFETs can be calculated. In this calculation, the worst-case values should be chosen[6] for the signal level v_{in} and for the chip temperature which also affects R_{on}.

In order to illustrate the effect of the nonzero R_{on} on the operation of a simple switched-capacitor stage, we will next consider the integrator of Fig. 5.4, reproduced in Fig. 7.5a. Including the on-resistance of both switches, the

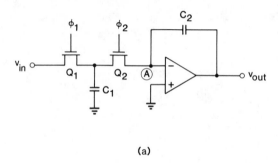

(a)

(b)

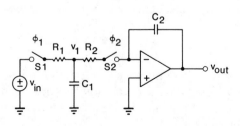

(c)

FIGURE 7.5. Switched-capacitor integrator: (*a*) circuit diagram; (*b*) clock waveforms; (*c*) equivalent circuit including switch resistances.

equivalent circuit of Fig. 7.5*c* results. In the ideal case, when $R_1 = R_2 = 0$, the transfer function is

$$H(z) \triangleq \frac{V_{out}(z)}{V_{in}(z)} = \frac{-C_1/C_2}{z-1}. \qquad (7.4)$$

Assuming that the switch resistances R_1 and R_2 are linear and that no clipping takes place, at the end of clock period ϕ_1 when $t = nT$ the voltage across C_1 is given by

$$v_1(nT) = v_{in}(nT)(1 - e^{-T/2R_1C_1}).$$

(7.5)

Here, it was assumed that C_1 was fully discharged at $t = (n - 1/2)T$, and that switches S_1 and S_2 are both closed for periods of length $T/2$, where $T = 1/f_c$ is the sampling period. Neither of these assumptions is exactly valid, but for usual conditions they represent good approximations.

Next, during the time when ϕ_2 is high $[nT < t < (n + 1/2)T]$ C_1 discharges into C_2. The charge delivered is

$$\delta q(nT + T/2) = C_1 v_1(nT)(1 - e^{-T/2R_2C_1}).$$

(7.6)

Hence, from $v_{out}(nT + T) - v_{out}(nT) = \delta q(nT + T/2)/C_2$, using z-transformation, we obtain for $R_1 = R_2 = R$

$$H(z) = \frac{-(1 - e^{-T/2RC_1})^2 C_1/C_2}{z - 1}$$

(7.7)

as the new transfer function. Thus, as a comparison with (7.4) shows, on-resistances reduce the effective value of the capacitance ratio C_1/C_2. The relative error is

$$\varepsilon = 1 - (1 - e^{-T/2RC_1})^2 \cong 2e^{-T/2RC_1}.$$

(7.8)

If ε is much less than the achievable tolerance (typically, 0.1%) of C_1/C_2, then this effect will not be noticeable. Thus, we may require, say,

$$\varepsilon \cong 2e^{-T/2RC_1} \leq 10^{-4}$$

(7.9)

so that

$$\frac{RC_1}{T} = RC_1 f_c \leq \frac{1}{2 \ln 20{,}000} \cong 0.05.$$

(7.10)

Thus, $RC_1 \leq T/20$ is a reasonable requirement. For a clock frequency of, say, $f_c = 500$ kHz and a capacitance value $C_1 = 5$ pF, this gives $R \leq 20$ kΩ which is easily achieved even for minimum-sized switches as Fig. 7.4 shows. It should be noted, however, that R_{on} is both voltage and temperature dependent. Hence, the designer must make sure that the condition on the time constants is satisfied for all possible values of v_{in} and T. If the nonzero charging time effect is *not* negligible, then it is also a source of nonlinear distortion, due to the dependence of R on v_{in}. Hence, it should be avoided whenever possible.

Capacitive Coupling of the Clock Signals

As briefly mentioned already in Section 3.6, some important nonlinear effects associated with MOS switches are due to their parasitic capacitances. As an illustration, Fig. 7.6 shows the crucial gate-to-source and gate-to-drain stray capacitances of the switches in the integrator of Fig. 7.5a. Each of these capacitances contains a voltage-independent (linear) component C_{ov} due to the overlap of the gate electrode and the source or drain diffusion, and a voltage-dependent (nonlinear) component C_{ch} due to the gate-to-channel capacitance. We shall next consider the dynamic behavior of this nonideal circuit when the clock signals are present at the gates of Q_1 and Q_2.

When ϕ_1 is "high" and hence Q_1 is conducting but Q_2 is cut off, C_1 is connected to v_{in}, and the stray capacitances play no role. However, as ϕ_1 goes "low" and ϕ_2 rises, a complicated transient involving the strays occurs. The top plate of C_{g4} acquires a positive charge $C_{g4}V_{\phi2}$ from the clock ϕ_2. This requires a matching charge $-C_{g4}V_{\phi2}$ at the bottom plate, which must come from C_2. Similarly, C_{g3} draws a charge from C_1. As $V_{\phi2}$ rises, Q_2 forms a channel and now C_{g3} also draws some of its charge from C_2. If the net positive charge delivered to both strays by ϕ_2 is δq, then v_{out} changes by $-\delta q/C_2$. In addition, as Q_1 cuts off, C_{g2} must recharge from C_2. These effects cause an error due to the strays.

At the end of the half-period during which ϕ_2 is high, $V_{\phi2}$ falls. Now C_{g3} and C_{g4} discharge and V_{out} is rising. At one point, when $V_{\phi2}$ falls below the threshold voltage V_{Tn}, Q_2 cuts off and isolates C_1. As $V_{\phi2}$ falls further, C_{g3} discharges only through C_1, and no longer through C_2. Hence, the total charge restored to C_2 will be *less* than that removed from C_2 when ϕ_2 was falling, and there will be a net change in v_{out} due to the clock feedthrough in every clock cycle. Thus, in the absence of dc feedback, after a number of such cycles the amplifier would saturate.

The above simplified discussion ignored the role which the channel charges of Q_1 and Q_2 play, and also the effect of ϕ_1 via C_{g2}. Nevertheless, it leads to some qualitatively correct conclusion, namely that a net change δV_{out}, proportional to the C_{gi} and the clock voltage but inversely proportional to C_2, is generated in each clock cycle, and will eventually saturate the op-amp in the absence of negative dc feedback.

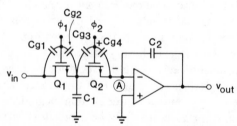

FIGURE 7.6. Switched-capacitor integrator with feedthrough capacitors.

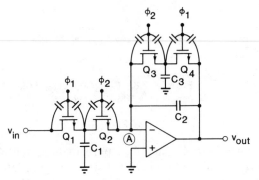

FIGURE 7.7. Switched-capacitor stage with feedback.

In the usual situation when a negative dc feedback does exist in the circuit, the cumulative effect of the clock feedthrough can be avoided. Consider again, for example, the "lossy" integrator of Fig. 7.7. Assuming that $v_{in} = 0$ and the op-amp is ideal, under steady-state conditions the only charges injected into node Ⓐ are δq_{clock} (the net charge due to clock feedthrough), and the charge $C_3 v_{out}$ conveyed by C_3. Hence, from charge conservation in the steady state when the charge in C_2 does not vary from cycle to cycle, we have

$$\delta q_{clock} + C_3 v_{out} = 0$$

and   (7.11)

$$v_{out} = -\frac{\delta q_{clock}}{C_3}.$$

Thus, a finite offset output voltage $-\delta q_{clock}/C_3$ results. Typical values are $\delta q_{clock} \approx 0.1$ pC and $C_3 \approx 1$ pF; hence the output offset voltage is around 0.1 V.

The clock feedthrough can clearly be reduced by using minimum-size transistors with small stray capacitances. MOSFETs fabricated with self-aligned gate techniques are especially useful as low clock feedthrough switches. Also, the feedback capacitances (C_2 and C_3 in Fig. 7.7) should be chosen as large as possible, so that for given clock charge injection the output voltage should not change significantly. In addition, the amplitudes of the clock voltages should not be larger than necessary for the reliable operation of the switches. In cascade circuits, alternating the roles of ϕ_1 and ϕ_2 from section to section may result in some cancellation of clock feedthrough effects.* Also, the autozeroing techniques used for op-amp offset cancellation may be used to cancel offset voltages due to clock feedthrough.

Another approach to reducing clock feedthrough effects is to use compensation (charge canceling) schemes. A circuit useful for NMOS technology[1] is shown in Fig. 7.8. Here, a "dummy" transistor Q_b with its drain and source shorted, and with a complementary gate signal $\bar{\phi}$, is added to the switching

---

*Only the dc offset will be cancelled by this arrangement, which will also reduce the settling time of the overall circuit.

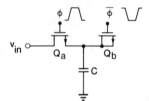

FIGURE 7.8.   Charge-canceling device for feedthrough compensation in NMOS technology.

transistor Q_a. Since both C_{gs} and C_{gd} participate in the clock feedthrough associated with Q_b, while only one of the strays of Q_a is always effective, the width W_b of Q_b should be approximately half of that of Q_a for equal lengths $L_a = L_b$.

A shortcoming of the scheme of Fig. 7.8 is that only a undefined part of the channel charge of Q_a flows into C, while all of that of Q_b does. An improved circuit[2] in which it is assured by circuit symmetry that only half of the channel charge of Q_a enters C, and hence the cancellation extends to the channel charges, is shown in Fig. 7.9. The cost is the extra capacitor C'.

For CMOS technology, for a first-order cancellation the transmission gate shown in Fig. 7.3 can be used. Now the dimensions of the two transistors should be about the same if they have complementary symmetry. For improved charge cancellation, the circuit of Fig. 7.8 may be used with both Q_a and Q_b realized by transmission gates. Thus, a single switch is realized using four MOSFETs. In all charge cancellation circuits the clock signal and its complement can be matched better, and hence better cancellation achieved, if the clock signals have trapezoidal (Fig. 7.8) rather than rectangular waveforms.

Charge cancellation is, of course, necessary only for those switches whose stray capacitances can inject some net charge into the signal path. For the circuit of Fig. 7.10a, with carefully controlled clock waveforms, the only sensitive switch is Q_5, that is, the device connected to the virtual ground node (A).

In fully differential circuits, to be discussed later, all offset effects (including those caused by clock feedthrough) cancel at least to a first-order approximation. Similar results can also be obtained by using partially differential circuits.[3] As an illustration, Fig. 7.10a shows an offset-free voltage amplifier. The only stray capacitance through which net clock feedthrough charges can enter node (A) and then C_1 and C_2 is C_{gds}, between the gate and drain (connected to (A))

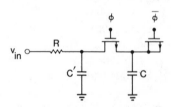

FIGURE 7.9.   Improved charge-canceling circuit. $C' = C$ for symmetry.

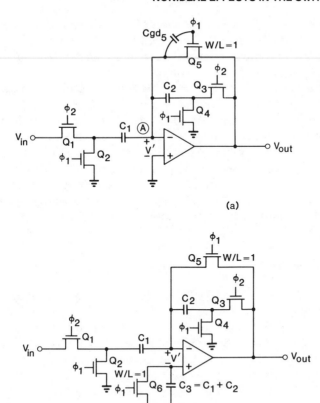

(a)

(b)

FIGURE 7.10. (a) Voltage amplifier. (b) Voltage amplifier with clock feedthrough compensation.

of Q_5. Thus, when ϕ_1 goes low and Q_5 cuts off, the charge in C_{gd5} changes by $\delta q \approx -C_{gd5}(V_T - V_{SS})$. This differential charge comes from C_1^* and C_2, and hence it results in a voltage drop $\delta q/(C_1 + C_2)$ at node Ⓐ. Figure 7.10b illustrates how this effect can be compensated for. By adding a transistor Q_6 and capacitor C_3 at the noninverting terminal of the op-amp, we can create a matching voltage drop at that terminal when ϕ_1 goes low. Thus, the voltage drop is canceled in the differential input voltage v' and hence also in v_{out}. Simulations indicate that a clock feedthrough reduction of over 40 dB is achievable for this circuit.[3] The technique can also be extended to more general SC stages.[3] As illustrated in Fig. 7.10b, the cost of the compensation is

---

*Note that in a practical implementation of the circuit of Fig. 7.10a, Q_5 should be cut off slightly before Q_2 to reduce clock feedthrough effects.[9] Hence, C_1 will still be grounded when Q_5 cuts off.

the doubling of the total capacitance needed for the circuit. Also, the symmetry (and thus the cancellation of the clock feedthrough) is not quite as good as for the fully differential circuit discussed later, in Section 7.4. The latter circuit is, however, more complicated, and requires a larger chip area.

In addition to the effect of the clock feedthrough capacitance connected to the inverting input terminal of the op-amp (e.g., in Fig. 7.6, C_{g4}), in stray-insensitive circuits there is also an additional clock coupling effect. This is due to the stray capacitances of the switches connected to the input capacitances of the integrators. It can be shown[9] that this effect is proportional to $(C_g/C_1)\Delta V$, where C_g is the gate-to-source (or gate-to-drain) stray capacitance, C_1 the input capacitance of the integrator, and ΔV the gate voltage range over which the switch remains open. This effect can be reduced by using slightly shifted clock signals for each switch, thus effectively replacing the two-phase clock by a four-phase one.[9]

Junction Leakage

The sources and drains of all MOS switching transistors are either diffused or ion-implanted regions, which are reversed biased with respect to the substrates. Thus, each such region is bordered by a reverse-biased p-n junction. Through each such junction, a small leakage current (due to minority-carrier generation, and thus strongly temperature dependent) flows. At room temperature, this current has a density of about 10 pA/mil^2. In an SC integrator, such as that shown in Fig. 7.5a, the leakage currents entering node Ⓐ will be integrated by the feedback capacitor C_2, and will eventually saturate the op-amp. As with the clock feedthrough effect, a negative dc feedback can prevent this saturation. Thus, consider the lossy integrator circuit of Fig. 7.11a, which also shows the leakage currents associated with the switches Q_1 and Q_2, modeled by the current sources I_{l_1}, I_{l_2}, and I_{l_3}. In the clock period when ϕ_1 is high, Q_1 is "on" while Q_2 is "off," and thus only the current I_{l_1} is integrated by C_1. Hence, at $t = nT - T/2^-$, the output voltage is given by

$$v_{\text{out}}(nT - T/2^-) = v_{\text{out}}(nT - T) + I_{l_1}T/2C_1. \qquad (7.12)$$

This voltage charges C_2, and thus when ϕ_2 turns high a charge $C_2 v_{\text{out}}(nT - T/2^-)$ enters C_1. By charge conservation at node Ⓐ, therefore,

$$-C_1[v_{\text{out}}(nT - T/2^+) - v_{\text{out}}(nT - T/2^-)] = C_2 v_{\text{out}}(nT - T/2^-). \qquad (7.13)$$

During the half-period when $nT - T/2 < t < nT$, the current $-(I_{l_1} + I_{l_2})$ charges C_1. Hence,

$$v_{\text{out}}(nT) = v_{\text{out}}(nT - T/2^+) + (I_{l_1} + I_{l_2})T/2C_1. \qquad (7.14)$$

In the steady state, $v_{\text{out}}(nT) = v_{\text{out}}(nT - T) = V_{os}$. Here, from (7.12)–(7.14),

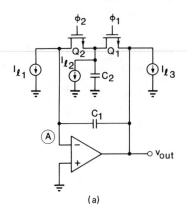

(a)

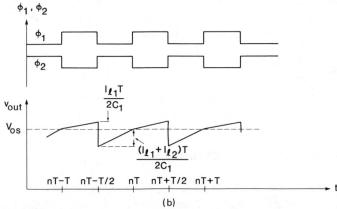

(b)

FIGURE 7.11. (*a*) Switched-capacitor lossy integrator with leakage sources. (*b*) Output voltage waveform of a lossy integrator due to leakage.

the offset voltage can be found to be

$$V_{os} = \frac{T}{2}\left[(2I_{l_1} + I_{l_2})/C_2 - I_{l_1}/C_1\right] \qquad (7.15)$$

(see Problem 7.2).

The waveforms of the parasitic signals introduced into v_{out} are illustrated in Fig. 7.11*b*. The resulting variation of v_{out} is a dual-slope ramp, with its dc offset given by (7.15). Since, as mentioned earlier, the density of I_l at room temperature is of the order of 10 pA/mil^2, for a 5-μm $\times$ 5-μm diffusion the leakage current will be around 0.4 pA. For capacitance values $C_1 = 10C_2 = 10$ pF and a clock frequency of 1 kHz, for example, (7.15) thus gives an output offset voltage of only a few mV. Thus this effect becomes important only at very low clock frequencies, say around 100 Hz or below. Such low clock rates seldom are needed. It should be noted, however, that the above calculations are

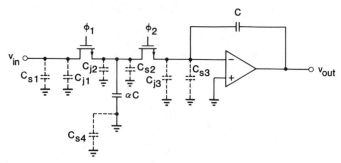

FIGURE 7.12.   Integrator circuit showing the important stray capacitances.

valid only at room temperature and below. Since the leakage current approximately doubles for every 10°C rise in temperature, at 100°C the minimum permissible clock rate rises to about 25 kHz!

Thermal and 1/f Noise

Another important nonideal effect is the noise generated internally by the switches. This will be discussed (along with the op-amp noise) in Section 7.4.

7.2.   CAPACITANCE INACCURACIES

As illustrated by the many SC circuits discussed in Chapters 5 and 6, the voltage gain of an SC two-port depends only on the *ratios* of the capacitance values of the circuit. As indicated in Problem 7.4, this statement is generally true for *any* SC network. Hence, only the factors affecting the achievable accuracy of these *ratios* are of primary interest to the designer. For circuits which do *not* have a fully stray-insensitive configuration, such as the simple integrator of Fig. 7.5a, the capacitance values are strongly influenced by the parasitic capacitances associated with the *p-n* junctions and the interconnects (leads). A circuit diagram illustrating these imperfections for the lossy SC integrator is shown in Fig. 7.12. Here, C_{S1} and C_{S2} are interconnect capacitances, while C_{S3} is the combination of the capacitances associated with the op-amp input, the corresponding interconnects, and the left-side plate* of the feedback capacitor C. Also, C_{j1}, C_{j2}, and C_{j3} are the junction capacitances between the drains and sources of the switches and the substrate. Finally, C_{S4} is the capacitance between the bottom plate of αC and ground. Neglecting all strays, the transfer function of the stage is

$$H(z) = \frac{V_{out}(z)}{V_{in}(z)} = \frac{-\alpha}{z-1}. \tag{7.16}$$

---

*As pointed out earlier, this electrode should be the top plate of C_1, to reduce noise injection into the circuit.

Considering now the parasitic capacitances, it is clear that for ideal op-amp performance only C_{j2} and C_{S2} contribute any net charge to the feedback capacitance C. They are, in fact, in parallel with the input capacitor αC and hence set a lower limit to the nominal size of this capacitor for a given accuracy. The resulting increase in the value of αC, and thus of C, can lead to a major increase in the total die area occupied by the stage. In addition, C_{j2} is a *p-n* junction capacitance, and, as such, it is a nonlinear element whose value depends on the applied voltage v. The formula relating C_j to v for a junction capacitance was given in Section 3.1, as Eq. (3.2). This relation can be rewritten in the form

$$C_j = \frac{C_j(0)}{\sqrt{|v|/\phi_i + 1}}, \tag{7.17}$$

where $C_j(0)$ is the capacitance when no voltage is applied. Assume that for the *n*-channel circuit shown in Fig. 7.12 the substrate is connected to the (negative) voltage V_{SS}. Then $|v| = |v_{in} - V_{SS}|$. Assuming, for simplicity, that $|v_{in}| \ll |V_{SS}|$, we obtain by series expansion from (7.17) the approximating relation

$$C_{j2}(v_{in}) = C_{j2}^0 \left(1 - \frac{v_{in}}{2(|V_{SS}| + \phi_i)} \right), \tag{7.18}$$

where

$$C_{j2}^0 \triangleq \frac{C_{j2}(0)}{\sqrt{|V_{SS}|/\phi_i + 1}}. \tag{7.19}$$

The actual capacitance ratio of the stage of Fig. 7.12 can therefore be approximated by

$$\alpha_{actual} \approx \alpha + \frac{C_{S2}}{C} + \frac{C_{j2}^0}{C} \left(1 - \frac{v_{in}}{2(|V_{SS}| + \phi_i)} \right). \tag{7.20}$$

The dependence of α on v_{in} will give rise to second harmonic distortion; its value can be estimated from Eq. (7.20).

As described in Section 5.3, the effects of the stray capacitances C_{Si} and C_{jk} can be almost completely eliminated by using the stray-insensitive integrators shown in Fig. 5.8. The only remaining stray capacitance in parallel with (say) αC is then that between the leads connecting the two plates of this capacitor to the rest of the circuit. This can be reduced to less than 1 fF by keeping the leads well separated, or (if that is impossible) by placing a grounded shielding ("guard") line between them.

In addition to the effects of the external stray capacitances, the capacitance ratios are, of course, also affected by the inaccuracies of the capacitances themselves. As discussed in Section 3.5, these inaccuracies can originate from

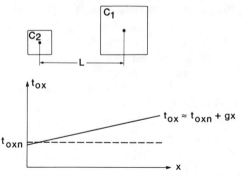

FIGURE 7.13. Capacitor ratio error due to thin-oxide gradient.

variations of the dimensions and the oxide thickness of the capacitors. Systematic variations, such as undercut, oxide thickness gradient, and so on, can be compensated for by using the strategies described in Section 3.5; the most effective technique is usually to construct all capacitors from smaller unit capacitors connected in parallel. To illustrate this technique, consider a circuit whose operation depends on the ratio of the capacitances C_1 and C_2. Placing these devices side by side, as shown in Fig. 7.13, leaves the ratio $\alpha = C_1/C_2$ vulnerable to a gradual variation of the oxide thickness t_{ox} along the chip. Assuming that this variation is linear (Fig. 7.13), a relative error

$$\frac{\Delta\alpha}{\alpha} \cong \frac{-g}{t_{oxn}} L \qquad (7.21)$$

results, where t_{oxn} is the average oxide thickness of C_2, g is the gradient $\partial t_{ox}/\partial x$, and L is the distance between C_1 and C_2. Experimentally, $|g/t_{oxn}| = 10 \sim 100$ ppm/mil has been found.[5]

Figure 7.14 shows the so-called common centroid geometry[4,5] realization of C_1 and C_2. For a linear variation of t_{ox} in any direction in the plane, the ratio C_1/C_2 remains (to a first-order approximation) unaffected. The same considerations hold for any parameter (temperature, etc.) which affects the capacitance and whose variation can be considered linear over the area of the capacitor array.

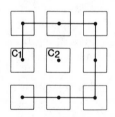

FIGURE 7.14. Typical common-centroid layout technique.

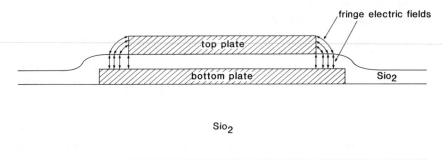

FIGURE 7.15.   Field fringing effect in a capacitor.

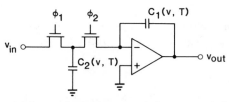

FIGURE 7.16.   Effects of voltage and temperature variations on a switched-capacitor integrator.

Since generally C_1/C_2 is not an integer value, one of the unit elements of the larger capacitor (here C_1) must be somewhat larger or smaller than the rest. To retain insensitivity to edge effects, such as undercut or field fringing (illustrated in Fig. 7.15), the procedure described in Section 3.5 can be used. Thus, the last element should be chosen rectangular (not square), and its dimensions determined as given by Eq. (3.31), where α is the ratio of the capacitance of the last element to that of a unit element, and the dimensions of the unit elements are assumed to be $W_2 = L_2$. Normally, $1 < \alpha < 2$.

The values of the MOS capacitors are somewhat dependent also on the applied voltage and the temperature. The voltage coefficient

$$\gamma_v^C \triangleq \frac{1}{C} \frac{\partial C}{\partial v} \qquad (7.22)$$

of a capacitor formed by a metal plate over heavily doped diffusion is of the order of -10 ppm/V* for a doping density of 5×10^{20} cm^{-3}. It is much lower than this value, and hence usually negligible, for metal-over-polysilicon, or polysilicon-over-polysilicon constructions.

To show how γ_v^C affects the performance of a SC stage, consider again the simple SC integrator shown in Fig. 7.16. If C_i^0 denotes the value of the capacitor i when no voltage is applied, then during the interval when ϕ_1 is high

---

*ppm stands for "parts per million," that is, it represents a factor 10^{-6}.

the capacitances are given by

$$C_1 = C_1^0\left(1 + \gamma_v^C v_{\text{out}}\right)$$

and   (7.23)

$$C_2 = C_2^0\left(1 + \gamma_v^C v_{\text{in}}\right).$$

Defining the nonlinear capacitance by the relation $C = \partial q / \partial V$, when ϕ_1 goes low and ϕ_2 goes high, a charge

$$\Delta q(nT) = \int_{v_{\text{in}}}^{0} C_2(v)\, dv = -C_2^0 v_{\text{in}}\left(1 + \gamma_v^C v_{\text{in}}/2\right)   \qquad (7.24)$$

is transferred from C_2 into C_1. [Here, v_{in} denotes $v_{\text{in}}(nT - T)$.] This causes v_{out} to change from $v_{\text{out}}(nT - T)$ to $v_{\text{out}}(nT)$. Hence, $\Delta q(nT)$ also satisfies the relation

$$\Delta q(nT) = \int_{v_{\text{out}}(nT-T)}^{v_{\text{out}}(nT)} C_1(v)\, dv = C_1^0\left[v_{\text{out}}(nT) - v_{\text{out}}(nT - T)\right]$$

$$\times \left\{1 + \frac{\gamma_v^C}{2}\left[v_{\text{out}}(nT) + v_{\text{out}}(nT - T)\right]\right\}. \qquad (7.25)$$

The factor multiplying γ_v^C in (7.25) can be approximated by $v_{\text{out}}(nT - T)$ if v_{out} does not change much from cycle to cycle. Using this approximation, and combining Eqs. (7.24) and (7.25),

$$v_{\text{out}}(nT) - v_{\text{out}}(nT - T) \cong -\frac{C_2}{C_1}\frac{1 + \gamma_v^C v_{\text{in}}/2}{1 + \gamma_v^C v_{\text{out}}(nT - T)} v_{\text{in}}. \qquad (7.26)$$

The ideal transfer relation, for $\gamma_v^C = 0$, is

$$v_{\text{out}}(nT) - v_{\text{out}}(nT - T) = -\frac{C_2}{C_1} v_{\text{in}}. \qquad (7.27)$$

Comparison of (7.26) and (7.27) shows that the effect of nonzero γ_v^C can be interpreted approximately as a change of $\alpha \triangleq C_2/C_1$ to an effective value

$$\alpha_{\text{actual}} = \alpha \frac{1 + \gamma_v^C v_{\text{in}}/2}{1 + \gamma_v^C v_{\text{out}}(nT - T)} \cong \alpha\left(1 + \gamma_v^C v_{\text{in}}/2\right)\left[1 - \gamma_v^C v_{\text{out}}(nT - T)\right].$$

$$(7.28)$$

Equation (7.28) shows that, as could be expected, the voltage dependence of the capacitors introduces a voltage-dependent capacitance ratio error. This is also a source of harmonic distortion. As long as the relative error

$$\frac{\Delta\alpha}{\alpha} \cong \gamma_v^C\left[v_{\text{in}}/2 - v_{\text{out}}(nT - T)\right] \tag{7.29}$$

is much less than the unavoidable random relative error (approximately 0.001), both effects are usually negligible. This condition normally holds for metal–polysilicon or polysilicon–polysilicon capacitors, but not always for capacitors which use a diffusion layer as a bottom plate.

As mentioned earlier, temperature effects also effect the MOS capacitances. However, the corresponding temperature coefficient

$$\gamma_T^C \triangleq \frac{1}{C}\frac{\partial C}{\partial T} \tag{7.30}$$

is very small, only about 20 ppm/°C. Also, it causes an error in a capacitance *ratio* (say C_1/C_2) only if there is a temperature *difference* between C_1 and C_2. This effect can be avoided by using common-centroid layout, which cancels the first-order effects of temperature gradients. Also, all capacitors should be placed as far away from high-current (and hence hot) output transistors as feasible. With such precautions, the temperature effects in the capacitors can usually be reduced to negligible proportions.

After all systematic errors have been eliminated, there still remains some unavoidable uncertainty in the values of the capacitance ratios due to random effects. One such effect, called *large-scale distortion*, is illustrated in Fig. 7.17a. It may be caused by the saturation of the etchant solution with the etched material in some areas, and/or by regional temperature gradients. Both effects result in some local etching-rate variations. To reduce these effects, each active edge (i.e., edge whose location affects the performance substantially) should be located at the same distance from adjacent edges to equalize the etching conditions and hence the etching rates.[5] This may necessitate the use of guard rings or leads.

A second reason for the random edge variation is the grainy, jagged boundary of the capacitor plates (Fig.7.17b). This may be caused by light interference patterns during the photolithograph process, or by localized temperature differences, or by gas bubbles which retard the etching process, and so on. Finally, the granular nature of the material being etched can also cause localized etching rate variations and hence edge graininess. The resulting capacitance variation is called *local edge effects*.[8]

Yet another source of nonuniform edge variation is *corner rounding* (Fig. 7.17c). This is caused by the inherent difficulty in realizing sharp corners with conventional photolithography methods. As Fig. 7.17c illustrates, a compensation of this effect can be achieved by having an equal number of 90° and 270°

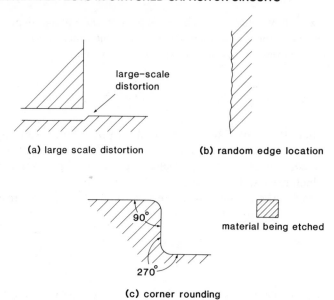

(a) large scale distortion

(b) random edge location

(c) corner rounding

FIGURE 7.17. Different forms of nonuniform undercut.

corners for each layer. However, the excess area gained due to a 270° corner does not exactly equal that lost at a 90° one, so this method is not very accurate.

The uncertainty in edge definition, illustrated in Fig. 7.17*b*, can be reduced by using improved processing methods. For example, using plasma etching rater than "wet" chemical etching can reduce the nonuniform undercut by as much as 50%.

The random *large-scale* variation of the mean locations of the edges can be represented by a random displacement of two entire edges, resulting in a change ΔW in the width and a change ΔL in the length of the capacitor (Fig. 7.18). The analysis of such a structure has already been briefly discussed in Section 3.5. The following results have been given there. Denoting the standard deviation (root-mean-square error, RMS error) of a parameter x by σ_x, the

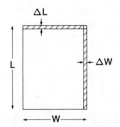

FIGURE 7.18. Model for random edge variations.

expected relative error of the capacitance C for a given $\sigma_L = \sigma_W$ is*

$$\frac{\sigma_C}{C} = \sigma_L\sqrt{W^{-2} + L^{-2}}. \tag{7.31}$$

For a given C, σ_C/C is minimized if $L = W$. Then $\sigma_C/C = \sigma_L\sqrt{2}/L$. Since $L \propto \sqrt{C}$, the expected relative error is proportional to $C^{-1/2}$.

Similar results were obtained when the ratio $\alpha \triangleq C_1/C_2$ of two capacitors was considered in Section 3.5. We have found there that for $\sigma_{L_1} = \sigma_{W_1} = \sigma_{L_2} = \sigma_{W_2} = \sigma_L$, the expected relative ratio error was

$$\frac{\sigma_\alpha}{\alpha} = \sigma_L\sqrt{L_1^{-2} + W_1^{-2} + L_2^{-2} + W_2^{-2}}. \tag{7.32}$$

For given C_1 and C_2, this error was minimized when $L_1 = W_1 = L_2\sqrt{\alpha} = W_2\sqrt{\alpha}$ was chosen. Then, the expected relative error was

$$\frac{\sigma_\alpha}{\alpha} = \frac{\sigma_L}{L_2}\sqrt{2(1 + \alpha)}. \tag{7.33}$$

Again, the expected relative error is proportional to $C^{-1/2}$.

For the case of edge variations caused by *local effects*, a derivation based on statistical considerations[7,8] reveals that the resulting RMS capacitance error σ_C is proportional to $C^{1/4}$, and hence the relative error to $C^{-3/4}$.

For a uniform (i.e., constant) undercut Δx along the perimeters of both capacitors, the actual capacitance ratio is

$$\alpha_{actual} \cong \frac{W_1 L_1 - 2(W_1 + L_1)\Delta x}{W_2 L_2 - 2(W_2 + L_2)\Delta x}. \tag{7.34}$$

It is possible to achieve $\alpha_{actual} = \alpha = W_1 L_1/W_2 L_2$, by choosing $L_2 = W_2$ and

$$W_1 = W_2\left(\alpha - \sqrt{\alpha^2 - \alpha}\right)$$

and $\tag{7.35}$

$$L_1 = L_2\left(\alpha + \sqrt{\alpha^2 - \alpha}\right).$$

(Here, $\alpha > 1$ is assumed, so that the *smaller* capacitor C_2 is made square!) Then the relative RMS ratio error is

$$\frac{\sigma_\alpha}{\alpha} = \frac{\sigma_L}{L_2}\sqrt{6 + 2/\alpha}. \tag{7.36}$$

As before, σ_α/α is proportional to $C^{-1/2}$.

*Typically, $\sigma_L \approx 0.1 \ \mu\text{m}$.

Another edge-related phenomenon, *field fringing* (Fig. 7.15), is similar in nature to uniform undercut, in that it also introduces a capacitance error which is proportional to the perimeter $2(W + L)$ of the capacitor plates. (Here, however, the effective area is *increased* by the effect.) Thus, all considerations valid for uniform undercut hold also for field fringing.

In addition to random variations of the edges, the capacitance is also affected by the random variations of the *thickness* of the oxide layer. The capacitance is given by

$$C = \varepsilon \frac{WL}{t}. \tag{7.37}$$

Here, $\varepsilon = K_{ox}\varepsilon_0$ is the permittivity of the oxide layer, and t is its thickness. Due to the granularity of the polysilicon layers (if they form the plates), or the surface defects of crystalline silicon or metal, and so on, there are *local small-scale* random variations of t within each capacitor. In addition, there is a *global* effect, due to the slow variation of t caused by the lack of absolute surface flatness, wafer warping, changes in the oxide growth rate, and so on. As discussed above, this latter effect can be compensated for by using unit-capacitor/common-centroid layout. The former, small-scale thickness variations, however, cannot be balanced out. It is especially serious for capacitors constructed from two polysilicon layers, where the grain boundaries of both plates "modulate" the oxide thickness. By contrast, the surface of the heavily doped crystalline–silicon substrate used as a capacitor plate is quite smooth.

It can be shown[7,8] that the small-scale thickness variations within a capacitor result in an RMS capacitance error σ_C which is proportional to $C^{1/4}$, while global ones cause $\propto {}_c\alpha C^{1/2}$. Hence, the expected relative error σ_C/C due to all these effects, when plotted as a function of C on a log–log scale, has the general shape shown in Fig. 7.19. The crossover point between edge and oxide

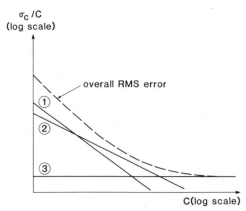

FIGURE 7.19. Expected relative capacitance error due to random effects. Line ① illustrates the local edge effects (slope $= -\frac{3}{4}$), line ② the global edge effects and local oxide variations (slope $= -\frac{1}{2}$), and line ③ the globe oxide effects (slope $= 0$).

effects usually occurs for $W = L = 20 \sim 50 \, \mu\mathrm{m}$, corresponding to $C = 0.2 \sim 1$ pF.

The reader is referred to Ref. 6 for a detailed comparative evaluation of the different MOS capacitor structures.

7.3.   NONIDEAL OP-AMP CIRCUIT EFFECTS[11,15]

As described in Section 4.1, the MOS op-amp is subject to various nonideal effects. The sources of these effects, as well as some methods for reducing them, were discussed in Chapter 4. In this section, we shall show how these effects influence the operation of the circuits in which the amplifier is used.

The Effect of the dc Offset Voltage

We shall illustrate this effect on the example of the lossy integrator of Fig. 7.20, where the added dc source V_{off} (typically, $5 \sim 20$ mV for MOS op-amps) represents the offset voltage of the op-amp. To isolate the effect of the offset voltage, we set $v_{\mathrm{in}} = 0$ and assume that the circuit is in steady state, so that v_{out} remains unchanged from clock interval to clock interval. Then, the net charge entering C via C_1 and C_2 at node $\textcircled{A}$ when ϕ_2 rises must be zero. Thus,

$$-C_1 V_{\mathrm{off}} + C_2 (v_{\mathrm{out}} - V_{\mathrm{off}}) = 0 \tag{7.38}$$

giving

$$v_{\mathrm{out}} = \left(1 + \frac{C_1}{C_2} \right) V_{\mathrm{off}}. \tag{7.39}$$

Note that v_{out} is independent of the value of C; it only depends on the ratio of

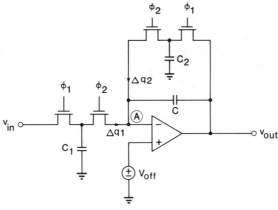

FIGURE 7.20.   A lossy integrator with op-amp offset voltage.

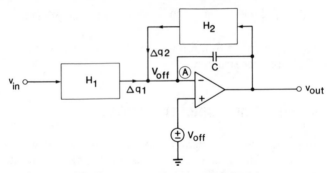

FIGURE 7.21. Integrator in a dc feedback loop.

the switched capacitors C_1 and C_2. Clearly, the above proof can readily be extended to the more general situation shown in Fig. 7.21. Here again if v_{in} is zero (or a constant) in the steady state $\Delta q_1 + \Delta q_2 = 0$ must hold,* where Δq_1 is the charge entering via input two-port H_1, which by linearity is

$$\Delta q_1 = H_{1a}v_{in} + H_{1b}V_{off} \tag{7.40}$$

and Δq_2 is the charge coming from the feedback circuit H_2:

$$\Delta q_2 = H_{2a}v_{out} + H_{2b}V_{off}. \tag{7.41}$$

Equating $\Delta q_1 + \Delta q_2$ to zero, v_{out} can be found. Problems 7.6–7.8 illustrate the use of formulas (7.40) and (7.41).

As described in Chapter 6, there exist several autozeroing methods which can be used to compensate for the dc offset effect, if it cannot be tolerated in a specific application.

As also mentioned earlier, in addition to the op-amp offset voltages, such effects as clock feedthrough noise and *p-n* junction leakage currents also contribute to the overall dc offset voltage of the complete circuit.

The Effects of the Finite dc Gain

This nonideal effect will be demonstrated on the stray-insensitive inverting integrator, discussed in Chapter 5 and reproduced in Fig. 7.22. If the op-amp has a finite dc gain A_0 (but an infinite bandwidth), its input voltage is $-v_{out}/A_0$, and hence at the instance $t = nT$, by the KVL

$$v_{out}(nT) = v_{C2}(nT) - \frac{1}{A_0}v_{out}(nT). \tag{7.42}$$

*It is implicitly assumed here that the circuit is stable, so that it possesses a steady state in its linear region of operation.

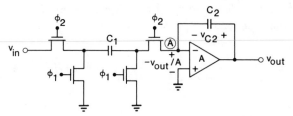

FIGURE 7.22.   A stray-insensitive inverting integrator.

Also, from the KCL at node Ⓐ,

$$C_2[v_{C2}(nT) - v_{C2}(nT - T)] + C_1\left[v_{in}(nT) + \frac{1}{A_0}v_{out}(nT)\right] = 0. \quad (7.43)$$

Solving (7.42) and (7.43) using z-transformation, we obtain

$$H(z) = \frac{V_{out}(z)}{V_{in}(z)} = \frac{-C_1}{C_2 + (C_1 + C_2)/A_0 - C_2(1 + 1/A_0)z^{-1}}$$

$$= \frac{-(C_1/C_2)[1 + (1 + C_1/C_2)/A_0]^{-1}z}{z - (1 + 1/A_0)/[1 + (1 + C_1/C_2)/A_0]}. \quad (7.44)$$

Comparing the $H(z)$ of (7.44) with the ideal transfer function valid when $A_0 \to \infty$, namely with

$$H_i(z) = \frac{-(C_1/C_2)z}{z - 1} \quad (7.45)$$

it is clear that the gain of the integrator has been reduced from C_1/C_2 to a somewhat smaller value. Also, the pole (earlier at $z_p = 1$) has now a smaller positive value. Substituting $z = \exp(j\omega T)$ into $H_i(z)$, the ideal frequency response can be written as

$$H_i(e^{j\omega T}) = \frac{-(C_1/C_2)e^{j\omega T/2}}{j2\sin(\omega T/2)}. \quad (7.46)$$

Similarly, the frequency response of the integrator with finite-gain op-amp is, from (7.44),

$$H(e^{j\omega T}) = H_i(e^{j\omega T})\frac{1}{1 + (1/A_0)(1 + C_1/2C_2) - j(C_1/C_2)/2A_0\tan(\omega T/2)}. $$

$$(7.47)$$

Here, $H_i(e^{j\omega T})$ is the ideal frequency response given in (7.46) (Problem 7.9). Clearly, the actual frequency response can be written in the form

$$H(e^{j\omega T}) = \frac{H_i(e^{j\omega T})}{1 - m(\omega) - j\theta(\omega)} = F(\omega)H_i(e^{j\omega T}), \qquad (7.48)$$

where

$$m(\omega) = -\frac{1}{A_0}\left(1 + \frac{C_1}{2C_2}\right)$$

and, for $\omega T \ll 1$,

$$\theta(\omega) = \frac{C_1/C_2}{2A_0\tan(\omega T/2)} \simeq \frac{C_1/C_2}{A_0\omega T}. \qquad (7.49)$$

The extra factor $F(\omega)$ introduced by the finite-gain effect into the frequency response can be written in the polar form $|F(\omega)|\exp[j\angle F(\omega)]$, where

$$|F(\omega)|^2 = \frac{1}{[1 - m(\omega)]^2 + [\theta(\omega)]^2} \simeq \left(\frac{1}{1 - m(\omega)}\right)^2 \simeq [1 + m(\omega)]^2$$

$$(7.50)$$

and

$$\angle F(\omega) = -\tan^{-1}\left(\frac{-\theta(\omega)}{1 - m(\omega)}\right) \simeq \tan^{-1}\theta(\omega) \simeq \theta(\omega). \qquad (7.51)$$

Here, the indicated approximations are usually justified, since $m(\omega)$ and $\theta(\omega)$ contain $1/A_0$ and hence (for $A_0 > 1000$ and usual values of C_1/C_2 and ωT) $|m| \ll 1$ and $|\theta| \ll 1$.* In conclusion, the extra factor $F(\omega)$ caused by $A_0 < \infty$ can be written as

$$F(\omega) \simeq [1 + m(\omega)]e^{j\theta(\omega)}, \qquad (7.52)$$

where $m(\omega)$ and $\theta(\omega)$ are given by (7.48) and (7.49). Thus, as Eq. (7.52) illustrates, $m(\omega)$ represents the relative magnitude error, while $\theta(\omega)$ the phase error in radians caused by the finite-gain effect. Note that the phase angle of $H_i(e^{j\omega T})$ is 90° for all values of ω, while that of $H(e^{j\omega T})$ is 180° at $\omega = 0$, decreasing to 90° as $\omega \gg 2/A_0T$ is reached.† Hence, $\theta(\omega)$ is large for $\omega < 2/A_0T$, as indicated earlier.

---

*Except near $\omega = 0$, where $\theta(\omega)$ becomes very large.
†Here, the linear phase $\omega T/2$ is disregarded for simplicity.

It is useful to introduce the *unity-gain frequency* ω_i of the integrator, defined by the condition

$$\left| H_i(e^{j\omega_i T}) \right| = 1. \tag{7.53}$$

(The frequency range of interest is usually $0 \le \omega \le \omega_i$ for the overall circuit.) From (7.46), we find then

$$\frac{C_1}{C_2} = 2 \sin(\omega_i T/2). \tag{7.54}$$

The values of $m(\omega)$ and $\theta(\omega)$ at ω_i are, from (7.49),

$$m(\omega_i) = -\frac{1}{A_0}[1 + \sin(\omega_i T/2)]$$

and

$$\theta(\omega_i) = \frac{1}{A_0}\cos(\omega_i T/2). \tag{7.55}$$

For the usual case when $\omega_i T/2 = \pi f_i/f_c \ll 1$, the simple result

$$\theta(\omega_i) \simeq -m(\omega_i) \simeq 1/A_0 \tag{7.56}$$

is thus obtained. Note that, as (7.46) illustrates, the magnitude error $m(\omega)$ can be simply regarded as the departure of C_1/C_2 from its nominal value by a relative error $m(\omega) \simeq 1/A_0$. Since normally $A_0 > 1000$, this is usually a negligible effect. By contrast, $\theta(\omega)$ can have important influence on the overall circuit response, as will be demonstrated later in this section.

The derivation can be repeated also for the stray-sensitive *noninverting* integrator (Fig. 7.23). It is left as a problem for the reader (Problem 7.10). Once again, the result is described by Eqs. (7.46)–(7.56). Here, however, the ideal response is

$$H_i(e^{j\omega T}) = \frac{(C_1/C_2)e^{-j\omega T/2}}{j2 \sin(\omega T/2)}. \tag{7.57}$$

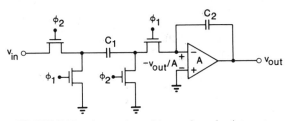

FIGURE 7.23.   A stray-insensitive noninverting integrator.

Assume next that the finite op-amp gain integrators of Figs. 7.22 and 7.23 are used in the *biquads* discussed in Section 5.4. Referring to Fig. 5.10a, if the ideal integrators with transfer functions $\pm 1/s$ are replaced by their finite-dc-gain equivalents, and only the magnitude errors $m(\omega) \simeq -1/A_0$ are considered, the effect on the natural modes (poles) is to replace $1/s$ by $(1 - 1/A_0)/s$ in the unexcited system.* Hence, the value s_p of a pole is replaced by $s_p/(1 - 1/A_0)$. This scaling effect, by Eqs. (5.31) and (5.32), does not affect the pole-Q, but it shifts the pole frequency $\omega_0 = |s_p|$ to $\omega_0/(1 - 1/A_0)$. This has usually negligible effect if $A_0 > 1000$.

To find the effect of the phase error, assume (as on earlier occasions) that $|sT| \ll 1$, and hence that we can replace z in (7.44) using the formula

$$z \triangleq e^{sT} \simeq 1 + sT. \tag{7.58}$$

Thus, the s-domain transfer function of the integrator is

$$H(s) \simeq \frac{-(C_1/C_2)[1 + (1 + C_1/C_2)/A_0]^{-1}(1 + sT)}{sT + (C_1/C_2A_0)/[1 + (1 + C_1/C_2)/A_0]} \tag{7.59}$$

$$\simeq \frac{-C_1/C_2T}{s + C_1/C_2A_0T}.$$

Hence, the s-domain pole of the integrator is

$$\sigma_i = \frac{C_1/C_2}{A_0T} \tag{7.60}$$

instead of $s = 0$. Thus, $H(s) \propto 1/(s + \sigma_i)$.

Referring back to Fig. 5.10a and changing the integrators' transfer functions to $1/(s + \sigma_i)$, the natural modes s_p can now be found from the relations

$$V_1(s) = -\frac{1}{s + \sigma_{i1}}\omega_0 V_{\text{out}}(s)$$

and

$$V_{\text{out}}(s) = -\frac{1}{s + \sigma_{i2}}\left(\frac{\omega_0}{Q}V_{\text{out}}(s) - \omega_0 V_1(s)\right). \tag{7.61}$$

This gives the condition

$$\left((s_p + \sigma_{i1})(s_p + \sigma_{i2}) + \frac{\omega_0}{Q}(s_p + \sigma_{i1}) + \omega_0^2\right)V_{\text{out}}(s_p) = 0 \tag{7.62}$$

*Where $V_{\text{in}}(s) \equiv 0$, or, equivalently, $K_0 = K_1 = K_2 = 0$, is set.

for the poles. Since $V_{\text{out}}(s_p) \neq 0$, (7.62) gives

$$s_p^2 + \left(\frac{\omega_0}{Q} + \sigma_{i1} + \sigma_{i2}\right)s_p + \left(\omega_0^2 + \frac{\omega_0\sigma_{i1}}{Q} + \sigma_{i1}\sigma_{i2}\right) = 0. \qquad (7.63)$$

Comparison with the $\sigma_{i1} = \sigma_{i2} = 0$ case shows that the new pole-Q, Q', is given by the relation

$$\frac{\omega_0}{Q'} \approx \frac{\omega_0}{Q} + \sigma_{i1} + \sigma_{i2}. \qquad (7.64)$$

Using (7.60),

$$\frac{1}{Q'} = \frac{1}{Q} + \frac{1}{A_0\omega_0 T}\left[\left(\frac{C_i}{C_F}\right)_1 + \left(\frac{C_i}{C_F}\right)_2\right] \qquad (7.65)$$

results, where the C_i are the input capacitors and the C_F the feedback capacitors of the two integrators. Alternatively, using (7.49), we obtain

$$\frac{1}{Q'} = \frac{1}{Q} + \theta_1(\omega_0) + \theta_2(\omega_0) \qquad (7.66)$$

where $\theta_i(\omega_0)$ is the phase error of integrator i at the pole frequency. Clearly, the Q is reduced by the phase-error effect. Since, as shown in Section 5.4, $C_i/C_F \simeq \omega_0 T$ for both integrators,* from (7.65)

$$\frac{1}{Q'} \simeq \frac{1}{Q} + \frac{2}{A_0}. \qquad (7.67)$$

[The same conclusion follows from Eqs. (7.66) and (7.56)!] For $A_0 = 1000$ and $Q = 15$, (7.67) gives $Q' \simeq 14.56$. Thus, for high-Q biquads, the effect is not negligible.

Another pertinent question is the change in the biquad's frequency response due to the finite-gain op-amps. Near the peak of the gain response, where $s \simeq j\omega_0$, the change in the magnitude of the denominator polynomial $s^2 + (\omega_0/Q)s + \omega_0^2$ due to the change in Q is simply

$$\Delta\alpha = 20\log_{10}\frac{Q}{Q'} \simeq 20\log_{10}(1 + 2Q/A_0) \qquad (7.68)$$

in dB. For $Q = 15$ and $A_0 = 1000$, $\Delta\alpha \simeq 0.257$ dB, a significant gain change. Note that the gain (along with the Q) is reduced by this effect.

As Eq. (7.63) shows, not only the pole-Q but also ω_0 is affected by the σ_i. Normally, however, this effect is negligible.

*For the usual case of $|H_a(0)| \cong 1$.

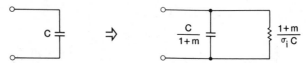

FIGURE 7.24. Finite op-amp gain effect on a simulated capacitor.

Consider now the effect of the finite op-amp gains on an SC *ladder filter*. As described in Section 5.5, in such a circuit each integrator simulates a reactive (L or C) element of the passive "prototype" filter. If the element is, for example, a capacitor C, the ideal integrator transfer function is

$$H_i(s) = \frac{-1/C}{s}. \tag{7.68}$$

In fact, as can be seen by combining Eqs. (7.52) and (7.59), the relation realized by the finite-gain integrator is

$$H_i(s) = \frac{-(1+m)/C}{s + \sigma_i}. \tag{7.69}$$

Thus, the passive branch actually simulated has the impedance

$$Z(s) = \frac{1}{sC/(1+m) + \sigma_i C/(1+m)}. \tag{7.70}$$

This branch is illustrated in Fig. 7.24. It is a lossy capacitor. The capacitance is $C/(1 + m)$ and the quality factor of the capacitor is

$$Q_C \triangleq \omega C R_C = \frac{\omega}{\sigma_i} = \frac{A_0 \omega T}{C_1/C_2} = \frac{1}{\theta(\omega)} \tag{7.71}$$

where (7.60) and (7.49) were used.

The change in element value, represented by $C \to C/(1 + m)$ has normally a negligible effect, since the doubly terminated passive two-port is insensitive to such variations.* The Q_C introduced, however, can result in significant changes in the gain response, especially for narrow-bandpass filters with high pole-Q's.

For an integrator simulating an inductor L, a similar derivation (Problem 7.12) shows that the branch simulated by an integrator containing an op-amp with finite dc gain is in fact the lossy inductor illustrated in Fig. 7.25. Its

---

*In fact, if the effect is uniform in all integrators, it simply corresponds to a frequency scaling by $1 + m$, which is usually a minor change.

FIGURE 7.25. Finite op-amp gain effect on a simulated inductor.

quality factor Q_L is given by

$$Q_L = \frac{\omega L}{R_L} = \frac{\omega}{\sigma_i} = \frac{A_0 \omega T}{C_1/C_2} = \frac{1}{\theta(\omega)}. \tag{7.72}$$

Thus, $Q_L = Q_C$, as a comparison with (7.71) shows.

For a reactance two-port with a uniform distribution of quality factors in all reactances, the loss distortion (in dB) can be approximated by[13]

$$\Delta\alpha \simeq 8.7\omega\tau(\omega)/Q(\omega). \tag{7.73}$$

Using our results, therefore

$$\Delta\alpha \simeq 8.7\theta(\omega)\omega\tau(\omega) \simeq 8.7\omega\tau(\omega)/A_0. \tag{7.74}$$

Here, $\tau(\omega)$ is the group delay of the prototype filter. For a typical fifth-order elliptic filter, for example, $\omega\tau(\omega)$ peaks just outside the passband limit, and has a maximum value around 10. Hence, for $A_0 = 1000$, $\Delta\alpha \simeq 0.087$ dB. This may be acceptable in a noncritical application. For a narrow-band bandpass filter, by contrast, $\omega\tau(\omega) > 100$ is not unusual. Then, $\Delta\alpha > 0.87$ dB, which can seldom be tolerated. Then, very-high-gain op-amps are needed, or alternatively, the N-path filter configuration discussed in Section 5.8 may be used.

The Effects of Finite Op-Amp Bandwidth[10-12]

As discussed in Section 4.8, stability considerations require that the op-amp response $A_V(s)$ have widely separated poles, so that we may assume that only the dominant pole s_1 affects the response in the frequency range of interest.* Then, Eq. (4.104) can be rewritten in the simplified form

$$A_V(s) = \frac{V_{out}(s)}{V(s)} \simeq \frac{-\omega_0}{s - s_1} \simeq \frac{-1}{1/A_0 + s/\omega_0}. \tag{7.75}$$

Here, A_0 is the dc gain and ω_0 the unity-gain bandwidth of the op-amp. (Note

*It is possible to treat the case of two poles also; see Problem 7.16 and Ref. 15.

that $\omega_0 \simeq A_0 |s_1| \gg |s_1|$.) Also, V_{out} is the output, while V the input voltage of the amplifier. In the time domain, Eq. (7.75) can be rewritten in the form[10]

$$\frac{1}{A_0} v_{out}(t) + \frac{1}{\omega_0} \frac{dv_{out}(t)}{dt} = -v(t). \qquad (7.76)$$

On the basis of Eq. (7.76), the transients occurring during both the $\phi_1 = \text{"1"}$ and $\phi_2 = \text{"1"}$ intervals in an SC integrator (Figs. 7.22 and 7.23) can be analyzed in the time domain.[10-12] The output voltage $v_{out}(t)$ is an exponential function of time during both half clock cycles: when C_1 is connected to the op-amp, the time constant entering the exponent is $\tau_1 = (1 + C_1/C_2)/\omega_0$; when C_1 is disconnected, it is $\tau_2 = 1/\omega_0 < \tau_1$. For the usual case when $C_1 \ll C_2$, $\tau_1 \simeq \tau_2 = 1/\omega_0$. When C_1 is connected to the amplifier (charging transient), $v_{out}(t)$ changes by a considerable amount. When C_1 is disconnected, (settling transient), $v_{out}(t)$ changes very little.

From $v_{out}(t)$, the samples $v_{out}(nT)$, $v_{out}(nT + T)$, and so on can then be calculated in terms of $v_{in}(nT)$. Finally, using z-transformation, the actual transfer function $H(z) = V_{out}(z)/V_{in}(z)$ can be calculated and evaluated for $z = \exp(j\omega T)$. This gives the sampled-data frequency response of the integrator using the band-limited op-amp characterized by Eqs. (7.75) and (7.76). The calculations, while straightforward, are lengthy and hence will not be repeated here. The results can be conveniently expressed[12] in terms of the magnitude error $m(\omega)$ and phase error $\theta(\omega)$ introduced earlier in Eqs. (7.48) and (7.52). Specifically, for the *inverting integrator* of Fig. 7.22,

$$m(\omega) = -e^{-k_1}[1 - k \cos \omega T]$$

and $\qquad\qquad\qquad\qquad\qquad\qquad\qquad\qquad\qquad\qquad\qquad (7.77)$

$$\theta(\omega) = -e^{-k_1} k \sin \omega T$$

where the notations

$$k \triangleq \frac{C_2}{C_1 + C_2}$$

and $\qquad\qquad\qquad\qquad\qquad\qquad\qquad\qquad\qquad\qquad\qquad (7.78)$

$$k_1 \triangleq k\omega_0 T/2$$

have been used. At the unity-gain frequency ω_i of the integrator, defined in (7.53) and (7.54), both error functions have the same value

$$m(\omega_i) \simeq \theta(\omega_i) \simeq -\omega_i T e^{-\omega_0 T/2} \qquad (7.79)$$

where $\omega_i T \ll 1$ was assumed.

Thus, if

$$\omega_0 T/2 = \pi \omega_0 / \omega_c \gg 1, \tag{7.80}$$

then both $m(\omega_i)$ and $\theta(\omega_i)$ become negligible. Calculations indicate that for a two-phase clock, $\omega_0 \simeq 5\omega_c$ is usually adequate; that is, the *unity-gain band-width* ω_0 *of the op-amp should be* (*at least*) *five times as large as the clock frequency* ω_c. To avoid unnecessary noise aliasing, ω_0 should *not* be chosen much larger than this value. This issue will be discussed in Section 7.4.

For the noninverting integrator of Fig. 7.23, the error functions are

$$m(\omega) \simeq -(1 - k)e^{-k_1}$$

and   (7.81)

$$\theta(\omega) \simeq 0$$

and at the integrator unity-gain frequency $\omega_i \ll \omega_c$, we have

$$m(\omega_i) \simeq -\omega_i Te^{-\omega_0 T/2} \tag{7.82}$$

as before. Thus, the condition $\omega_0 \geq 5\omega_c$ applies for the noninverting integrator also.

In the (omitted) derivations leading to Eqs. (7.77)–(7.82), it was assumed that $v_{in}(t)$ is an ideal sampled-and-held signal. In fact, typically it originates from a preceding SC stage also containing a finite-bandwidth op-amp. Hence, the input signal $v_{in}(t)$ is itself an exponential function of time. This need not be a serious problem, provided that $v_{in}(t)$ has settled by the time the input switches connect it to the op-amp. This time is at the *beginning* of the $\phi_2 = $"1" interval for the circuit of Fig. 7.22, and the *end* of the $\phi_2 = $"1" interval for the circuit of Fig. 7.23. Clearly, the latter is hence less sensitive to this effect. In many cases, there are several equivalent SC circuits realizing the same transfer function $H(z)$, which have, however, greatly different behaviors with regards to op-amp settling time. An important example is the high-Q biquad circuit of Fig. 5.13c, reproduced here in Fig. 7.26. With the indicated clock phases, both op-amps receive inputs during the interval when $\phi_2 = $"1." Also, during this same time period the output of OA2 is one of the inputs of OA1. Hence, the output voltage v_1 of OA1 will be affected by the charging transients of *both* op-amps. This will slow down v_1, and hence even if the condition $\omega_0 > 5\omega_C$ is satisfied by both op-amps, the settling time may be still too long. Changing the phasing of the four switches ($\phi_1 \leftrightarrow \phi_2$) marked by asterisks, the overall response $V_{out}(z)/V_{in}(z)$ of the biquad remains unaffected, but now OA2 receives input when $\phi_1 = $"1," and hence its charging transient does not overlap with that of OA2. Then, the two transients are essentially decoupled, and $\omega_0 > 5\omega_C$ guarantees a satisfactory settling behavior.*

---

*Note that in this circuit the output voltage v_1 of OA_1 changes *twice* during a clock period. This should not cause any difficulties, however.

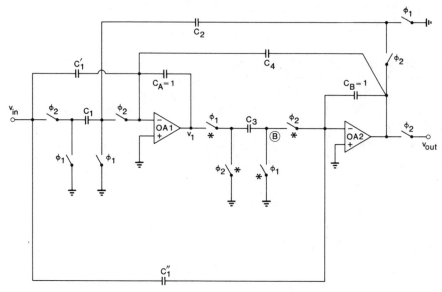

FIGURE 7.26.   High-Q switched-capacitor biquad.

It is also possible to modify the clock phasing of the low-Q biquad circuit of Fig. 5.10c to decouple the charging transients of its two op-amps. The details are left to the reader, as an exercise (Problem 7.13).

In ladder filters with finite nonzero loss poles, there may also be *loops* containing two op-amps and coupling capacitors. An example is the low-pass SC filter shown in Fig. 5.69b, where the first and last op-amps form a closed loop with capacitors C_1 and C_2. In such a case, these elements form a *second-order* system capable of generating an underdamped oscillatory response. Fortunately, if this cannot be tolerated, it is possible to apply some simple circuit transformations[14, 24] which break such loops.

On the basis of the above discussions, we can make the following observations concerning the relative values of ω_0 and ω_c.

1.   For given op-amps, and hence given ω_0, the clock frequency ω_c should be chosen low enough so that the individual op-amps have enough time to settle. However, ω_c should not be too low, or the noise aliasing effect (discussed later in this chapter) becomes serious, and also the antialiasing and smoothing filters needed in the system must be too selective and hence too complex.

2.   For a given clock frequency ω_c, the op-amp bandwidth ω_0 should be just high enough to assure that the stage can settle within each clock phase. Any higher value worsens unnecessarily the noise aliasing effect, and raises the dc power drain and chip area requirements of the op-amps.

In the above discussions, it was assumed that only the dominant pole $s_1 = -\omega_0/A_0$ of the op-amp has a significant effect on the settling time of the integrator. It is, however, possible to extend the analysis to the case where the transfer function of the op-amp contains *two* dominant poles s_1 and s_2, so that it is in the form

$$H(s) = \frac{-A_0}{(1 - s/s_1)(1 - s/s_2)}$$

where usually $|s_2| \gg |s_1|$ for a compensated op-amp. The detailed analysis of the circuit of Fig. 7.27 under these conditions is given in Ref. 15, and implied in Problem 7.18. The main results of the analysis are the following. The transient voltage $v_{out}(t)$ will be oscillatory if the condition $|s_2/s_1| < 4A_0C_2/(C_1 + C_2)$ holds; otherwise, it will contain two real exponential terms. If the response is oscillatory, it will be damped with a time constant $T_d \simeq 2/|s_2|$. If the response is exponential, the two terms will be $V_1\exp(-t/T_1)$ and

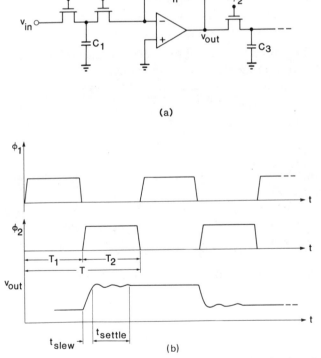

(a)

(b)

FIGURE 7.27. (*a*) Switched-capacitor integrator section; (*b*) Clock signals and output waveform.

$V_2 \exp(-t/T_2)$, where

$$T_{1,2} = \frac{2}{|s_2|} \left[1 + x \mp \sqrt{(1+x)^2 - 4A_0 x C_2/(C_1 + C_2)} \right]^{-1}.$$

Here, $x \triangleq |s_1/s_2| \ll 1$. The longer time constant T_2 is, for $x \rightarrow 0$, $T_2 \simeq (1 + C_1/C_2)/A_0|s_1| = (1 + C_1/C_2)/\omega_0$. This is the same result as derived earlier for the single-pole op-amp transfer function.

The Effects of Finite Op-Amp Slew Rate

As described in Section 4.9, the finite slew rate of an op-amp is usually caused by the inability of the input stage to charge or discharge the compensation capacitor fast enough, so that the output voltage cannot follow the input voltage. If the slew rate is so slow that the output voltage does not reach its final value within the available time slot, then nonlinear distortion will take place. The situation is illustrated for an integrator stage in Fig. 7.27. Since the sampling of the output voltage $v_{out}(t)$ ends at the end of $\phi_2 = $ "1" interval, the condition

$$t_{slew} + t_{settle} < T_2 \tag{7.83}$$

must hold. The previous discussions covered the calculation of t_{settle} and the conditions which it imposes on the op-amp bandwidth. The slew time t_{slew} is determined by the slew rate of the op-amp, defined as

$$S_r = \left| \frac{dv_{out}}{dt} \right|_{max} \tag{7.84}$$

and the change $\Delta v_{out} = v_{out}(nT) - v_{out}(nT - T)$ which $v_{out}(t)$ must undergo in each step. Clearly, $t_{slew} = \Delta v_{out}/S_r$.

To obtain an estimate of the maximum step Δv_{out} possible, assume that the highest passband frequency of the SC circuit in the $0 \leq \omega \leq \omega_c/2$ baseband is ω_B. Then, in the worst case, $v_{out}(t)$ will be a sampled-and-held sine wave with a frequency ω_B and a hold time T (Fig. 7.28). It can thus be regarded as the result of sampling and holding a continuous-time voltage

$$v(t) = V_{max} \sin \omega_B t. \tag{7.85}$$

Here, to obtain worst-case results, the amplitude of the sine wave was chosen as the largest voltage swing V_{max} of the op-amp. (Typically, for balanced dc supplies $V_{DD} = -V_{SS}$, $V_{max} \simeq V_{DD}$ may be used.) Then, the maximum slope

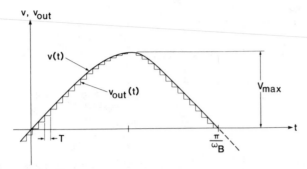

FIGURE 7.28. Output voltage v_{out} and its envelope $v(t)$.

possible for $v(t)$ is given by

$$\left| \frac{dv(t)}{dt} \right|_{max} = \omega_B V_{max}. \tag{7.86}$$

The maximum step which $v_{out}(t)$ must take every $T = 1/f_c$ seconds is therefore (Fig. 7.28)

$$\Delta v_{out,max} \simeq \left| \frac{dv(t)}{dt} \right|_{max} T = \omega_B V_{max}/f_c. \tag{7.87}$$

We can assign a certain portion x (say, 10%) of the available time slot T_2 to t_{slew}. Then, $t_{slew} = xT_2$, and the required op-amp slew rate is

$$S_r \geq \frac{\Delta v_{out,max}}{t_{slew}} \simeq \frac{\omega_B V_{max}}{f_c x T_2} \simeq \frac{2\omega_B V_{max}}{x}. \tag{7.88}$$

In the last part of (7.88), we assumed that $T_1 = T_2 \simeq T/2 = 1/2f_c$, so that $f_c T_2 \simeq \frac{1}{2}$. If, for example, $V_{max} = 5$ V, $\omega_B = 2\pi\,10^4$ rad/s, and $x = 0.1$, then (7.88) requires $S_r \geq 2\pi10^6$ V/s $\simeq 6.3$ V/μs, a fairly high slew rate. This is due to the worst-case assumption implicit in analyzing a single high-frequency sine wave. Other possible assumptions are discussed in Problem 7.15; they result in maximum slope value estimates $|dv(t)/dt|_{max}$ which are much lower (by about a factor of 4) than that predicted by (7.86).

In fact, there are of course also signals of higher frequencies than ω_B present internally in the SC circuit. However, they and their harmonics fall into the stopbands of the SC filter, and hence the resulting distortion will normally not affect the output signal of the complete circuit.

The Effects of Nonzero Op-Amp Output Resistance

As discussed in Chapter 4, the op-amp (which is ideally a voltage-controlled voltage source) has in fact a nonzero output resistance R_0. This may be of the order of only a few kΩ if a buffer output stage is used, but it can be much

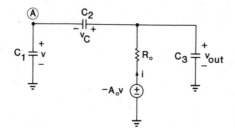

FIGURE 7.29. Equivalent circuit of the integrator of Fig. 7.27a during the $\phi_2 = $"1" interval.

larger (of the order of several MΩ) if there is no buffer stage in the op-amp. The effect of R_0 may again be illustrated on the integrator stage shown in Fig. 7.27a. During the $\phi_2 = $"1" interval, when the op-amp is discharging C_1 and recharging C_2 and C_3, the equivalent circuit of the stage is that shown in Fig. 7.29. Here, he op-amp was simulated simply by a voltage-controlled voltage source with a constant gain A_0 and an output resistance R_0. It was also assumed that the on-resistance of the switches is low, so that the initial redistribution of charges between C_1, C_2, and C_3 when ϕ_2 goes high requires a negligible amount of time. We then analyze the subsequent transient caused by the nonzero op-amp output resistance.

By the KVL

$$v_{\text{out}} = -A_0 v - i R_0 = v_c + v \tag{7.89}$$

and the KCL and branch relations give

$$i = C_3 \dot{v}_{\text{out}} + C_1 \dot{v} = C_3 \dot{v}_{\text{out}} + C_2 \dot{v}_c \tag{7.90}$$

where the dot denotes derivatives with respect to time. Also, if the initial values of $v(t)$ and $v_c(t)$ (after the charge redistribution) are v^0 and v_c^0, respectively, then by charge conservation at node Ⓐ,

$$C_1(v - v^0) = C_2(v_c - v_c^0). \tag{7.91}$$

Combining Eqs. (7.89)–(7.91), we obtain for $A_0 \gg 1 + C_1/C_2$ the differential equation

$$-\dot{v}_{\text{out}} \simeq (v_{\text{out}} - V^0)/T_0 \tag{7.92}$$

where the abbreviations

$$V^0 = v_c^0 - \frac{C_1}{C_2} v^0$$

and

$$T_0 = (R_0/A_0)(C_1 + C_3 + C_1 C_3/C_2) \tag{7.93}$$

were used.

Hence, solving (7.92), we obtain

$$v_{\text{out}}(t) = V^0 + \left[v_{\text{out}}(0) - V^0\right]e^{-t/T_0}$$
$$= V^0 + (1 + C_1/C_2)v_c^0 e^{-t/T_0} \tag{7.94}$$

where $v_{\text{out}}(0) = v^0 + v_c^0$ is the initial value of $v_{\text{out}}(t)$ *after* ϕ_2 went high (at $t = 0^+$).

As Eqs. (7.93) and (7.94) show, the time constant T_0 associated with the exponential change of $v_{\text{out}}(t)$ is related to R_0/A_0, as well as the circuit capacitances. If $C_2 \gg C_1$ and C_3, which is often the case, this time constant is approximately $T_0 = (R_0/A_0)(C_1 + C_3)$. If the op-amp has no buffer output stage, for the typical values $R_0 \simeq 10^6 \ \Omega$ and $A_0 \simeq 3 \times 10^3$, we get $R_0/A_0 \simeq 333 \ \Omega$ which is much less than the usual on-resistance of a minimum-sized switch. We can hence conclude that for a purely capacitive load even an unbuffered (but otherwise ideal) op-amp can operate sufficiently fast. If the load is resistive, however, the load current requirements usually make a buffer output stage necessary. Also, the above derivation assumed that the op-amp has infinitely wide bandwidth, and is hence somewhat unrealistic.

The analysis becomes considerably more complicated if both the nonzero output resistance and finite-gain-bandwidth characteristics of the op-amp are taken into account simultaneously. Then a second-order differential equation can be used to describe $v_{\text{out}}(t)$, which (for unbuffered op-amps) has a damped sine-wave function as its solution (Problem 7.17). The time constant T_1 of the exponential damping is, for usual parameter values, determined almost completely by the time constant associated with R_0; its value is

$$T_1 \simeq 2R_0\left(\frac{C_1 C_2}{C_1 + C_2} + C_3\right). \tag{7.95}$$

In the derivation of (7.95), it was assumed that the condition

$$T_1 \ll \frac{A_0}{\omega_0} = \frac{1}{|s_1|} \tag{7.96}$$

held. This will usually be true even for a relatively fast op-amp with a fairly high output impedance R_0.

Note that under the above conditions, the transient is much slower than the value T_0 derived for the case of an infinite-bandwidth op-amp. In fact $T_1 \simeq 2A_0 T_0$, as a comparison of Eqs. (7.95) and (7.93) reveals. Of course, in order to obtain sufficiently complete settling, T_1 should be much smaller* than the time interval during which $\phi_2 =$ "1."

---

*Say, by a factor of 7 or more in order to obtain a 0.1% settling accuracy.

The frequency of the damped oscillation of the output voltage $v_{out}(t)$ can be obtained from the approximating formula

$$\omega_1 \simeq \sqrt{2\omega_0/T_1} \,. \tag{7.97}$$

Here, ω_0 is the op-amp's unity-gain frequency, and T_1 the output time constant given in Eq. (7.95).

It is also possible to include in the analysis the effects of the on-resistance R_{on} of the input switches. It can then be shown[16] that for an integrator settling error of 0.1% or less the following conditions must hold:

$$A_0 \geq 5000,$$

$$\omega_0/\omega_c \geq 4, \tag{7.98}$$

$$T/R_{on}C_1 \geq 40.$$

The proof is lengthy[16] and is hence omitted here.

Thermal and 1/f Noise

As discussed in Chapter 4 (see Sections 4.10 and 4.13) both thermal and $1/f$ noise are generated internally by the MOSFETs of the op-amps. This noise can be reduced by appropriate device design techniques. Its effect on the overall circuit noise, as well as some circuit design techniques aimed at minimizing this effect, will be discussed in Section 7.4, below.

7.4. NOISE GENERATED IN SWITCHED-CAPACITOR CIRCUITS

The output voltage of a switched-capacitor circuit is always contaminated by noise, originating from a variety of sources. This noise is much larger than what is usually found in comparable active-RC filters, and hence it is important to understand its origin and dependence on circuit parameters.

There are three main sources of noise in an SC circuit:

1. Clock feedthrough noise.
2. Noise coupled directly or capacitively from the power, clock, and ground lines, and from the substrate.
3. Thermal and flicker $(1/f)$ noise generated in the switches and op-amps.

The first of these noise sources has already been discussed extensively in Section 7.1; the second will be discussed in the next section. Here, we will deal with the analysis and minimization of internally generated noise effects.

As described in Section 3.7, there are two important mechanisms for noise generation in an MOS transistor. One is the *thermal noise* with a constant (frequency-independent) power spectral density (PSD) given by

$$S_T(f) = \frac{\overline{v_{nT}^2}}{\Delta f} = 4(kT)R = 4\theta R. \tag{7.99}$$

Here, $\overline{v_{nT}^2}$ is the mean square of the noise voltage in the frequency range Δf. Note that S_T is a *one-sided* distribution, so that Δf is the difference of two *positive* frequencies. (For a two-sided representation along the negative as well as positive frequency axis, S_T must be divided by 2.) Also, k is Boltzmann's constant, T is the absolute temperature, and $\theta = kT$ is a notation introduced to prevent confusion in this section between the identical symbols used to denote clock period and temperature. Finally, R is the incremental channel resistance.

The other dominant noise source is the *flicker* $(1/f)$ *noise*, with a frequency-dependent PSD given by

$$S_f(f) = \frac{\overline{v_{nf}^2}}{\Delta f} = \frac{K}{C_{ox}WLf}. \tag{7.100}$$

Here, $C_{ox}WL$ equals the gate-to-channel capacitance in the triode region, and K is a process- and temperature-dependent parameter. Since the two noises in the same device are uncorrelated, their PSDs can be added directly. Thus, the overall noise generated in a MOSFET has the spectral density $S(f) = S_T + S_f$ illustrated in Fig. 7.30. Here, f_{cr} (the *corner frequency*) separates the two frequency regions in which the two different noise effects dominate.

Since the noise generated by an op-amp originates from its MOSFETs (usually mostly from the input devices), its spectrum also has the general shape shown in Fig. 7.30. The value of f_{cr} for an MOS op-amp depends on the device

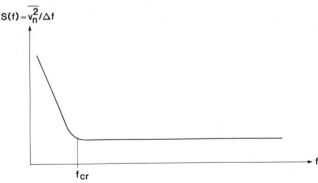

FIGURE 7.30. Op-amp noise power spectral density showing $1/f$ and thermal noise and the corner frequency f_{cr}.

dimensions, the fabrication process, the temperature, and so on; for typical parameters, it is usually in the range of $1 \sim 50$ kHz, with 10 kHz a typical value.

As described in Sections 6.1 and 6.3, there exist simple circuit techniques for the cancellation of the dc offset voltages of the op-amps. These operations are equivalent to subtracting the instantaneous values of the noise existing at times nT and $nT - T/2$. Hence, the overall effect can be represented by the transfer function $H_{CDS}(z) = 1 - z^{-1/2}$ or, in the frequency domain, by

$$H_{CDS}(e^{j\omega T}) = 1 - e^{-j\omega T/2} = e^{-j\omega T/4} 2j \sin\left(\frac{\omega T}{4}\right). \qquad (7.101)$$

(Here, the subscript CDS stands for *correlated double sampling*, the often-used name for this process.) Thus, the PSD of the noise is multiplied by

$$\left| H_{CDS}(e^{j\omega T}) \right|^2 = 4 \sin^2\left(\frac{\omega T}{4}\right). \qquad (7.102)$$

This function suppresses noise not only at dc, but also at low frequencies as well as around the frequencies $2f_c, 4f_c, \ldots$, where $f_c = 1/T$ is the clock frequency. Hence, for the usual case when $f_{cr} \ll f_c$, the $1/f$ noise is essentially eliminated near $f = 0$ by the offset compensation (Fig. 7.31).

A different strategy for the elimination of the $1/f$ noise is to use a fully differential circuit combined with chopper stabilization.[17] The basic principle of chopper stabilization is illustrated in Fig. 7.32a, which shows the block diagram of a chopper-stabilized amplifier.[17] The power spectrum of the input voltage is illustrated in Fig. 7.32b; the input-referred noise spectrum $S(f)$ $= \overline{v_n^2}/\Delta f$ of the input stage A_1 in Fig. 7.32c. As these spectra illustrate, a direct amplification of v_{in} would result in a severe degradation of the dynamic range at low frequencies, since the peaks of the signal and noise spectra overlap.

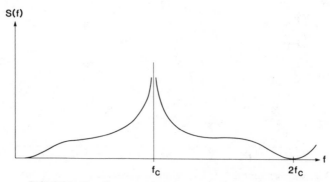

FIGURE 7.31. Op-amp noise after correlated double sampling.

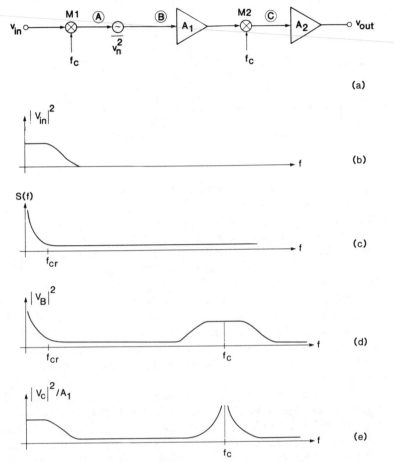

FIGURE 7.32. Chopper-stabilized amplifier circuit: (a) block diagram; (b) input spectrum; (c) noise spectrum; (d) the spectrum at node Ⓑ; (e) the spectrum at node Ⓒ.

Hence, before entering v_{in} into A_1, it is multiplied by a clock signal of frequency f_c, using the balanced modulator M1. This operation shifts the input signal spectrum, so that it is now centered at f_c. The combined spectra of the modulated signal and the noise at node Ⓑ are illustrated* in Fig. 7.32d. Clearly, the peaks are now well separated. After amplification by A_1, a second balanced modulator M2 can be used to return the signal spectrum to the baseband. Figure 7.32e shows the resulting spectrum at node Ⓒ.

The resulting low-noise amplified signal can now be further amplified and/or buffered by the output stage A_2, to obtain v_{out}.

*In Fig. 7.32, the noise amplitudes are greatly exaggerated for easier comprehension of the basic concepts of the system.

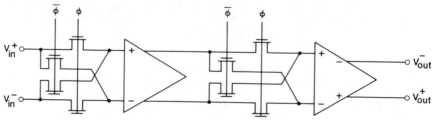

FIGURE 7.33. Differential chopper-stabilized op-amp.

Figure 7.33 shows how the choppers shown in Fig. 7.32a can be implemented using MOSFET switches. Note that the amplifiers A_1 and A_2 must be fully differential at both input and output terminals.

In a switched-capacitor stage (such as the integrators in Figs. 7.5, 7.22, and 7.23), the use of the fully differential op-amp requires a duplication of all capacitors and switches. Corresponding to the integrator of Fig. 7.5, for example, we now have the circuit of Fig. 7.34. The general scheme, which can realize also the stray-insensitive integrators of Figs. 7.22 and 7.23, is shown in Fig. 7.35. Here, v_{in1}^+ and v_{in1}^- form one of the differential input signals and v_{in2}^+ and v_{in2}^- the other. Assuming that the common-mode value of the two differential input signals is zero, the common-mode voltage of the op-amp input terminals will be V_{bias}. This voltage can be set to a convenient value by an on-chip circuit.

Clearly, any low-frequency noise, whether caused by $1/f$ noise, clock feedthrough noise aliased to dc, or op-amp offset voltage, and so on is suppressed by the combination of the differential configuration and chopper stabilization, as long as the layout as well as the dimensions are fully symmetrical. In addition, if the choppers are operated at one-half of the clock frequency f_c, then the noise peaks occurring near dc, as well as near f_c, $2f_c$, and so on will be shifted to $f_c/2$, $3f_c/2$, and so on, and cannot be aliased into the low-frequency passband range by subsequent sampling at the clock rate.

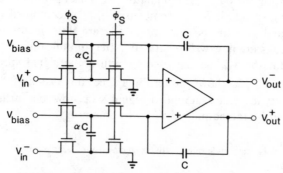

FIGURE 7.34. Differential switched-capacitor integrator.

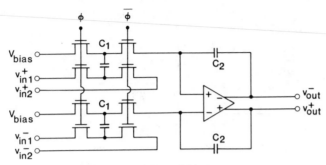

FIGURE 7.35. Differential integrator stage.

The use of the fully differential circuit illustrated in Fig. 7.35 has additional advantages. Since for a symmetrical layout many of the noise voltages (power-supply noise, clock feedthrough noise, offset voltages) appear as common-mode signals, they are to a considerable extent canceled in the differential output voltage v_{out} at all frequencies.

On the negative side, the design of the fully differential op-amps needed in these circuits is somewhat more complicated than of their simpler single-ended counterparts; this topic was briefly discussed in Section 4.16 of Chapter 4. Also, the total capacitance, as well as the number of switches, clock, and other lines are effectively doubled. In addition, differential-to-single-ended conversion may be needed in some applications. Finally, the thermal noise will be increased by the added components and switching operations.

The maximum voltage swing of the differential op-amp for the same supply voltages is twice that of a single-ended one. Hence, in addition to the $1/f$ and supply noise reduction, the maximum signal level of the filter is doubled.

Using the integrator of Fig. 7.35 as the basic building block with very large integrating capacitance ($C_2 \simeq 100$ pF), Hsieh et al.[17] were able to design and fabricate a fifth-order Chebyshev SC filter with a dynamic range of 102 dB, an exceedingly high value.

While the reduction of $1/f$ noise is very useful, in switched-capacitor circuits the effect of the wide-band thermal noise generated in the switch transistors and op-amps is often much more important than that of the $1/f$ noise. To understand the reason for this phenomenon, consider the simple switched-capacitor stage[18] shown in Fig. 7.36a. Assume that the input signal v_{in} is zero, and that the switch transistor Q has an on-resistance R_{on} during the duty cycle when ϕ is high (Fig. 7.36b). The noise waveform v_{cn} across C has then the waveform shown in Fig. 7.36c. During $\phi = $"1," the noise v_n from Q appears directly in v_{cn}. As ϕ goes low, the last value of v_n remains "frozen" in v_{cn}, due to the sample-and-hold (S/H) character of the circuit. The waveform of Fig. 7.36c may be partitioned into a "direct" waveform v_{cn}^d which equals v_{cn} when $\phi = $"1" but is zero otherwise (Fig. 7.36d), and a S/H waveform $v_{cn}^{\text{S/H}}$ which contains the piecewise-constant segments of v_{cn} during the $\phi = $"0" intervals.

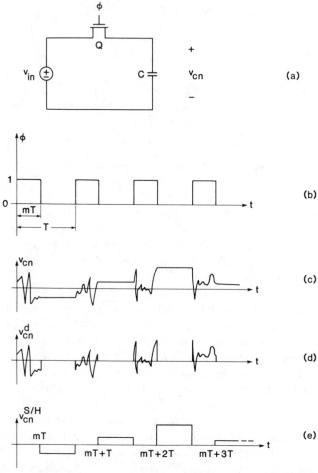

FIGURE 7.36. Switched-capacitor noise: (*a*) circuit diagram; (*b*) clock signal; (*c*) output noise waveform; (*d*) direct noise component; (*e*) sampled-and-held noise component.

To calculate the RMS value of v_{cn}^d, we note that when $\phi =$ "1," the equivalent circuit of the SC stage is that shown in Fig. 7.37. Clearly, the power spectral density $S(f)$ of v is

$$S(f) = \frac{\overline{v^2}}{\Delta f} = S_T(f) \left| \frac{1/j\omega C}{R_{on} + 1/j\omega C} \right|^2 = \frac{4\theta R_{on}}{1 + (2\pi f T_{on})^2}. \quad (7.103)$$

The mean-square value of v for all frequencies is hence

$$\overline{v^2} = \int_0^\infty S(f)\, df = \frac{4\theta R_{on}}{2\pi} \int_0^\infty \frac{d\omega}{1 + \omega^2 T_{on}^2}, \quad (7.104)$$

where $T_{on} = R_{on}C$ is the time constant of the switch during its "on" time.

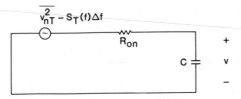

FIGURE 7.37.   Equivalent circuit of the switched capacitor when ϕ = "1."

The integral in (7.104) is readily evaluated and gives $\pi/2T_{on}$. Hence, $\overline{v^2} = \theta/C$ results. However, the resistor R_{on} feeds the capacitor C only during the duty cycles when ϕ = "1"; hence the noise power must be multiplied by $m < 1$, giving

$$\overline{\left(v_{cn}^d\right)^2} = \frac{m\theta}{C}. \tag{7.105}$$

Similarly, the spectral density $S^d(f)$ of v_{cn}^d is $mS(f)$, where $S(f)$ is given by (7.103).

Next, the spectral density $S^{S/H}(f)$ of the sampled-and-held wave $v_{cn}^{S/H}(t)$ will be given. As shown in Fig. 7.36, $v_{cn}^{S/H}(t)$ is obtained by sampling $v_{cn}^d(t)$ at time instances mT, $mT + T$, $mT + 2T$, $mT + 3T$, and so on, with a frequency $f_c = 1/T$, and holding these samples for intervals of $\tau = (1 - m)T$. Since at these instances $v_{cn}^d(t) = v(t)$, we can assume that it is $v(t)$ which is being sampled and held. By using an argument similar to that presented in connection with Eqs. (2.33)–(2.56), and also assuming that the values of $S(f)$ at different frequencies are uncorrelated, it can be shown that $S^{S/H}(f)$ and $S(f)$ are related by

$$S^{S/H}(f) = \frac{\overline{\left(v_{cn}^{S/H}\right)^2}}{\Delta f}$$

$$= \left(\frac{\tau}{T}\right)^2 \left(\frac{\sin(\pi\tau f)}{\pi\tau f}\right)^2 \sum_{k=-\infty}^{\infty} S(f - kf_c). \tag{7.106}$$

Here, by Eq. (7.103),

$$S(f) = \frac{\overline{v^2}}{\Delta f} = \frac{1}{2} \frac{4\theta R_{on}}{1 + (\omega T_{on})^2}. \tag{7.107}$$

In (7.107), as compared to (7.103), there is an additional factor $\frac{1}{2}$ transforming $S(f)$ into a two-sided ($-\infty \leq \omega \leq \infty$) PSD function.

The general shape of the $S(f)$ curve is shown in Fig. 7.38. The total area under the curve, as derived earlier in (7.104), is θ/C. Hence, the curve may be

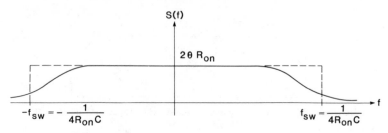

FIGURE 7.38.   Noise spectrum for the circuit of Fig. 7.37.

approximated by the rectangle shown in broken line in the figure. The bandwidth associated with the idealized response is thus $f_{SW} = 1/4T_{on}$. According to (7.106), this response must be shifted by integer multiples of $f_c = 1/T$, and the resulting replicas added.

To draw meaningful conclusions from the above theoretical results, some practical assumptions must also be made. In dimensioning the switching transistor Q in the circuit Fig. 7.36a, its aspect ratio W/L is presumably chosen such that it makes a (nearly) complete charging of C from the input voltage v_{in} possible during the interval $0 \le t \le mT$. Since the charging time constant is $T_{on} = R_{on}C$, for a 0.1% settling accuracy the condition

$$mT \ge 7T_{on} \tag{7.108}$$

must be satisfied. Hence, the direct noise bandwidth f_{SW} satisfies

$$f_{SW} = \frac{1}{4T_{on}} \ge \frac{7}{4mT} = \frac{7}{4m}f_c > 3.5f_c \tag{7.109}$$

where $m < 0.5$ was used. In practice, usually $f_{SW} \ge 5f_c$. Some of the replicas $S(f - kf_c)$ appearing in (7.106) are illustrated, for $N = f_{SW}/f_c = 5$, in Fig. 7.39, where two contiguous replicas are shown on each f-axis line to save space. Clearly, after adding all the (uncorrelated) replicas, the power spectral density of $S(f)$ is magnified by a factor $2N = 2f_{SW}/f_c = 10$. Hence, in general, the summation in (7.106) gives a PSD

$$\frac{2f_{SW}}{f_c}S(f) \simeq \frac{1/2R_{on}C}{f_c}2\theta R_{on} = \frac{\theta/C}{f_c}. \tag{7.110}$$

The total noise power in the baseband $-f_c/2 \le f \le f_c/2$ due to all replicas is hence θ/C. The aliasing due to the sampling of the noise thus concentrates the full noise power of the switch resistor into the baseband. This result shows that it is futile to reduce R_{on} below the value required by the settling-time condition (7.108), since while this reduces the direct thermal noise PSD $S^d(\omega)$, it

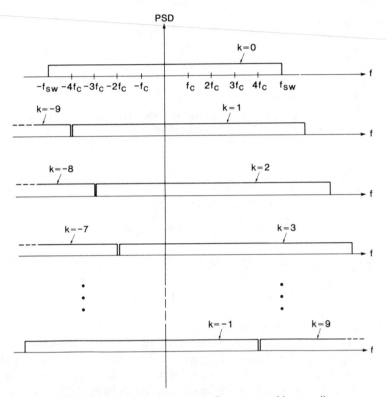

FIGURE 7.39.   Noise spectrum replicas generated by sampling.

increases the aliasing, and the two effects cancel. By contrast, increasing C reduces both the direct and the aliased thermal noise PSDs.

Substituting Eq. (7.110) into (7.106), and recalling that $\tau = (1 - m)T$, we obtain the sampled-and-held noise density:

$$S^{\text{S/H}}(f) = \frac{\overline{\left(v_{cn}^{\text{S/H}}\right)^2}}{\Delta f} = (1 - m)^2 \frac{\sin^2\left[(1 - m)\pi f/f_c\right]}{\left[(1 - m)\pi f/f_c\right]^2} \frac{\theta}{f_c C}. \quad (7.111)$$

Figure 7.40 illustrates the shape of $S^{\text{S/H}}(f)$ for $m = 0.25$, in decibels. For low (passband) frequencies where $f \ll f_c$, the two-sided S/H noise density is

$$S^{\text{S/H}}(f) = \frac{\overline{\left(v_{cn}^{\text{S/H}}\right)^2}}{\Delta f} \simeq \frac{(1 - m)^2 \theta}{f_c C} \quad (7.112)$$

while the two-sided direct noise density is for low frequencies

$$S^d(f) = \frac{\overline{\left(v_{cn}^d\right)^2}}{\Delta f} \simeq m \frac{\overline{v^2}}{\Delta f} \simeq 2m\theta R_{\text{on}}. \quad (7.113)$$

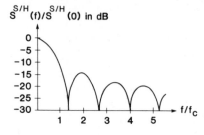

$S^{S/H}(f)/S^{S/H}(0)$ in dB

FIGURE 7.40. The density function of the sampled-and-held noise.

Their ratio is

$$r = \frac{S^{S/H}(f)}{S^d(f)} \simeq \frac{(1-m)^2}{2m} \frac{1}{f_c R_{on} C}. \tag{7.114}$$

Since, by (7.108), $f_c R_{on} C \leq m/7$ and $m < 0.5$,

$$r \geq 3.5(1/m - 1)^2 > 3.5. \tag{7.115}$$

In fact, for $m = 0.4$, $r \simeq 8$; for $m = 0.25$, $r = 31.5$. We can hence conclude that the S/H noise dominates the direct noise effect at low frequencies in the circuit of Fig. 7.36a. For the usual case when the corner frequency f_{cr} is less than $f_c/2$, the $1/f$ noise contributes significantly only to the direct noise in the baseband. Hence, it is usually much less important than the S/H noise.

Note that due to the $\sin^2 x/x^2$ factor in $S^{S/H}(f)$, as given in (7.111) and illustrated in Fig. 7.40, the PSD of the sampled-and-held noise is most important in the $|f| < f_c$ frequency range. This can be verified in the time domain from Fig. 7.36 which illustrates that while the RMS value of the S/H noise $v_{cn}^{S/H}$ is of the same order of magnitude as that of the direct noise v_{cn}^d, its PSD is mostly at low frequencies while that of v_{cn}^d is spread over a much wider frequency range. Thus, while the PSD of the direct noise occupies a broad frequency band, that of the S/H noise is a narrow-band spectral density.

Since the direct and S/H thermal noises are essentially uncorrelated, their PSDs may be simply added to obtain the overall thermal noise PSD $S^T(f)$. From Eqs. (7.103) and (7.111), therefore

$$S^T(f) = S^d(f) + S^{S/H}(f) = mS(f) + S^{S/H}(f)$$

$$= \frac{4\theta m R_{on}}{1 + (2\pi f T_{on})^2} + \frac{(1-m)^2 \theta}{f_c C} \frac{\sin^2[(1-m)\pi f/f_c]}{[(1-m)\pi f/f_c]^2} \tag{7.116}$$

gives the total thermal noise of the switched capacitor.

The concepts of the preceding analysis can now be extended to the complete integrator of Fig. 7.5.[19] For $v_{in} = 0$, the noise-equivalent model of the circuit is

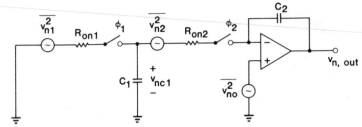

FIGURE 7.41.   Noise-equivalent circuit of the integrator of Fig. 7.5.

shown in Fig. 7.41. In this circuit, each switch is modeled by its on-resistance R_{on}, the associated thermal noise source $\overline{v_n^2}$, and an ideal switch.* The noise voltage of the op-amp will be modeled by the noise source $\overline{v_{no}^2}$ connected to the noninverting input terminal of the op-amp. The op-amp will again be described by the one-pole model, so that its output–input relation is given by Eq. (7.75). Its voltage gain is hence

$$|A_V(j\omega)| \simeq \frac{A_0}{\sqrt{1 + (A_0\omega/\omega_0)^2}}. \qquad (7.117)$$

As for the circuit of Fig. 7.36a, discussed above, the output noise will again contain some broad-band direct noise, and also some S/H noise due to noise charges trapped on the switched capacitor C_1 when it is disconnected from a noise source. To calculate the direct noise, the equivalent circuits drawn for the $\phi_2 = $ "1" and $\phi_2 = $ "0" periods in Figs. 7.42a and 7.42b, respectively, can be used. Note that during the $\phi_2 = $ "1" interval, the direct noise sources are both switch 2 and the op-amp, while during $\phi_2 = $ "0," only the op-amp contributes direct noise. Switch 1 is, of course, never connected by a direct path to the output, and therefore does not add to the direct output noise.

The calculation of the direct PSDs from the circuits of Fig. 7.42 is straightforward if laborious, and is left to the reader as an exercise (Problem 7.19). For $A_0 \gg 1$ and $\omega/\omega_0 \ll 1$, the PSD at the output of the op-amp in Fig. 7.42a is

$$S_1^d \simeq 2m\theta\left[\alpha^2 R_{on2} + (1 + \alpha)^2 R_{eq}\right] \qquad (7.118)$$

where $\alpha \triangleq C_1/C_2$, and R_{eq} is the (hypothetical) resistor which would generate as much thermal noise as $\overline{v_{no}^2}$, so that (for two-sided noise density) $\overline{v_{no}^2} =$

---

*The off-resistance of an MOS switch is so large that the corresponding time constant satisfies $R_{off}C_1 \gg T$. Hence, the noise due to the off-resistance is essentially a small slowly varying dc offset voltage, which will be neglected here.

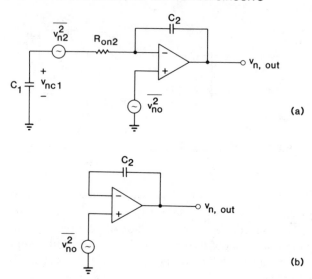

FIGURE 7.42.   Direct noise generation in the circuit of Fig. 7.41: (*a*) when $\phi_2 =$ "1"; (*b*) when $\phi_2 =$ "0."

$2\theta R_{eq}\Delta f$. Usually, $R_{eq} \gg R_{on}$. For the circuit of Fig. 7.42*b*,

$$S_2^d \simeq (1 - m)2\theta R_{eq}. \qquad (7.119)$$

The reader is referred to Ref. 19 for the exact formulas.

As mentioned earlier, the S/H noise is caused by the sampling-and-holding operation which C_1 and one or the other ideal switches perform on the noise voltage v_{nc1} across C_1. This causes an aliasing of the spectrum, and concentrates the noise power in the baseband. The wider the bandwidth of v_{nc1}, the more serious the aliasing effect becomes. For the noise due to $\overline{v_{n1}^2}$, the band limiting is performed by the $R_{on1} - C_1$ section. For $\overline{v_{n2}^2}$ and $\overline{v_{no}^2}$, as illustrated in Fig. 7.42, both the $R_{on2} - C_1$ section and the finite unity-gain bandwidth ω_0 of the op-amp contribute to the band-limiting operation. Usually, the op-amp band limits the signal since $\omega_0 < \omega_{sw} = \pi/2R_{on}C$.

The S/H noise voltage across C_1 due to switch 1 can be calculated in the same way as in the previous example. The result must then be multiplied by the sampled-data power transfer function

$$\left| H(e^{j\omega T}) \right|^2 = \frac{(C_1/C_2)^2}{4\sin^2(\omega T/2)} \qquad (7.120)$$

of the integrator. The contribution of switch 2 can be found from Fig. 7.42*a*, by setting $v_{no} \equiv 0$ and calculating the RMS noise voltage v_{nc1} across C_1. Then, the aliasing due to the sampling is taken into account, and the $(\sin x/x)^2$ factor representing the holding operation appended. Finally, the result is multiplied by $|H(e^{j\omega T})|^2$ to obtain the output PSD. The same process can be used to obtain the S/H noise due to the op-amp noise source v_{no}^2: now v_{n2} is set to zero in Fig. 7.42a, and v_{nc1} found, and so on. Assuming also that the time constants $R_{on}C_1$ associated with the two switches are equal to each other, but much less than $1/\omega_0$ so that the key frequency limiting element is the op-amp, the result can be written in the form[19]:

$$S^{S/H}(f) = (1 - m)^2 \frac{\theta}{f_c C_1}\left(1 + \frac{(R_{on} + R_{eq})C_1\omega_0}{\alpha + 1}\right)\left(\frac{\alpha}{\omega T}\right)^2. \quad (7.121)$$

Here, the first term represents the S/H noise due to v_{n1}, while the term containing ω_0 represents the S/H noise due to v_{n2} and v_{no}. It is directly proportional to ω_0, indicating that the band-limiting performed by the op-amp is important in reducing the aliasing which enhances the S/H noise. Hence, ω_0 should be chosen as low as possible while still allowing the adequate settling of the op-amp; as discussed in the previous section, this means $\omega_0 \simeq 5\omega_c$. Normally, $R_{on} \ll R_{eq}$, so that $S^{S/H}$ is essentially independent of R_{on} as long as the above-stated conditions on $R_{on}C_1$ hold. The last factor of $S^{S/H}(f)$ represents the integrator transfer function $|H|^2$ for $\omega T \ll 1$. Note that the broad-band direct noise may also have an important effect if the output noise $v_{n,\,out}$ is undersampled by the input switch of the following SC stage.

From our two simple examples the following general conclusions, valid for most SC circuits, may be drawn. The internally generated noise in an SC circuit consists of the flicker ($1/f$) and thermal noises. Due to the internal sampling and holding performed by the switches and capacitors this noise will be replicated in the frequency domain. For usual parameter values, this does not lead to appreciable aliasing of the $1/f$ noise, since the corner frequency f_{cr} is usually well below $f_c/2$; however, the thermal noise will get seriously undersampled and hence aliased. As a result of the aliasing, the PSD of the low-frequency thermal noise will be multiplied by $2f_{noise}/f_c$, where f_{noise} is the bandwidth of the thermal noise after being band-limited by the circuit. The noise bandwidth is determined by the time constants of the on-resistance of the switches multiplied by the switched capacitances, and by the op-amp bandwidths. Hence, to minimize noise aliasing, these time constants should be as large and the op-amp bandwidths should be as low as possible. Typically, $R_{on}C \simeq 0.1mT$ and $\omega_0 \simeq 5\omega_c$ may be used. All noise effects, along with the clock feedthrough noise, are reduced if the circuit capacitances are increased; the noise power is proportional to $1/C_{min}$.

7.5.   LAYOUT CONSIDERATIONS IN SWITCHED-CAPACITOR CIRCUITS[6, 21, 22]

While any integrated circuit is to some extent sensitive to the physical arrangement of its components and their interconnections, analog ICs are more sensitive than most others. Hence, some key points concerning the optimal arrangement (layout) of switched-capacitor circuits will be discussed in this section. The level of sophistication of this discussion will be low, since it is aimed primarily at the inexperienced designer. However, the basic principles to be described are of great practical importance, and should not be ignored even in simple circuits.

Some of the performance parameters which are often strongly affected by the layout are the following:

1. Noise injection from power lines, clock lines, and ground lines.
2. Noise injection from the substrate.
3. Clock feedthrough noise.
4. Accuracy of matched elements.
5. Rise time and overshoot, or (equivalently) the high-frequency response of the circuit.
6. Nonlinear distortion.
7. Sensitivity to process variations.
8. Op-amp offset voltages.

These effects will be briefly discussed next.

To prevent noise injection from the power, ground, and clock lines, two precautions of equal importance must be observed. First, these lines must, as far as possible, be kept free from noise. Second, the noise coupling from the lines into the signal path should be minimized. The first requirement is especially important if the chip contains both analog and digital circuitry, which is an increasingly common situation.* Consider the system shown in Fig. 7.43a where the analog and digital sections are fed by the same positive (V^+) and negative (V^-) bias voltages, and where both sections use the same bonding pads and power lines. Clearly, the positive supplies of both sections share a common resistance R^+, which carries the supply current of the analog section (i_a) as well as that of the digital one (i_d). The digital supply current i_d typically contains some asynchronously recurring large noise spikes, which are due to the switching transients of the digital logic circuits. These spikes are especially large if some of the digital inputs or outputs are connected to bipolar transistor/transistor logic (TTL) circuits on adjacent chips. The common resistance R^+ contains the output impedance of the outside power supply providing V^+, the resistance of the bonding wire and bonding contact, as well as the shared portion of the on-chip power line. The line voltage at the analog section will be therefore $v^+ = V^+ - (i_a + i_d)R^+$. Hence, the spikes in i_d will appear (multiplied by $-R^+$) in v^+. The resulting noise is often intolerable for the analog circuit, since it has both a large amplitude and a wide bandwidth,

*See, for example, the systems discussed in Chapter 8.

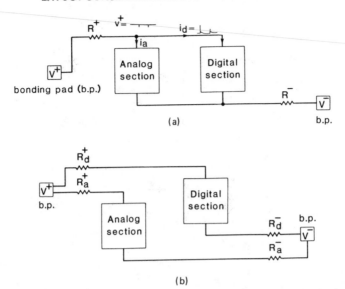

FIGURE 7.43. (*a*) Biasing arrangement for an integrated circuit containing both analog and digital sections. (*b*) Improved biasing arrangement.

leading to noise aliasing effects. Similar considerations hold, of course, for the negative power supply.

The values of the shared resistances R^+ and R^- may be considerably reduced, and hence the analog supply noise suppressed, by using separate ("dedicated") power lines for the analog and digital circuitry (Fig. 7.43*b*). Further improvement can be achieved by using also separate bonding pads for these dedicated lines, connected to the same outside pin; or, in an optimal arrangement, using separate pins as well, which are short-circuited externally. In the latter arrangement, the only remaining common impedance of the analog and digital lines is the output impedance of the external supply. Using external decoupling capacitors at the pins, this residual impedance can be reduced to a very low value and the spike noise essentially eliminated. Under these circumstances, the bias voltage lines for the substrates and wells can also be connected to the analog supply pads without introducing any digital noise into the substrate or wells. These substrate bias lines must have as many contacts to the substrate (or well) as possible. These contacts will collect the electrons or holes injected into the substrate (or well), keep the substrate (or well) at a fixed potential, and will thus prevent the occurrence of latch-up, a fatal problem which can affect CMOS integrated circuits.

To prevent some crucial signal or clock lines either from picking up or injecting noise, they can be shielded from their environments. A possible shielding arrangement for a metal line is shown in Fig. 7.44, where the shield consists of two grounded metal lines and a grounded polysilicon layer.[21,22]

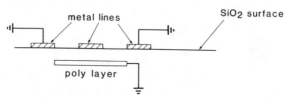

FIGURE 7.44.   Shielding arrangement for a metal line.

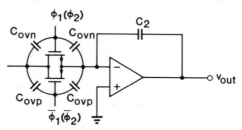

FIGURE 7.45.   Coupling paths from the clock sources to the output.

Such shielding can also be used to separate the analog and digital circuitries, and to prevent noise from coupling into or out of the substrate.

Noise injection *from* the power-supply lines can easily occur in the op-amps; the resulting noise depends on the PSRR of the op-amp. This subject was briefly discussed in Section 4.14, item 8, where it was shown that the low-frequency noise usually enters only via the bias circuit and through the asymmetries of the input stage, while high-frequency noise may enter through the stray capacitances and the output stage as well. We have also discussed techniques (such as fully differential circuitry, supply-independent biasing, modified compensation circuits, cascode input stage, etc.) for reducing these effects.

Power-line or clock-line noise can also enter into the signal path through the switches. To illustrate this situation, consider the stray-insensitive integrator of Fig. 7.22. The part of the circuit containing the op-amp and the input switch is shown in Fig. 7.45, where the switch is assumed to be realized by a CMOS transmission gate. The gate-to-source and gate-to-drain overlap capacitances C_{ovn} and C_{ovp} are also shown.* Clearly, any noise in the four clock signals ϕ_1, $\bar{\phi}_1$, ϕ_2, and $\bar{\phi}_2$ will be coupled to the output voltage v_{out} of the op-amp, with weighting factors C_{ovn}/C_2 or C_{ovp}/C_2. The clock voltage is here (as in a majority of applications) a two-phase nonoverlapping signal, which may typically be generated by the circuit of Fig. 7.46. In this clock generator, the delay needed to avoid the overlap of ϕ_1 and ϕ_2 as well as $\bar{\phi}_1$ and $\bar{\phi}_2$ is realized by the gate delays. The circuit is normally fed only by the digital supply lines;

---

*These include also the gate-to-channel capacitances of the conducting devices.

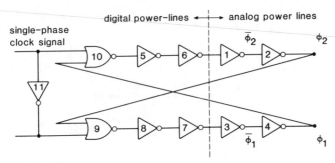

FIGURE 7.46. Nonoverlapping clock signal generator, with dedicated power-line connection scheme.

however, it is a better arrangement to use the analog power lines for biasing the last four inverters (numbered 1 through 4 in the figure), which actually generate the four clock signals needed. This arrangement will reduce the digital noise in the signals ϕ_1, ϕ_2, $\bar\phi_1$, and $\bar\phi_2$, and hence (as Fig. 7.45 shows) also in v_{out}. Naturally, the switch sizes should all be chosen as small as the voltage settling and other considerations permit. Also, if possible, silicon-gate switches with self-aligned geometries should be used to reduce the gate-overlap capacitances.

Another important noise source in the circuit is due to coupling from the substrate. Again, this effect can be reduced by minimizing the noise coupling *into* the substrate as well as *out* of it. The former can be achieved by using a clean power supply to bias the substrate; by establishing a good bond (via gold bonding, if necessary) between the back surface of the substrate and the package header; and by shielding the substrate from the bottom plates of all capacitors by placing grounded wells under them. It is also a good practice to shield the substrate from all noisy lines, such as digital clock lines, using grounded polysilicon lines or diffusion wells. Perhaps the best technique for reducing substrate noise is to use an epitaxial process; the resulting n^+-layer then acts as a large ground plane. To reduce noise coupling *from* the substrate, a number of techniques may be used. These include the use of fully differential circuitry[17] (with special attention paid to the symmetry of the layout), as well as shielding and careful selection of the topology of the circuit. The largest capacitances coupling the substrate to the circuit are by far those between the substrate and the bottom plates of the capacitors. Depending on the ratio of the field-oxide and thin-oxide thicknesses, this bottom-plate-to-substrate capacitance may be 20% or more of the nominal top-plate-to-bottom-plate one. Hence, the bottom plate should *not* be connected or switched to the inverting input terminal of any op-amp, since these terminals are the most noise-sensitive ones in the circuit (i.e., the noise gain from these terminals to the output is the highest). Thus, for example, in the circuit of Fig. 7.22 the left plate of C_1 and the right plate of C_2 should be chosen as the bottom plates. Then, at low

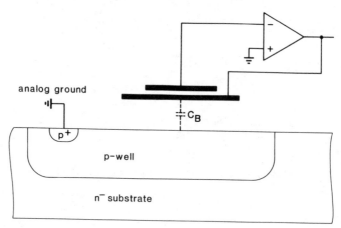

FIGURE 7.47. Shielding technique using a p-well connected to an analog ground line.

frequencies, the low output impedance of the op-amps operating under closed-loop conditions will prevent noise injection through the bottom-plate capacitances. At high frequencies, however, the op-amp gain is reduced; hence the closed-loop output impedance increases, and high-frequency noise can thus be coupled into v_{out}. To prevent this, in CMOS technology often a p-well is established in the substrate to shield all capacitances from the bottom. This well can be connected to a low-impedance/low-noise ground line (Fig. 7.47), and will greatly reduce high-frequency noise coupling into the circuit via the bottom-plate capacitance C_B. It will also, as mentioned above, reduce the coupling of noise spikes *into* the substrate from the switched capacitors.

Other steps should also be taken to protect the inverting input terminals of the op-amp from substrate noise. The lines connecting the input nodes to any capacitors and switches should be as short as possible. They should be fabricated using either polysilicon or (preferably) metal; the use of diffusion lines must be avoided here because they are too closely coupled capacitively to the substrate. The crossing of input-node lines with any other signal-carrying lines (digital or analog) should be avoided. Whenever possible, the input lines should be shielded (although this may not be necessary for very short lines), and guard rings should be used to shield the input devices of the op-amp.

The amount of circuitry connected to an input node should be minimized;[6] only one switch should be used at the input terminal, realized with minimum-area devices. If clock feedthrough and bias considerations permit, this switch should be a single-channel one rather than a transmission gate. Also, a separate p-well may be used for those NMOS switches which are connected to the input nodes, biased by a separate low-noise negative supply voltage. To illustrate the noise coupling mechanism from the substrate to the op-amp input nodes, Fig. 7.48*a* shows a part of the integrator circuit of Fig. 7.22, with CMOS

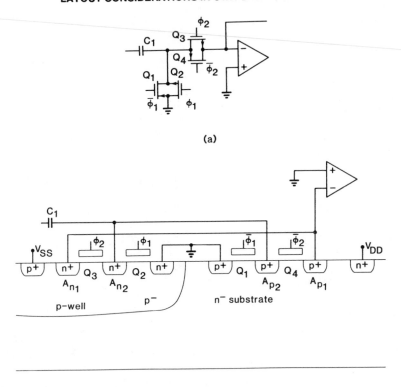

(a)

(b)

FIGURE 7.48.   (*a*) Circuit of the input branches. (*b*) Cross-sectional view of the source–drain junctions connected to the inverting input of the op-amp in an integrator stage.

transmission gates used to realize the switches. Figure 7.48*b* shows a schematic picture of the cross section of an actual realization. Clearly, when $\phi_1 =$ "0" and $\phi_2 =$ "1" so that Q_3 and Q_4 conduct, the junction area which couples noise to the op-amp includes the diffusion junction areas A_{n1}, A_{n2}, A_{p1}, and A_{p2}, as well as the channel areas of Q_3 and Q_4. By contrast, when $\phi_1 =$ "1" and $\phi_2 =$ "0" so that Q_1 and Q_2 conduct, the active area for noise coupling is only $A_{n1} + A_{p1}$. (In calculating the corresponding noise coupling capacitances, it is important to include the areas of the sidewalls as well as the planar bottoms of the diffusions; the sidewall areas may in some cases dominate!) The above discussion and Fig. 7.48*b* shows that a considerable capacitance exists between the substrate and the op-amp input nodes, and hence illustrates the importance of using dedicated low-noise bias voltage supplies for the *p*-wells and the substrate.

A typical layout of the four devices Q_1, Q_2, Q_3, and Q_4 shown in Fig. 7.48 is illustrated in Fig. 7.49. Here a self-aligned process is assumed, so that the

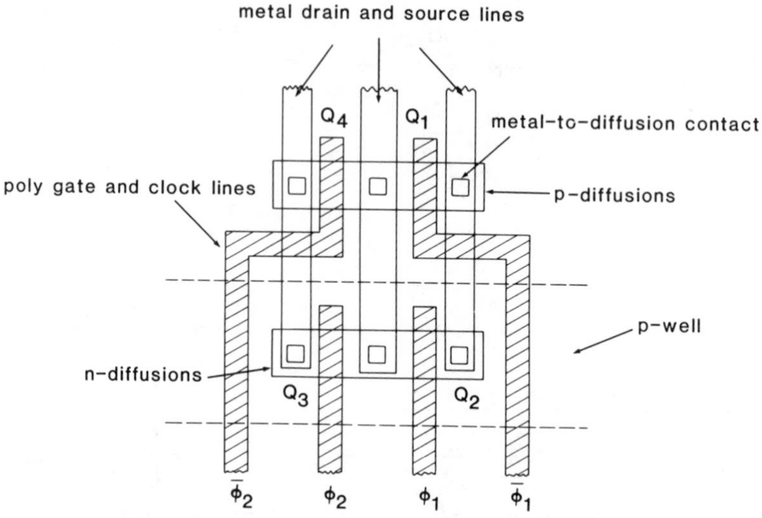

metal drain and source lines

Q_4 Q_1 metal-to-diffusion contact

poly gate and clock lines

p-diffusions

p-well

n-diffusions

Q_3 Q_2

$\bar{\phi}_2$ ϕ_2 ϕ_1 $\bar{\phi}_1$

FIGURE 7.49.   Typical layout of a CMOS switch arrangement.

channel region is defined by the common areas of the diffusion and polysilicon-gate masks. Hence, the source and drain diffusions, shown as single units, are in fact interrupted under the gate lines as can be seen from the cross-sectional view of Fig. 7.48b. As discussed before, ideally all dimensions of the switches should have the minimum value allowed by the resolution of the technology. Sometimes, however, settling-time considerations require a smaller on-resistance for the switch than that provided by the minimum-width device, and hence the width must be increased. Also, often the metal-to-diffusion contact rule specifies the minimum extension e of the diffusion beyond the contact opening (Fig. 7.50), which would require a wider width for the channel. To reduce the channel width, the bone-shaped geometry of Fig. 7.50 may be used. Now, however, the distance d between the polysilicon and diffusion edges has a minimum allowable value, and hence the area and periphery of the diffusion is increased as compared to the simpler geometry. Figure 7.51 shows the layout of the CMOS switches $Q_1 - Q_4$ using the geometry of Fig. 7.50.

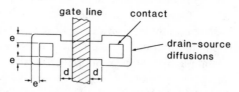

gate line   contact

e

e

drain-source diffusions

d d

e

FIGURE 7.50.   MOSFET geometry for a gate width which is narrower than the allowed drain/source width.

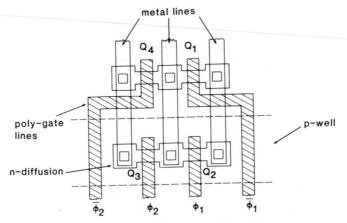

FIGURE 7.51. Alternative layout of the CMOS switch arrangement.

The layout geometry of the *op-amp* has important effects on its rise time, overshoot, and high-frequency response, as well as on its sensitivity to process variations and its offset voltage. To achieve good step response and high-frequency response, the components of the amplifier must be arranged so as to minimize the line lengths, especially for lines connecting high-impedance nodes. The devices and lines of the input and output stages should be well separated, to avoid spurious feedback effects. If the nominal design of the op-amp is done carefully, so that no systematic voltage offset exists, some random offset may still occur due to errors in the ratios of (nominally) matched components. For example, as discussed in Section 4.14 [see Eq. (4.187)], a relative error ε in the matching of the aspect ratios of the input devices results in an input-referred offset voltage $v_{off} = \varepsilon(v_{GS} - V_T)/2$, where v_{GS} and V_T are the gate-to-source and threshold voltages of the input devices, respectively. For example, if $v_{GS} - V_T = 1$ V and $\varepsilon = 1\%$, then an offset voltage of 5 mV results from this single imperfection. This may be unacceptable in some circuits. Similarly, a matching error ΔV_T between the threshold voltages of the input devices results in an offset voltage equal to ΔV_T. Again, this may give an impractically large offset. To minimize these errors, both the input devices and their loads should be placed side by side, with identical geometries, including all connecting lines. Matched devices requiring wells should share the same well. Only straight-line channels should be used, since the geometry of corners of bent channels is poorly controlled. If high accuracy and a low offset voltage is required, it is advisable to split both input and load devices into two or more unit transistors connected in parallel, and arrange them in a common centroid geometry similar to that used for accurate capacitance matching. If a heat source (e.g., high-current output stage) is near the matched elements, the latter should be located symmetrically with respect to it, to ensure matched temperatures.

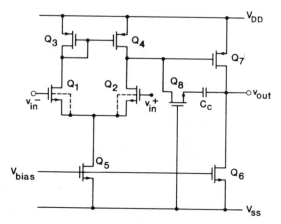

FIGURE 7.52. A two-stage CMOS op-amp.

The circuit geometry also has some effect on the internally generated noise. Thus, as already mentioned, the $1/f$ noise may be reduced by using larger input devices in the op-amps; the wideband and aliased noise can be reduced by choosing larger capacitors. Also, it was observed that very short devices introduce "hot electron" noise if operated at large voltages.[22] This should be avoided, by using increased channel lengths for such devices.

As an illustration of efficient layout procedure, Fig. 7.52 shows the circuit diagram of a simple two-stage compensated CMOS op-amp. Its layout for self-aligned silicon-gate p-well process is illustrated in Fig. 7.53. Note the symmetry of the matched devices (Q_1/Q_2 and Q_3/Q_4), the short connecting lines, the separation of the input and output lines, and the compact arrangement of the overall structure resulting in a small total chip area.

In laying out op-amps with previously untested circuitry and/or fabrication process, it is a useful practice to include test-circuit versions of the amplifier on the first chip. These test circuits may contain the individual sections of the op-amp separately, and/or the full circuit but with large test pads added at all important internal nodes. These pads enable the designer to probe and measure the dc bias voltages, the low-frequency signal voltages, the saturation conditions of the individual devices, and so on, and to discover the reasons for any deficiency of the op-amp response. Such diagnostic tests may otherwise be difficult to perform on a cramped integrated circuit.

On a chip containing many op-amps, switches, and capacitors, it is usually efficient to arrange these components in separate areas on the chip. Thus, all op-amps may occupy one row, all capacitors a second one, and all switches a third, with the supply and clock lines running in parallel along the edges. A possible arrangement is shown in Fig. 7.54. Here, all capacitors are placed over a shielding p-well connected to the analog ground, and all n-channel switches

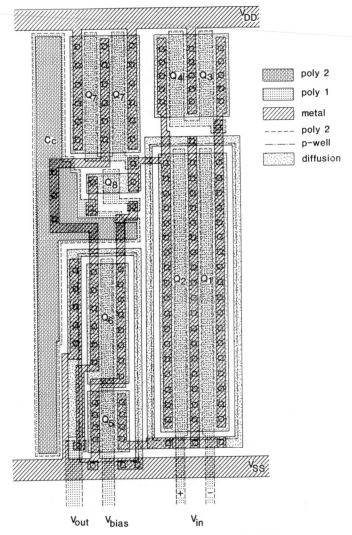

FIGURE 7.53.   Layout of the two-stage CMOS op-amp.

share a p-well biased by a separate analog negative supply voltage V^-. Separate ground lines are used for the noninverting op-amp input terminals and the capacitor p-well on the one hand, and the switches on the other. The voltage supply lines (V^+, V^-, ground) used by the op-amps run along the upper and lower edges of the op-amp row.

The layout of Fig. 7.54 (or some variation thereof) is useful not only for switched-capacitor filters, but for any switched-capacitor circuit including gain

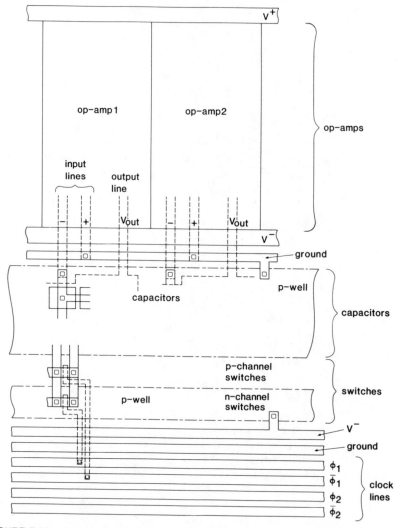

FIGURE 7.54. A typical layout for a switched-capacitor circuit. Lines drawn in solid are metal and those indicated with broken lines polysilicon.

stages, rectifiers, oscillators, modulators, and so on. An alternative arrangement is described in Ref. 23.

PROBLEMS

7.1. Derive Eq. (7.3) for the circuit of Fig. 7.1, assuming that $v_{\text{out}}(0) = 0$, and that v_{in} is a constant. Also, ignore the dependence of V_{Tn} on v_{out}, that is, treat V_{Tn} as a constant.

7.2. Derive Eq. (7.15) from Eqs. (7.12)–(7.14).

7.3. Let the leakage currents in the circuit of Fig. 7.11a be $I_{l2} = 2I_{l1} = 0.5$ pA, and let $C_1 = C_2 = 3$ pF. What is the lowest clock frequency $f_{c\,min}$ if the maximum permissible output offset due to leakage is 0.1 V?

7.4. Prove that the z-domain voltage gain

$$H(z) = \frac{V_{out}(z)}{V_{in}(z)} = \frac{a_n z^n + a_{n-1} z^{n-1} + \cdots + a_1 z + 1}{b_m z^m + b_{m-1} z^{m-1} + \cdots + b_1 z + b_0}$$

of an SC circuit depends only on the *ratios* C_i/C_j of the capacitance values in the circuit. [*Hint:* The coefficients a_l and b_k depend only upon the C_i values. What are the dimensions of $H(z)$, z, a_l, and b_k?]

7.5. Show how the lossy integrator of Fig. 7.55 can be compensated for clock feedthrough by using (a) a fully differential circuit configuration and (b) a partially differential configuration (cf. Fig. 7.10).

7.6. Calculate the steady-state output voltage v_{out} of the circuit of Fig. 5.15c if v_{in} is constant, and the op-amp has an input-referred offset voltage V_{off}.

7.7. Calculate the steady-state output voltage v_{out} of the circuit of Fig. 5.13c if $v_{in} = 0$, and both op-amps have equally large input-referred offset voltages $V_{off1} = V_{off2}$.

7.8. Repeat the calculation of Problem 7.7 for the biquad of Fig. 5.10c.

7.9. Replace z by $\exp(j\omega T)$ in Eqs. (7.45) and (7.44). Verify that the results can be written in the forms given in Eqs. (7.46) and (7.47).

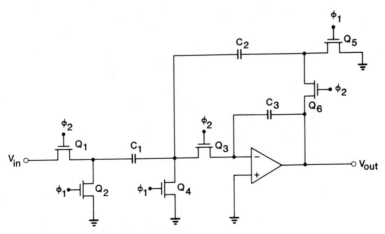

FIGURE 7.55. Lossy SC integrator (Problem 7.5).

7.10. Analyze the noninverting integrator circuit of Fig. 7.23 to prove that Eqs. (7.47)–(7.56) hold also for this circuit, in the case of finite dc op-amp gain.

7.11. Derive the frequency response of the damped integrator of Fig. 5.20 if the op-amp has a finite dc gain A_0. How large must A_0 be if the change introduced by $A_0 < \infty$ is to be less than the changes due to 0.1% capacitance ratio errors?

7.12. Consider an integrator with a transfer function $H(s) = -1/sL$, simulating the voltage–current relation of an ideal inductor. Let the op-amp of the integrator have a finite dc gain A_0, but infinite bandwidth. Show that the branch actually simulated is that shown in Fig. 7.25. [*Hint:* Use Eq. (7.69).]

7.13. As indicated in Fig. 7.26 for the high-Q biquad, by changing the phasing of the four switches marked by asterisks, the charging transients of the two integrators are decoupled from each other. Which four switches should be differently clocked in the low-Q biquad of Fig. 5.10c to achieve the same result?

7.14. The condition on the unity-gain op-amp bandwidth $\omega_0 \geq 5\omega_c$ was derived for two-phase clock signals. What are the equivalent conditions for four-phase and six-phase clocks?

7.15. To obtain some reasonable estimates for the highest slope occurring in a band-limited signal, and hence for the required slew rate of the op-amps, assume that the signal $v(t)$ is a continuous-time step function $Au(t)$ which is then passed through an analog low-pass filter with a cut-off frequency ω_B. Calculate the maximum slope $|dv(t)/dt|_{\max}$ for the following analog filter characteristics:

   1. *Ideal Low-Pass Filter.*

$$H_{\mathrm{LP}}(\omega) = \begin{array}{ll} 1, & |\omega| \leq \omega_B \\ 0, & |\omega| > \omega_B \end{array}$$

   2. *Gaussian Filter.* $H_{\mathrm{LP}}(\omega) = e^{-(\tau\omega)^2}$.
   3. *RC Filter.* $H_{\mathrm{LP}}(\omega) = 1/(1 + j\omega\tau)$.

   In cases 2 and 3, calculate the value of τ such that the gain of the antialiasing filter is down to -40 dB at $\omega = \omega_B$. Assume (for simplicity) $A = A_{\max}$, and compare your results with (7.86). [*Hint:* Note that the impulse response $h(t)$ of the analog filter is identical to the slope $dg(t)/dt$ of its step response!]

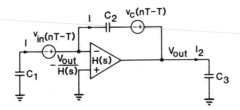

FIGURE 7.56. Initial state of the integrator of Fig. 7.27 at the beginning of the $\phi_2 = $ "1" interval (Problem 7.18).

7.16. Derive Eqs. (7.92)–(7.94) for the equivalent circuit of Fig. 7.29. Compare the time constant T_0 with the charging/discharging time constants of C_1 and C_2 via the on-resistance of a minimum size switch!

7.17. Carry out the analysis of the charging transient ($\phi_2 = $ "1" interval) for the case when the op-amp has a nonzero output resistance R_0 *and* its input–output relation is given by Eq. (7.76). Assume typical values for ω_0, A_0, R_0, and the capacitances, and make reasonable approximations to obtain a simplified expression for $v_{out}(t)$. Under what conditions will v_{out} be an oscillatory function of time? Is the condition satisfied for an unbuffered op-amp? What is its frequency and damping coefficient?

7.18. Carry out the transient analysis during the $\phi_2 = $ "1" inteval of the SC intergrator of Fig. 7.27, under the following conditions: (a) The op-amp transfer function $H(s)$ contains two real poles, s_1 and s_2, where $|s_1| \ll |s_2|$; (b) The op-amp had adequate time to settle during the $\phi_1 = $ "1" period; and (c) The initial conditions when ϕ_2 goes high can be represented by the equivalent circuit of Fig. 7.56.

7.19. Calculate the direct-coupled output noise densities for the two circuits shown in Fig. 7.42.

7.20. The offset-compensated noninverting SC voltage amplifier of Fig. 6.5a is shown again in Fig. 7.57. Analyze the performance in the presence of the following nonideal effects: (a) Finite constant op-amp gain A; (b) stray capacitances between all nodes and the substrate; (c) stray capacitances C_{gs} and C_{gd} for all switch transistors; (d) an input-referred offset voltage V_{off} for the op-amp. Show that the ideal output voltage $v_{out} = av_{in}$ is changed to

$$v'_{out} = \frac{av_{in} + (V_H - V_L)(C_{gs}/C) + V_{off}\left[(1 + \alpha)C + C_{gs} + C_A\right]/(1 + A)C}{\left[(1 + \alpha)C + C_{gs} + C_A\right]/(AC) + 1},$$

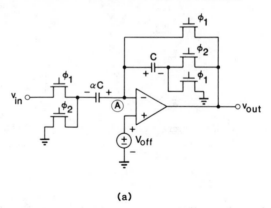

(a)

FIGURE 7.57. An offset-compensated SC amplifier (Problem 7.20).

where C_{gs} is the gate-to-source stray capacitance of the reset switch between node Ⓐ and the output node, and C_A is the stray capacitance between node Ⓐ and ground. Also, V_H is the high level and V_L the low level of the clock signal. Find v'_{out} for the following values: $\alpha = 0.1$, $V_H = -V_L = 8$ V, $C_{gs} = 10$ fF, $C = 10$ pf, $V_{off} = 5$ mV, $C_A = 15$ fF, and $A = 1000$.

REFERENCES

1. R. E. Suarez, P. R. Gray, and D. A. Hodges, *IEEE J. Solid-State Circuits*, **SC-10**, 379–385 (1975).
2. L. A. Bienstman and H. DeMan, Proceedings of the International Solid-State Circuits Conference, pp. 16–17 (1980).
3. K. Martin, *Electron. Lett.*, **18**, 39–40 (1982).
4. J. L. McCreary, *IEEE J. Solid-State Circuits*, **SC-16**, 608–616 (1981).
5. J. L. McCreary and P. R. Gray, *IEEE J. Solid-State Circuits* **SC-10**, 371–379 (1975).
6. D. J. Allstot and W. C. Black, Jr., *Proc. IEEE*, **71**, 967–986 (1983).
7. J. B. Shyu, G. C. Temes, and K. Yao, *IEEE J. Solid-State Circuits*, **SC-17**, 1070–1076 (1982).
8. J. B. Shyu, G. C. Temes, and F. Krummenacher, *IEEE J. Solid-State Circuits*, **SC-19**, 948–955 (1984).
9. D. G. Haigh and B. Singh, Proceedings of the International Symposium on Circuits and Systems, pp. 586–589 (1983).
10. G. C. Temes, *IEEE J. Solid-State Circuits*, **SC-15**, 358–361 (1980).
11. K. Martin, *Switched-Capacitor Networks*, Ph.D. dissertation, University of Toronto, 1980.
12. K. Martin, and A. S. Sedra, *IEEE Trans. Circuits Systems*, **CAS-28**, 822–829 (1981).
13. M. L. Blostein, *IEEE Trans. Circuits Systems*, **CT-14**, 21–25 (1967).
14. A. Hedayati, *New Techniques for the Design of Bilinear Switched-Capacitor Ladder Filters*, Ph.D. dissertation, UCLA (1984).

15. D. J. Allstot, *MOS Switched-Capacitor Ladder Filters*, Ph.D. dissertation, University of California, Berkeley, 1979.

16. G. Fischer and G. S. Moschytz, *IEEE J. Solid-State Circuits*, **SC-19**, 510–518 (1984).

17. K. C. Hsieh, P. R. Gray, D. Senderowicz, and D. C. Messerschmidt, *IEEE J. Solid-State Circuits*, **SC-16**, 708–715 (1981).

18. Cl. A. Gobet and A. Knob, *Electron. Lett.*, **16**, 734–735 (1980).

19. Cl. A. Gobet and A. Knob, *IEEE Trans. Circuits Systems*, **CAS-30**, 37–43 (1983).

20. J. H. Fischer, *IEEE J. Solid-State Circuits*, **SC-17**, 742–752 (1982).

21. R. Gregorian, K. W. Martin, and G. C. Temes, *Proc. IEEE*, **71**, 941–966 (1983).

22. K. W. Martin, unpublished lecture notes, UCLA.

23. P. E. Fleischer, K. R. Laker, D. G. Marsh, J. P. Ballantyne, A. A. Yiannoulos, and D. L. Fraser, *IEEE Trans. Circuits Systems*, **CAS-27**, 552–559 (1980).

24. T. H. Hsu, *Improved Design Techniques for Switched-Capacitor Ladder Filters*, Ph.D. dissertation, UCLA (1982).

SYSTEM
CONSIDERATIONS
AND APPLICATIONS

Switched-capacitor circuits are usually fabricated as parts of a larger signal processing system, although some programmable filters have also been made commercially available as stand-alone units. As system components, SC circuits must be compatible with the other parts of the system, and must also satisfy its overall specifications.

In this chapter, we first discuss the specific design considerations associated with the sampled-analog character of SC circuits, discussed earlier in Chapter 2. Then, a number of fully integrated signal processing systems incorporating SC components are described. In selecting these examples, the authors gave preference to devices which they (especially R.G.) were instrumental in designing. This was motivated both by familiarity and emotional involvement. We hope that the reader excuses this bias.

8.1. PREFILTERING REQUIREMENTS FOR SWITCHED-CAPACITOR FILTERS

As discussed in Section 2.5, both the sampling or sample-and-hold (S/H) operations result in the replication of the spectrum of the input signal, and (unless Nyquist's criterion is satisfied) introduce aliasing distortion. Therefore, SC circuits which perform a S/H function are vulnerable to aliasing. To prevent this, in general, a low-pass continuous-time analog antialiasing filter (AAF) is required before a continuous-time signal can be processed by an SC circuit (Fig. 8.1). Since only that part of the input spectrum which extends

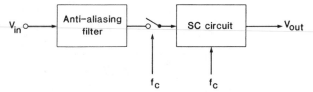

FIGURE 8.1.   SC circuit with antialiasing prefilter.

beyond $f_c/2 = 1/2T$ is responsible for the aliasing (cf. Fig. 2.9 of Section 2.5), a continuous-time low-pass AAF with a stopband limit frequency $f_{stop} = f_c/2$ can prevent all aliasing. Naturally, the passband limit f_{pass} of the AAF should be at a high enough frequency so that all important signal frequencies are contained in the passband $0 \leq f \leq f_{pass}$.

In most SC systems, the antialiasing filter is integrated on an MOS chip which is shared with the rest of the system. The AAF is then typically an active-RC filter,[1] containing thin-oxide capacitors and polysilicon resistors. Since the polysilicon layer has a sheet resistivity of only about 50 $\Omega/\square$, the sizes of such resistors (or of the associated capacitors) become very large when RC time constants of the order of 10 μs or more are required.[2] In addition, the tolerance of the resistivity is often as large as $\pm 50\%$, and this variation must not affect the flatness of the passband, nor permit aliasing. Fortunately, the *ratios* of matched resistors can be controlled quite closely, using symmetrical layouts, common-centroid geometry, and so on. The same statement concerning tolerances holds, of course, for the capacitors. It can be shown that under these conditions the effect of the R and C variations is simply to shift the frequency response of the AAF up or down along the frequency axis. The resulting relative errors of f_{pass} and f_{stop} are at most equal to the sum of the relative errors of R and C. Since the passband and stopband specifications must be met even under worst-case tolerance conditions, the AAF is usually greatly overdesigned.

To reduce the area requirement of the AAF, it is expedient to design the system so as to allow the use of very low degree ($n = 2$ or 3) AAFs. As an illustration, Fig. 8.2 shows a second-degree Sallen–Key active-RC filter,[1] often used as an AAF. It is easy to show (Problem 8.1) that the transfer function of

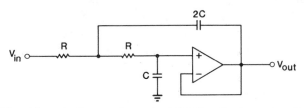

FIGURE 8.2.   Schematic diagram of an active-RC Sallen–Key low-pass antialiasing filter.

this circuit is

$$H(s) = \frac{\omega_0^2}{s^2 + (\omega_0/Q)s + \omega_0^2},\qquad(8.1)$$

where the pole frequency is $\omega_0 = 1/\sqrt{2}\,RC$ and the pole-Q is given by $Q = 1/\sqrt{2}$; the latter is independent of R and C, as long as the two resistors track each other, and also the two capacitors are matched.

The response provided by this circuit is a maximally flat (Butterworth) one.[1] Since the pole-Q is independent of R and C, so is the flatness of the passband. To illustrate the selectivity which can be achieved by the Sallen–Key filter, let the 3-dB frequency limit chosen to be at 36 kHz; then at $f_{\text{stop}} = 512$ kHz the loss is approximately 46 dB (Problem 8.2). Furthermore, if the accuracy of the RC product is $\pm 50\%$, then a minimum loss of 32 dB can be guaranteed at f_{stop}, and a maximum loss of 0.01 dB at $f_{\text{pass}} = 3.4$ kHz under all tolerance conditions. The circuit can also provide a peaking response, which may be useful to equalize the gain droop introduced by S/H distortion, finite op-amp gain, and so on (Problem 8.9.).

From the above numbers, some generally valid conclusions can be drawn. For usual specifications, in order to be able to use a low-degree AAF, the $f_{\text{stop}}/f_{\text{pass}}$ ratio should be large: typical values range from 10 to several hundreds. Also, the circuit must be overdesigned, so that its response continues to satisfy both passband and stopband specifications even under worst-case tolerance conditions; this must be checked by the designer.

To achieve a sufficiently large $f_{\text{stop}}/f_{\text{pass}}$ ratio for the AAF, the clock frequency f_c must be chosen much higher than f_{pass}. This, however, often introduces new complications. As discussed in the previous chapter, the op-amps' unity-gain frequency needs to be about five times higher than f_c, otherwise distortions occur in the frequency response. Hence, for a large f_c, the high-speed specifications of the op-amp may become hard to satisfy. In addition, as shown, for example, by Eq. (5.37) of Section 5.4, for high f_c (i.e., small T), the capacitance spread $C_{\text{max}}/C_{\text{min}}$ of the SC circuit becomes very large.

There are at least two ways of avoiding the impasse of these conflicting considerations. One is based on the observation that the SC circuit following the AAF usually has a low-pass or bandpass response. Assuming the former, let us consider the responses shown in Fig. 8.3. In the figure, for simplicity, the S/H effect ($\sin x/x$ response) is neglected, and the SCF response is hence considered periodic. As the figure indicates, the stopband limit frequency f_{stop} of the AAF is chosen here *not* as $f_c/2$, but as $f_c - f_1$, where f_1 is the cutoff (stopband limit) frequency of the SCF. Hence, the input spectrum of the SCF will contain energy in the $f_c/2 \leq f \leq f_c - f_1$ range, which will be therefore aliased into the $f_1 \leq f \leq f_c/2$ range, as illustrated in the figure. Since, however, the $f_1 \leq f \leq f_c/2$ frequency range falls into the *stopband* of the SCF response,

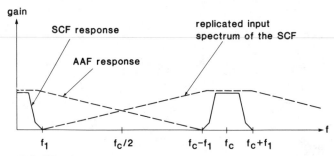

FIGURE 8.3.   Frequency responses of the antialiasing and SC filters.

the resulting aliasing distortion is suppressed by the SCF, and will not appear in the output v_{out} of the overall system.

Since normally $f_1 \ll f_c$, specifying $f_{stop} = f_c - f_1$ rather than $f_{stop} = f_c/2$ for the AAF nearly doubles the f_{stop}/f_{pass} ratio, and hence allows the use of a lower f_c for a given AAF circuit, or of a lower-order AAF for a given clock frequency f_c.

Another technique involves the use of multiple clock frequencies.[43] Consider the modified system shown in Fig. 8.4. It contains an additional building block—a *decimator* (DEC). This is a low-pass SCF clocked at a higher harmonic nf_c of the main clock rate f_c, where n is an integer. As will be shown, the DEC can usually be realized by a simple modification of the input stage of the SCF of the system of Fig. 8.1. Assume that the passband limit of the DEC is the same as that of the SCF, but that its stopband limit is at $f_c - f_1$ where f_1 is the stopband frequency of the SCF. The resulting frequency responses are schematically shown (for $n = 4$) in Fig. 8.5a. Note that the first replica of the DEC response is centered at nf_c, which is the clock rate of the DEC. The AAF response is also shown; it now cuts off only at $f_{stop} = nf_c - f_1 \approx nf_c$. Thus, the selectivity of the AAF is reduced by a factor close to $2n$.

When the output of the DEC is next sampled at the main clock frequency f_c (Fig. 8.4), its spectrum gets replicated (Fig. 8.5b). There is considerable aliasing occurring in this sampling operation; all, however, takes place in frequency regions which subsequently fall into the stopbands of the main filter SCF response illustrated in Fig. 8.5c. Hence, the resulting distortion does not appear in v_{out}.

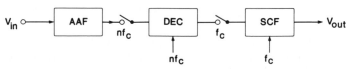

FIGURE 8.4.   Modified SCF system containing a decimator stage.

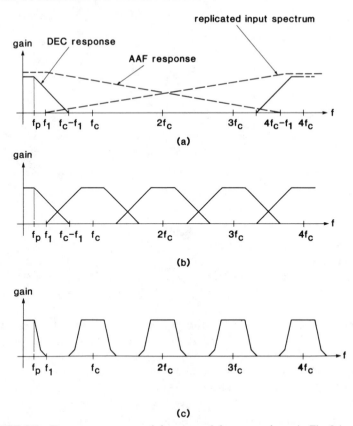

FIGURE 8.5.   Frequency responses of the stages of the system shown in Fig. 8.4.

Using this decimation strategy, the selectivity of the AAF is reduced by a factor $2n$. Hence, its degree can be much lower for the same f_c than that of the AAF in the original system of Fig. 8.1.

To find an economical implementation of the DEC, consider, as an example, the SCF input stage shown in Fig. 8.6a. The charge flow q in the input branch can be described by a sequence $q_0, q_1, q_2, \ldots$, where q_k is the charge which flows into C_1 from the input capacitor $\alpha_3 C_1$ during the kth clock interval $(k-1)T < t < kT$. Clearly, $q_k = \alpha_3 C_1 v_{in}(kT - T/2)$ where it is assumed that the $t = kT$ instance occurs at the end of the kth clock interval, when ϕ_1 goes high (Fig. 8.6b). Hence, the z-transform of the q_k sequence is simply

$$Q(z) = \alpha_3 C_1 V_{in}(z) z^{-1/2}, \tag{8.2}$$

where $V_{in}(z)$ is the z-transform of the $v_{in}(kT)$ sequence.

Consider next the modified input stage shown in Fig. 8.7. In this circuit, the continuous-time input voltage v_{in} is sampled n times during each clock pulse

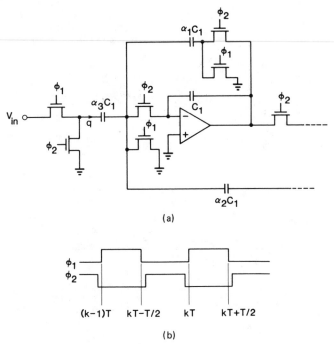

FIGURE 8.6.   (*a*) Input stage of an SCF. (*b*) Clock waveforms.

(in the figure, $n = 6$). Hence, the total charge q_k entering through $\alpha_3 C_1$ during the kth clock pulse is now

$$q_k = \alpha_3 C_1 \sum_{i=0}^{n-1} v_{\text{in}}(kT - iT/n - T/2n). \tag{8.3}$$

Therefore, in the z-domain (Problem 8.7),

$$Q(z) = \alpha_3 C_1 \sum_{i=0}^{n-1} z^{-i/n} V_{\text{in}}(z) z^{-1/2n}$$

$$= \alpha_3 C_1 V_{\text{in}}(z) \frac{1 - z^{-1}}{1 - z^{-1/n}} z^{-1/2n}. \tag{8.4}$$

Comparison with (8.2) shows that the multiple sampling introduces a factor $H_n(z)$ into the transfer function, where

$$H_n(z) = \frac{1 - z^{-1}}{1 - z^{-1/n}} z^{(1-1/n)/2}. \tag{8.5}$$

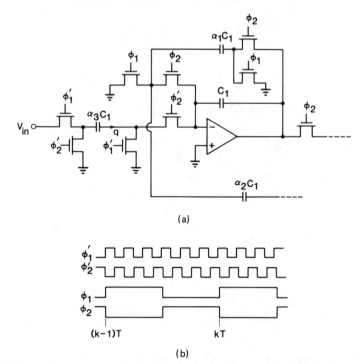

(a)

(b)

FIGURE 8.7. (*a*) Input stage of filter which samples the input signal at nf_c. (*b*) Timing diagram.

In the frequency domain, with z replaced by $\exp(j\omega T)$, the amplitude of H_n is readily found to be

$$\left| H_n(e^{j\omega T}) \right| = \left| \frac{\sin(\pi f / f_c)}{\sin(\pi f / n f_c)} \right|. \tag{8.6}$$

Figure 8.8 shows this response for $n = 8$. As required (cf. Fig. 8.5*a*), the response has passbands around the frequencies $f = 0$, nf_c, $2nf_c$, and so on, and stopbands between f_c and $(n - 1)f_c$, between $(n + 1)f_c$ and $(2n - 1)f_c$, and so on. Hence, it performs the required function of the DEC stage in the system of Fig. 8.4.

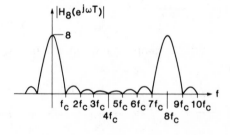

FIGURE 8.8. The amplitude response of $H_n(e^{j\omega T})$ for $n = 8$.

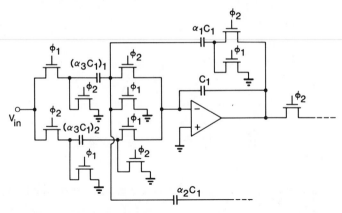

FIGURE 8.9. Modified input stage which incorporates a cosine filter.

A disadvantage of the input circuit of Fig. 8.7 is that it requires a multiple-phase clock signal. For $n = 2$, the input circuit of Fig. 8.9 can be used instead.[3] Here, two input capacitors, switched alternatively, are used, and no extra clock phases are required.* It is also possible to use a single input capacitor and a single two-phase clock to implement the DEC for the $n = 2$ case[3]; the resulting circuit is shown in Fig. 8.10. Its analysis is left to the reader (Problem 8.4). This circuit is, however, *not* stray insensitive; the response is affected by the stray capacitance loading node Ⓐ (Problem 8.8).

For the $n = 2$ case, (8.6) gives

$$\left| H_2(e^{j\omega T}) \right| = \left| \frac{\sin(\pi f/f_c)}{\sin(\pi f/2f_c)} \right| = 2 \left| \cos \frac{\pi f}{2f_c} \right|. \tag{8.7}$$

Hence, the $n = 2$ DEC is often called a *cosine filter*. Its amplitude response is illustrated in Fig. 8.11.

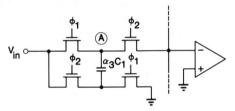

FIGURE 8.10. Simplified input branch for implementing a cosine filter.

*However, it is important that ϕ_1, and ϕ_2 be symmetrical.

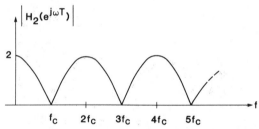

FIGURE 8.11. Cosine filter amplitude response.

As Fig. 8.5 shows, the necessary performance conditions on the DEC are the following:

1. Its response should be large and flat up to at least f_p, the passband limit of the SCF.

2. Its response should be small between $f_c - f_1$ and $(n - 1)f_c + f_1$. For a cosine filter, if (say) we choose $f_c = 30f_p = 20f_1$, at the passband limit f_p the response is down by only about 0.012 dB, while at $f_c - f_1$ it is down by over 22 dB and at $f_c - f_p$ by about 25.6 dB. The higher f_c is for a given SCF response, and the higher n is, the flatter the DEC response will be in its passband and the sharper will be its cutoff.

The described techniques can be modified and/or combined to achieve further relaxation of the AAF specifications. As an illustration, consider the system of Fig. 8.12. Here, a separate decimator filter SCF1 precedes SCF. As Fig. 8.13 illustrates for $n = 4$, the cutoff frequency of the AAF needs to be here at most $nf_c - f_1$. If now, in addition, the input branch of the *decimator* filter is modified so as to implement a cosine filter, then the responses shown in Fig. 8.14 result. Now the AAF needs to cut off only at $2nf_c - f_1$. More generally, if the input stage of Fig. 8.7 is used in a SCF with an input clock rate of mnf_c, then the stopband limit of the AAF can be raised to $mnf_c - f_1$. Thus, at the cost of using more complicated SCFs and multiple-phase clock signals, the specifications of the AAF can be drastically reduced. Since the area of the AAF on the chip (as discussed earlier) can be very large, such a tradeoff may be very attractive in many systems.

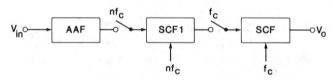

FIGURE 8.12. Filter system with multiple clock frequency.

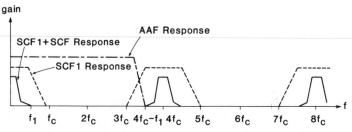

FIGURE 8.13.   Switched-capacitor filter responses and the required antialiasing filter response for $n = 4$.

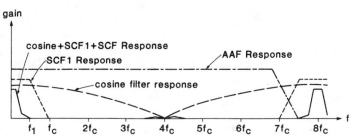

FIGURE 8.14.   Cosine, switched-capacitor, and required antialiasing filter responses for $n = 4$.

8.2.   POSTFILTERING FOR SWITCHED-CAPACITOR FILTERS

The output voltage of an SCF is usually a sampled-and-held (S/H) signal. As discussed in detail in Section 2.5, the S/H operation causes the spectrum of the continuous-time input signal to be replicated at $\pm f_c$, $\pm 2f_c$, and so on, and also to be multiplied by the characteristic S/H function given in Eq. (2.55) and reproduced below:

$$H_{SH}(j\omega) = T e^{-j\omega T/2} \frac{\sin(\omega T/2)}{\omega T/2}. \tag{8.8}$$

Note that (as mentioned earlier)

$$|H_{SH}(j\omega)| = T\frac{\sin(\omega T/2)}{\omega T/2} \tag{8.9}$$

giving rise to the commonly used name "sin x/x response" for H_{SH}. Figure 8.15 illustrates $|H_{SH}(j\omega)|$.

The effect of the S/H operation on the spectrum $V(f)$ of a continuous-time signal $v(t)$ is illustrated in Fig. 8.16. For a sine-wave signal, the waveforms are as shown in Fig. 8.17. If the sampling rate is much higher than the signal frequency (which is usually the case) then the granularity due to the S/H

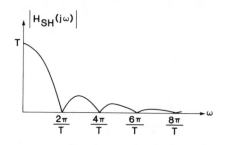

FIGURE 8.15. The amplitude response of the sample-and-hold operation.

nature of the signal is slight. In many applications, however, even this small effect is intolerable, and a continuous-time output signal is required. Since in the frequency domain the S/H effect is represented by the sidebands centered at f_c, $2f_c$, and so on (cf. Fig. 8.16b), by removing these sidebands the desired continuous-time signal is obtained. This can be achieved by using a continuous-time analog low-pass filter (usually an active-RC filter) which passes signals in the $0 \geq f \geq f_1$ frequency range, but cuts off at or below $f_c - f_1$ (Fig. 8.18). Such an analog postfilter is often called *smoothing filter* (SMF). Its specifications are usually identical to those of the AAF, and hence the SMF and the AAF are normally realized by similar circuits. Whenever possible, the SMF is integrated on the same chip as the SCF which precedes it. Hence, it is important to keep its degree low (2 or 3) and thus its structure simple. This will

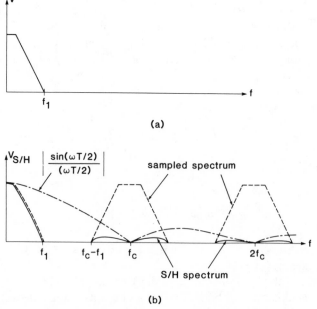

FIGURE 8.16. Spectra of a continuous-time signal before (a) and after (b) the sample-and-hold operation.

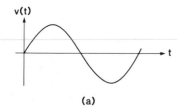

(a)

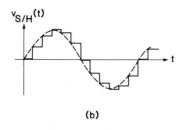

(b)

FIGURE 8.17. Sine-wave signal before (a) and after (b) the sample-and-hold operation.

only be the case if $f_c \gg f_1$ (say, $f_c > 10f_1$), as for the AAF. If this is not true, the use of an additional SCF may be expedient, as illustrated in Fig. 8.19. Here, the main filter SCF1 is followed by an SC postfilter SCF2 which has a higher clock frequency than SCF1 (in the figure, $f_{c2} = 4f_{c1}$ is illustrated), and only then by the SMF. As Fig. 8.19b illustrates, SCF2 suppresses the sidebands of the output of SCF1 around f_{c1}, $2f_{c1}$, and $3f_{c1}$. The sidebands around $4f_{c1}$, $5f_{c1}$, and so on are then suppressed by the SMF.

Unfortunately, unlike the decimator filter used to relax the specifications of the AAF, the auxiliary SCF2 (often called an *interpolator filter* and abbreviated IF) usually cannot be realized by a minor modification of SCF1. It thus requires an additional filter stage, typically with one or two op-amps.[3,6] One conceptually simple realization of the interpolator filter can be obtained using linear interpolation. The concept is illustrated in Fig. 8.20, which shows the input and output signals for an interpolation ratio $r \triangleq f_{c2}/f_{c1} = 3$. (Note that a delay of $1/f_{c1}$ is necessary for the interpolation, since the IF cannot predict the next value of v_1. This delay was omitted, for clarity, in Fig. 8.20.) To obtain

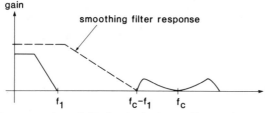

FIGURE 8.18. Spectrum of an S/H signal (continuous line) and suitable smoothing filter response.

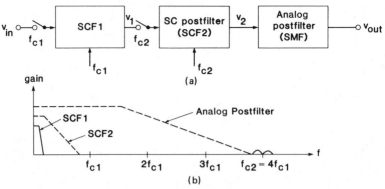

FIGURE 8.19. Postfiltering system: (a) block diagram; (b) frequency responses.

the transfer function of the IF, we note that the samples of v_1 and v_2 satisfy

$$v_2(nT_2) - v_2(nT_2 - T_2) = \frac{1}{r}[v_1(nT_2) - v_1(nT_2 - rT_2)], \quad (8.10)$$

where $T_2 = 1/f_{c2} = T_1/r = 1/rf_{c1}$. Defining z as $\exp(sT_2)$, and performing z-transformation in (8.10) gives

$$(1 - z^{-1})V_2(z) = \frac{1}{r}(1 - z^{-r})V_1(z). \quad (8.11)$$

Hence, the required transfer function of the linear interpolator filter is

$$H_{IN}(z) = \frac{1}{r}\frac{1 - z^{-r}}{1 - z^{-1}} = \frac{1}{r}\sum_{k=0}^{r-1} z^{-k}. \quad (8.12)$$

A simple realization of $H_{IN}(z)$, based on (8.10), is shown for $r = 4$ in Fig. 8.21a. In the circuit, $C_1 = C_2 = C_3 = C_4 = C_a/r$. Each time the ith switch ϕ_i closes, a charge $C_i[v_1(nT_2) - v_1(nT_2 - rT_2)]$ enters C_a. Hence, the corresponding change in v_2 is given (apart from a negative sign) by (8.10), as required.

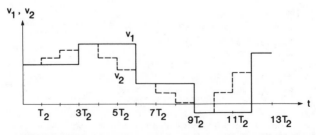

FIGURE 8.20. Linear interpolation for a sampled-and-held signal.

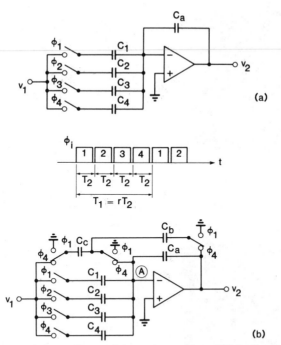

FIGURE 8.21. (*a*) A four-step linear SC interpolator without DC feedback, (*b*) A four-step linear SC interpolator with DC feedback.

While the stage of 8.21*a* is simple, it is of limited practical usefulness, since the op-amp functions without any dc feedback, and is thus vulnerable to saturation due to leakage currents. Fortunately, by adding two switched capacitors C_b and C_c satisfying $C_b = C_c = C_a/r$ (Fig. 8.21*b*), a negative dc feedback is introduced and the response otherwise left unchanged.[7] During the first three clock cycles, the circuit functions the same way as that of Fig. 8.21*a*. When ϕ_4 goes high, C_b and C_c feed equal charges of opposite signs into C_a, and hence the net change in v_2 is found to be once again given by the RHS of (8.10), as required.

The amplitude response associated with $H_{\rm IN}(z)$ is, from (8.12),

$$\left| H_{\rm IN}\!\left(e^{j\omega T_2}\right) \right| = \frac{1}{r}\left| \frac{\sin(r\omega T_2/2)}{\sin(\omega T_2/2)} \right|. \tag{8.13}$$

This transfer function does *not* include the S/H functions associated with SCF1 and SCF2. It can be shown[6] that the overall amplitude response including all functions is, for $\omega T_2/2 \ll 1$,

$$\left| H_{\rm total}(\omega) \right| \simeq \left(\frac{\sin(r\omega T_2/2)}{r\omega T_2/2} \right)^2. \tag{8.14}$$

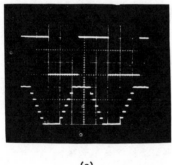

(a)

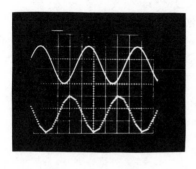

(b)

FIGURE 8.22. Waveforms of the linear interpolator system for (a) a square wave and (b) a sine-wave input signal.

Hence, it has the general shape indicated in Fig. 8.15, but with smaller side lobes. The zeros are at ω_{c1}, $2\omega_{c1}$, and so on, where $\omega_{c1} = 2\pi/T_1 = 2\pi/rT_2$. The stopband of the response starts at ω_{c1}. Hence, as an inspection of Fig. 8.19 reveals, $|H_{\text{total}}(\omega)|$ can be used to realize the required SCF2 response.

As an illustration, Fig. 8.22 shows the observed waveforms of the system of Fig. 8.19 to a square wave and a sine-wave input signal.[8] Here, the SCF1 used was simply a S/H stage with $f_{c1} = 4$ kHz, while SCF2 was a linear interpolator with $f_{c2} = 24$ kHz (i.e., $r = 6$). The top curve shows $v_{\text{in}}(t)$ and the bottom curve shows $v_2(t)$. The frequency response is illustrated in Fig. 8.23. Since the sidebands to be suppressed are centered around $f_{c1} = 4$ kHz, $2f_{c1} = 8$ kHz, and so on where the attenuation is very high, this design technique has proved to be efficient for obtaining smooth output signals when combined with a simple continuous-time analog output filter.

Reference 6 describes some alternative realizations for the linear interpolator. At the cost of using two op-amps rather than one, it is possible to use a circuit[6] which requires only four clock phases, independent of the value of r. Figure 8.24 illustrates the circuit. Its analysis is left as an exercise to the reader (Problem 8.6).

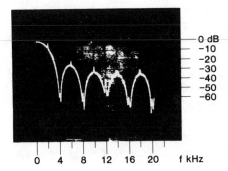

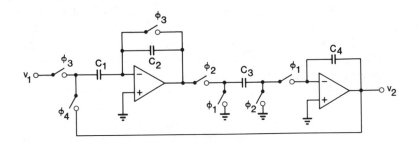

FIGURE 8.23.   The gain response of the S/H linear interpolator system.

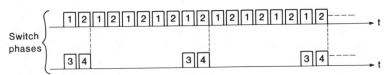

FIGURE 8.24.   Switched-capacitor interpolator using two op-amps. The element values are $C_2 = C_4 = \sqrt{r}\,C_1 = \sqrt{r}\,C_3$. The input need *not* be a sampled-and-held signal.

8.3.   THE APPLICATION OF SWITCHED-CAPACITOR CIRCUITS IN DIGITAL VOICE TRANSMISSION SYSTEMS

The use of digital methods for the transmission of analog signals is becoming increasingly common in telecommunication systems. There are two major reasons for this. First, if digital rather than analog signals are transmitted, then the system remains nearly immune to noise as long as it is below a threshold level. This is due to the fact that for long-distance transmission the digital signal can be regenerated at each repeater, creating a new, noise-free signal. Thus, noise does not accumulate as it would in a comparable analog system. *Second*, the components of a digital system lend themselves well to integrated implementation using large-scale-integrated (LSI) circuits.

Even in such a digital transmission system, however, many signals which are being processed are usually analog in nature. These analog signals need therefore to be converted into digital format. There are various modulation schemes to accomplish this, including pulse-width modulation (PWM), pulse amplitude modulation (PAM), and pulse code modulation (PCM). PCM is currently the most commonly used modulation system in digital telecommunication systems.[9] The PCM signal is generated by sampling, quantizing, and coding an analog signal. The result is a stream of binary digits (bits), that is, an alternation of high and low voltage levels in the signal. This stream can then be applied to the line, either directly or after some additional modulation steps, for example frequency modulation.

The process of sampling an analog voltage $v(t)$, as discussed in Chapter 2, consists of developing a pulse train in which the amplitude of the nth pulse equals the amplitude of $v(t)$ at $t = nT$. As was also shown in Chapter 2, in order for the pulse train to represent uniquely the information contained in $v(t)$, the sampling rate $f_c = 1/T$ must be at least twice as large as the highest frequency component in the spectrum of $v(t)$. The amplitudes of the pulses in the sampled signal are then quantized and converted into a set of bits, called a *digital word*.

The transmission of digital words in a practical PCM telecommunication system often utilizes time-division multiplexing (TDM). In a TDM system, the digital words taken from several channels are transmitted interlaced over the same line. In voice-frequency (telephone) systems, usually 24 or 32 voice channels are thus multiplexed onto a single pair of wires. The International Telegraph and Telephone Consultative Committee (CCITT) has recently recommended two main architectures ("hierarchies"), which now many nationwide PCM transmission networks use. The first one, shown schematically in Fig. 8.25, has been adopted mostly in Europe, Africa, Australia, and South America. In this system, the input multiplexers (MUX) interleave 32 digital words. Of these, 30 words represent the signal amplitudes in 30 voice channels, while the remaining two time slots contain signaling and synchronization information. The second hierarchy, used mainly in the United States, Canada, and Japan, is illustrated in Fig. 8.26. In this system, 24 voice channels are multiplexed at the input section.

There are two main tasks performed in a digital telecommunication system: transmission and switching. Transmission involves sending the digitally coded voice signal from one location to another, while switching consists of establishing the desired connection between two voice channels which carry digital signals.

The unit which performs the pulse code modulation and demodulation of the signals in a digital transmission system is called a *co*der–*dec*oder, or, in abbreviated form, a *codec*. In early digital systems, the switching was done in an analog form, using such analog devices as the electromechanical crossbar mechanism. Then, the analog voice signal channels were time-multiplexed into groups of 24 or 30 channels. Each group was next converted into digital form

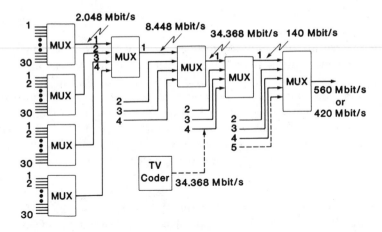

| First order | Second order | Third order | Fourth order | Fifth order |

FIGURE 8.25.   The PCM hierarchy used in Europe, Africa, Australia, and South America.

by a single high-speed codec. This arrangement resulted in some savings in terms of the number of components; however, the design of the shared codec with a sufficiently low line-to-line crosstalk and noise, as well as the required high speed in a fully integrated form represented a very difficult design problem. In addition, any failure of the codec resulted in the loss of service to all lines serviced by that unit. Also, the system required a large number of analog switches, which were bulky and slow. Finally, the analog multiplexing

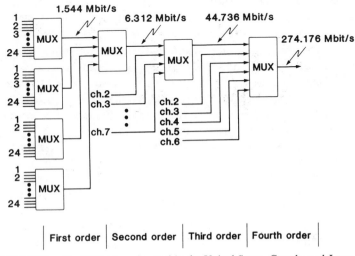

| First order | Second order | Third order | Fourth order |

FIGURE 8.26.   The PCM hierarchy used in the United States, Canada, and Japan.

needed was more difficult to perform and less flexible than digital multiplexing would have been.

With the recent availability of low-cost and high-performance integrated circuits, it became feasible to allocate one codec to each voice channel, rather than to a group of channels. The voice signal in each channel is thus first digitized, and the subsequent switching and multiplexing are performed by digital systems which can use low-cost digital logic and memory circuits. Such a "per-channel codec" system results in a considerable reduction of crosstalk and noise. Also, there is a significant saving in size and cost, due to the elimination of most of the bulky and expensive electromechanical components.

The standard sampling rate f_c recommended by the CCITT for the usual 300–3400 Hz voice channel is 8000 Hz. In the per-channel codec system, the analog signal is thus sampled at intervals of $1/8000$ Hz $= 1.25 \times 10^{-4}$ s $= 125$ μs, and the amplitude of the sample is converted into an eight-bit digital word which is then transmitted serially to the multiplexer. The 125-μs-long time interval available for the conversion is called a *frame*.

The digital multiplexer places all the bits corresponding to one sample of each of 24 or 30 channels (depending on the hierarchy used) into a single frame, and transmits the resulting serial bit stream. To mark the beginning and the end of each frame at the receiver, some identifying signal should be added to the transmitted bit stream. For example, in the 24-channel eight-bit PCM system called D2 developed by AT & T, an additional bit is appended to the bits of each frame. Thus, a frame in the D2 system contains $24 \times 8 + 1 = 193$ bits. Hence, the data rate of the transmitted signal is 193 bits per 125 μs, which corresponds to 1.544×10^6 bits/s. Figure 8.27 shows two groups of the system, including handsets, codecs, and multiplexers.

The components of a typical switching or transmission path are illustrated in Fig. 8.28. In the transmit direction, the subscribed line is connected to the *s*ubscriber-*l*ine *i*nterface *c*ircuit (SLIC). The SLIC performs the two-to-four wire conversion, the battery feed, the line supervision, and also provides ringing access and overvoltage protection. The signal then passes through the transmit filter, which limits its spectrum to (approximately) 3400 Hz. The transmit filter thus removes the part of the spectrum which extends over 4 kHz, and thus makes it possible to sample the signal at an 8-kHz rate without introducing aliasing. In addition, the low-frequency portion (below 300 Hz) of the spectrum is also suppressed by the transmit filter, to prevent power-line frequency (50 or 60 Hz) noise from being transmitted.

The filtered signal, now band limited to the 300- to 3400-Hz range, is then sampled at an 8-kHz rate, and then converted (encoded) into eight-bit PCM data. This conversion is performed *nonlinearly*; that is, the resolution of the resulting digital signal is finer for small signals than for large ones. The reasons for this are the following. Studies performed on telephone systems indicated that the amplitude probability density of the speech waveform has a Gaussian distribution centered at a value which corresponds to a small amplitude. Thus, by making the slope of the digital output versus analog input curve (Fig. 8.29)

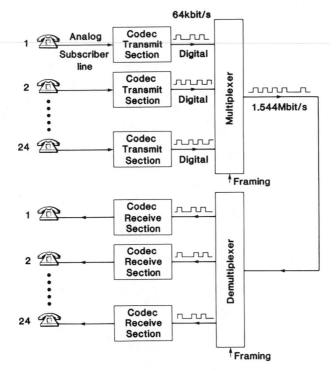

FIGURE 8.27. A 24-channel time-division multiplexed PCM telephone system.

steep for small signals and reducing the slope for large ones improves the system performance. This also ensures that the system functions in a satisfactory manner for both soft and loud talkers. It can be shown[10,11] that if the step size Δx_k needed in the analog input $x(t)$ to change the digital output by one bit

$$\Delta x_k = x(kT + T) - x(kT) \tag{8.15}$$

is chosen to be proportional to the average value of $x(kT + T)$ and $x(kT)$, then the signal-to-quantizing-noise ratio is independent of the probability density of the signal x. Equation (8.15) holds for a logarithmic analog-to-digital conversion.

The most commonly used compression characteristics used in the analog-to-digital conversion are hence based upon the logarithmic function $y = \log_{10} x$. Here, y is the value of the binary output and x is that of the analog input. To achieve a linear characteristic for the overall path, the digital-to-analog converter (decoder) in the "receive" section of the system must then have an exponential expanding characteristic $x = 10^y$, where y is the digital input and x the analog output of the encoder. (The use of *comp*ressing transmitters and exp*anding* receivers is often called "companding.")

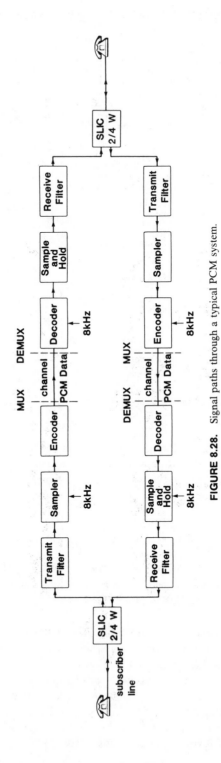

FIGURE 8.28. Signal paths through a typical PCM system.

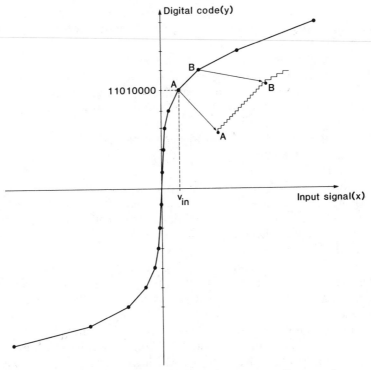

FIGURE 8.29. The transfer curve of the μ-law nonuniform encoder.

Since the logarithmic characteristic becomes unusable for $x = 0$, the actual compression function used is only an approximation to the logarithmic one. In North America, the commonly used conversion function is instead the so-called μ-*law*

$$|y| = \frac{\log(1 + \mu|x/x_{max}|)}{\log(1 + \mu)} \qquad (8.16)$$

where $\mu = 255$. In Europe, the *A-law* given by

$$|y| = \frac{A|x/x_{max}|}{1 + \log A} \qquad \text{for} \quad 0 \le |x/x_{max}| \le 1/A$$

$$\qquad (8.17)$$

$$= \frac{1 + \log(A|x/x_{max}|)}{1 + \log A} \quad \text{for} \quad 1/A \le |x/x_{max}| \le 1$$

is used, with $A = 87.6$. In these relations, x_{max} is the maximum instantaneous value of the analog input signal $x(t)$. Using either of these characteristics, the average quantizing noise power becomes essentially independent of the amplitude density function of the signal $x(t)$.

The functions given in (8.16) and (8.17) can be approximated by a piecewise linear characteristic, as shown in Fig. 8.29 for the μ-law. Within each segment, there are 16 steps used in the analog-to-digital conversion. The horizontal step sizes Δx_k are constant within each segment, but they double as one goes from segment to segment towards larger $|x|$. In the μ-law curve, there are 16 segments, and the ratio of the largest-to-smallest step sizes is 128.

The output of the nonlinear A/D converter is a specially coded eight-bit digital word. The first bit indicates the sign of the input x, with "1" representing positive and "0" negative polarity. The next three bits provide the number of the linear segment in which x lies, where "000" corresponds to the segment at the origin, and "111" to the last one. Finally, the last four bits indicate the step number within the segment: "0000" represents the first step, and "1111" the last one.

Returning to the block diagram of Fig. 8.28, the digital output of the encoder is then time-division multiplexed with the outputs from the other lines, and the resulting bit stream is transmitted over the channel. At the receiver side, the incoming data are demultiplexed, that is, distributed among the various channels, and decoded (i.e., D/A converted). The analog output is then sampled-and-held, and passed through the receive filter. The latter is a low-pass smoothing filter, which removes the high-frequency "side lobes" of the signal spectrum, and hence smoothes out the staircase noise in the signal. It is also often used to equalize for the $\sin x/x$ amplitude distortion introduced by the sample-and-hold stage.

There have been several approaches to the realization of fully integrated low-cost per-channel codecs. Among the first ones was a codec using charge-coupled device (CCD) filters.[12] With the evolution of switched-capacitor filters, several codecs using SC filters appeared. Initially, a two-chip approach was used, with the transmit and receive filters realized on one chip, and the PCM encoder and decoder on the other.[4,13-18] In a somewhat different approach, the transmit filter and encoder were integrated on one chip, and the receive filter and encoder on the other.[19] This latter approach had the advantage that the transmit and receive functions were physically separated, and thus the crosstalk and noise in the asynchronous operation of the codec were reduced.

Among the commercially available MOS integrated PCM channel filters, we mention the Intel 2912 and the Motorola 14413 and 14414. A simplified block diagram[14] of the Intel 2912 is shown in Fig. 8.30. Here, the signal to be transmitted is first applied to an op-amp which (using two external resistors) provides accurately controlled voltage amplification. The signal is then passed through an active-RC antialiasing filter, which band limits it to allow its sampling by the subsequent SC filter without aliasing. The circuit diagram of this filter is the same as that shown in Fig. 8.2; the element values were $R = 160$ kΩ and $C = 25$ pF, resulting in a flat response in the 0–3 kHz range

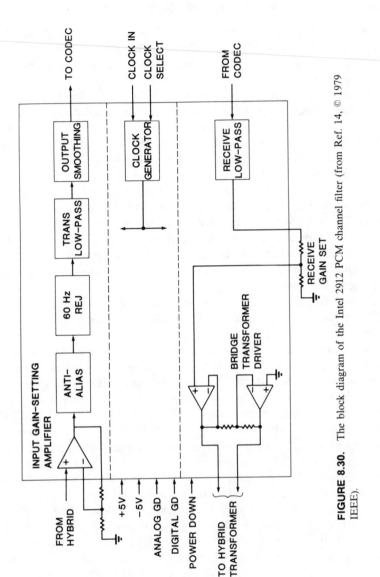

FIGURE 8.30. The block diagram of the Intel 2912 PCM channel filter (from Ref. 14, © 1979 IEEE).

and a nominal cutoff frequency of 12 kHz. This gave a loss of about 52 dB at the clock frequency ($f_c = 256$ kHz) of the SC filter, thus reducing aliasing. The capacitors used two polysilicon plates, and the resistors were made from polysilicon lines.

The antialiasing filter is followed by the 50/60 Hz hum rejection filter, which is here a second-order notch filter, containing two cascaded SC integrators and a resistor-op-amp summing stage.

The main transmit low-pass filter was realized by a fifth-order elliptic SC filter, including, however, also some polysilicon resistors which were used to generate the sum of some signal voltages. The output of this filter is then smoothed by another active-RC filter, similar to the input antialiasing one.

In the receive direction, a low-pass filter is used which is identical to the transmit low-pass filter, except that a single-pole pre-emphasis network is added at the input. This provides a 2-dB peak which compensates for the $\sin x/x$ rolloff contributed by the decoder. The transmit filter is followed by an (external) attenuator, and a transformer driver circuit containing two high-current amplifiers.

The chip contains a total of 400-pF capacitance realized between two polysilicon layers, a total of 600-kΩ polysilicon resistance, and 20 op-amps. The die size was 150×160 mil^2.

In order to reduce cost, power dissipation, and space requirements, the next generation of codecs integrated all functions required for voice-to-digital data conversion on a single chip.[20-23] Several single-chip MOS codecs are commercially available, including the AMI S3506/S3507, the Hitachi HD44231A/HD44233A, National Semiconductor's TP3052, TP3053, TP3054, and TP3057, and Motorola's MC14400, MC14401, and MC14402.

In the following, we shall briefly discuss the organization and some of the circuitry[23] of the AMI S3506/S3507 codec. The block diagram of this circuit is shown in Fig. 8.31. Starting at the top left, the voice signal can be entered through an on-chip op-amp which (with external resistors) can also be used to adjust the gain of the signal path. The amplifier is followed by the antialiasing filter, which is again a Sallen–Key circuit, as shown in Fig. 8.2. For nominal values, the 3-dB frequency is 36 kHz; the loss at the input sampling frequency of the transmit low-pass filter is about 46 dB. Even with a $\pm50\%$ variation of the resistance values, the passband droop remains below 0.01 dB and the 512-kHz loss stays over 32 dB.

As indicated in Fig. 8.32, the input branch of the transmit low-pass filter incorporates the "cosine" filter described in Section 8.1 (cf. Fig. 8.9). Since the clock rate of ϕ_1 and ϕ_2 is $f_c = 256$ kHz, the effective sampling rate of the filter is (as mentioned earlier) 512 kHz. As can be seen from Eq. (8.7) of Section 8.1, such an input branch provides a transmission zero at f_c. In the range 256 ± 3.4 kHz, the loss due to the cosine filter is over 33.5 dB. It is thus a reasonably effective decimator filter. The main low-pass filter is a fifth-order elliptic ladder SC filter, which limits the transmitted signal spectrum to 3.4 kHz.

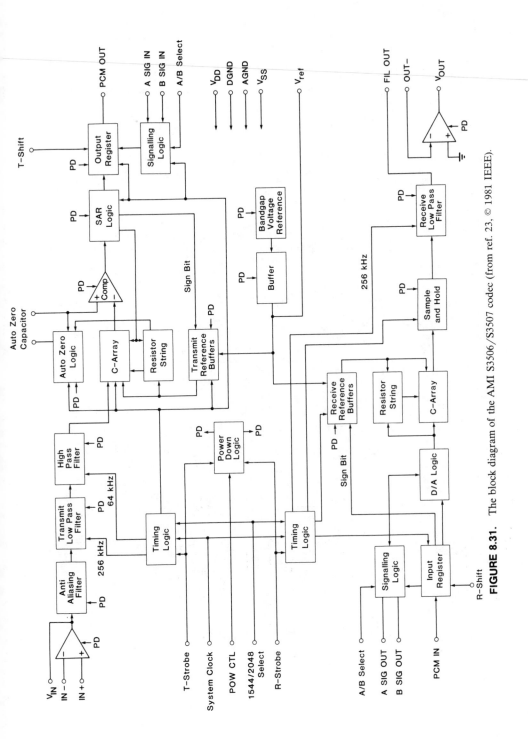

FIGURE 8.31. The block diagram of the AMI S3506/S3507 codec (from ref. 23, © 1981 IEEE).

555

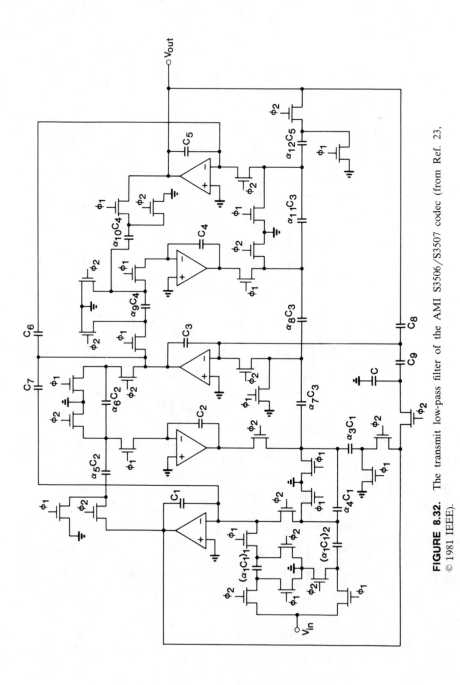

FIGURE 8.32. The transmit low-pass filter of the AMI S3506/S3507 codec (from Ref. 23, © 1981 IEEE).

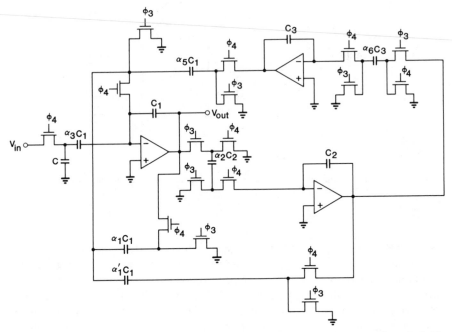

FIGURE 8.33. The transmit high-pass filter of the AMI S3506/S3507 codec (from Ref. 23, © 1981 IEEE).

The high-pass filter shown in Fig. 8.33 is a third-order "state-variable"-type[1] SC filter, which is used to remove the power-line frequency (50 or 60 Hz) noise from the signal. Its passband starts at 300 Hz. Since the poles of this filter are at low frequencies, a high clock rate would result in a large capacitance spread. On the other hand, a very low clock rate (say, 8 kHz) would result in an unacceptably large signal delay through the filter. Therefore, $f_c = 64$ kHz was used, as a compromise value.

The output of the high-pass filter is sampled and held at 8 kHz and then applied to a nonlinear (μ-law) PCM A/D converter (encoder). This circuit uses a charge redistribution conversion technique (as described in Chapter 6, in connection with Fig. 6.13) to determine the linear segment corresponding to the analog input. The step within the segment is then found using a chain of 16 equal-valued resistors.[24] Both the segment and step bits are obtained using successive approximation, at a 128-kHz clock rate, in nine steps. In the first step, the sign of the input signal is determined; then the seven bits of the PCM data identifying the segment and step numbers are found, and clocked out serially. The clock frequency can be varied in the 64-kHz to 2.048-MHz range.

Returning to the block diagram of Fig. 8.31, in the receive direction the incoming PCM word is forwarded to the μ-law D/A converter (decoder). A simplified circuit schematic of the decoder is shown in Fig. 8.34. The operation

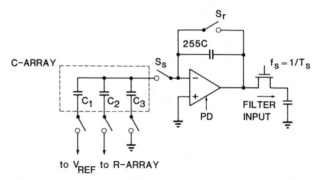

FIGURE 8.34. The simplified schematic of the μ-law decoder used in the AMI S3506/S3507 codec (from Ref. 23, © 1981 IEEE).

is as follows. At the beginning of the conversion period, the sign bit of the PCM data is fed to the receive reference generator, which then provides the required positive or negative reference voltage V_{ref} to the decoder. An appropriate combination of switches, controlled by the PCM word, is then closed to develop the correct analog voltage at the top plates of the capacitor array. During this same time interval, the reset switch S_r discharges the feedback capacitor. Subsequently, S_r opens and S_s closes, transferring the charge stored in the top plates of the capacitance array into the feedback capacitor. As a result, the appropriate analog voltage appears at the op-amp output.

The converted analog signal at the decoder output is then sampled and held at an 8-kHz clock rate, and fed to the SC receive low-pass filter. The receive filter removes the signal frequencies above 3.4 kHz. It also incorporates some passband gain equalization: it has a peak near the passband edge which compensates for the $\sin x/x$ droop caused by the 8-kHz sample-and-hold operation. This filter is similar in structure to the fifth-order elliptic low-pass SC filter used as the transmit low-pass filter; however, it does not have the cosine-filter input branch, and its capacitance values have been modified to achieve the peaking passband gain response. The clock rate of the receive filter is 256 kHz. It is followed by an uncommitted op-amp, which can be used to drive a hybrid transformer.

The codec needs an accurate and temperature-stable voltage reference V_{REF} to be used for the analog-to-digital and digital-to-analog conversions. Any changes in V_{REF} affect the gain stability of the whole PCM codec; hence, the reference voltage must have an accuracy of 0.1 dB, or about 1.15%. Such accuracy can be obtained using the precision bandgap reference voltage generator[25] shown in Fig. 8.35a. Here, the bipolar transistors Q_1 and Q_2 are obtained using the parasitic devices of the p-well CMOS process: the n-doped substrate acts as the collector, the p-well as the base, and an n^+ diffusion as the emitter. The emitter areas of the two devices have the ratio $A_2/A_1 = 10$. By contrast, the two NMOS devices Q_3 and Q_4 are matched, operate in satura-

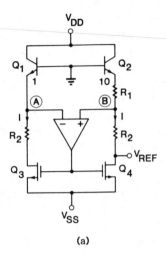

(a)

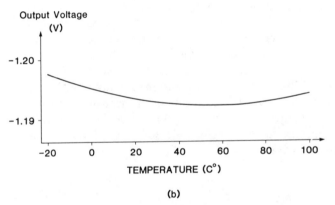

(b)

FIGURE 8.35. Bandgap voltage reference (from Ref. 23, © 1981 IEEE), (a) circuit diagram; (b) voltage vs temperature characteristics.

tion, and have identical gate and source voltages. They force therefore Q_1 and Q_2 to carry the same emitter current I. Since the op-amp forces both nodes (A) and (B) to be at the same potential, we have

$$V_A = V_B = -V_{BE_1} = -V_{BE_2} - IR_1. \qquad (8.18)$$

Hence, the common current I equals $(V_{BE_1} - V_{BE_2})/R_1$. Therefore, the output voltage is

$$V_{\text{REF}} = V_B - IR_2 = -V_{BE_1} - (V_{BE_1} - V_{BE_2})R_2/R_1. \qquad (8.19)$$

Hence, V_{REF} is the weighted average of V_{BE_1} and $(V_{BE_1} - V_{BE_2})$. It can readily be shown (Problem 8.11) that the temperature coefficients of these two terms have opposite signs. Hence, it is possible to find a value for R_2/R_1 which makes the temperature coefficient of V_{REF} nearly zero over a range of temperatures. The resulting temperature characteristics for the reference generator is shown in Fig. 8.35b. The required 1.15% accuracy is clearly satisfied over the whole temperature range -20 to $+100°C$.

Other types of temperature-stable voltage references can also be used; see, for example, Refs. 13 and 26 for an NMOS and a CMOS voltage reference, respectively.

The overall size of the codec chip was approximately 212×212 mil^2. The required low voltage-coefficient capacitors were realized using two layers of polysilicon. Polysilicon lines were also used to fabricate the required resistors.

The requirements which the per-channel codecs must satisfy include specified limits on the idle-channel noise, crosstalk coupling, signal-to-quantization-noise ratio, gain tracking, and the gain-frequency response. Typically, most of the idle-channel noise (i.e., the noise measured in the absence of a signal in the channel) as well as most of the crosstalk is contributed by the filters and the analog circuits which precede the encoder. The noise power is measured by first sending the noise through a filter with a "C-message weighting" frequency response.[27] This response emphasizes the noise in the 1- to 3-kHz range, and gradually suppresses it outside of this frequency band.

To satisfy the idle-noise specification, the dynamic range achieved by the AMI S3506/S3507 was 85 dB. The measured signal-to-quantization-noise ratio is compared with the CCITT specifications in Fig. 8.36 for various levels of the input signal. Figure 8.37 shows the gain-tracking performance, with the encoder and decoder in cascade, for different output levels, compared with the CCITT requirements.*

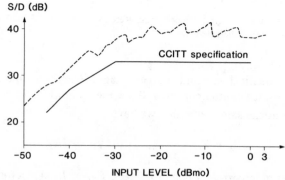

FIGURE 8.36.   The measured signal-to-quantization/noise ratio of the AMI S3506/S3507 codec (from Ref. 23, © 1981 IEEE).

---

*In these figures, the unit dBmo is used to measure signal levels. By definition, this is the logarithmic value of the signal power in mW, at the reference point in the system.[27]

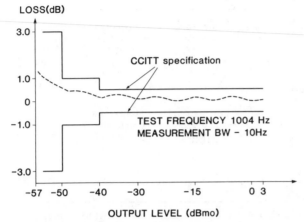

FIGURE 8.37.   The measured gain-tracking performance of the AMI S3506/S3507 codec, with its encoder and decoder looped. The test frequency was 1004 Hz and the bandwidth 10 Hz. (From Ref. 23, © 1981 IEEE).

Other specifications on the codec include the passband flatness and minimum stopband loss. These requirements were also fully met by the codec.[23]

8.4.   THE APPLICATION OF SWITCHED-CAPACITOR CIRCUITS IN DATA MODEMS

With the increasing use of telephone transmission facilities for data transmission (e.g., for communicating between a data terminal and a remote computer), a new high-volume application developed for switched-capacitor circuits. Voice-graded telephone lines are designed to carry voice signals from the subscriber to the central office, but are not suitable for the direct transmission of a pulse stream representing the bits of data.* To permit the transmission of such a signal over voice-graded telephone facilities, it is necessary to convert the digital data signal into a signal whose spectrum is well within the voice-frequency range. The unit which performs this conversion is called a data "modem," an abbreviation of "*mo*dulator–*dem*odulator." Among the available modulation schemes, the ones which are most widely used in current modems are frequency-shift keying (FSK) and phase-shift keying (PSK) modulation. In FSK modems, the two binary states are represented by two different voice frequencies. In the two-phase PSK system, by contrast, there is a single carrier frequency. One phase of the carrier signal then represents one binary state, and the other phase (shifted by 180°) represents the other state.

*Note that the bit streams used in digital telephony are transmitted through wide-band trunk lines capable of direct pulse transmission.

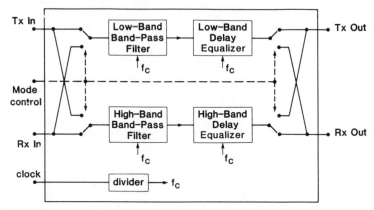

FIGURE 8.38. The block diagram of the AMI S3522 modem filter set.

A considerable part of the cost of a high-speed modem is due to the required filters. Earlier, these filters were implemented using discrete active-RC circuitry. Such circuits required a lengthy and costly trimming process in order to meet the accuracy requirements of the system. Recently, it was found that SC circuits provide an efficient and cost-effective realization for modem filters. One commercially available PSK modem filter set, intended for 600 baud* data transmission, is the AMI S3522. It will be briefly described next.

The block diagram of the S3522 modem filter set is shown in Fig. 8.38. This unit can be used in a four-wire (two-line) "full-duplex" system, that is, in a system where the modem can transmit and receive data simultaneously. The chip contains two SC bandpass filters with added allpass delay equalizers. The "low-band" bandpass filter has a response centered at 1200 Hz, and the "high-band" bandpass filter has one centered at 2400 Hz. The bandwidth of both filters is 800 Hz. Both filters are based on ladder simulation, while both delay equalizers are realized using cascaded second-order sections. The low-band filter is of order 14; the corresponding delay equalizer is of order 6. The high-band filter is of order 10, and its equalizer of order 4. The specified amplitude and group delay responses of the filters incorporate equalization for the line. The maximum overall passband ripple is 1 dB, and the maximum group-delay variation from the ideal equalized response is ± 100 μs for the low-band path and ± 50 μs for the high-band one. The measured gain responses of the two paths are shown in Fig. 8.39. Clearly, the frequencies centered around 1200 and 2400 Hz are well separated by the two bandpass filters.

---

*Bauds measure the speed of data transmission. For binary data, 600 bauds indicate 600 bits transmitted each second.

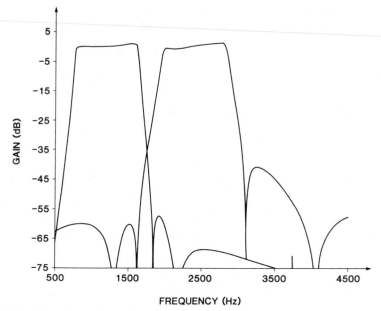

FIGURE 8.39. Typical frequency response of the high-band and low-band filters in the AMI S3522 modem filter set.

The chip realizing the filter set was fabricated using 5-μm CMOS technology. The die size was 189×138 mil^2.

Other commercially available integrated 600-baud PSK modem filter sets are the Reticon R5632 and the AMI S35212. The former uses NMOS technology,[28] while the latter CMOS. In both of these units, the required amplitude and group-delay responses are all realized by circuits consisting of cascaded second-order sections. In each path, five of these sections provide the selective amplitude response, and five allpass sections the necessary delay equalization.

For low-speed (300-baud) data transmission, the Reticon R5630 and R5631 filter sets are available. The former is compatible with the Bell Systems' specifications 103/113 for modems, the latter with the CCITT V.21 specifications. The block diagram of these units is shown in Fig. 8.40. The two bandpass filters contained on the chip are both 10th-order SC circuits. Switching the filters between the "originate" and "answer" modes is accomplished by using an external TTL-compatible signal entered via the "MODE" pin. Also included on the chip are a receive gain control stage whose gain is externally adjustable between 0 and 30 dB; a separate limiter for use with the receive output signal; an on-chip clock oscillator (which, however, requires an external crystal); and low-pass filters in each path. The latter act as smoothing filters, and remove the staircase (S/H) distortion from the output signal.

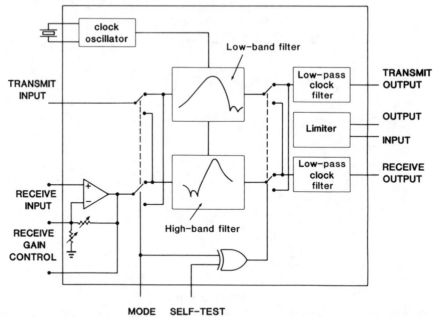

FIGURE 8.40.   The block diagram of the Reticon R5630 and R5631 300-baud modem filter sets.

8.5. THE APPLICATION OF SWITCHED-CAPACITOR FILTERS IN TONE RECEIVERS

Signaling with single-frequency sine waves ("tones") is used commonly in telecommunication systems. While the generation of such tones can be performed relatively economically and simply,[29] their detection and decoding is more complicated. Typical applications of tone signaling in telephony include the multifrequency (MF) detection system used for signaling between switching machines, and the dual-tone multifrequency (DTMF) detection system used for signaling between the subscriber and the central office.[30] Other applications are paging and multifrequency code signaling (MFC). The largest-volume application is currently the DTMF detection system used with touch-tone receivers. This performs the dialing function which earlier used a rotary mechanism and now pushbuttons. This system, and its realization using SC circuits, will be described in what follows.

The DTMF dialing system uses eight frequencies, which are selected so as to avoid harmonically related interference from speech signals. These tones are divided into a low group and a high group, each containing four tones. When a pushbutton is depressed, a dual-tone signal containing one tone from each group and representing the digit on the pushed button is generated. This signal is then transmitted to the central office, where it must be correctly detected even in the presence of noise and dial tone on the line. Moreover, the detection must be correct with a high probability even if a speech signal is simultaneously present. This is especially difficult, since all DTMF tones are in the voice-frequency range.

There are several possible approaches to the detection of a DTMF signal. The early technique[30] was to use analog low- and high-pass filters to separate tones which belonged to the low- or high-frequency groups. The output voltages of these filters were then hard limited to make their peak amplitudes essentially independent of input level variations. The amplitude-limited output voltage of each band-separating filter was then applied to a bank of four simple bandpass filters. Each filter was tuned to one of the tone frequencies, and was followed by an envelope detector. The detected digit was considered valid whenever the outputs of exactly two bandpass filters were sufficiently large and obeyed the appropriate timing relations.[31]

Digital methods can also be used for tone detection. In a simple approach,[32] the analog band-separating filters and hard limiters described above were followed by zero-crossing detectors and counters. The number of zero crossings counted over a fixed period gave an estimate of the tone frequency. It is also possible to implement an all-digital tone receiver, using spectral moment estimation,[33] or a digital tone detector based on correlation.[34]

Recently, switched-capacitor circuits have been used extensively in the integrated implementation of tone receivers. An example of these devices is Silicon Systems Inc.'s single-chip receiver SSI 201 (also 202, 203, and 204). This unit essentially implements the classical analog system using switched-capacitor building blocks.[35] The basic system is illustrated by the block diagram shown in Fig. 8.41. The input signal (ANALOG IN) is prefiltered by the preprocessor, and then applied to the band-separating filters BS1 and BS2. These are both six-pole Chebyshev band-reject filters, realized by cascading three second-order SC filter sections. Each filter passes one tone group and rejects the other. The outputs of these filters are hardlimited by the zero-crossing detectors, which replace the tones by square waves. Thus, the input to the bandpass filters which follow will not depend very much on the level of the tones. The bandpass filters form two groups, each containing four filters; each filter is implemented by a second-order SC bandpass section. For a valid tone pair, one filter in each of the two groups will have a high output signal. The two high outputs are identified by the two amplitude detectors. These find the difference between the maximum and minimum values of the signals, and are thus insensitive to their dc offsets. From the two identified tones, the four bits of the originally transmitted digit are found in the digital decoder, and passed on to the output register.

The complete receiver contains 37 op-amps, and includes circuits realizing 36 filter poles in addition to precision zero-crossing and amplitude detectors, as well as digital logic, clock generator, voltage reference, and so on. Its dimensions are 211×223 mil^2.

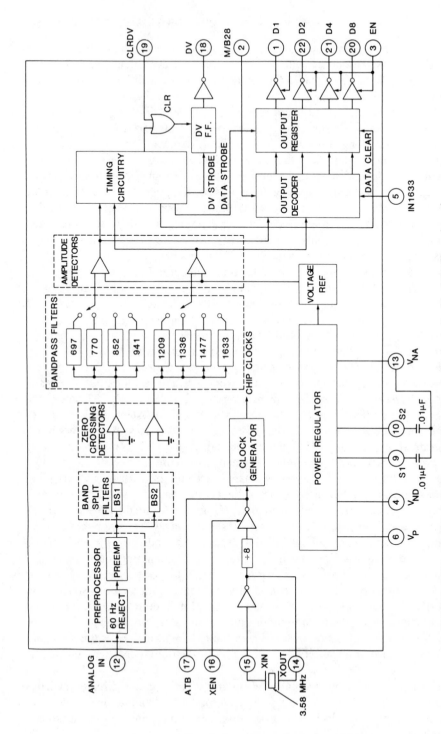

FIGURE 8.41. The block diagram of Silicon Systems Inc.'s single-chip DTMF receiver (from Ref. 35, © 1979 IEEE).

Another class of tone receivers uses SC band-separating filters to perform the front-end functions, and then employs digital zero-crossing algorithms to detect the tones.[31, 36, 37] Several integrated band-separating filters are commercially available for such systems. These include Mitel's MT 8865 and AMI's S3525. The block diagram of the latter is shown in Fig. 8.42. The dual-tone input signal is entered directly or via the input op-amp OA1 to the low-pass filters. OA1 can be used as a unity-gain buffer, a voltage amplifier, or a differential amplifier, by appropriately connecting external resistors. The input low-pass filter is a third-order elliptic SCF, clocked at $f_1 = 150$ kHz. It limits the input spectrum to about 3 kHz; this allows the subsequent high-selectivity filters to operate at a low clock rate without introducing aliasing. The output of the input low-pass filter is then sampled and held at a low ($f_2 = 28$ kHz) clock rate, and applied to the dial-tone-reject filter. The latter is also clocked at 28 kHz; it is a fifth-order general-stopband high-pass filter. Its main function is to suppress the 450-Hz dial tone in the input spectrum, along with any 60-Hz power-line noise and dc offset. It provides the low-frequency skirt of the overall frequency response of the low-group path. The output of the dial-tone-reject filter is then applied to the main low-pass/high-pass band-separating filter pair. This pair carries out the actual separation of the two tones in the signal. Both of its filters are fifth-order elliptic SC circuits, clocked at 28 kHz. Each of the two output signals of the pair is then resampled at the high clock rate $f_1 = 150$ kHz, and filtered by the third-order smoothing filter. The resulting tones can be hard limited by the on-chip op-amps OA2 and OA3. The necessary timing and clock signals are generated by on-chip circuitry, which (as in all such circuits) requires an external crystal.

The complete chip uses 29 op-amps, and realizes 25 filter poles. It is fabricated in CMOS technology; the total chip area is 123×186 mil^2. The measured gain responses for the two signal paths (including all on-chip filters) is shown in Fig. 8.43.

In addition to multifrequency signaling, single-frequency (SF) signaling is also used in telephone systems, mostly for supervision. The signaling frequency may be in the voice band ("in-band signaling") or outside of it ("out-of-band signaling"). The most commonly used in-band SF signaling frequency is 2600 Hz. This tone can be used to indicate the condition of a line. Thus, when the line between two switching sites is idle (unused), the 2600-Hz tone is present; when the line is in use, the tone is removed.

The functional block diagram of a 2600-Hz SF signaling circuit is shown in Fig. 8.44. In normal operation, the four-wire transmission line between switching sites A and B is continuously monitored by the system. If the line is idle, the "energy detect" circuits operate the switches marked C, and thus apply the 2600-Hz tone at both sites to the transmitting ports. Now the only signal present on the lines is the signaling tone. This causes the output voltages of the bandpass filters (tuned to 2600 Hz) to have much higher energy than those of the band-reject filters which suppress 2600 Hz. This condition is decoded by the energy detect circuit as an indication that the line is unused.

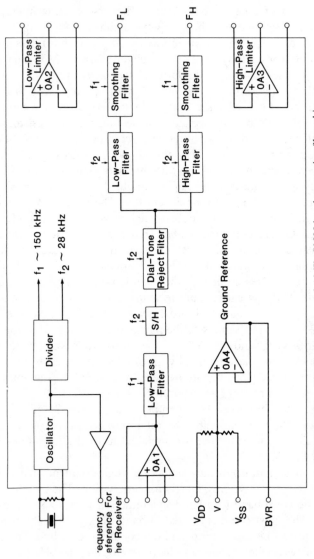

FIGURE 8.42. The block diagram of AMI's S3525 band-separating filter chip.

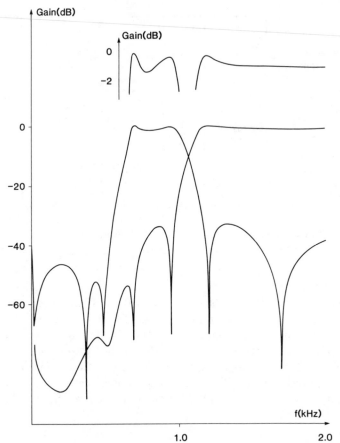

FIGURE 8.43.   Measured overall frequency responses of the low- and high-group channels in the AMI S3525 integrated band-separating filter (from Ref. 37).

When a call is initiated from (say) site A, after the supervisory unit indicates that the line is idle, the 2600 Hz is removed from the transmit port of site A by opening switch C at that port. Now in the signaling unit at site B neither filter has a detectable output voltage. This condition is interpreted by the supervisory circuits as an indication that the lines are about to be used, and the 2600-Hz tone is therefore removed at site B as well, freeing both lines for voice transmission. As long as the line is used, the output energy of the band reject filters is much larger than that of the bandpass ones; this is then interpreted by the energy detect circuits as an indication of line usage. These precautions avoid the possibility of premature activation, or deactivation, of the supervisory equipment, which may occur under some conditions for a less elaborate detection system.

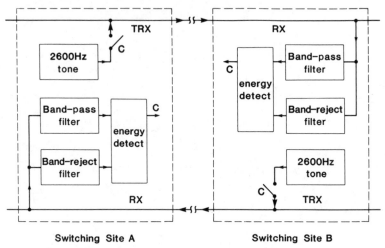

Switching Site A   Switching Site B

FIGURE 8.44. The functional block diagram of a SF signaling system.

A commercially available SC filter set for 2600-Hz SF signaling is the AMI S3526. Its block diagram is shown in Fig. 8.45. The unit contains a bandpass and a band-reject filter, both centered at 2600 Hz. The bandpass filter is a sixth-order SC circuit, with maximally flat (Butterworth) passband characteristics. It is realized as the cascade of three biquad sections. The band-reject circuit is a sixth-order elliptic SC filter. It is also implemented as the cascade of three biquads. Both filters use a 128-kHz clock frequency. The time base and

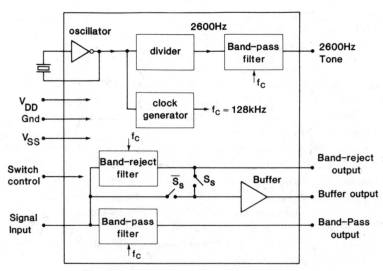

FIGURE 8.45. The AMI S3526 SF signaling filter unit.

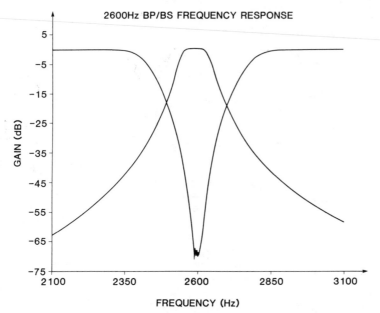

FIGURE 8.46. Typical frequency response of the 2600-Hz bandpass/band-reject filters in the AMI S3526 signaling filter unit.

clock signals are generated by an on-chip oscillator and frequency divider; as usual, an external crystal unit is required. In addition, the divider circuit generates a 2600-Hz square wave. This is then converted into the required 2600-Hz signaling tone by a bandpass filter, which rejects all odd harmonics of 2600 Hz, leaving only the fundamental tone. A buffer is also provided for driving 600-Ω lines. The input signal can be connected either directly to the filters, or through a notch filter, depending on the "switch control" input.

The chip is fabricated using CMOS technology. Its dimensions are 163×131 mil^2. The measured response of a typical unit is illustrated in Fig. 8.46. The complete circuit realizes 12 filter poles, and contains 16 op-amps.

8.6. PROGRAMMABLE SWITCHED-CAPACITOR FILTERS

There are several ways in which the frequency response of a switched-capacitor filter can be changed. Conceptually, the simplest method is to vary its clock frequency f_c. Multiplying f_c by a factor a results in the frequency response being scaled (multiplied) by the same factor along the frequency axis. Thus, any gain previously obtained at some frequency f will now be obtained at the scaled frequency af. If the scaling of f_c is digitally programmable, then so will be the scaling of the SCF response. This feature makes, say, a SC low-pass

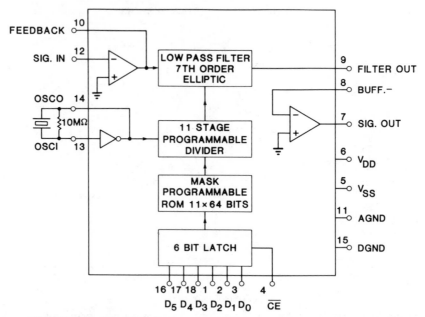

FIGURE 8.47. The block diagram of the AMI S3528 programmable low-pass filter.

filter essentially a general low-pass filter; its cutoff frequency can be varied to satisfy the specifications of the system at hand.

Because of their flexibility and general usefulness, there are numerous such clock-programmable SCFs commercially available. Reticon markets the R5609, which is a seventh-order elliptic low-pass SCF, with less than 0.5-dB passband ripple and over 75 dB minimum stopband loss. The Reticon R5613, by contrast, is a linear-phase low-pass filter with 60-dB minimum stop-band attenuation. Both are fabricated using an NMOS double-polysilicon process; both are tunable by changing their clock frequencies.

Reticon's R5611 is a five-pole Chebyshev high-pass filter, with less than 0.6-dB passband ripple and a 30-dB/octave rolloff. The R5612 is a four-pole notch filter, with over 50-dB rejection at the (tunable) notch frequency. All these filters are essentially self-contained, and require only the clock signal and power-supply voltages for their operation. The inherent stability and accuracy of switched-capacitor filters, as compared to their active-RC counterparts, eliminate alignment and trimming problems, as well as the need for low-tolerance temperature-stable external components.

Another type of clock-programmable filter is represented by the AMI S3528 switched-capacitor low-pass filter. The block diagram of the device is shown in Fig. 8.47. The analog input signal "SIG. IN" can be entered through an op-amp which, with appropriate external resistors and capacitors, can perform amplification and/or antialiasing filtering. The output of the op-amp is then

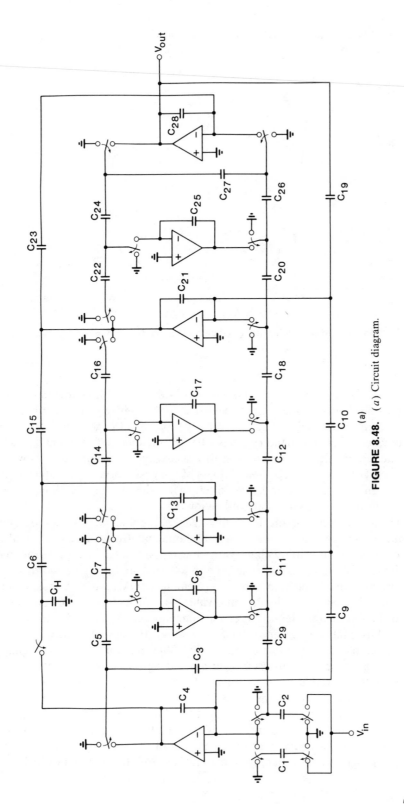

FIGURE 8.48. (a) Circuit diagram.

(a)

573

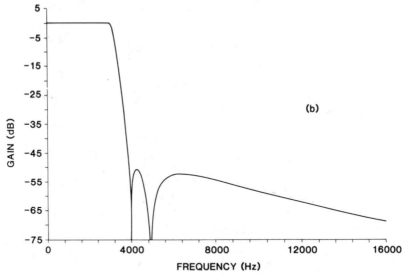

FIGURE 8.48. (*b*) Measured frequency response of the AMI S3528 programmable low-pass filter.

applied to the main SC filter, which is a seventh-order elliptic low-pass filter. The passband edge of the filter is located at $f_c/40$, where f_c is the (programmable) clock frequency. The stopband edge/passband edge ratio is 1.3. The design value of the passband ripple is 0.05 dB; the minimum stopband loss 51 dB. The circuit diagram and a typical measured gain response are shown in Fig. 8.48. To achieve the required flatness of the passband response, the circuit was designed as a bilinear ladder filter, using the synthesis techniques described in the second subsection of Section 5.5.

The output signal of the main filter can be applied to a buffer op-amp capable of driving a 600-Ω load. The programmable-frequency clock signal is derived from an on-chip oscillator with an external crystal, and a programmable divider. The latter is controlled by an external six-bit digital word, via a mask-programmable on-chip read-only memory (ROM) which translates the input word into an 11-bit divider-ratio value. Thus, the clock frequency f_c, and with it the filter cutoff frequency, can assume any one of $2^6 = 64$ values.

The chip is fabricated using CMOS technology. It measures 103×130 mil^2.

A different strategy for making an SC filter tunable is to use digitally programmable capacitor arrays instead of fixed capacitors in the circuit. While this leads to somewhat more complicated circuitry, it also provides increased flexibility. For example, in the biquad shown in Fig. 5.13c of Chapter 5, the capacitors C_2 and C_3 can be programmed to adjust the pole frequency ω_0, and C_4 to change the pole-Q, essentially independently. Changing C_1, C_1', and C_1'' by the same factor k merely changes the gain of the stage by k. A somewhat different programmable filter[38] is shown in Fig. 8.49. It is a biquad which uses

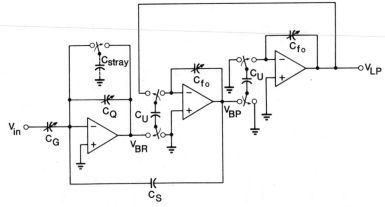

FIGURE 8.49.   A programmable biquad (from Ref. 38, © 1979 IEEE).

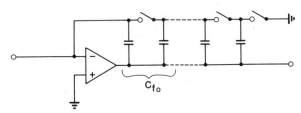

FIGURE 8.50.   The realization of the C_{fo} programmable array (from Ref. 38, © 1979 IEEE).

three, rather than two, op-amps in order to reduce the capacitance spread and to have a single capacitor C_G determine the gain. Using the approximations described in Section 5.4, the relations between the programmable capacitance values and the filter response parameters can be shown to be (Problem 8.12)

$$Q \simeq C_Q/C_S,$$

$$G \simeq C_G/C_S,$$ (8.20)

$$\omega_0 \simeq f_c C_U/C_{fo}.$$

Here, ω_0 is the pole frequency and Q the pole-Q, while G is the peak gain.

The circuit produces a low-pass (V_{LP}), a bandpass (V_{BP}), and a band-reject (V_{BR}) response.* Since the first op-amp has no negative dc feedback, a small (0.01 pF) stray capacitance is switched between the input and output of the op-amp to prevent latch-up.

---

*Strictly speaking, the circuit does not transmit dc, since C_G blocks it. However, since C_{stray} is very small (~ 0.01 pF), it does transmit down to very low frequencies.

In the actual circuit fabricated, the capacitors C_G and C_Q were realized as six-bit binary-weighted capacitor arrays similar to those discussed in Chapter 6. However, in order to vary ω_0 logarithmically, the C_{fo} array was realized using the circuit of Fig. 8.50. In this circuit, all but one of the switches are closed. The open switch determines the value of C_{fo}; the other capacitors (to the right of the open switch) are grounded. This arrangement gives convenient element values in the array. The circuit was used in a formant speech synthesis system.

Another programmable filter realization, even more flexible, is offered by the *mask-programmable* chips. Such a device may contain a large number of building blocks, such as op-amps, resistors, and switches, which are not interconnected in any way. In addition, an area may be set aside for the realization of capacitors. The chip may also contain the clock oscillator, logic and drivers, buffers, voltage reference, and the devices necessary to realize the antialiasing and smoothing filters. The appropriate interconnection of these components, as well as the realization of the capacitors, is accomplished using the last one or two masks in the process. This concept is similar to that of the gate arrays used in digital ICs. It results in low-cost fabrication for SC systems, after the initial expenses of the general-purpose chip have been recovered.

An example of this approach is an NMOS chip recently developed at the Bell Laboratories.[5] This system contains 22 op-amps, an oscillator, clock circuitry, buffers, a voltage reference, resistors, 88 switches, and prefilters and postfilters. The capacitors and interconnections are realized by the customized polysilicon I and II masks. This circuit can implement up to 22 filter poles, of which six can be allocated to the antialiasing and smoothing filters. The chip size is 120×240 mil^2; when all parts are functioning the total dc power consumption is 150 mW.

8.7. THE APPLICATION OF SWITCHED-CAPACITOR CIRCUITS IN SPEECH PROCESSING

An important area where switched-capacitor circuits have been extensively applied recently is *speech processing*. This field can be divided into *speech synthesis* (the production of synthetic speech) and *speech analysis* (the characterization, verification, or recognition of speech). There are many possible economical reasons for applying speech processing. An early motivation was the need to reduce the bandwidth of the speech signal transmitted over telephone or radio channels, by transmitting only a few essential parameters rather than the complete waveform. A more recent application was the storage of speech for computer voice response systems such as the ones used in telephone systems, talking cars, and so on. Yet another application (still largely in the future) is the recognition of spoken commands by "smart" appliances, and/or the verification of the speaker's identity for security purposes.

The most effective techniques for storing, producing, analyzing, or transmitting speech signals are based on the modeling of the mechanism used by the human vocal apparatus. The human vocal tract can be modeled as a lossy acoustic resonator with a large number of natural frequencies called *formants*. The first one of these frequencies is at a few hundred hertz, the second usually around 1 kHz, the third around 2 kHz, and so on. Their exact values depend upon the shape and size of the vocal tract, and they change with time as different sounds are produced by the speaker. For *voiced* sounds, such as vowels, the vocal tract resonator is excited by a periodic stream of pulses (called *pitch*) produced by the vibration of the vocal cord. For *unvoiced* sounds, such as the consonants f, th, s, sh, the excitation is a turbulent air flow created by a constriction in the vocal tract. This excitation is similar to a white noise. Finally, by completely closing and then suddenly opening the vocal tract, transient excitations can also be produced. This process can be used to create the "stop" consonants such as b, d, p, t, and so on. The reader is referred to Ref. 39 for a comprehensive discussion of speech production and digital speech processing methods.

Based on the simple model described above, it is possible to derive the block diagram of a system which can simulate (synthesize) the speech signal. Figure 8.51 shows the basic scheme. Two methods based on this concept, which differ only in their simulation of the vocal tract, will be discussed next. The first one, called *formant synthesis*, is illustrated in Fig. 8.52. Here, the vocal tract is represented by its three lowest-frequency formants, each of which may be realized by a simple second-order bandpass resonator. The resonant frequency f_{oi}, the pole-Q Q_i, and the gain G_i ($i = 1, 2, 3$) of each resonator is independently controlled by the digital "parameter interpolation" circuit. In one recent implementation,[38] the resonators were realized by the programmable SC bandpass biquad described in the previous section (Fig. 8.49). The response of these biquads was readjusted every 1 to 10 ms under external control to

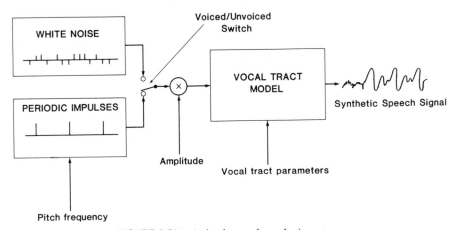

FIGURE 8.51. A simple speech synthesis system.

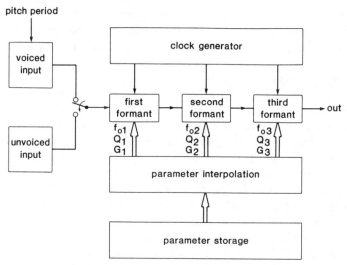

FIGURE 8.52. Block diagram of a formant synthesis system.

provide the desired formants. In addition to changing the programmable capacitors, the clock frequencies of the biquads were also controlled: the first formant stage was operated either at 10 or at 20 kHz, the second stage either at 30 or at 60 kHz, and the third one either at 60 or at 120 kHz. The resulting vowel reproduction was quite realistic. The filters used in the SC formant synthesizer based on this concept each occupied an 88×105 mil^2 area, including the bonding pads.

A different method of speech synthesis is *linear productive coding* (LPC). In this process,[39] the vocal tract and vocal cords are simulated by a time-variable all-pole sampled-data filter with a transfer function

$$H(z) = \frac{G}{1 - \sum_{k=1}^{p} a_k z^{-k}}. \qquad (8.21)$$

The system represented by $H(z)$ is then excited either by a periodic pulse train (for voiced speech), or by a random noise signal (for unvoiced sounds). Thus the parameters of this speech model include the voiced/unvoiced classification, the period of the impulse train (called *pitch period*) for voiced sounds, the gain G, and the coefficients $a_1, a_2, \ldots, a_p$ of the all-pole filter. All these parameters vary slowly with time. In the analysis phase, these speech parameters can be determined by a computation which minimizes the RMS error between the actual speech samples and the values predicted by the model.[39] Because of the time-varying nature of speech, all parameters must be estimated from a fairly short segment, called *frame period*, of the speech signal. The typical frame period may last 20 ms.

The block diagram of a single-chip speech production system[40] based on the LPC model is shown in Fig. 8.53. The time-varying LPC filter needed in the system is expediently implemented by the *lattice* structure illustrated (for

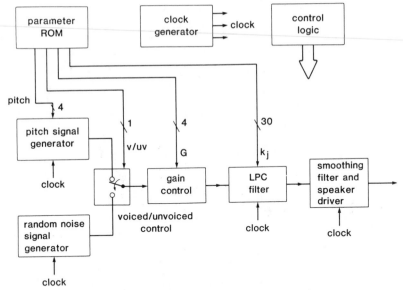

FIGURE 8.53. Block diagram of a LPC speech synthesizer.

$p = 10$) in Fig. 8.54.* Here, the multipliers k_i (often called *reflection coefficients*) replace the coefficients a_i of $H(z)$ in Eq. (8.21). Clearly, the structure requires the repeated performance of operations in the form

$$e_j(i) = e_{j+1}(i - 1) - k_j b_j(i - 1)$$

and $\qquad\qquad\qquad\qquad\qquad\qquad\qquad\qquad\qquad\qquad$ (8.22)

$$b_j(i) = b_{j-1}(i - 1) + k_{j-1} e_{j-1}(i).$$

The required delay and multiply/add operations can be performed efficiently using switched-capacitor stages.[40] In particular, the delay (and gain) required can be implemented by the offset-free amplifier stage discussed in Chapter 6 (cf. Fig. 6.5) and reproduced in Fig. 8.55. The multiplication/addition operations of the general form

$$v_3 = v_1 + k_j v_2 \qquad\qquad\qquad\qquad\qquad\qquad (8.23)$$

as required by (8.22) can be performed using the circuit of Fig. 8.56. In this stage, v_1, v_2, and v_3 are sampled-data *analog* voltages, while each factor k_j is represented by a nine-bit *digital* word with a sign bit b_0 and magnitude bits b_1 to b_8. This multiplier/adder circuit is similar in principle to the multiplying DAC discussed in Section 6.2 (see Fig. 6.12), but with some modifications. The present circuit contains *two stages* to reduce the capacitance spread $C_{max}/C_{min} = 2^n$ of the MDAC of Fig. 6.12, which for $n = 8$ gives $C_{max} = 256 C_{min}$. The first stage converts the least significant bits b_8, b_7, b_6, and b_5; its output is

---

*This implementation is unconditionally stable, and its coefficients can readily be computed.[41]

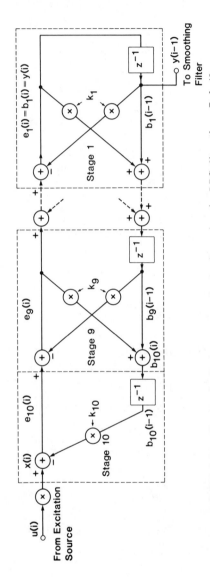

FIGURE 8.54. Tenth-order lattice structure for implementing the LPC filter (from Ref. 40, © 1983 IEEE).

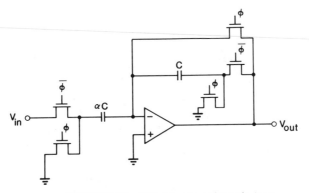

FIGURE 8.55.   Offset-free inverting gain stage.

attenuated by 16 when it is entered into the second stage which processes the most significant bits $b_1 - b_4$. The resulting capacitance spread is only 16. The sign bit b_0 determines the polarity of the gain for both stages, via an arrangement similar to that used for the circuit of Fig. 6.12. (It is also possible to use a two's-complement format for negative values of k_j; the corresponding form of the multiplier/adder is described in Ref. 40.)

The LPC parameters stored in the parameter ROM shown in Fig. 8.53 are obtained b₁ using a mainframe computer to extract these values from the actual speech signal sampled at a rate of 8000 samples/s. The frame period used in this process is 20 ms. In the reproduction of speech, the LPC filter parameters are interpolated, so that the new effective frame period is reduced to 5 ms. This is fast enough for the ear to perceive the change as continuous rather than abrupt.

The input and output circuits of the LPC filter are operated at a clock rate of 8 kHz. During each sample period of 1/8000 Hz = 125 μs, the filter accepts an input sample, performs sequentially the necessary $2p = 2 \times 10 = 20$ multiply/add operations as given in Eq. (8.22), and produces an output sample.

The signal generators used to excite the LPC filter with an 8-kHz S/H signal are digitally controlled switched-capacitor circuits. During a *voiced* frame, the input signal is a train of odd-symmetry pulses, with an appropriate energy spectrum and a repetition period equal to the pitch period of the sound. For an *unvoiced* frame, the input is a constant-amplitude signal, whose sign is controlled by a pseudorandom bit generator. Figure 8.57 shows the sampled version of the two input signals. The circuit generating these excitations is illustrated in Fig. 8.58. For *voiced* sound, the pitch-generator logic controls the MOS switches via the logic signals a_0, a_1, and a_2 so as to produce the input pulse train as the output voltage v_{out}. For an *unvoiced* frame, the pseudorandom bit generator changes the polarity of the input signal $v_{in} = \pm V_{ref}$. The repetition period of the bit sequence equals 2^{15} sample periods, so that it can be considered random.

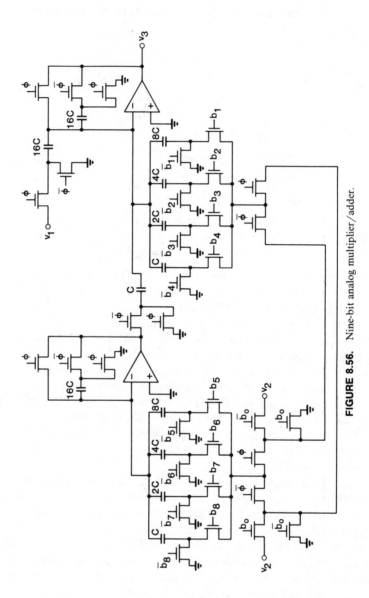

FIGURE 8.56. Nine-bit analog multiplier/adder.

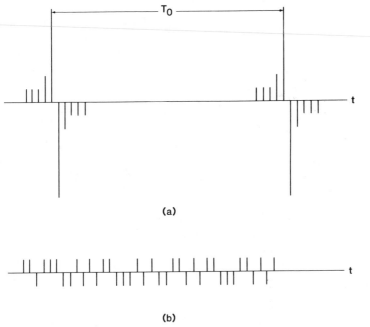

FIGURE 8.57. (*a*) Excitation pulse train for voiced speech. (*b*) Excitation pulse train for unvoiced speech (from Ref. 40, © 1983 IEEE).

The output of the LPC filter is an 8-kHz sampled-and-held signal. To remove the staircase distortion of this voltage, it is entered into a third-order smoothing low-pass filter clocked at 160 kHz, which attenuates the signal frequencies above 4 kHz. The output of this filter drives a balanced on-chip power amplifier capable of delivering over 30 mW audio power into a 100-Ω loudspeaker.

The system was fabricated using 5-μm CMOS technology. The chip size was 280×168 mil^2. It is capable of storing and reproducing speech for a 17-s duration, using an on-chip ROM containing about 20,500 bits. This illustrates the efficiency of the LPC algorithm; direct storage of the speech samples represented by, say, four-bit words would only permit a speech duration of about 0.64 s. Thus, the efficiency of the storage is increased by a factor of over 26, thanks to the LPC coding.

As an illustration of the use of switched-capacitor circuits in an integrated *speech analysis* system, we shall discuss next a single-chip 20-channel speech spectrum analyzer.[42] In this system, the incoming voice spectrum is split into 20 frequency bands, and the energy in each channel is measured and stored in a digital form. The results can be used to achieve high-quality speech recognition.

The block diagram of the system[42] is shown in Fig. 8.59. The automatic gain controller (AGC) is digitally adjusted over a 46.5-dB range in 1.5-dB steps. It

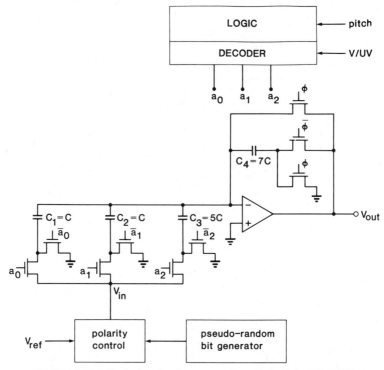

FIGURE 8.58. Excitation signal generator (from Ref. 40, © 1983 IEEE).

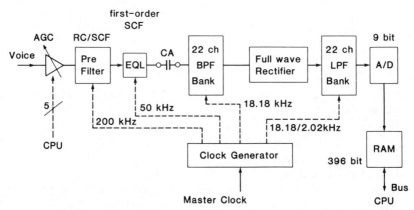

FIGURE 8.59. Speech spectrum analyzer block diagram (from Ref. 42, © 1984 IEEE).

maintains the signal level of the rest of the system at a desirable value. The prefilter contains the cascade combination of a second-order active-RC filter and a 10th-order SCF with a high (200 kHz) clock frequency. It enables the use of a low (18.18 kHz) clock rate in the following stages. This is followed by a first-order SC equalizer (EQL) which introduces a 6-dB/octave slope to emphasize higher frequencies. The main part of the analyzer is the 20-channel bandpass filter (BPF) channel bank. Each BPF is a sixth-order SCF; the whole bank is realized by a single time-division multiplexed (TDM) filter. The filter outputs are full-wave rectified. The rectified signal is then low-pass filtered in a 20-channel low-pass filter (LPF) bank. The LPFs are again obtained by using a single eighth-order TDM SCF, with programmable cutoff frequencies. The resulting 20 signals are then converted into nine-bit digital words and stored in a buffer memory from which they can be read out asynchronously.

The key component in this complicated circuit (which must realize 308 filter poles) is the multiplexed SCF biquad. The biquad used is the "low-Q" stage described in Section 5.4, and shown there in Fig. 5.10c.* Since it is used to realize a notch response, the input capacitance C_1' is omitted. In multiplexing the section, the *unswitched* capacitors C_A, C_B, and C_1'' must be replicated n times (where n is the number of channels sharing the section) since each replica must store the charge for a different channel. The *switched* capacitors, which are discharged each time when the filter is used by a new channel, need not be replicated. To obtain the different capacitor ratios required by the different channels, it is possible to use different values for each element of the unswitched capacitor arrays. This approach is illustrated in Fig. 8.60a, which shows one of the integrators used in the multiplexed biquad. In this circuit, the values of the memoried feedback capacitors C_{fi} as well as the values used for the memoryless switched capacitor C are, in general, different for each channel i. The disadvantage of this arrangement is that, due to the different element values, the dc offset voltages caused by the op-amp dc offset as well as by the clock feedthrough and channel-charge injection effects will be different from channel to channel. This will introduce a "fixed-pattern" error into the spectrum, unless it is canceled individually for each channel, which is a difficult task. To avoid either of these unpleasant alternatives, the circuit parameters must be uniform for all channels, but yet provide different responses for each one. These contradictory requirements may be resolved by adding a variable resistive divider to the circuit (Fig. 8.60b) and using the same capacitance values for all channels. If the voltage division provided by the resistive divider for channel i is k_i, then this arrangement is equivalent to changing the value of C to $k_i C$ or that of C_f to C_f/k_i (Problem 8.13). All offset effects will then be the same for all channels, and a single offset-voltage cancellation will be sufficient.

The application of this principle to the full biquad is illustrated in Fig. 8.61. This contains five resistive dividers (realized by using three shared resistor strings) to determine the five coefficients of the response $H(z)$ of the biquad

---

*There is a minor change in the switch phasing which improves the settling speed of the biquad.

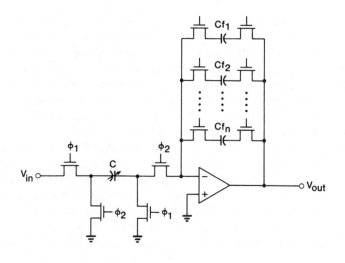

(a)

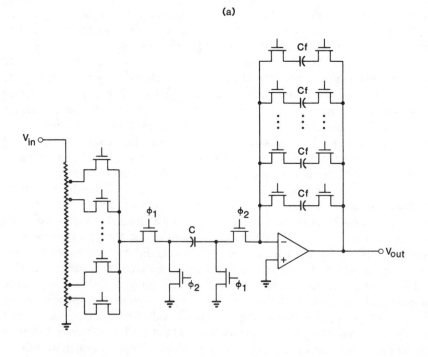

(b)

FIGURE 8.60. TDM SC integrator structures: (*a*) conventional type; (*b*) new type (from Ref. 42, © 1984 IEEE).

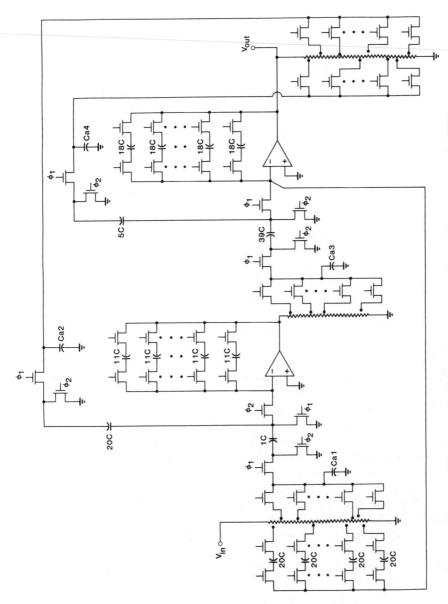

FIGURE 8.61. TDM SCF biquad with resistor strings (from ref. 42, © 1984 IEEE).

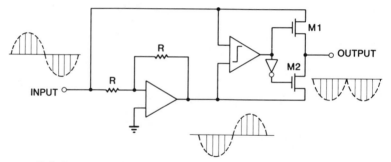

FIGURE 8.62. Full-wave rectifier (from Ref. 42, © 1984 IEEE).

[cf. Eq. (5.27) of Chapter 5] for each channel. In this TDM biquad, the dc *output* offset voltage depends on the division factor k_i associated with the 20C capacitor, and hence varies from channel to channel. However, the *input-referred* offset is the same for all channels, and hence it can be eliminated at the first op-amp input with a single offset-canceling circuit.[42] To find the compensatory voltage needed to balance out the offset, two dummy channels have been added to the channel bank, making the total number of the channels equal to 22.

For the bandpass filters, the dc offsets introduced by the first two biquads into the filter output can be reduced by having transmission zeros at dc ($z = 1$) in the response of the third (output) biquad.

The extra capacitors C_{a1}, C_{a2}, C_{a3}, and C_{a4} shown in Fig. 8.61 are used to improve the efficiency of the charge-canceling clock feedthrough compensation. This arrangement was discussed in Section 7.1 (see Fig. 7.9).

The bandpass filters, as well as the two input biquads of the low-pass filters, operate at a clock rate of 18.18 kHz. This value results in a relatively small capacitance spread in the biquads. Since there are 22 channels, the switched capacitors must discharge at a relatively fast ($22 \times 18.18 \simeq 400$ kHz) rate. To save area and increase the switching speed, a very small (0.2 pF) unit capacitor was used in these filters. Since it is difficult to maintain a reasonable accuracy for the realized response with such small capacitances, a discrete optimization technique was used to find the nominal capacitance values. The two output stages of the LPF are operated at the reduced clock rate of 2.02 kHz. This is made possible by the antialiasing response provided by the first two LPF stages. Since the last two biquads have very low cutoff frequencies (in the range of 12.5–400 Hz), such low clock rate is necessary to keep the capacitance spread reasonably low.

The rectifier used in the system is shown schematically in Fig. 8.62. In this circuit, the amplifier compares the input v_{in} with $-v_{in}$. If $v_{in} > -v_{in}$ (so that $v_{in} > 0$), the comparator closes the switch $M1$ which allows v_{in} to appear unchanged at the output. Otherwise, $M2$ will conduct, and $-v_{in}$ appears at the output. As was the case with the SCFs, a single rectifier circuit is shared by all 22 channels.

The nine-bit A/D converter which follows the rectifier uses the charge redistribution method described in Section 6.2 (see Fig. 6.13). It has slightly larger unit capacitors (0.3 pF) than the SCFs.

The overall chip contains a total of 30 op-amps, two comparators, a $2 \times 22 \times 9 = 396$-bit RAM which permits the asynchronous readout of the spectral samples, and 500 digital gates. Thanks to the multiplexing of the op-amps, the total number of filter poles realized can equal $22 \times (6 + 4 + 4) = 308$. The chip was fabricated using 3-μm CMOS technology. Its die size is 276×256 mil^2.

PROBLEMS

8.1. Analyze the circuit of Fig. 8.1 to verify Eq. (8.1). Express Q and ω_0 in terms of R and C.

8.2. Calculate RC for the filter of Fig. 8.1 so that the response is down by 3 dB from its dc value at $f_{3dB} = 36$ kHz. How much is the loss at $f_{pass} = 3.4$ kHz and at $f_{stop} = 512$ kHz?

8.3. Assume that the accuracy of the time constant RC of the filter discussed in Problem 8.2 is $\pm 50\%$. What are the worst-case values of the loss at f_{pass} and f_{stop}?

8.4. Show that replacing the input branch of the circuit of Fig. 8.7 by that shown in Fig. 8.10 implements a cosine decimator filter.

8.5. For a decimator filter ($n = 8$), it is required that the response be flat to within 0.01 dB for $0 \le f \le 4$ kHz, and the loss be down by at least 40 dB within the $f_c - 4$ kHz $\le f \le f_c + 4$ kHz range. What is the minimum value of f_c?

8.6. Show that the circuit of Fig. 8.24 satisfies the input–output relations given in Eqs. (8.10)–(8.12).

8.7. Prove Eq. (8.4) for the input circuit of Fig. 8.7.

8.8. Analyze the effect of a stray capacitor C_A loading node (A) on the transfer function of the circuit of Fig. 8.10. Under what conditions is the effect significant?

8.9. The active-RC antialiasing filter shown in Fig. 8.2 can easily be generalized to provide a peaking, rather than maximally flat passband gain response. To achieve this, the capacitor $2C$ must have the value rC instead, where $r > 4$. The applicable design equations are then

$$r = 2A_{max}\left(A_{max} + \sqrt{A_{max}^2 - 1}\right),$$

$$C \quad \text{arbitrary},$$

$$R = \frac{\sqrt{r - 2}}{rC 2\pi f_{max}}.$$

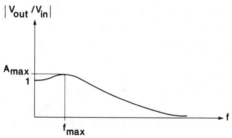

FIGURE 8.63. A possible gain response for the active-RC filter of Fig. 8.2 (used in Problem 8.9).

Here, $A_{\max} = |V_{\text{out}}/V_{\text{in}}|_{\max}$ is the voltage gain at the peak (Fig. 8.63), and $f_{\max}$ is the frequency at which it occurs. (a) Derive the above design equations. (b) Calculate the location and Q-value of the complex poles in terms of $A_{\max}$ and $f_{\max}$. (c) Design the circuit for $C = 10$ pF, $f_{\max} = 10$ kHz, and $A_{\max} = \sqrt{2}$ (i.e., for a 3-dB peak). Where are the poles, and how much is the pole-Q?

8.10. The unity-gain voltage follower shown in Fig. 8.64 is often used as a buffer stage in SC chips. Assume that the op-amp gain is $A(\omega) = |A(\omega)|e^{j\phi(\omega)}$. (a) Calculate $A_u(\omega) = V_{\text{out}}(\omega)/V_{\text{in}}(\omega)$. (b) Find the gain and phase of $A_u(\omega)$ at the unity-gain frequency ω_0 of the op-amp. (c) Under what conditions will $|A_u(\omega_0)| > 1$ hold?

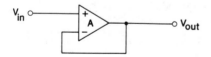

FIGURE 8.64. Unity-gain voltage follower (Problem 8.10).

8.11. Show that the two terms $-V_{BE_1}$ and $-(V_{BE_1} - V_{BE_2})R_2/R_1$ in V_{REF} as given in Eq. (8.19) have opposite temperature coefficients. (*Hints:* ①. The emitter-current relations of the bipolar transistors Q_1 and Q_2 are in the form

$$I_e = AJ_0\left(e^{\frac{qV_{BE}}{kT}} - 1\right)$$

where A is the emitter area and J_0 a fabrication-dependent current density, common to Q_1 and Q_2. ②. The temperature coefficient of V_{BE} is negative for a constant value of I_e.[25])

8.12. Prove Eq. (8.20) for the circuit of Fig. 8.49.

8.13. Analyze the circuits of Figs. 8.60a and 8.60b. What should be the values of the resistive voltage divisions k_i if the two circuits are to provide the same response in all channels?

REFERENCES

1. G. C. Temes and J. W. LaPatra, *Introduction to Circuit Synthesis and Design*, McGraw-Hill, New York, 1977.

2. D. J. Allstot and W. C. Black, Jr., *Proc. IEEE*, **71**, 967–986 (1983).

3. R. Gregorian and W. E. Nicholson, Jr., *IEEE Trans. Circuits Systems*, **CAS-27**, 509–514 (1980).

4. R. Gregorian and W. E. Nicholson, Jr., *IEEE J. Solid-State Circuits*, **SC-14**, 970–980 (1979).

5. P. E. Fleischer, K. R. Laker, D. G. Marsh, J. P. Ballantyne, A. A. Yiannoulos, and D. L. Fraser, Jr., *IEEE Trans. Circuits Systems*, **CAS-27**, 552–559 (1980).

6. M. B. Ghaderi, G. C. Temes, and S. Law, *IEE Proc.*, **128**, Pt. G, 213–215 (1981).

7. T. H. Hsu, Improved Design Techniques for Switched-Capacitor Ladder Filters, Ph.D. dissertation, UCLA (1982).

8. T. H. Hsu, unpublished work.

9. F. E. Owen, *PCM and Digital Transmission Systems*, McGraw-Hill, New York, 1982.

10. J. Max, *IEEE Trans. Information Theory*, **IT-6**, 7–12 (1960).

11. G. M. Roe, *IEEE Trans. Information Theory*, **IT-10**, 384–385 (1964).

12. J. T. Caves, C. H. Chan, S. D. Rosenbaum, L. P. Sellars, and J. B. Terry, *IEEE J. Solid-State Circuits*, **SC-14**, 65–73 (1979).

13. M. E. Hoff, J. Huggins, and B. M. Warren, *IEEE J. Solid-State Circuits*, **SC-14**, 47–53 (1979).

14. P. R. Gray, D. Senderowicz, H. Ohara, and B. M. Warren, *IEEE J. Solid-State Circuits*, **SC-14**, 981–991 (1979).

15. I. A. Young, *IEEE J. Solid-State Circuits*, **SC-15**, 997–1006 (1980).

16. W. C. Black, D. J. Allstot, and P. A. Reed, *IEEE J. Solid-State Circuits*, **SC-15**, 929–939 (1980).

17. R. Gregorian and W. E. Nicholson, *IEEE J. Solid-State Circuits*, **SC-14**, 970–980 (1979).

18. K. B. Ohri and M. J. Callahan, *IEEE J. Solid-State Circuits*, **SC-14**, 38–46 (1979).

19. Y. A. Haque, R. Gregorian, R. W. Blasco, R. A. Mao, and W. E. Nicholson, *IEEE J. Solid-State Circuits*, **SC-14**, 961–969 (1979).

20. K. Yamakido, T. Suzuki, H. Shirasu, M. Tanaka, K. Yasunari, J. Sakaguchi, and S. Hagiwara, *IEEE J. Solid-State Circuits*, **SC-16**, 302–307 (1981).

21. D. G. Marsh, B. K. Ahuja, T. Misawa, M. R. Dwarakanath, P. E. Fleischer, and V. R. Saari, *IEEE J. Solid-State Circuits*, **SC-16**, 308–315 (1981).

22. A. Iwata, H. Kibuchi, K. Uchimura, A. Morino, and M. Nakajima, *IEEE J. Solid-State Circuits*, **SC-16**, 315–321 (1981).

23. R. Gregorian, G. A. Wegner, and W. E. Nicholson, *IEEE J. Solid-State Circuits*, **SC-16**, 322–333 (1981).

24. Y. P. Tsividis, P. R. Gray, D. A. Hodges, and J. Chacko, *IEEE J. Solid-State Circuits*, **SC-11**, 740–747 (1976).

25. R. J. Widlar, *IEEE J. Solid-State Circuits*, **SC-6**, 2–7 (1971).

26. Y. P. Tsividis and R. W. Ulmer, Digest of Technical Papers, International Solid-State Circuits Conference, pp. 48–49 (1978).

27. Bell Laboratories Staff, *Transmission Systems for Communications*, Western Electric Technical Publications, Winston-Salem, North Carolina, 1970.

28. L. T. Lin, H. F. Tseng, and L. Querry, Digest of Technical Papers, International Solid-State Circuits Conference, pp. 148–149 (1982).

29. M. J. Callahan and C. B. Johnson, Ref. 28, 64–65 (1977).

30. R. N. Battista, C. G. Morrison, and D. H. Nash, *IEEE Trans. Commun. Electron.*, 9–17 (1963).

31. M. J. Callahan, *IEEE J. Solid-State Circuits*, **SC-14**, 85–90 (1979).

32. K. Niwa and M. Sato, *Proc. Int. Conf. Commun.*, **18 F1–5** (1974).

33. N. J. Denenberg, *Bell Syst. Tech. J.*, **55**, 143–155 (1976).

34. T. A. C. M. Claasen and J. B. H. Peek, *IEEE Trans. Commun.*, **COM-24**, 1291–1300 (1976).

35. B. J. White, G. M. Jacobs, and G. F. Landsburg, *IEEE J. Solid-State Circuits*, **SC-14**, 991–997 (1979).

36. T. Foxall, R. Whitbread, L. Sellars, A. Aitken, and J. Morris, Digest of Technical Papers, International Solid-State Circuits Conference, pp. 90–91 (1980).

37. R. Gregorian, W. E. Nicholson, and G. C. Temes, *Microelectron. J.*, **11**, 5–12 (1980).

38. D. J. Allstot, R. W. Brodersen, and P. R. Gray, *IEEE J. Solid-State Circuits*, **SC-14**, 1034–1041 (1979).

39. L. R. Rabiner and R. W. Schafer, *Digital Processing of Speech Signals*, Prentice-Hall, Englewood Cliffs, New Jersey, 1978.

40. R. Gregorian and G. Amir, *IEEE J. Solid-State Circuits*, **SC-18**, 65–75 (1983).

41. J. Makhoul, *IEEE Trans. Acoust. Speech and Signal Process.*, **ASSP-25**, 423–429 (1977).

42. Y. Kuraishi, K. Nakayama, K. Miyadera, and T. Okamura, *IEEE J. Solid-State Circuits*, **SC-19**, 964–970 (1984).

43. K. Martin and A. S. Sedra, *Electron. Lett.*, **16**, 613–614 (1980).

INDEX